THE SCIENCE OF HARMONICS IN CLASSICAL GREECE

The ancient science of harmonics investigates the arrangements of pitched sounds which form the basis of musical melody, and the principles which govern them. It was the most important branch of Greek musical theory, studied by philosophers, mathematicians and astronomers as well as by musical specialists. This book examines its development during the period when its central ideas and rival schools of thought were established, laying the foundations for the speculations of later antiquity, the Middle Ages and the Renaissance. It concentrates particularly on the theorists' methods and purposes and the controversies that their various approaches to the subject provoked. It also seeks to locate the discipline within the broader cultural environment of the period; and it investigates, sometimes with surprising results, the ways in which the theorists' work draws on and in some cases influences that of philosophers and other intellectuals.

ANDREW BARKER is Professor of Classics in the Insititute of Archaeology and Antiquity at the University of Birmingham.

THE SCIENCE OF HARMONICS IN CLASSICAL GREECE

ANDREW BARKER

CAMBRIDGE
UNIVERSITY PRESS

CAMBRIDGE UNIVERSITY PRESS
Cambridge, New York, Melbourne, Madrid, Cape Town,
Singapore, São Paulo, Delhi, Tokyo, Mexico City

Cambridge University Press
The Edinburgh Building, Cambridge CB2 8RU, UK

Published in the United States of America by Cambridge University Press, New York

www.cambridge.org
Information on this title: www.cambridge.org/9780521289955

First published 2007
First paperback edition 2011

A catalogue record for this publication is available from the British Library

ISBN 978-0-521-87951-4 Hardback
ISBN 978-0-521-28995-5 Paperback

O dear white children, casual as birds,
Playing among the ruined languages,
So small beside their large confusing words,
So gay against the greater silences . . .

W. H. Auden, *Hymn to Saint Cecilia*

Contents

Figures

Preface

I did most of the research for this book and wrote the first draft during my tenure of a British Academy Research Professorship in the Humanities in 2000–2003. It was a great privilege to be awarded this position, and I am deeply indebted to the Academy for its generous support of my work, which would otherwise have been done even more slowly or not at all. I am grateful also to the University of Birmingham for freeing me from my regular duties for an extended period. In that connection I should like to offer special thanks, coupled with sympathy, to Matthew Fox, for uncomplainingly taking over my role as Head of Department at a particularly difficult time, and to Elena Theodorakopoulos, Niall Livingstone and Diana Spencer for shouldering a sack-full of other tasks that would normally have come my way. Many others have been splendid sources of help, encouragement and advice. I cannot mention them all, but here is a Mighty Handful whose members have played essential parts, whether they know it or not: Geoffrey Lloyd, Malcolm Schofield, David Sedley, Ken Dowden, Carl Huffman, Alan Bowen, André Barbera, Franca Perusino, Eleonora Rocconi, Donatella Restani, Annie Bélis, Angelo Meriani, David Creese, Egert Pöhlmann, Panos Vlagopoulos, Charis Xanthoudakis. My sincere thanks to all these excellent friends. Jim Porter and another (anonymous) reader for the Press read two versions of the entire typescript in draft; without their comments, to which I have done my best to respond, the book would have been a good deal less satisfactory than it is. I appreciate the magnitude of the task they generously undertook, and though they added substantially to my labours I am exceedingly grateful for theirs. This is the fourth book of mine to which the staff of Cambridge University Press have served as midwives, and they have amply lived up to the standards of efficiency, courtesy and patience which I have come to expect and appreciate. My thanks to all concerned on this occasion, and especially to my admirable copy-editor, Linda Woodward, both for her careful work on the lengthy typescript and for the gratifying interest she took in its contents. Thanks, too, to my oldest son, Jonathan Barker, who showed me how to solve certain vexing mathematical conundrums; and as always, to the rest of my family and especially my wife, Jill, for their continuing patience and encouragement.

I can only regret that David Fowler is no longer here to be thanked. His untimely death has deprived me and many others of a friend and colleague whose enthusiasm and insatiable curiosity were infectious and inspiring, and whose lively and sympathetic humanity put some warmth and light into this cynical world. He was one of the most charming people who ever trod the earth, and he will be sadly missed.

PART I

Preliminaries

Introduction

Few books have more splendidly informative titles than Theon of Smyrna's *Mathematics useful for reading Plato*. A title modelled on his, perhaps *Harmonic theory useful for reading classical Greek philosophy and other things* would have given a fair impression of my agenda here. But that's a little cumbersome; and for accuracy's sake, I would have had to tack the phrase '*and indications of the converse*' onto the Theonian title, since I shall be trying to show not only how harmonics can be 'useful' to students of other fields, but also how the preoccupations of Greek writers who tilled those fields can shed light on the development of harmonics itself, and can help us to understand its methods and priorities. More importantly, this hypothetical title would have been dangerously hubristic; it has the air of presupposing a positive answer to one of the book's most serious questions. Leaving one or two exceptional passages aside (the construction of the World-Soul in Plato's *Timaeus*, for example), does a knowledge of the specialised science of harmonics, and of its historical development, really give much help in the interpretation of texts more central to the scientific and philosophical tradition, or in understanding the colourful environment inhabited by real Greek musicians and their audiences, or indeed in connection with anything else at all? Can such knowledge be 'useful', and if so, in which contexts, and how? I intend to argue that it can, though not always in the places where one would most naturally expect it.

There is a point I should like to clarify before we begin, to avoid misunderstandings and to help explain some of this book's unavoidable limitations. Specialists in the ancient musical sciences may be few (though there are many more swimmers in these tricky waters now than there were when I took my first plunge over twenty-five years ago); but they are nevertheless various. By and large, they fall into two main groups. Some are professional musicologists, who may have worked their way upstream into these reaches from a starting point in the Middle Ages or the Renaissance. Others set out from a training in Classics, within which broad church I include devotees of ancient philosophy and science. Musicologists, of course, are sometimes proficient in Greek and Latin, and some classicists are excellent musicians; but when tackling their professional work, each group brings to it the equipment, the presuppositions and the puzzlements of their own academic tribe. I am no exception, and I make no bones about the fact. I am a classicist and a student of Greek science and philosophy. As it happens, I have made a good deal of music in my time, but I am not a trained musicologist. American colleagues have sometimes chided me, no doubt rightly, for my lack of a properly musicological perspective. So be it; each of us does what he or she can.

Most work published nowadays in this field is written by specialists for specialists. From time to time, over the years, I have contributed my own penny-worth to these esoteric conversations; but I have always had another objective in mind. Like the other branches of ancient 'musical theory' (and indeed all other serious forms of enquiry), harmonics was not a water-tight, self-contained enterprise, ring-fenced from its cultural and intellectual surroundings. In some of its guises it drew extensively on the concepts, methods and doctrines of other fields of intellectual study, and fed them, in turn, with its own; in others, or so I shall argue, its relations with philosophy and the natural sciences are more distant and its interactions with real-world music-making and musical appreciation much closer than is often supposed. Its exponents wrote and taught in ways, and for purposes, that responded to the wider controversies of the day, and to the specific intellectual, cultural and educational demands of their environment. Most of the authors I shall consider in this book did not compose free-standing treatises on musical topics, but pursued the subject as an element in some other philosophical, literary, scientific or artistic enterprise. Even when these external points of reference are put aside, experience with my own students has convinced me that one does not need an unusually eccentric turn of mind to find harmonic theory as delightfully fascinating in its own right as any other discipline, once one has been lured into the labyrinth. In other publications, and in lectures and seminars here and there around the globe, I have therefore tried to find ways of advertising its charms to people who work in other, intersecting areas, to musicians, to mathematicians, to classicists in general, and especially to students of ancient philosophy and science, and I shall continue with that attempt in this book. I hope that the musicological cognoscenti will find things in it to interest and perhaps to infuriate them, but I would like to show others as well that forays into this little jungle will not be a waste of their time.

This will involve a delicate balancing act between intricacies of detail and the larger perspective; no doubt from time to time I shall fall off the tight-rope on one side or the other. Too bland and generalised an approach would disguise the subject's intellectual meat and sophistication; equally, I do not want to face readers with impenetrable thickets of minutiae. The writer of a book of this sort must also decide whether the science's content and its contexts should be allowed to intermingle, enriching and informing one another in a seamless exposition, or should be addressed in separate compartments for the sake of clarity. I have adopted a mixed strategy; some chapters are principally concerned with one or the other, and in others, for various reasons which I hope will become apparent, I have done my best to weave the two together. But of course the division is thoroughly artificial. The internal agenda that drove the discipline's development was in many cases a response to pressures from outside its borders, and one can make little sense of its changes by considering it in isolation. The separate chapters on 'contexts' are not just titbits for non-specialists. Neither should the more technical parts of the book be treated as if they were labelled 'For Experts Only'. Each depends on the other. Anyone who pursues the history of Greek harmonics beyond the period covered here will find that in later times the situation is even more acute. The story spans more than a thousand years; and though significant developments in the methods and doctrines of harmonic theorists are confined, with some minor and a very few major exceptions, within the compass of its first two centuries, that is not to say that nothing happened for the rest of the millennium. A great deal did. But the history of changes in those later centuries is to a large extent a history of shifting contexts. It is a story about the ways in which inherited ideas were used, abused, recombined and inserted into new settings, new forms of discussion and new patterns of enquiry. In the earlier period, while the discipline was inventing itself, there is much more to be said about its internal debates and transformations, but processes of the sort which take the limelight later were crucially involved from the start.

Greek harmonics in general, and in this period in particular, is not the easiest of topics. This is not only, or even principally, because it involves esoteric technicalities. The most obstructive difficulty is one that it shares with other, more familiar fields of study, Presocratic philosophy for instance; no extensive texts on the subject survive from the fifth century and very few from the fourth, and much of its history has to be reconstructed out of fragments and reports embedded in other people's writings, of various kinds and dates. By no means all the evidence we have can be taken at face value. Later reports and even contemporary ones are commonly coloured or distorted by their authors' own agendas; some are plainly anachronistic or otherwise inaccurate; a considerable number are bare-faced fictions. This does not mean that the project is impossible. Modern studies in other areas affected by these problems have done a great deal to illuminate them and to show how they can, to some extent, be resolved; and harmonics and its history are now much better understood and more widely known than they were twenty-five years ago. But a great deal remains to be done, both in interpreting the theorists' work and (still more) in unravelling its contexts, and again in trying to communicate the significance of sometimes arcane researches to a wider readership.

THE AGENDA OF GREEK HARMONICS

Non-specialist readers will be getting impatient with my repeated and unexplained references to 'harmonics'. It is high time I said something to explain what the subject is. It is one of three sister-sciences which share a strong family resemblance; the others are rhythmics and metrics.[1] They deal, plainly enough, with different aspects of the subject. But each, in its own sphere, has a similar goal: it is to identify, classify and describe, with the maximum of objectivity and clarity, the regular and repeated patterns of form and structure which underlie the bewildering diversity of melodic, rhythmic or metrical sequences found in musical compositions themselves. Metrics studies the patterns formed by the lengths of syllables in verse, whether or not it is set to 'music' in our sense of the word. I shall say little about it here; all students of Greek poetry in its literary guise are already familiar with it, and its mysteries have been expounded, time and again, by scholars much better qualified in its black arts than I am. Rhythmics (when it is distinguished from metrics, which is not always the case either in ancient or in modern treatments[2]) is a more strictly 'musical' discipline. It examines the patterns within which, when poetry becomes song (or when purely

[1] The names of the three sciences appear first in fourth-century sources. Harmonics is *ta harmonika* at Plato, *Phaedrus* 268e6 (the *Republic* identifies it only by reference to its subject-matter, *harmonia*, 531a1), *harmonikē* in several passages of Aristotle, e.g. *Metaph.* 997b21, and often in later writers; rhythmics is *rhythmikē* at Aristox. *El. harm.* 32.7, where it is distinguished from *harmonikē*, *metrikē* and *organikē* (the study of instruments); Aristotle had earlier referred to metrics as *ta metrika* and *metrikē* at *Poetics* 1456b34, 38.

[2] Plato, for example, rarely marks a clear distinction between metre and rhythm; but for explicit instances of the three-part classification which later became common, see *Gorg.* 502c5–6, *Rep.* 601a8. There is an earlier indication of it at Aristoph. *Clouds* 635–50.

instrumental music is in question), the singers' sequences of long and short syllables (or the instrument's sequences of long and short notes) are divided and grouped into repeated rhythmical structures, not necessarily identical with metrical 'feet' and analysed rather differently, and roughly analogous to the 'bars' of more modern music. This form of enquiry will make occasional appearances in this book, but only fitfully and in a supporting role. Composers themselves, of course, may well have found that its analyses were sometimes helpful to them in the practice of their craft, and it is true that its findings sometimes surface in the work of philosophers, scientists and other non-musical writers. In the period we are considering, however, they do so less frequently and less significantly than those of harmonics; it is harmonics, out of the three central musical disciplines, that lives the most vigorous life outside its own specialised sphere, and interacts most intimately with patterns of thought characteristic of other intellectual domains. Greek writers themselves commonly take the view that harmonics is the first and most important of the musical sciences, whereas rhythmics becomes visible to us as a substantial discipline, and one into which serious philosophical issues have been absorbed, in the surviving work of only one author.[3]

The other essential ingredient of all Greek music, alongside rhythm, was melody; and it is the structures underlying melody that are the concern of harmonics. 'Harmony' and 'harmonic progression', as we understand such things, had no place in Greek musical practice, and the concepts would have meant nothing to their theorists.[4] Any sequence of sounds recognisable as a melody depends for its musical coherence on a pattern of relations between the notes and intervals on which it draws, one that can be set out, formally and abstractly, as a scale of some specific type (or, in more complex cases, as a combination of two or more such scales). More concretely, when a Greek lyre-player set out to play a melody, it was essential that the strings of his instrument were already tuned to a pattern of intervals which would make such a melody possible. But from the perspective of most Greek theorists, though perhaps not all, this puts the relation between melody and

[3] The author is Aristoxenus; the remnants of his work in rhythmics and other pieces of evidence about it are collected and discussed in Pearson 1990. Two other disciplines will be discussed from time to time in our reflections on harmonics itself, as distinct from its contexts. One is mathematics, especially the branch of it known as *arithmētikē* or 'number-theory'. The other is physical acoustics, a science of broader scope than harmonics since it deals with sounds in general, not only those relevant to music. But it seems to have originated as an accessory to one form of harmonic research, and will be considered here only in that role.

[4] In practical music-making, accompanists sometimes – perhaps often – played notes other than those currently sounding in the melodic line. But we know all too little of this practice, and there are few traces of Greek attempts to study it from a theoretical point of view. For further discussion see Barker 1995.

attunement the wrong way round. In their view the status of a sequence of notes as a genuine melody depends on its being rooted in a scale or attunement which is itself formed in a properly musical way.[5] Melodies are infinitely various, but the structures from which they draw their musical credentials are not. Not just any arrangement of notes and intervals can form the basis of a melody, and according to the Greek theorists those that can do so can be sharply distinguished from those that cannot. The central task to which they set themselves was to identify and analyse the varieties of scale and the systems of attunement which could be reckoned as musical, and which could transmit their musicality to melodies constructed on their foundations.

Put like this, the harmonic theorists' project may seem simple, even trivial. Our melodies, by and large, are built either on a major or on a minor scale (with one or two variants), and a seven-year-old child can learn to describe them. But even before other complexities arise, as they will, there are at least two reasons why the Greek theorists' task was more demanding than the modern analogy suggests. First, as is well known, the Greeks used many more types of scale than we do, and included among their elements a much more various repertoire of intervals than our scales contain, restricted as they are to permutations of the tone and the semitone. Tiny differences between the intervals used in two scales – the difference, for instance, between a quarter-tone and one third of a tone – might mark the borderline between radically distinct musical systems, credited with strongly contrasting aesthetic properties. Other differences, equally small, could amount – or so the theorists assure us – to the distinction between a musically acceptable scale and a meaningless and melodically useless jumble of noises. Much larger differences, in certain contexts, were construed as generating no more than variants of the same type of scale. The theorists, furthermore, were far from unanimous in their analyses of the various scale-systems they considered. This is evidently a much more intricate field of study than we might initially have suspected.

Secondly, we should not underestimate the importance of the fact that in the fifth century this enterprise was entirely new. Musicians, of course, were

[5] Scales differ from attunements in two principal ways. A scale is a series of notes set out in order of pitch, while an instrument's strings need not always be arranged with the highest note at one end and the lowest at the other, and the remainder set out in pitch-order between them. Secondly, to think of a set of notes as a scale is to think of it as a sequence of steps, unfolding successively in time; an attunement is simply a structure or pattern, in which no element is temporally prior to any other. In some Greek approaches it is attunements and in others it is scales that are the main focus of attention, and sometimes at least there are philosophically and musically interesting reasons for their difference in emphasis. But these distinctions and complications need not yet concern us.

familiar with the systems they used; they could recognise the distinctions between one pattern of attunement and another, and could construct them in practice. But there is a world of difference between the capacity to recognise, create and use a system of notes and intervals, and the capacity to analyse and describe it in clear and objective terms. There is no evidence to suggest that musicians of the earlier period had a vocabulary of the sort that such descriptions demand, or even that they thought of the relations between elements of their systems in ways that could, even in principle, be made 'precise' in anything like a scientific sense. When theorists began to tackle the task, most of them (perhaps not quite all) took the view that it could be achieved only if the relations between notes could somehow be represented quantitatively, and measured; no other approach would allow the intervals in each scale or attunement to be precisely specified and compared in a way that the mind could grasp. That is all very well; but how are musical intervals to be measured? No appropriate metric existed. It had to be invented from scratch (in fact two quite different methods of measurement were devised, as we shall see in Chapter 1); and there were difficult obstacles to be negotiated both in the invention of any such metric and in its application to the musical phenomena.

Once the harmonic enterprise was well under way, in at least two quite different forms, the theorists began to engage with issues of more complex and abstract sorts. Given that there are many different scales, are they related to one another in an intelligible way? Are all scales that span (let us say) an octave or more constituted out of sub-systems of identical or analogous types, and if so, are there constraints on the ways in which these sub-systems can be combined? Are there orderly procedures which permit the transformation of one kind of scale into another? Is it possible to identify all the musical systems there can be, and to show that the tally is complete? Given that the two approaches to measurement I have mentioned formed the basis of enquiries which differed quite radically in their methods and results, what grounds were there for preferring one or the other? Most crucially of all, are all the schemes of relations which harmonics identifies unified and governed by some fundamental principle or some coherent group of principles, so that all structures conforming to those principles, and no others, are thereby constituted as properly musical? If so, what kinds of principle are involved? What gives them their authority? Are they somehow rooted in human nature, or in the nature of something independent of humanity, or in mathematics, or are they merely products of social convention and tradition? Are they peculiar to the musical sphere, or do they have wider application?

Questions of these sorts are first raised explicitly by fourth-century writers. They answer them in various ways, but in one fundamental respect they are unanimous. Greek musicians, as I have said, used a number of different kinds of scale and attunement, considerably more than are familiar to most modern 'Western' ears, and there were ways in which they could be varied, transformed and combined with one another. Greek musical historians commonly credit this or that composer with having pioneered some new variety of scale. But it is axiomatic for all the theoretical writers whose views can be clearly pinned down that there is an objective and discernible line of demarcation, independent of human whim, decision or ingenuity, between musically well-ordered relations and transformations on the one hand, and on the other the indeterminate chaos of the non-musical. The distinction is not one of convention or taste, but is somehow fixed in the order of things, awaiting discovery, and from this perspective the innovative composers discussed by the historians are 'discoverers' rather than 'inventors'. (Sometimes, indeed, they are represented as 'perverters' of genuine music, and it is implied that their productions should not really be thought of as musical at all.) The task of harmonics, then, is to identify the structures on which melodies *must* be based if they are truly to be melodies, the ways in which they can *properly* be related to one another, modified, recombined, and so on, and to uncover the unchanging principles which govern them and determine the immutable boundaries of the melodic realm.

When theorists have come to regard the subject in this light, we can say with some assurance that they are treating it as a full-blooded scientific discipline, a branch of investigation dedicated to the discovery and demonstration of a body of truths, regardless of whether they assimilate it to the mathematical sciences or to the 'sciences of nature', the realm of *physiologia*. Students of this branch of reality must therefore adopt as reliable a methodology, as rigorous an approach to the evidence and as meticulous standards of reasoning as those of any other. But the appearance of these views in the major fourth-century writings should not tempt us to run away with the idea that harmonics had any such pretensions from the start. In subsequent chapters I shall try to show that its original aspirations were much less ambitious, and that this fact has an important bearing on the way in which its pronouncements were treated by people outside the ranks of the theorists themselves. The relation between harmonics and other matters is not a constant. If harmonics is to be 'useful for reading' texts of any other sort or to help us in understanding the dynamics of Greek culture, it is imperative that each stage of its development should be located as exactly as possible in its own historical environment. In practice some of its phases can be dated only very approximately, but we must do what we can. In this connection as in almost any other, generalisations which ignore chronology are almost bound to mislead.

When we come to these contextual matters, some of my conclusions will be familiar, at least in outline, to specialists in the adjacent fields in question. It will come as no surprise, for instance, to students of Plato or Aristotle or the Pythagoreans, that ideas drawn from harmonics had a significant role in their arguments and speculations; and the fact that it contributed to the theories in astronomy and medicine is almost equally well known. Precisely which form of harmonics and which of its aspects were involved is not always so clear, nor is it always easy to say whether the non-musical writers represent elements of harmonic theory accurately, or have misunderstood them or deliberately modified them for their own purposes. These issues need some attention if we are to understand what harmonics had to offer to natural scientists and philosophers; and we need to consider also the extent to which ideas flowed back into harmonics from these other directions. But at a general level, my comments in these areas will follow fairly well-trodden paths. Suggestions I make elsewhere may be more unexpected. It is often supposed, for example, that however intriguing harmonics may be as a body of abstract thought, and however important its contributions to philosophy and the sciences, it had little or nothing to do with the realities of Greek musical practice. Statements of this sort can be understood in two ways. They may mean either that the theorist's analyses had no basis in the facts and regularly misrepresented them, or that whether they did so or not, they had nothing to offer to musicians themselves or to connoisseurs in their audiences; they revealed nothing about individual compositions, and made no contribution to the skills of composition and musical appreciation. Except in certain very special cases, I shall argue, and perhaps even there, all these judgements are false. Another point at which my contentions may not match expectations is in the territory where ideas about music intersect with ethics, and where music is credited with a powerful influence on its hearers' emotions, dispositions and characters, a theme we meet repeatedly in philosophical writings and in more colourful terms in plays for the comic stage. Modern scholars have written copiously on this fascinating topic, especially in connection with Plato, and have often drawn on harmonic theory in the course of their interpretations.[6] I shall treat it a good deal more briskly and briefly than

[6] See for instance Moutsopoulos 1959, Lippman 1964 ch. 2, Anderson 1966, Gamberini 1996, Rossi 1988, 2000, Pagliara 2000, Boccadoro 2002.

readers might anticipate, since in the Greek writings themselves as I read them (with one very notable exception), what needs to be explained about harmonics in these contexts is not how it contributes to the discussions but why it is so remarkably absent.[7]

A NOTE ON THE 'PERFECT SYSTEMS'

At various points in this book I refer to notes by their Greek names, and to structures such as tetrachords which form parts of a larger system. I shall explain some of these references as we go along, but it seems sensible to give readers some guidance here, to which they can turn at need. From the later fourth century onwards, all Greek writers on harmonics were in broad agreement about the basic shape of a structure, or a group of structures, which contained within itself all the patterns of relations they set out to examine. The systems described by earlier theorists do not always fit exactly into those structures, but the picture developed by the later theorists still gives a useful point of reference and comparison. Certain constructions mapped out by Aristoxenus and his successors also subject the systems to more or less complex manipulations which I shall ignore for the present. All I offer here is a sketch of the regular scheme which formed the background to Aristoxenian analysis, together with a very few comments about the ways in which some earlier conceptions are related to it.

The system within which most of Aristoxenus' simpler constructions find a place is a scale spanning two octaves. It inhabits no particular range of pitch; in certain contexts (which we shall meet later in connection with the theory of *tonoi* and modulation) writers may refer to several instances of it, set at different pitch-levels. What gives each of its notes its identity is not its pitch but the relations in which it stands to others in the system. Within this fundamental scale, the principal sub-structures are the tetrachords, groups of four notes of which the outermost are a perfect fourth apart. The whole system, in fact, is a continuous sequence of such tetrachords. Some of them are linked in 'conjunction' (*synaphē*), where the highest note of the lower tetrachord is also the lowest note of the tetrachord above. Others are in 'disjunction' (*diazeuxis*); that is, there is an interval (but no note) between them, and this interval is always a tone (in modern parlance, a major second). Each note of the system has its own name. When written in full, most of the notes' names contain two words, of which the second

[7] This ceases to be true in the Roman imperial period, when writings on music and ethics became permeated with ideas borrowed from the exceptional case to which I have just alluded, that is, from Plato's *Timaeus*.

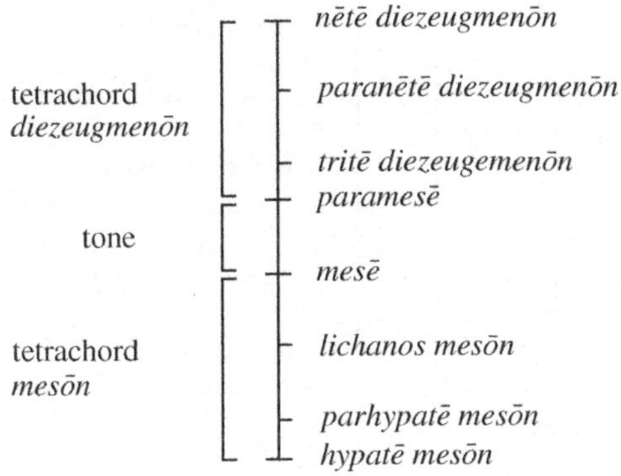

Figure 1 The central octave

identifies the tetrachord to which the note belongs. Thus *lichanos mesōn* is the '*lichanos* of the tetrachord *mesōn*'; and this distinguishes it from the corresponding note in the next tetrachord down, *lichanos hypatōn*, the '*lichanos* of the tetrachord *hypatōn*'.

The original core of the system, as theorists understood it, was a single octave, made up of two tetrachords disjoined by a tone. Most discussions before the later fourth century confined themselves to this octave, and references to individual named notes in writers such as Plato are to be construed as alluding to it. Its tetrachords are framed, from the top down, by the notes called *nētē diezeugmenōn*, *paramesē*, *mesē* and *hypatē mesōn*; see Figure 1. By the later fourth century, two further tetrachords had been added to this octave, one above it and one below, both in conjunction with their neighbours. A single note called *proslambanomenos* (the note 'taken in addition') was placed at the bottom of the system, a tone below the lowest tetrachord, to complete the double octave. The system used by Aristoxenus is thus made up of two identically formed octaves, each of which has a tone at the bottom, followed by a pair of tetrachords linked in conjunction (see Figure 2). For reasons that will appear shortly, this structure became known as the Greater Perfect System.

The notes bounding the tetrachords and the tones of disjunction are 'fixed' notes; the relations between them are invariable, and they form an unchanging framework for the whole. The two notes inside each tetrachord, by contrast, are 'moveable'. The higher and more structurally significant of the two, according to Aristoxenus, may lie anywhere between a tone and two tones below the tetrachord's upper boundary, and the lower at any distance between a semitone and a quarter-tone above its lower boundary.

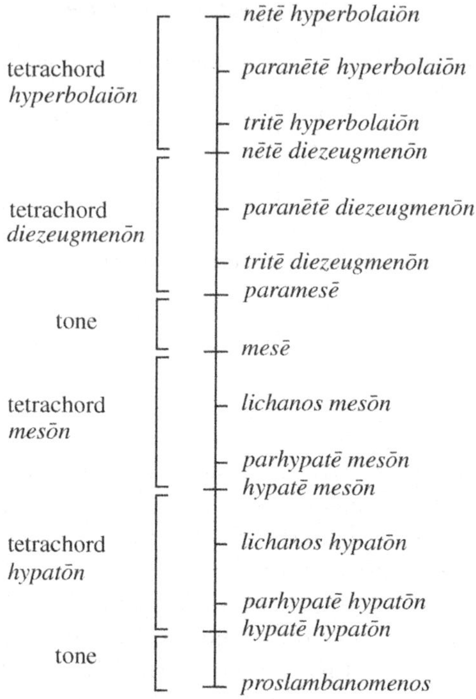

Figure 2 The Greater Perfect System

In Aristoxenian language, the most important changes created by their shifts of position are called changes of *genos* or 'genus', and there are three such genera, diatonic, chromatic and enharmonic. Less significant shifts produce changes from one variant or 'shade' of a genus to another (for more details see p. 129 below). But in any straightforward scale, every tetrachord contains the same pattern of intervals as every other. As a consequence, where two tetrachords are conjoined, every note in the lower of them lies a perfect fourth below its counterpart in the higher; and where they are disjoined by a tone, the corresponding interval is always a perfect fifth.

All this is fairly straightforward. The situation is complicated by the theorists' recognition of another tetrachord, which is not added at the top or the bottom, but runs upwards from *mesē* as an alternative to the tetrachord *diezeugmenōn*. Instead of being disjoined from the tetrachord *mesōn* by a tone (*diezeugmenōn* means 'of disjoined notes'), this one is conjoined with it at *mesē*. Its name is the tetrachord *synēmmenōn* ('of conjoined notes'). Thus the system is conceived as branching into alternative pathways as one

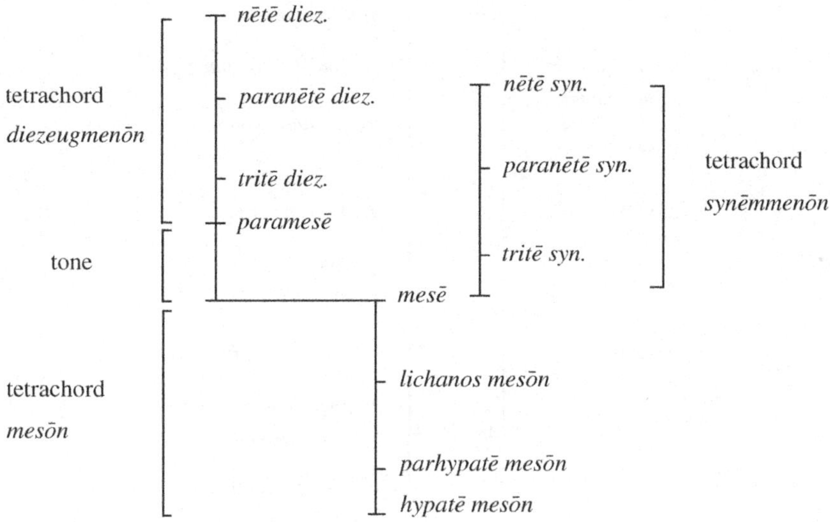

Figure 3 Disjunction and conjunction

passes upwards through *mesē*; at this point a melody may take either route (see Figure 3).

The origins of this curious appendage probably lie in ancient procedures for tuning a seven-stringed lyre, some of which, according to our sources, gave attunements falling short of the octave by a tone. Such attunements were represented in the form of two conjoined tetrachords. They appear in the theorists' scheme as the tetrachords *mesōn* and *synēmmenōn*; the tetrachord *diezeugmenōn* was conceived as an alternative, introduced by later musicians to complete the octave.[8] In putting the two structures together in a single system, the theorists seem still to have been broadly in tune with contemporary musical practice, since there is evidence that melodies in Aristoxenus' time often took a course that could be described as modulating between the tetrachords *diezeugmenōn* and *synēmmenōn*. Such modulations were apparently so common that the two pathways were felt as equally natural, and hence both were accommodated within the one, compendious system.[9]

[8] For references and brief discussion see West 1992a: 176–7.

[9] This type of modulation is given a special name, 'modulation of *systēma*', at Cleonides 205.5–6 and in several other sources of the Roman period. Cf. Aristox. *El. harm.* 5.9–14. A modulation of this sort occurs, for example, in the Delphic paean of Athenaeus, in bar 24 of the transcription of Pöhlmann and West 2001: 63.

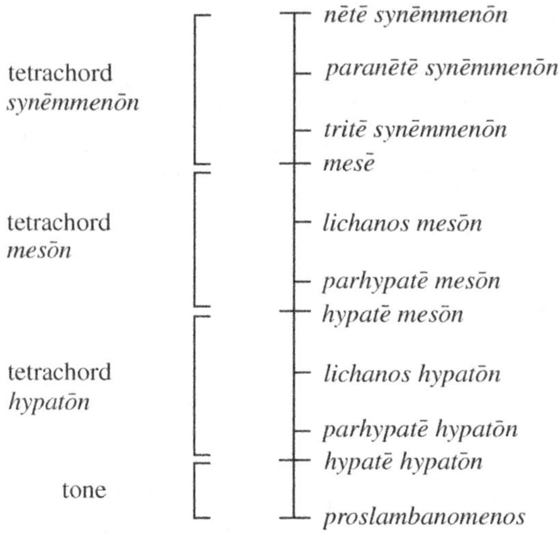

Figure 4 The Lesser Perfect System

The title 'perfect system' was assigned both to the complex structure as a whole and to two of its components. The straightforward double-octave running from *proslambanomenos* via *mesē* and the tetrachords *diezeugmenōn* and *hyperbolaiōn* to *nētē hyperbolaiōn* is the Greater Perfect System (see Figure 2 above). The Lesser Perfect System is the scale spanning an octave and a fourth which proceeds from *proslambanomenos* to *mesē* and then into the tetrachord *synēmmenōn*, ending with *nētē synēmmenōn* (Figure 4).[10] The Unchanging (or 'Non-modulating', *ametabolon*) Perfect System is the complete structure combining both the others (Figure 5).

Melodies with a compass of two octaves were rare at any period, and the primary role of the perfect systems was not to make room for the analysis of such prodigies. It was to make it possible to locate all acceptable melodic patterns and structures and to identify the relations between them within a single, integrated scheme.[11] I should emphasise again that the bald account

[10] The identification of the LPS as a structure in its own right seems artificial, corresponding to nothing significant in musical reality. Ptolemy (whose terminology differs from Aristoxenus' in other ways too) denies it the status of a 'perfect system', distinct from the GPS, though on theoretical rather than historical grounds. He argues that any melodic shift from the GPS into what we know as the tetrachord *synēmmenōn* should be understood as a modulation of *tonos* or 'key', involving a temporary transposition of the regular GPS through the interval of a fourth. But this complication need not concern us.

[11] Thus, for instance, the scales upon which two melodies, each spanning an octave, were based might differ in the order in which they placed the intervals within the octave. Every acceptable arrangement of the octave's intervals (the 'species of the octave') will be found in some stretch of the GPS; and according to the theory of *tonoi*, the distances in the system between the locations of the various arrangements govern the possibilities for modulation between them; see pp. 215–28 below.

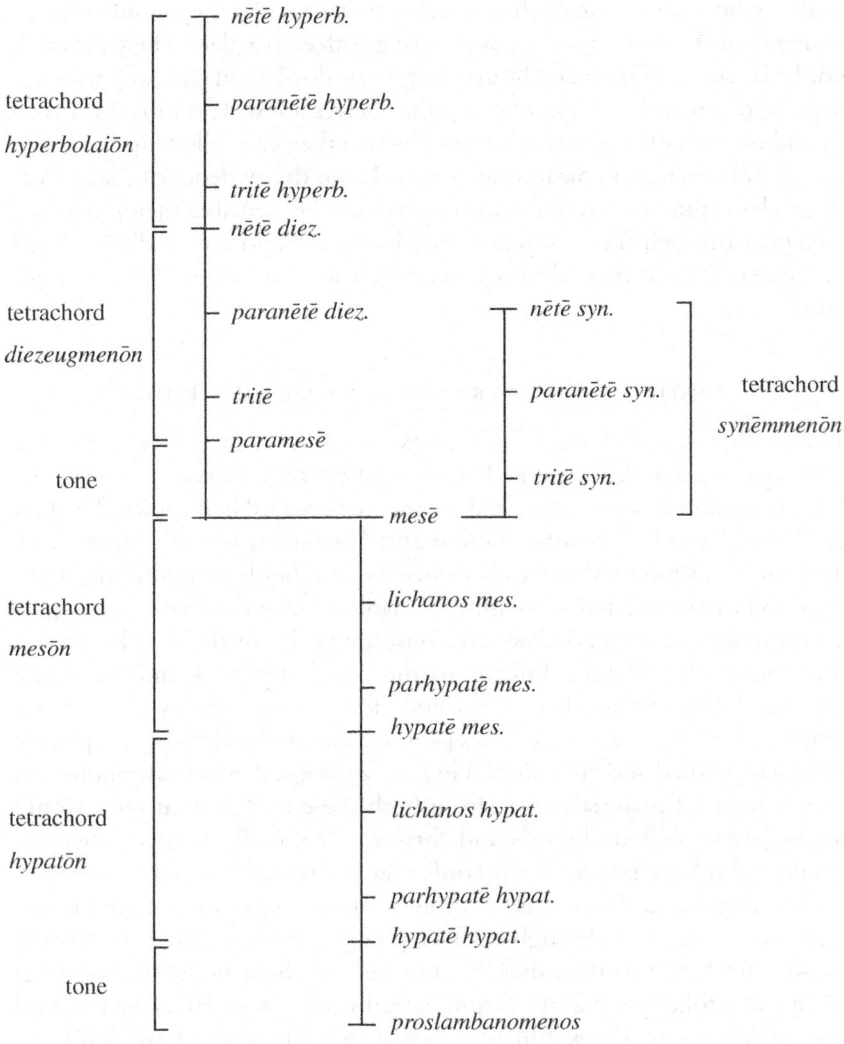

Figure 5 The Unchanging Perfect System

I have given here is only a sketch of that framework, calculated to provide some help with discussions elsewhere in this book. Even in those discussions many details have been elided. There is a great deal more to be said about the patterns of notes and intervals that were identified within the perfect systems and assigned musical roles, about the theory of *tonoi*, in which several instances of the GPS are located in different ranges of pitch, and about the perfect systems' historical precursors. The best advice I can offer readers who want to pursue these matters further is that they should retreat, armed with West 1992a and a pot of strong coffee, to a place where they will not be disturbed for several hours, and work slowly and carefully through his chapters 6 and 8. After allowing themselves a few days to recover they should do it all over again, in a library where they can follow up the lavish supply of references in his footnotes and absorb the evidence of the sources themselves (though acceptable coffee may not be available there). Anyone who goes through these exercises conscientiously and repeatedly 'will end up knowing about these things as accurately as anyone does' (Plato, *Meno* 85c11–d1)

A NOTE ON THE ARRANGEMENT OF THIS BOOK

The book is divided into three parts, together with a brief tail-piece which pulls a few threads together and glances for a moment at what the Hellenistic and Roman future had in store for classical harmonics. The first part includes only this Introduction and Chapter 1, where I introduce a division of harmonic theorists into two main schools or traditions. They seem to have originated at roughly the same time and to have run in parallel throughout the period we are considering. From the very beginning they apparently had quite different methods and objectives, and only rarely interacted with one another; by the end of the fourth century (though not much earlier) they were sometimes portrayed as implacable rivals, a picture routinely painted and embellished in later writings. A purely chronological arrangement of materials taken from both these traditions at once would require us to shift backwards and forwards repeatedly between them; it would, I think, be unnecessarily confusing and would highlight mere temporal succession at the expense of more essential continuities. I have therefore chosen to work through each of them separately. Part 2 is devoted to one and Part 3 to the other. Within each of those parts, the sequence of chapters follows a mainly chronological route (in so far as we can find one, which is sometimes difficult), except that wherever I have devoted a separate chapter to questions about 'contexts', I have placed it after my exposition of the 'contents' of the relevant theorists' work. It would make rather little sense to try to contextualise something without first finding out what it is.

CHAPTER 1

Beginnings, and the problem of measurement

The arrangement of this book is only partly chronological, as I have said. Still, one would like to begin at the beginning. But it is hard to know what would count as a solid historical starting point. New intellectual enterprises do not spring into being in a single bound, fully armed and fully recognisable; typically they trickle together as a confluence of tributaries which are themselves side-shoots of other, pre-existing traditions of thought, none of which 'is' this new thing, but whose developing combination can be seen, in retrospect, gradually to have become it. In any case, such information as we have about the earliest explorations of territory eventually claimed by this science is vague, unreliable and hopelessly entangled with legends and misconceptions that grew up in later times. Pythagoras, of course, is often identified as the fountain-head, and if that were true it would take us back to the late sixth century. There are also reports that his younger contemporary, the poet and composer Lasus of Hermione, wrote the first 'treatise' (*logos*) about music,[1] and allusions in two other writers seem to suggest that Lasus' work had some connection with issues in harmonics. But one of these latter references is too enigmatic to be interpreted safely;[2] in the second, which hints at connections with Pythagorean speculations in acoustics, there is a serious gap in the surviving text, and we do not know precisely what the author was asserting.[3] If Lasus did indeed write a work of the sort with which he is credited, it may not have survived for long after its author's lifetime; even Aristoxenus, who apparently had some inkling of its contents, may not have known it as a whole or at first hand. We are in no position to reconstruct Lasus' ideas or

[1] See Martianus Capella 9.936, and the entry under Lasus' name in the *Suda*. The latter gives no details; the former assigns to Lasus a set of distinctions between branches of musical theory which certainly originated in later times. See Privitera 1965: 37–46.

[2] Aristox. *El. harm* 3.21–4, which indicates that Lasus offered a rather curious view about the nature of a musical note. The passage does at least do something to confirm that a musical treatise by Lasus existed, and we shall revisit it briefly in Chapter 6.

[3] Theo Smyrn. 59.7–12.

to identify the relation, if any, in which his *logos* stood to the writings of the later theorists.[4]

The familiar figure of Pythagoras, mystic, mathematician, philosopher and scientist, is almost entirely a construct of the Pythagorean revival of Roman times, though its creators found some of the material from which it was built in writings of the fourth century. Very little of the colourful information retailed by Nicomachus, Porphyry, Iamblichus and others can be taken at face value, and the amount of reliable evidence they offer to link Pythagoras with harmonic science is vanishingly small. Much the same is true of the accounts we possess of the work of Pythagoras' followers down to the late fifth century, at least in point of detail. What we know, reasonably securely, about Pythagorean harmonics around 400 BC certainly presupposes an earlier tradition, which may go back to Pythagoras himself. But we are deceiving ourselves if we think that we can pin down its contents with any precision, let alone attribute specific ideas with justifiable confidence to particular individuals and dates.[5]

In the light of these dismal reflections I shall abandon the search for a historical beginning. I shall take as my starting point remarks made about their predecessors by writers of the fourth century, which seem to refer to ideas and activities current in the later years of the fifth. It may be possible, on the basis of a study of these reports, to work our way with due caution a little further into the past. But there is a general issue that needs to be faced first. In the rest of this short chapter I shall consider its broad outlines; more intricate details will be examined in later parts of the book.

A melody, a scale or an attunement is a complex of pitched sounds, or notes, whose pitches stand to one another in particular relations. It is in virtue of these relations that they are melodies, scales or attunements; and it is these relations that must form the principal focus of any attempt to analyse them. Students of these matters must therefore be able to conceptualise relations between pitches in a way that allows them to compare one relation with another, and to map groups of such relations into intelligible patterns; and they need linguistic resources – and perhaps others too, as we shall see – which will allow them to express, with a high degree of precision, the way in which one relation differs from another, and to specify the different

[4] For some fascinating discussion and suggestions see Porter (2007).

[5] The foundation-stone of all recent scholarship on the early Pythagoreans is Burkert 1962, with its English translation of 1972. Though some of Burkert's views have been disputed and others are still under debate, the book remains indispensable. For an admirable, brief and recent study see Kahn 2001, whose bibliography gives a useful guide to the modern literature and refers to other, more compendious bibliographical surveys.

characteristics of the patterns they form. In our own language (if we set aside for the present the perspective of scientific acoustics) we speak of notes as 'higher' or 'lower', as if they were placed as points on a vertical continuum of 'up' and 'down', and we talk of the relations between them as larger and smaller 'intervals', as if they were spatial gaps between these points of pitch, and could be measured and compared like distances along a line. This way of depicting the phenomena is simple and convenient, and to us it seems obvious. But of course we know that it is entirely metaphorical. It is not a direct, objective representation of the facts, but a way of thinking and talking which arises from the contingencies of our own culture and its history. Other peoples have other metaphors, and those to whose language this particular piece of imagery is alien may have no easy access to the mode of thought that it invites.

Just occasionally, a Greek writer speaks of notes as 'above' and 'below', *anō* and *katō*; but the usage is very rare.[6] Where we would call a note 'high', a Greek would most commonly describe it as *oxys*; where we would call it 'low' it is *barys*. But *oxys* and *barys* do not mean 'high' and 'low'; they mean 'sharp' and 'heavy'. I have argued elsewhere that these designations are not conceptually neutral, amounting to nothing more than another way of labelling the same distinctions of pitch that we make, but that they radically condition the way in which the Greeks experienced and envisaged the phenomena.[7] I shall not labour the point here. It is obvious, however, that these terms provide no ready access to a metric within which pitches can be precisely compared and the relations between them pinned down. 'Sharp' and 'heavy' are not even direct contraries, and there is no metric through which the 'sharpness' of one sound can be measured against the 'heaviness' of another.

The standard Greek word for 'pitch' is *tasis*, which literally means 'tension'; and another, rather more erudite way of calling a note 'high' or 'low' was to describe it as *syntonos*, 'tense', or *aneimenos* (sometimes *chalaros*), 'relaxed' or 'slack'. This usage too is in a certain sense metaphorical, though it need be none the worse for that; its source is probably in the observation that an increase in tension raises the pitch of an instrument's string. At first sight it seems more promising, since degrees of tension can be measured and compared with mathematical exactitude. But that is true only in principle. The Greeks had no reliable way of measuring tension, and the one method to which writers on harmonics refer, that of suspending larger or smaller

[6] It occurs at Hippocr. *De victu* I.18, [Ar.] *Problems* XIX.37, 47, and very occasionally in later sources.

[7] See Barker 2002b.

weights from strings, introduces unfortunate complications; pitch does not vary directly with the weights of the suspended objects.[8] Some relatively complex mathematics would be required in order to express accurately the relations between the pitches of an attunement by reference to tensions measured in this way. No Greek theorist seems to have understood the difficulties fully;[9] and no Greek theorist outside Pythagorean legend seems even to have attempted to measure musical relations by this procedure.

Nor did the words used by early musicians to designate particular intervals, so far as we can recover them, give any purchase to a system of measurement. What we call the perfect fourth, for instance, was called *syllabē*, 'grasp', or *dia tessarōn*, 'through four [strings]'. The former term seems to refer to the span of strings that the fingers can comfortably grasp on an instrument such as the lyre, and the second merely expresses the number of strings one passes across, in a regular form of attunement, in order to complete this interval. Neither tells us, in any relevant sense, 'how big' the interval is. The terms for the perfect fifth and for the octave are no more helpful. The former is *di' oxeiōn*, 'through the high [strings]', or *dia pente*, 'through five'; the octave is *harmonia* ('attunement', indicating that the normal compass of an attunement was an octave), or *dia pasōn*, 'through all'. None of this terminology has anything to do with the measurement of the relations between the pitches of notes standing, as we might put it, a fourth, a fifth or an octave 'apart' from one another.

How, then, are the relations between pitches to be identified objectively and compared with scientific precision? The issue, as my previous remarks have suggested, is essentially one of measurement. We can specify precisely the system of relations which gives a well designed building, for instance, its pleasing form, because these relations hold between measurable distances of height, length and breadth. (Of course this is an over-simplification, but measurement has an essential part to play.) It is much harder, and impossible without a sophisticated array of equipment and techniques, to give a comparable account of the relations between colours and tones in a skilfully balanced painting, since we have no way, outside a modern laboratory, of measuring shades and intensities of green, blue and the rest of them, and comparing them with one another on a single, objective scale.

[8] For references to this procedure see e.g. Theo Smyrn. 57.2–4, 60.7–9, Porph. *In Ptol. Harm.* 119.29–120.7, and the famous story of Pythagoras and the 'harmonious blacksmith', told most elaborately (and misleadingly) in Nicomachus, *Harm.* ch. 6.

[9] Difficulties involved in the procedure are noted by Ptolemy at *Harm.* 17.7–17 (cf. Porph. *In Ptol. Harm.* 120.33–121.10); but even he does not mention the fundamental problem, that the pitches of the notes do not vary directly with the weights attached to the strings, but with their square roots.

Serious harmonic analysis cannot begin without a system of measurement which allows relations between pitches to be expressed in quantitative terms. (Since the science focuses on the relations between pitches, rather than on the absolute pitches of the notes themselves, the capacity to measure pitches absolutely is unnecessary.) I do not mean that everything in harmonic science reduces to quantification; Aristoxenus, as we shall see later, does his best to minimise its role. But even he cannot proceed without it.

Methods of measurement in harmonics figure prominently in a well-known passage of Plato's *Republic* (530c–531c), which sketches the procedures of adherents to two very different schools of thought. The dramatic setting of the dialogue (itself written around 385 BC) would place their activities in the later decades of the fifth century. We shall consider the work of one group in more detail in Chapter 2 and that of the other in Chapter 10; for the present let the issue of measurement take centre-stage.

MUSICAL INTERVALS AS LINEAR DISTANCES

Responding (inappropriately, as it turns out) to a remark of Socrates, Glaucon describes the procedures of one set of theorists as follows.

> What they do is ridiculous, when they call certain things '*pyknōmata*', and bend their ears to the task as if trying to catch a sound from next door, some of them declaring that they can still just hear a sound in between, and that this is the smallest interval, by which measurement should be made, while others disagree, claiming that the notes sounded are already the same. (*Rep.* 531a4–8)

The task these people have set themselves, then, is to identify a unit 'by which measurement should be made'. It is to be the smallest gap between pitches that the human ear can pick out; larger ones will be 'measured' as multiples of that unit. Their procedure, as Socrates' sarcastic metaphors in his next speech make clear (531b2–6), involves adjusting the pitches of strings on an instrument by twisting the tuning-pegs, until two strings give notes so nearly identical with one another that they can approach no closer without reaching an apparent unison. When that situation is achieved, the unit of measurement has been found.

The crucial tools of the trade here are the ears. Under this procedure the unit involved is accessible to the hearing and to nothing else; there is no question, for example, of measuring the tensions exerted by the tuning-pegs. Since acuity of hearing varies from one individual to another, it is hardly surprising that this approach generated disagreements of the sort that Glaucon describes; and no system of measurement based directly upon it can be fully objective, or remain demonstrably constant when it is deployed by different researchers. There will be substantial difficulties, too, even for a single scientist who is undisturbed by others' doubts and has established such a unit of measurement to his own satisfaction, in using it for the purposes for which it was intended. In determining how many times the unit fits into the 'gap' between two notes of a scale, for example, he will apparently have laboriously to repeat the procedure by which the unit was established as many times as it takes to fill up the vacant space.

Another crucial implication of the passage is that these people were working with – or perhaps fumbling towards – a quasi-linear conception of pitch, in which the relations between pitches are thought of as gaps or spaces, some larger and some smaller, and hence measurable. I have already suggested that this mode of representation does not arise naturally from imagery inherent in the Greek language; it demands a degree of conceptual detachment from inherited cultural norms. Here we must consider the noun *pyknōma*, which Glaucon evidently regards as a piece of pretentious jargon. The adjective *pyknos* means 'compressed' or 'dense', and is applied to things whose constituents are packed closely together. Its usual converse is *araios* (sometimes *manos*), 'loosely packed', 'diffuse'. It plays significant roles in mainstream harmonic science, as we shall see later in this book; and in Chapter 2 we shall encounter another way in which cognate expressions are linked specifically to the people we are considering here (see p. 42 below). *Pyknōmata*, then, are 'densifications', complexes of items stacked tightly up against one another. The word entered the language of harmonics, if we may judge by Glaucon's reaction to it, as an abstruse technicality, not as part of the common coinage of every-day musical talk. Even the adjective *pyknos*, which is common in literature of every kind from Homer onwards, makes its first surviving appearance in connection with music in the technical writings of the fourth century. Despite their many descriptions of music, the poets of the preceding period never exploit, in this context, a contrast between 'dense' and 'diffuse'.[10] But the contrast does play important roles in the cosmological and scientific speculations of Presocratic philosophers

[10] Although the first occurrence of *pyknos* in harmonic works that have come down to us is in Aristoxenus' *Elementa harmonica* (late fourth century), where it is common, the present passage of the *Republic* allows us to infer its use a century earlier; it seems unlikely that theorists who talked of *pyknōmata* failed to describe these same complexes of notes, adjectivally, as *pykna*. The noun *pyknōma* itself occurs here for the first time in Greek of any sort (with the doubtful exception of a disputed reading at Aeschylus, *Suppl.* 235). Subsequently, like other nouns from the same stem, *pyknotēs* and *pyknōsis*, and the related verb *pyknoein*, it is used almost exclusively in philosophical, scientific or technical contexts, though the adjective was still used frequently in other settings. (*Pyknotēs* is contrasted with *manotēs* in a musical context at Plato, *Laws* 812d6–7, but in this instance is more likely to refer to the 'close packing' of notes in time than in pitch.)

and in the medical tradition;[11] and I suspect, though of course I cannot prove, that it was from them that it passed into the language of harmonics. If so, the contrast and the conceptual associations that it brings with it mark a link between the musical theorists and those who pursued rarefied intellectual research in other fields. We shall find in due course that it would not be the only stowaway from Presocratic philosophy to have embarked on a new career in harmonics.

There is no great distance between a representation of pitches as densely or loosely packed together, and a more explicitly linear conception of the 'dimension' of pitch. But I shall postpone any further examination of this issue, and of other evidence about this group of theorists, until Chapter 2. Let us now take a preliminary look at the other group mentioned in this passage of the *Republic*.

MUSICAL INTERVALS AS RATIOS

The direct information Plato provides here about their methods of measurement is slight and enigmatic. We are told that they 'measure audible concords against one another' (531a1–2), and that they 'search for numbers in those audible concords' (531c1–2). If we had to interpret those remarks in isolation we would be hard pressed to do it. We could tell that these people were engaged in attempts at measurement; that the objects measured were audible 'concords' (*symphōniai*) and notes; that the measurements were relative, not absolute, (they are measured 'against one another'); and that these relative measurements were somehow expressed in terms of numbers. From these modest certainties some inferences might hesitantly be drawn. Fortunately, however, we have other information that helps us to construe Plato's allusive remarks. We can identify it because the *Republic* identifies this group of harmonic specialists by name. They are the Pythagoreans (530d8, with e1–2 and 531b7–8).

One thing that is securely known about Pythagorean harmonics, throughout its history from the fifth century onwards, is that it did not represent relations between pitches as variable distances between points, but as ratios of numbers. Let us give this mode of expression some general consideration before buckling down to evidence and to detail. In Pythagorean harmonics, as in modern physical acoustics, the octave, for example, is represented by the ratio 2:1, the perfect fifth by the ratio 3:2 and the perfect

11 See e.g. DK 13B1 (Anaximenes), 28B8.59 (Parmenides), 30B7.8 (Melissus), 31B27.3, B75.1 (Empedocles), Hippocr. *Vet. med.* 22, *De victu* III.78, *Epid.* VI.3.1, *Mul.* I.1. The contrast also appears repeatedly in Plato's *Timaeus* and in Aristotle's scientific writings.

fourth by the ratio 4:3. The first question to be asked is how these ratios are to be understood as expressions of the relevant musical relations. They cannot be interpreted as reflections of ways in which the intervals might present themselves to our experience once we were in the right conceptual gear, as it were, in the way that the 'distances' of the linear representation can. No such conceptual gear seems to be available; no one, so far as I know, and certainly no extant Greek theorist, claims to perceive – say – the fourth note of 'Greensleeves' as one-and-a-half times as high-pitched as the first note, whatever that might mean (they are a perfect fifth apart, to which the ratio 3:2 corresponds). The relation between the phenomenon and its expression as a ratio must be of some other and less direct variety.

The route to understanding it, at least in the Greek context, runs through an examination of certain properties of the physical devices by which notes standing in these relations are produced. The simplest case is that of a taut string, of even thickness and constitution. If we pluck it to produce a note, and then stop the string with a bridge at its mid-point and pluck the half-length, the pitch of the second note will be an octave above the first. In this case, then, it is the lengths of string that stand in the ratio 2:1, the ratio assigned to the octave. Similarly, adjustments of the bridge's position to give lengths in the ratios 3:2 and 4:3 will produce notes in the relations of a perfect fifth and a perfect fourth respectively. It is usually and plausibly assumed that the Pythagoreans' practice of representing intervals as ratios began from just such observations as these.

Two problems of different sorts complicate this simple story. The first is that no early source mentions this way of using a string to demonstrate the correspondences between intervals and ratios. There is no record before the late fourth century of the instrument standardly used for this purpose in later times, the monochord or *kanōn*.[12] When Archytas, some seventy or eighty years earlier, discusses the means by which higher and lower notes are produced by different sound-sources, he does not mention strings at all. He does refer to the reed-blown pipes called *auloi*, and to Panpipes, on both of which the same correspondences can in principle be generated; but it is a great deal harder to produce and measure them precisely and consistently.[13] Writers of the Roman period tell us that Pythagoras and his followers 'demonstrated the ratios of the concords' on instruments of

[12] The earliest reference is in Duris frag. 23, though it asserts that the instrument was invented very much earlier, alleging that the inventor was really Pythagoras, and that the person who claimed it as his own, a certain Simos (whose date is not known) was doing so fraudulently.

[13] See Archytas frag. 1, a passage to which we shall return. For some of the difficulties in using wind-instruments for these purposes see Ptol. *Harm.* 16.32–17.7.

many different sorts, including *auloi*, Panpipes, sets of metal disks, strings subjected to different tensions by the attachment of weights, vessels filled to various levels with water, and so on.[14] Not all these attributions are believable, if only because several of them will not in fact yield the required results; and though such authors sometimes include the monochord in their list, they do not assert, in this context, that the early Pythagoreans assigned it special importance. Yet it is the only one that will perform the required task straightforwardly and reliably.[15]

Leaving that difficulty unresolved for the moment, let us turn to the other. Up to the point we have reached, the ratios describe only relations between lengths of string or pipe, and we have been given no way of understanding the notion that they characterise relations between notes themselves, let alone any reason for believing it. Yet Plato tells us that what the Pythagoreans measured were notes and 'concords', not pipes or strings, and that they are searching for numbers 'in' those audible phenomena. Many later writers in this tradition explicitly treat the ratios as descriptions of musical intervals as such, not just of the relative dimensions of the physical objects that produce them.[16] In this context the diversity of 'experiments' attributed to the early Pythagoreans begins to make sense, however unreliable the details offered by our authorities may be. If the same ratios are found to correspond to the same intervals, no matter what kind of physical agency – strings, pipes, disks or whatever – is used to create them, it is easily if not quite securely inferred that the ratios must somehow belong to the intervals in their own right. They are evidently not attached only to specific modes of sound-production. The ratio 2:1 seems to follow the octave around wherever it occurs; and this fact is most readily explained if its constituent pitches themselves can somehow be thought of as standing to one another in that ratio. Two strings, pipes, disks or other agencies whose relevant dimensions are related in the same way will, on this interpretation, emit pitches between which the same relation is reproduced.

What is needed, then, is an account of sound itself which allows its pitches to be construed as values of some property that varies quantitatively, values which are altered by changes in the dimensions of the sound-sources, and which vary in exactly the same ratios as those dimensions do. Here we return to the long passage of Archytas nowadays labelled as fragment 1. The key point he makes is that sound involves movement, and that – so Archytas

14 E.g. Theo Smyrn. 57.1–10, 59.4–61.17, Nicom. *Harm.* 248.13–18.

15 On the history of the monochord from earliest times to Ptolemy see Creese 2002.

16 Thus the Euclidean *Sectio canonis*, for example (see Ch. 14 below), consistently assigns ratios to the musical intervals (*diastēmata*) themselves.

argues – it is those sounds which move from their source more swiftly and vigorously, and which strike our ears with greater speed and force, that are perceived by us as higher in pitch. He underwrites his thesis with a barrage of examples drawn from observation, claiming support for it from the cases of human voices, *auloi*, *rhomboi* ('bull-roarers') and Panpipes; and the passage concludes with the statement: 'Thus the fact that high-pitched notes move more quickly and low-pitched ones more slowly has become clear to us from many pieces of evidence.'

Archytas' account involves a good many difficulties which need not trouble us here; other fourth-century and later writers modify and develop his ideas in a variety of ways.[17] Whatever its weaknesses and obscurities, however, it succeeds in representing sound and pitch in such a way that the expression of the difference between two pitches as a ratio becomes intelligible. The higher of two sounds pitched an octave apart moves (in some sense) twice as vigorously and twice as fast as the lower, and it is their speeds and vigours that constitute their pitches. We do not know exactly how Archytas interpreted this 'equivalence' between a movement's speed or force and the pitch of a sound. It seems easiest to understand, however, as the claim that pitch, as we perceive it, is not a feature of sound conceived as a physical event independent of the perceiver. We might go further, and say that what occurs as an independent physical event is not even, strictly speaking, a sound; alternatively we might say that in describing it as a movement we are penetrating behind mere 'appearance' and explaining what a sound 'really is'. (Archytas does not discuss this esoteric issue, but so far as we can judge from the evidence of frag. 1, he seems to have taken the latter approach.) At any rate, if sound or its physical counterpart is a particular kind of movement, any instance of it, like instances of movements of other sorts, has a speed. We distinguish it from other varieties of movement and classify it as a 'sound' merely because we can detect it only through our auditory apparatus, not through the visual and tactile organs which are capable of recognising movements and their speeds for what they 'really' are. Its 'pitch' is simply the guise in which its speed represents itself to our hearing.

Here, then, is a second and quite different way of measuring relations between musical pitches. A physical theory such as that expounded by

[17] See for instance Plato, *Tim.* 67b–c, 80a–b, Ar. *De anima* 420a–b, *De gen. an.* 786b–787a, [Eucl.] *Sect. can.* 148.3–149.24. The later tradition contains a multitude of repetitions of fourth-century theories and variations upon them. The most intricate and distinctive is Ptol. *Harm.* 7.17–9.15; Porphyry's commentary on the passage (*In Ptol. Harm.* 45.22–78.2) incorporates extensive quotations on the subject from other writers.

Archytas will allow the ratios of string-lengths and the like to be transferred to a property of the sounds themselves; and in this way we can understand Plato's contention that it is in the notes and their relations that the Pythagoreans search for 'numbers'.[18] Plato's implication, through the dramatic context of the *Republic*, that these approaches date back to the fifth century is supported, in some degree, by Archytas' account, since he introduces at least part of the theory as the work of his predecessors – how much of it is not altogether clear.[19]

THE TWO SYSTEMS OF MEASUREMENT COMPARED

There are four noteworthy differences between the two methods of measurement we have unearthed.

(i) One of them represents intervals as larger or smaller 'gaps' between items to which no features seem to be assigned other than audibility, position and separation within a special dimension of their own; they gradually come to be imagined as points strung out along a line. The other makes no use of this linear perspective. It portrays intervals not as distances but as ratios between the speeds of movements that travel through the familiar space inhabited by ordinary physical objects.

(ii) The first, but not the second, demands the identification of an auditory unit of measurement. Pythagoreans measured nothing by ear, though they identified by ear the intervals (octave, fifth, fourth and so on) that they were measuring.

(iii) Thirdly, in both cases the measurements were purely relative in the sense that they measured relations between notes and not absolute pitches. But where numbers appear in analyses of the first kind, larger numbers indicate larger intervals, reckoned as containing more of the elementary

[18] When transferred from string-lengths to pitches the terms of the ratios are usually reversed, the larger number being assigned to the higher pitch (corresponding to its supposedly greater speed, or to the greater value of whatever other variable is held to be responsible for pitch) instead of to the longer string, whose pitch is lower. This minor complication is not discussed in the early sources, and is mentioned only occasionally in later ones, e.g. by Thrasyllus, as quoted at Theo Smyrn. 87.9–18. For an intriguing exception – a theorist who argues that the larger number should be assigned to the lower pitch – see Theo's report on Adrastus at 65.10–66.11; but Adrastus wrote around AD 100, and his position corresponds to nothing known from the earlier period.

[19] The first twenty-three lines of the fragment, as printed in DK, are explicitly presented as a report about the work of Archytas' predecessors; the bulk of this passage is in indirect speech. The change to direct speech in the sequel may indicate a transition to views originating with Archytas himself. On this and other issues surrounding the fragment see Burkert 1972: 379 n. 46, Bowen 1982, and especially Huffman 2005: 103–61. But it would be unsafe to draw firm conclusions. The respectful attribution of important ideas to unnamed predecessors may be merely conventional; conversely, the shift into direct speech in line 24 may be no more than a literary convenience.

'units'; whereas in Pythagorean theory the sizes of the intervals do not correspond to the sizes of the numbers in the ratios. The interval correlated with the ratio 2:1, for instance, is larger than the one whose ratio is 4:3; and the expressions '2:1' and '4:2' are equivalent.

(iv) Finally and crucially, the project of the first group of theorists was an attempt to apply measurement to the auditory phenomena directly. They were trying to find a way of representing, in quantitative terms, the relations between items that appear in our auditory landscape; and although this could not be done without introducing new ways of conceptualising the phenomena, the new 'map' still had to be a recognisable representation of the topography of that landscape itself, as it presents itself to us. Pythagorean ratios, by contrast, give direct descriptions, in the first place, of relations between lengths of string or pipe or the dimensions of other such agencies, and secondly, by inference, of features of the physical events that are the immediate causes of our perceptions. They do not describe what we perceive as we perceive it. It can therefore be argued, as it was, later, by Aristoxenus,[20] that their approach is irrelevant to music and to harmonic science. On his view the relations that matter in music, and hence in harmonics, are those accessible to our hearing. The characteristics of the physical events underlying them, and of the mathematical relations between those events, may be of interest to a physicist, but have nothing to contribute to the analysis of musical structures as such; and if the ratios revealed by Pythagorean researches had been completely different, or even if the physical facts had been wholly incapable of being described in terms of ratios, the patterns within our auditory field that we perceive as well-formed scales, attunements and so on would have remained completely unaffected. It is clear that there is a major argument brewing.

[20] Aristox. *El. harm.* 12.4–34, 32.19–28.

PART II

Empirical harmonics

CHAPTER 2

Empirical harmonics before Aristoxenus

The study of Pythagorean thought, including harmonics, is firmly entrenched in the agendas of several scholarly disciplines, and the writings devoted to it in the last two centuries alone would fill a respectable library. The other style of harmonic analysis mentioned in the *Republic*, by contrast, has rarely been examined except – briefly – in commentaries on the *Republic* itself, and by a handful of specialists in ancient musical theory. The force of its exponents' claim to more generous treatment should not be exaggerated; they did not, like the Pythagoreans, pioneer a system of ideas which played a major part in the formation of scientific, philosophical and popular thought for over two millennia. But they deserve more than the occasional learned footnote. Their work broke new ground in the study of music; it casts valuable light on the history of the empirical sciences in the period before Aristotle; and it adds, in small but significant ways, to our knowledge of the environment and the practices of Greek culture between about 450 and 350 BC. In the present chapter I shall be exploring such evidence as we have about the character and content of their contributions to musical knowledge; the discussion will inevitably involve some technicalities, though none of them are alarmingly abstruse. In Chapter 3 I shall offer some thoughts about the broader context within which their researches took place, about the nature of the audience to which they were addressing their ideas and the purposes to which they put them.

None of these people's writings survives, even as fragments, and there is no single, definitive ancient source of information about them, though once, perhaps, there was.[1] There are enough allusions in texts written within a century or so of their time, and in one or two later sources which are probably reliable, to encourage the hope that we might be able to piece together a tolerably coherent and informative picture; and I think we can. But the

[1] At *El. harm.* 2.28–30 Aristoxenus mentions a previous work in which 'we examined the opinions of the *harmonikoi*'. The passage is quoted more fully on p. 39 below.

task is not a straightforward matter of locating factual reports about them in the available Greek writings. The evidence is patchy and fragmented. Any attempt to make sense of it as a whole will have to rely on more or less plausible guesses about the missing pieces, and to fasten the *disiecta membra* together with some rather speculative glue. Most of the surviving reports appear, furthermore, in the writings of hostile or even contemptuous witnesses (we shall consider some of the reasons for their disdain in the next chapter), and in order to arrive at an unbiased interpretation we shall have to work our way through layers of prejudice and distortion.

THE EVIDENCE OF PLATO

In the *Republic*, as we have seen, Socrates and Glaucon are more interested in denouncing these theorists' work than in expounding its details or examining its credentials, but at least one solid piece of information emerges. They are represented as seeking to identify the smallest interval accessible to the human ear, and they are doing so in order to establish a unit of measurement, such that any larger interval will be measured by the number of these units it takes to fill up the 'space' between its boundaries.

There is nothing in our texts to suggest that exponents of this method regarded the unit as minimal or indivisible in principle. Though their procedures may stand in some distant relation to debates among mathematicians and philosophers about the infinite division of spatial magnitudes, their approach by-passes the abstract issues involved in these controversies. What they were listening for is the smallest interval that human hearing can distinguish; it is the interval at the limit of our powers of perception, not at the limit of theoretical possibility, and the problems which beset their attempts to identify it are not of an abstract sort but arise straightforwardly from variations in the acuteness of different people's ears. They were concerned with the business of making measurements on a basis which all listeners could accept as reliable for practical purposes, and though agreement on this matter was hard to come by, there is no indication that they tried to explore or resolve the difficulties by philosophical or mathematical means, or that they saw any relevance in intellectual conundrums about the division of continua. Their unit is simply the smallest interval that we can identify and re-identify by ear, and as such it must also be the smallest interval that could be recognised as a step in a musical melody or scale.[2]

[2] In addition to the passage of the *Republic*, see Aristotle, *De sensu* 446a. For the contrast between units that are indivisible in principle and those that are so only so far as perception can judge, see *Metaph.* 1087b–1088a, where the example of the 'unitary' musical interval again appears; and cf. Aristox. *El. harm.* 13.30–14.11.

The theorists' apparent inference that every larger interval must be regarded as an exact multiple of this unit may seem unsafe, since the fact, if it is one, that our ears can pick out no smaller interval than (for instance) a quarter-tone does not at first sight guarantee that an interval such as one spanning one-and-a-third tones is unidentifiable and musically unusable; but their position can be defended. If the quarter-tone and its multiples can be identified reliably, we can identify the intervals of one-and-a-quarter and of one-and-a-half tones; and if no smaller interval than the quarter-tone can be recognised, we cannot recognise the difference between those intervals and any interval that lies between them. We may indeed be able to hear that the interval presented to our ears is larger than the one and smaller than the other, but we cannot say by how much. It has no determinable identity.

The fact that the function of this minimal interval was to provide an auditory basis for measuring others allows us to infer that their project involved quantifying the intervals which the ear encounters, and which are reckoned significant in the context of our experience of differently pitched sounds. Since it is only in music that clearly distinguishable intervals between audible pitches play an important role, it is only in connection with real music, that is, the music that is actually performed and heard, that their method can have any plausible application. It is more than a reasonable guess, then, that their aim was to specify, as precisely as possible, the various sizes of the auditory 'spaces' that separated the notes of the scales and attunements used by contemporary musicians, and perhaps also, more ambitiously, those of the melodies they composed and performed. The final product of such an investigation would be a fully quantified description of the patterns of intervals that made up these musical systems and sequences (perhaps in the form 'a step of n units followed by a step of m units', and so on). It would also allow the sizes of different intervals, and the sequences of intervals in different musical constructions, to be compared directly with one another.

Socrates seems to be alluding to analyses of these sorts, perhaps among others, in a later Platonic dialogue, the *Philebus* (17b–e). In order to be a 'musical expert' (*sophos tēn mousikēn*), he says, a person must understand much more than merely the differences between higher, lower and equal pitches. They will deserve this description only when they have grasped 'the intervals of sound in respect of high pitch and low – how many they are in number, and what qualities they have – and the boundaries of the intervals, and the number of systems (*systēmata*) which have come into being from those intervals, which our predecessors examined and passed down to us, their successors, under the title "*harmoniai*"'; and when, in addition, they have reached a comparable understanding of rhythms and metres. I shall not comment here on the more enigmatic features of this passage.[3] What seems clear is that this 'musical expert' will be in command of information arising from an extensive and detailed analysis of musical structures. He will be able to describe and enumerate all the intervals of which music makes use, to locate their boundaries, and to give an account of the *systēmata* ('scales') or *harmoniai* ('attunements') of which they are components. It is worth noting Socrates' indication that this knowledge was established at least a generation before the time at which he is speaking; and though the dramatic date of the *Philebus* is indeterminate, it must at least be prior to Socrates' death in 399 BC. The dialogue itself was probably written several decades later, towards the end of Plato's life (perhaps around 355 BC); but if its author is playing fair with his readers, Socrates' apparently unmotivated reference to their 'predecessors' suggests that they were already at work by about 420 BC, and perhaps substantially earlier.

In the *Republic* Plato uses no single word as a title for studies of this sort, or for those who pursue them. He speaks only of people who conduct enquiries 'about *harmonia*' (*peri harmonias*, 531a1, cf. 531b8).[4] The *Philebus*, similarly, has no title for a harmonic theorist which would distinguish him from any other sort of musical specialist; the person who knows about intervals, scales and so on is described simply as a *mousikos*, or as *sophos tēn mousikēn* ('expert in music'). In a brief passage of the *Phaedrus*, however (268d–e), Socrates names a specific field of enquiry as *ta harmonika* (though he does not identify its defining features), and a specialist in it as a *harmonikos*. In Chapter 3 we shall need to consider the characteristics assigned to such a person in the *Phaedrus*, and his relation to the other hypothetical individual, described as a *mousikos*, with whom he is confronted in Socrates' imagination. All we need to note for the present is that the term *harmonikos* was available for use by the mid-fourth century, and that Plato saw no need to explain its meaning; he assumes that his readers will know quite well, at least in outline, what a *harmonikos* was and what he did.[5] In view of the word's uses in writings of the next generation, we can fairly assume (and I shall argue further in Chapter 3) that his activities would have fitted the description provided in the *Philebus*, despite the difference in terminology, and probably that the word by itself, as Plato employs it, does not pick out

[3] Some of them are discussed in Barker 1996.

[4] *Harmonikē*, the feminine form of the adjective *harmonikos*, regularly used by later authors as the name of the science we call 'harmonics', in fact appears nowhere in Plato's writings.

[5] Plato had already used the adjective in its neuter form in an earlier dialogue, to refer to the contents of a branch of knowledge which is acquired through expertise in *mousikē*; see *Laches* 170c.

harmonic theorists of any one specific type. Aristotle uses *harmonikos* and related terms to refer to exponents of harmonics in both its mathematical and its empirical versions, where necessary adding a qualifying phrase to make it clear which he has in mind.

THE EVIDENCE OF ARISTOXENUS

(a) The harmonikoi *and the enharmonic genus*

Aristoxenus, however, towards the end of the fourth century, uses the word in a more specialised way. He repeatedly refers to earlier theorists as 'the *harmonikoi*', but he does so only when alluding to those whom he regards as his own legitimate predecessors, that is, to those who adopted an empirical rather than a mathematical approach to the subject. His *harmonikoi* turn out to be theorists of just the same sort as the non-Pythagorean students of harmonics mentioned in the *Republic*. He draws, in fact, a distinction which apparently marks off exactly the same two groups as Plato does, though his perspective on them is different. Among earlier theorists, he tells us, there are those who 'stray into alien territory and dismiss perception as inaccurate, devising theoretical explanations and saying that it is in certain ratios of numbers and relative speeds that high and low pitch consist, stating propositions irrelevant to the subject and in total conflict with what we perceive' – that is, mathematical theorists including the Pythagoreans; and there are those who 'utter oracular pronouncements about various topics without explanation or proof, and without even properly enumerating the perceptual data' (*El. harm.* 32.20–31). Though Aristoxenus' description here of the latter group, taken by itself, is fairly uninformative and characteristically abusive, it at least suggests that they were attempting, however feebly, to 'enumerate the perceptual data', that is, to describe real musical structures in the guise in which the ear perceives them. Unlike the Pythagoreans, whose work he considers irrelevant to his concerns, he elsewhere treats the exponents of this second approach as engaged, however incompetently, in a similar project to his own; and his comments on them not only establish their connection with the 'empiricists' of the *Republic*, but provide a substantial body of additional information about their views and procedures. It is to them that he attaches the label 'the *harmonikoi*'.

Aristoxenus' first comment on these people in the *Elementa harmonica* incorporates a rather laboured joke, based on a pun; they showed that they really deserved the title '*harmonikoi*' (in the sense 'experts in the study of *harmonia*, musical attunement') because they dealt only with systems in the enharmonic genus (which in his writings, when he refers to it by a noun, is also called 'the *harmonia*'). 'Of the other genera', he says, 'they had absolutely no conception' (2.7–11). We shall discuss these 'genera' more fully in later chapters. For the present we need only remind ourselves that a scale in any of the three genera that Aristoxenus and others identify, enharmonic, chromatic and diatonic, divides up each interval of a perfect fourth between the fundamental, 'fixed' notes of the system in its own distinctive way. A scale in the enharmonic genus, for instance, locates the two 'moveable' notes between the boundaries of such a fourth low down in its span and very close together; on Aristoxenus' own analysis the sequence of intervals between the four notes of an enharmonic tetrachord is quarter-tone, quarter-tone and ditone, reading from the bottom up.[6] By contrast a familiar type of diatonic tetrachord (there were several others) follows the sequence semitone, tone, tone; and one type of chromatic tetrachord has the form semitone, semitone, tone-and-a-half (these quantifications too are those of Aristoxenus). It is obvious at a glance that melodies constructed from the resources of any one of these systems will have created a markedly different aesthetic impression from those based in either of the others.

Distinctions between enharmonic, chromatic and diatonic systems do not seem to have been formalised and assigned these names until the fourth century. Even then there are very few traces of them in the surviving literature before the time of Aristoxenus himself. Our major pre-Aristoxenian sources, Plato and Aristotle, do not mention them at all, and use an entirely different way of dividing melodies and attunements into types.[7] Just one piece of early fourth-century writing, an anonymous fragment dating probably from the 380s, mentions diatonic, chromatic and enharmonic music (not 'genera'), but its treatment suggests that distinctions between them in that period were still hazy (see pp. 70–1 below). Around the same time the Pythagorean Archytas, whom we shall consider in Chapter 11, apparently recognised distinctions corresponding broadly to those of Aristoxenus, and offered mathematical analyses of systems of all three types. But there is no

[6] A tetrachord, in this usage, is a sequence of four notes separated by three intervals jointly spanning a perfect fourth. The tetrachords most commonly and importantly discussed by the theorists are those whose highest and lowest notes are 'fixed' notes, for instance *mesē* and *hypatē mesōn*. Larger systems are regularly conceived as sequences of such tetrachords, either 'conjoined' with one another (so that the highest note of one is also the lowest of the next) or separated ('disjoined') by the interval of a tone. See the diagrams on pp. 13–17 above.

[7] The principal distinctions they draw are between the systems of attunement which they call '*harmoniai*'. There are difficulties in deciding precisely what these *harmoniai* were and how they differed from one another (see pp. 43–55 and 309–11 below), but they certainly did not correspond to the genera of Aristoxenian theory.

good evidence to indicate that this method of classifying scales and attunements was current in the fifth century, to whose second half some at least of Aristoxenus' *harmonikoi* probably belong, even though later theorists were perfectly prepared to apply their 'generic' distinctions to the compositions of that period as well as to those of their own.

When Aristoxenus announces that these theorists studied only 'the *harmonia*', that is, patterns of attunement in what he calls the enharmonic genus, he is therefore unlikely to mean that they themselves stated the fact in those terms, and explicitly disavowed any interest in chromatic or diatonic systems. What he means, as his next remarks make clear, is that a close inspection of their work would reveal no trace of anything that looked like an analysis of a chromatic or a diatonic scale. The patterns of intervals they specify are in his opinion inaccurately drawn; but they can only be interpreted as fumbling attempts at the representation of enharmonic structures. 'Here is the evidence', he says. 'The diagrams they set out are those of enharmonic systems only, and no one has ever seen any of systems in diatonic or chromatic' (2.11–14).

The evidence Aristoxenus had at his disposal, then was a set of diagrams. He continues:

> And yet the diagrams in which they spoke only of enharmonic octachord systems did represent the whole ordering of melody; but about the other magnitudes and arrangements in the enharmonic genus itself, and in the others, no one even attempted to learn anything. Instead, they cut off, from the whole of melody, just one magnitude, the octave, in just one of the three genera, and devoted all their attention to it. The fact that they worked incompetently even on the topics they happened to address became obvious to us in our earlier discussions, when we were examining the opinions of the *harmonikoi*. (2.15–30)[8]

Aristoxenus' confident assertion that his predecessors worked only on enharmonic systems is less surprising than it may appear. By the standards of our own culture, melodies whose scale divides the perfect fourth into two quarter-tones and a ditone would seem very strange, and such music was alien to musical practice in later phases of Greek antiquity.[9] In Aristoxenus' own time, though putatively enharmonic melodies were still performed, there was already a tendency, he tells us, to 'stretch' its tiny quarter-tones, and to make this style of music approximate to a form of the chromatic

[8] The earlier treatise or lecture to which this remark refers is lost; some traces of it may survive in passages of the Plutarchan *De musica*.

[9] See for instance Ptol. *Harm.* 38.2–9, Arist. Quint. 16.13–18. Of the surviving musical scores, only those preserving fragments of (probably) fifth-century music, that of Euripides, are recognisably enharmonic; see the fragments numbered 2 and 3 in Pöhlmann and West 2001: 12–21.

(*El. harm.* 22.30–23.22). Aristoxenus himself describes genuine enharmonic music as the most sophisticated kind of all, to which human perception becomes accustomed 'only with difficulty and after a great deal of work' (19.27–9). But he also makes explicit his opinion that such music, when properly performed, is the finest music there is, and that it is to be found in musical styles current in the distant past (23.3–12; see also p. 297 below). There is other evidence, too, that in fifth-century music of high cultural status, especially that of Athenian tragedy, enharmonic forms of melody predominated.[10] In restricting their studies to the patterns of enharmonic convention, the *harmonikoi* were simply concentrating on the music that had the most eminent credentials in their own environment. Aristoxenus' remarks therefore provide further grounds for the conclusion that their work was focused, though admittedly in a selective, even elitist way, on the practices of real musicians at work in contemporary culture. They were directly concerned with music, not, like some of the Pythagoreans, with the application of musically inspired ideas to researches in another domain.

One more preliminary point must be made before we examine other aspects of Aristoxenus' evidence. In the passage I have quoted, and sometimes elsewhere, he treats these predecessors as an undifferentiated group, as if all of them followed identical procedures and reached identical conclusions. This is, on the face of it, unlikely. I shall argue below that they were not all members of some single 'school' or organised tradition; and there are passages elsewhere in the *Elementa harmonica* which record disagreements between them and differences of approach, or which identify specific individuals as having examined topics or adopted views which others did not.[11] Hence when we come across a remark about *harmonikoi* in one part of Aristoxenus' work which seems to conflict with something he says about them in another, we need not assume that one of these statements or the other is mistaken. He may simply have had different people in mind. We must nevertheless interpret his accounts with caution, not because the evidence he possessed or his understanding of it was unreliable, but because he makes no attempt to offer a balanced report of his predecessors' work.

[10] See *P.Hib.* 1.13, discussed briefly below (pp. 69–73), [Plut.] *De mus.* 1137e–f, Plut. *Quaest. conv.* 645e, and ch. 5 of the Byzantine treatise *On Tragedy* sometimes attributed to Michael Psellus (the most recent edition is Perusino 1993); cf. the scores mentioned in n. 9 above.

[11] Some of them will be discussed more fully below. Summarily, the most important passages are these: 5.9–22, 6.12–31, where a certain Eratocles is credited with studies not undertaken by others; 7.29, referring to 'some of the *harmonikoi*'; 36.26–37.1, where the work of Pythagoras of Zakynthos, Agenor of Mytilene and unspecified others is said to differ in certain respects from that of another group (which may include Eratocles); 37.8–38.5, recording disagreements about relations between the structures known as *tonoi*; and a long stretch of polemical argument at 39.4–43.24, belabouring the exponents of two quite different conceptions of the nature, basis and aims of harmonic science.

He consistently adopts the posture of prosecuting counsel, and if we are to make sense of what they were doing, and why, we have to read rather carefully between the lines of his denunciations and complaints.

(b) The harmonikoi *and their diagrams*

When Aristoxenus claimed that the *harmonikoi* concentrated exclusively on enharmonic systems, the core of his evidence consisted, as we have seen, in a set of diagrams, which were evidently representations of forms of attunement or scale. They were presumably designed as economical and readily intelligible summaries of the main results of their authors' researches; and Aristoxenus' repeated assertion that his predecessors failed to 'demonstrate' their conclusions or to ground them adequately in 'principles' suggests that they did not accompany their diagrams with fully argued written explanations.[12] Aristoxenus does not tell us explicitly what form their diagrams took, but his hints, in passages which mention them directly or seem to be alluding to them, allow us to reconstruct their general outlines fairly confidently. Several times in such passages he refers to a procedure called *katapyknōsis*, roughly 'densification', and sometimes he relates it explicitly to the diagrams themselves. On one occasion he says that they insisted on the task (which he regards as pointless) of 'densifying the diagram' (*katapyknōsai to diagramma*, 7.32), and on another he speaks of their 'densifications of the diagrams' (28.1–2). His references are regularly linked to the comment that the procedure is quite useless for certain purposes that he himself has in mind, and it must therefore be one which might plausibly, though erroneously, be thought of as casting some light on the issues in question. They are connected with at least two quite different topics on Aristoxenus' own agenda, but there is a common thread. Both of them are concerned with the distances between notes, or between whole systems of notes, which are to be reckoned, in a musical sense, as being 'adjacent to' or 'successive with' one another. It is in the project of reaching conclusions on matters of that sort, so Aristoxenus asserts, that *katapyknōsis* is useless.

One of his tirades in this vein is particularly enlightening. He has raised the question how melodic 'succession' or 'adjacence' (*to hexēs*) is to be understood, that is, on what basis we can say that one note stands 'next

[12] James Porter has pointed out to me that Aristoxenus' silence does not prove that they provided no written explanations of their theses, and of course he is right; Aristoxenus may have had his own (polemical) reasons for suppressing them (see also p. 54 below). But in that case (with a very few exceptions which will be mentioned in due course), we have no clue to their contents. We know only that in Aristoxenus' view they did not amount to 'demonstrations' (see further pp. 104, 153–4 below).

to' another in a scale or similar system, and that no notes can properly be inserted between them. This is not a question in physics, or about the ear's capacity to discriminate very small intervals. In any scale there will be notes that are in this musical sense neighbours, but which are separated in pitch by a gap that plainly could contain others capable of being distinguished by the ear. Nevertheless, from the perspective of this particular form of scale, the notes bounding the gap are musically successive. Aristoxenus' question is what it is that makes them so:

> To put it simply, succession (*to hexēs*) is to be sought in accordance with the nature of melody, and not in the way in which those who had an eye on *katapyknōsis* used to represent continuity (*to syneches*). For they seem to neglect the sequence of melody, as is clear from the multitude of dieses which they place in succession, since the voice cannot string together [musically] even as many as three. Hence it is clear that succession is not to be sought always in the smallest intervals, or in their equality or inequality, but we must follow what is [melodically] natural. (53.2–15)

Katapyknōsis, then, involved setting out (in a diagram, as the other references show) an extensive sequence of the tiny intervals called 'dieses', and representing the pitches bounding them as 'successive'. This would appear to mean that no melody can place any note between them. In Aristoxenus' usage, except where he explicitly qualifies the term to change its application, a diesis is a quarter-tone; and it is always so in contexts dealing with the enharmonic genus. The term *katapyknōsis* is cognate with the expression *to pyknon*, which Aristoxenus uses regularly (in enharmonic contexts) to refer to the miniature structure formed by the two quarter-tones at the bottom of the tetrachord. The diagram was therefore divided into steps of a quarter-tone each. It may have been no more than a simple line marked off at equal distances representing these successive dieses, upon which the notes of a scale were then mapped. *Katapyknōsis*, of course, also recalls the word *pyknōma* which we met in the *Republic* (p. 24 above), in connection with the project of establishing a minimal interval 'by which measurement is to be made'. The theorists whose work Aristoxenus knew had evidently settled on the identification of this unit with one quarter of a tone.

We should not be bamboozled by Aristoxenus' insinuation that by the standards of these *harmonikoi* there could be a musical scale, a usable melodic sequence, consisting of a couple of dozen consecutive quarter-tones. It is in the last degree improbable that they supposed anything of the kind; their dieses are 'successive' only in the sense that no scale can introduce a note that falls between the boundaries of a 'unit'. What they have 'densified' with their thicket of quarter-tones is a diagrammatic grid upon which musical structures are projected. It is not itself a musical structure of any sort. Aristoxenus' own closing statement in this passage gives the game away, in fact, with its reference to (and rejection of) the idea that melodic succession can be defined in terms of equal and unequal intervals, pinned down, that is, to specific sequences of intervallic 'distances'. His own approach to the matter is complex and will be discussed in the appropriate place; but his remark indicates that his predecessors had tried to identify what counts as a melodically successive series of notes by specifying the sequence of distances, equal or unequal, that separate them in a continuous scale. In this context each such distance would be identified as a step of so many dieses, and would be marked as such on the diagrammatic grid.

The passage from which we began (2.7–30) speaks of 'diagrams' in the plural, and says that they represented 'enharmonic systems', not just one single structure. Given that they were all recognisable as enharmonic, they cannot have been distinguished from one another by the sizes of their elementary intervals. Each must have been made up, at least predominantly, of ditones and pairs of quarter-tones (together with the 'disjunctive' tone that may separate one tetrachord from another), or have approximated reasonably closely to that pattern. Nor did the analyses mentioned here differ by representing structures with different compasses; they were all 'enharmonic octachord systems', and each of them spanned an octave. One of Aristoxenus' references to a theorist named Eratocles might be construed as providing a clue about the ways in which they differed, and how they were related to one another. So it does; but I must issue advance warning that the passage's evidence may not bear on these questions quite as directly as at first appears. Eratocles' approach, in my view at any rate, was probably not typical of *harmonikoi* in general.

(c) Eratocles' systematisation of the 'ancient harmoniai*'*

What Aristoxenus says is that Eratocles attempted 'to enumerate the forms of one *systēma*, the octave, in one genus, without any demonstration, by moving the intervals around cyclically' (6.21–5). To this we may add a later reference to people who focused all their attention on 'the seven octachords which they called *harmoniai*' (36.30–2).[13] As writers of an Aristoxenian

[13] The MSS here have *heptachordōn*, 'heptachords'; the emendation *hepta oktachordōn*, 'seven octachords' was proposed by Westphal. It has been adopted by most modern editors and must surely be correct, in view of Aristoxenus' contention that his predecessors studied only 'octachord *systēmata*' (2.15–18), and his discussion of the seven octave-systems constructed by Eratocles (6.21–31).

persuasion in the Roman period confirm, these allusions are to a procedure for generating different patterns of attunement from a single basic structure by taking away (for example) the interval at the top of a given system and relocating it at the bottom.[14] Since there are seven intervals in a typical octave, the procedure generates seven octachords constituting seven different ways in which an instrument such as a lyre might be attuned. Later writers also tell us that the seven systems were assigned names closely related to (in some cases identical with) those that had traditionally been attached to the so-called *harmoniai*, the different ways of tuning an instrument which enabled executants to perform melodies in different melodic styles (these are the familiar names 'Dorian', 'Phrygian' and so on).[15]

But it is obvious that the collection of attunements produced in this way is desperately artificial. It is beyond belief that the Dorian, Phrygian and Lydian styles of melody, for instance, which entered Greek culture from different ethnic and geographical sources and retained quite distinct ethical and aesthetic connotations, were related to one another, in musical practice, in so neat and systematic a way. We are certainly dealing with a theorist's rather than a musician's perspective. Yet the results of Eratocles' project must have been conceived as standing in some tolerably close relation to the *harmoniai* that were in real contemporary use. Though the names used in the earlier musical tradition and those of the theorists do not match completely, there is substantial overlap; and names from the same stable could hardly have been applied both to the traditional attunements and to the scientifically 'rationalised' constructions if there were no affinities between them at all. To make any more precise judgement about the degree of distortion imposed by the Eratoclean system we would obviously need some detailed information about the attunements actually current among musicians of the time; and on this occasion the haphazard accidents through which such information is preserved or lost may have worked in our favour. There is a good chance that some such information survives.

The evidence is buried in the work of a much later writer, but a link is provided by Aristoxenus himself. In one of his polemical diatribes about the *katapyknōsis* of diagrams he specifies the number of dieses (quarter-tones)

[14] Thus (where q = quarter-tone, d = ditone and t = tone), if the first *harmonia* has the form q,q,d; t; q,q,d, the second would be d; q q d; t; q,q, the third would be q,d; q,q,d; t; q, and so on.

[15] Our sources sometimes record two different names for certain of the systems. In these cases one name is clearly a theorist's coinage, the other a name taken from the usage of earlier musicians. The implication is that the theorist's construction corresponds directly to an attunement used in practice in 'ancient' times; but this (as I shall argue immediately) may be prompted more by wishful thinking than by hard evidence. See e.g. Cleonides 197.4–198.13, Bacchius 77, 308.17–309.9, Arist. Quint. 15.9–19.

in the diagram as twenty-eight (28.6–7). But there are only twenty-four quarter-tones in an octave; and the 'seven octachords' can most conveniently be mapped out in relation to one another if they are all represented within the same twenty-four-unit span. What, then, is the role of the additional four? The answer may be provided in a passage of a treatise dating from the third century AD, the *De musica* of Aristides Quintilianus. The passage is at *De mus.* 18.5–19.10. What it purports to record are the structures of the *harmoniai* mentioned in Book 3 of Plato's *Republic*, to which Plato refers by their traditional 'ethnic' names, Dorian, Phrygian and so forth (*Rep.* 398e–399a). These *harmoniai*, as Aristides portrays them, are not all octachords, and not all of them have the compass of exactly an octave. Several are less extensive than that; but one exceeds the octave by a whole tone, so occupying the space represented by twenty-eight dieses in the diagrams of the *harmonikoi*.

No other writer mentions the strange and irregular *harmoniai* described by Aristides, and we must pause to consider the credentials and probable sources of his report. It is virtually certain that they are not deliberate forgeries, by Aristides or anyone else; ancient literary fabrications are rarely so technical, and are typically constructed out of familiar (if anachronistic) commonplaces. The unparalleled eccentricity of these *harmoniai* can have no such basis, and by itself effectively rules the hypothesis of forgery out of court. Where, then, did Aristides find them? He refers to them in two ways, first as systems 'used for the *harmoniai* by people of distant antiquity', and later as 'those that the noble Plato mentions in the *Republic*', with a direct citation of *Republic* 398e–399a (18.5–6, 19.2–7). The fact that the names given in Aristides' account correspond very closely to those mentioned at that point in Plato's text makes it probable that they were indeed linked to this dialogue in some quite intimate way;[16] and it has sometimes been suggested that Aristides' source was a commentary on the *Republic* itself. That is possible, but I think it unlikely. If the report had appeared in that context, in a work surviving in the intellectual world of the third century AD, it would be very surprising that no echo of its remarkable analyses reappears elsewhere in the writings of this and subsequent centuries, in which studies of Plato's dialogues multiplied, and his ghost stalked through the pages of every essay in scholarship or philosophy. Secondly, the structures of the *harmoniai* are not expressed in the manner one would expect of a Platonist writer, that is, in terms of numerical ratios. Their intervals are represented

[16] Plato mentions Dorian, Phrygian, Mixolydian, Iastian, 'tense Lydian' (*syntonolydisti*), and 'some forms of Lydian that are called "slack" (*chalarai*)'. Aristides' list includes all of these and no others, except that the last type is replaced by a simple 'Lydian'.

in the language of quarter-tones, ditones and the like, as linear 'distances'; their mode of presentation, in fact, is wholly Aristoxenian.

We should therefore consider the possibility that the immediate authority for Aristides' information was a work by Aristoxenus himself, or perhaps one by an intermediate source which summarised or excerpted Aristoxenian material. There is much to encourage this hypothesis. For one thing, the passage in Aristides' treatise, in Chapter 9 of Book 1, is embedded in a sequence of chapters retailing material drawn exclusively (though not necessarily directly) from Aristoxenus, including parts of Chapters 4–5 and the whole of 6–8, everything in Chapter 9 (excluding for a moment the passage under discussion), the whole of Chapters 10–11, and probably but less certainly Chapter 12. Secondly, Aristoxenus was by far the most prolific ancient writer on music. He touched on a wide range of topics, many of them dealing in one way or another with the music and musical ideas of earlier periods. As he tells us himself in the *Elementa harmonica* (n. 1 above), he had already composed a study of the work of previous musical theorists. Passages in another *De musica*, the one attributed conventionally but falsely to Plutarch, show that he was a passionate advocate of musical forms and styles which in his own time were already antique, and that his discussions of them included analyses, comparable to those found in Aristides, of the patterns of intervals that some of their structures employed. The scales described in the Plutarchan treatise are as irregular, by the systematic standards of later theory, as are those that Aristides records.[17]

Either Aristoxenus' essay on earlier theorists or a work of musical history such as that excerpted in the Plutarchan compilation could therefore have appropriately contained Aristides' information. More generally, it seems to me impossible to make out a persuasive case for any other known author as his source. The terms in which the account, as it stands, is set, betray the hand of a writer familiar with Aristoxenian usage; hence it cannot have been offered in this form before the late fourth century. Though several of Aristoxenus' contemporaries and near-contemporaries (Theophrastus, Heraclides of Pontus, the author of the Euclidean *Sectio canonis*, even Aristotle himself) had a lively interest in music and musical theory, none of them used his theoretical vocabulary, and none of them, so far as our evidence takes us, concerned himself with the details of such musical structures as these. Some writings of the Hellenistic period and the early centuries of the Roman empire (that is, before Aristides' time) set out harmonic analyses in Aristoxenian terminology, and others offer snippets of music-historical

[17] [Plut.] *De mus.* 1134f–1135b, 1137b–e; see also pp. 99–100 below.

information. But almost everything in these works on harmonics, and much of the historical data too, can be traced back to a single origin, Aristoxenus himself. The idea that any of these writers somehow unearthed pre-Aristoxenian material independently, and recast it for themselves in the form in which we have it, has nothing whatever to recommend it.

I think it overwhelmingly likely, then, that Aristides' authority for these *harmoniai* was Aristoxenus. In that case the pressing question is how Aristoxenus acquired his information about them. The *Republic* represents them as current in the later fifth century, and Plato himself provides no analysis of them. There is therefore a gap of about a hundred years between the time they are said to have been in use and the time at which Aristoxenus could have recorded them. The Socrates of the *Republic*, later in the same discussion, says that on certain other technical issues to do with music they should consult Damon, the much-discussed Athenian musician and associate of Pericles, whose work probably falls into the period between the 440s and the 420s. It is generally (though in my view wrongly) assumed, by both ancient and modern commentators, that Plato's list of the *harmoniai* and their emotional and ethical characteristics is also derived from him.[18] Aristides, in a later passage, mentions *harmoniai* 'handed down' by Damon, which may be identical with those set out in the analysis we are discussing (*De mus.* 80.25–81.3).[19] But even if I am wrong, and Damon really was Plato's source in this passage, we do not know of any written work which he left;[20] and if he gave detailed descriptions of the arrangements of intervals in the *harmoniai* he identified, we know nothing of the conceptual or linguistic resources they employed.

Yet on the hypothesis I am suggesting, an account of the eccentric systems which Aristides describes somehow survived through the next hundred years; and since (on that hypothesis) it was into Aristoxenus' hands that they arrived, and by him that they were recast into his own language and dispatched into a future where Aristides would find them, the most responsible guess we can make is that when they reached him, they were set out in

[18] For discussions of Damon and his work see Anderson 1955 and 1966 (especially 38–42, 74–81, 147–53), Wallace 1991 and 1995, which also explores many other issues relevant to the present chapter, West 1992a: 246–9. The issues are examined in detail by Wallace in a forthcoming book on Damon, which he has generously allowed me to see in typescript; I am happy to record that he too sees no grounds for thinking that Damon is the source of Plato's account of the *harmoniai*, though he is less sceptical than I am about his importance in other areas of musical thought. I shall attempt to justify my heretical views on this matter in another book, but I cannot pursue the issues here.

[19] Hence some scholars (e.g. West 1992a: 174 ff.) refer to the Aristidean *harmoniai* as the 'Damonian scales'.

[20] Wallace 1991: 32–44 examines the question whether he left any such works or not, and answers it, persuasively in my view, in the negative.

one of the forms of representation that he attributes to his predecessors. The only representations of musical systems which he mentions repeatedly are the diagrams of the *harmonikoi* (one isolated passage referring to another way of depicting them will be considered shortly). Since every interval in the Aristidean *harmoniai* is either a quarter-tone or one of its multiples, these diagrams could readily have displayed their structures. Overall, the supposition that it was they that gave Aristoxenus his information seems much the most likely and the most firmly attached to solid pieces of evidence.

This reconstruction of the route by which Aristides' 'ancient *harmoniai*' were transmitted must plainly remain hypothetical. It has the merit, however, of providing a lineage which allows them to be treated, with certain qualifications, as an authentic record of fifth-century patterns of attunement; and there are other reasons, which other scholars have identified, for regarding them in that optimistic light.[21] No process of transmission which excludes the *harmonikoi* and their diagrams is in my view sustainable. If that is accepted, it generates an important result; the *harmonikoi* of the early period were not all in the business of 'rationalising' the data confronting them, in an attempt to make them fit some systematic patterns like that of Eratocles' 'seven octachords'. From the perspective of an orderly minded theorist, the Aristidean *harmoniai* are chaotic. They can only be the fruits of an attempt to describe what musicians actually did, regardless of its disorderliness. At the same time, however, each of these *harmoniai* contains, as its core, a pattern which in Aristoxenus' terms is recognisably enharmonic, and will still allow us to accept his assertion that the *harmonikoi* studied enharmonic systems and no others.

The question which prompted our excursion into this terrain was how, if at all, the systematic set of octachords ingeniously devised by Eratocles was related to attunements that he might have found musicians using, or which other *harmonikoi* had transcribed on the basis of their own experience of performers' practices. I shall not set out here on a full examination of all the *harmoniai* anatomised by Aristides, or compare all of them, one by one, with their putative counterparts in Eratocles' scheme.[22] Three examples will be enough to sketch grounds for the conclusions I want to draw, and to exhibit some of the structural irregularities which these *harmoniai* display.

[21] See Winnington-Ingram 1936: 21–30; West 1981, and 1992a: 174–7; Anderson 1994: 154–8; cf. Barker 1982a, and 1984a: 165–8.

[22] For comparisons of that sort see Winnington-Ingram and Barker (n. 21 above), and West 1992a: 226–8.

An enharmonic octave in the Dorian *harmonia*, as it is portrayed in the 'cyclic' Eratoclean system, is made up of two tetrachords, separated or 'disjoined' by the interval of a tone. Each tetrachord is divided into intervals (from the bottom upwards) of quarter-tone, quarter-tone and ditone. Letting their initial letters stand for the intervals, then, its structure is q,q,d; t; q,q,d. The Dorian attunement in Aristides' collection is very similar; in fact it is identical, except that it adds one further interval, a tone, at the bottom. Thus it runs t; q,q,d; t; q,q,d. This is the *harmonia* which I mentioned earlier as the one that exceeds the octave by a tone, and so requires the grid of the diagram in which it is depicted to span twenty-eight dieses, as Aristoxenus says, and not the twenty-four that complete an octave.

Consider next the Mixolydian. In the cyclic system it is three steps away from the Dorian; that is, its structure is reached by removing the three highest intervals from the top of the Dorian octave structure, and replacing them in the same order at the bottom. Its sequence is q,q,d; q,q,d; t. In the pre-systematic version recorded by Aristides, its most immediately striking peculiarity is the enormous undivided interval of a tritone (three whole tones) at the top; it has the form q,q,t,t,q,q,tritone. The overall shapes of these two Mixolydians have clear affinities with one another. To convert the Aristidean structure into its orderly counterpart one need only combine the two adjacent tones of the former into a ditone, and divide its concluding tritone into a ditone followed by a tone. As it happens, we have a little more information about the history of this *harmonia*, in a discussion in the Plutarchan *De musica* of which at least part, and possibly the whole, is derived from Aristoxenus (1136c–d). The relevant point here is that according to 'the *harmonikoi* in their historical works', there was a controversy in the fifth century about the location of the disjunction in the Mixolydian *harmonia*, that is, about which of its intervals was to be identified with, or should be reckoned to include, the tone whose function is to separate one substructure from another. What these *harmonikoi* state is that 'Lamprocles the Athenian, realising that the disjunction in this *harmonia* is not where almost everyone imagined it to be, but at the top of its range, gave it the form of the [downwards] sequence from *paramesē* to *hypatē hypatōn*' (that is, the form it has in the regularised set of cyclically transformed *harmoniai*).

If the Mixolydian which 'almost everyone' had in mind followed the pattern recorded by Aristides, the nature of the problem about the disjunction is easily understood. The tone of disjunction normally separates two regularly formed tetrachords, each spanning a perfect fourth (as in the Dorian *harmonia* we have glanced at), or else it stands between one complete tetrachord and another interval or intervals that can be recognised as a fragment of a tetrachord of the same type. (In the regularised Lydian *harmonia*, for instance, it stands above a complete tetrachord, q,q,d, and below a single interval of a quarter-tone, which constitutes, in theory at least, the first interval of another tetrachord, not completed within the relevant octave; the remainder of this has been transferred to the bottom of the system.) Neither of the tones in Aristides' Mixolydian fulfils these conditions. What 'almost everyone' apparently did was to make the best of a bad job, and to construe the higher of the two successive tones as a disjunction between a four-note (and in that sense 'tetrachordal') structure above it spanning a perfect fifth, the structure q,q,tritone, and a structure below it that is in the same sense a tetrachord, but whose compass is only a tone and a half, q,q,t. Lamprocles' innovative analysis must have involved the insertion of a note a whole tone from the top, dividing the tritone into a ditone with a tone above it. If he also fused the pair of tones into a ditone, so keeping constant the number of notes in the octave, he will have arrived at the pattern assigned to the Mixolydian in the cyclic system. Since Lamprocles probably belongs to the early or mid-fifth century, we may guess (though it is only a speculation) that the *harmonikoi* who retailed this account in their 'historical works' were theorists of an Eratoclean persuasion writing a few generations later,[23] and that they were providing their own systematic analysis with a respectable ancestry.

For my third example, deliberately chosen as the most difficult to accommodate, I shall take the *harmonia* which Aristides calls 'Iastian', in line with the usage of the *Republic* (398e10). It sets off from the bottom in a regular way, with a straightforward enharmonic tetrachord, q,q,d. But its remaining intervals abandon that pattern. There are only two, a step of one and a half tones followed by a tone. It thus contains only six notes and five intervals, and its compass falls short of a complete octave by a whole tone. Neither the name 'Iastian' (or 'Ionian') nor any structure that is plainly a rationalisation of this one appears in our accounts of the cyclically organised scheme. The fact, I suggest, should be taken as evidence of good faith on both sides, on that of the analyst who recorded it, for not foisting upon it some more readily systematised form, and on that of the Eratocleans, for resisting the temptation to assign its well-known name to an artificially constructed pattern with which it had no real affinity. They simply set it

[23] Eratocles is undatable, but may most plausibly be thought of as active in the later decades of the fifth century.

aside. This is not to say that its structure is theoretically unintelligible. Each of its substructures, q,q,d (at the bottom), and tone-and-a-half, tone (at the top), spans the interval of a fourth. The former is an enharmonic tetrachord, and the latter can be construed as a 'defective' tetrachord of another sort (apparently diatonic, implicitly semitone, tone, tone, but with one note missing). The tetrachord and the quasi-tetrachord are linked directly, in conjunction, rather than being disjoined by a tone. Some account could therefore be given of it in terms, at any rate, of Aristoxenian theory. But we need not insist on this interpretation. What matters is that just at the point where it becomes impossible to find any real connection between an Aristidean *harmonia* and a construction in the systematic set, we find also that the former's name has disappeared from the systematisers' list.

If we approach the matter from the opposite direction, we can be reasonably confident that Eratocles' procedure produced some patterns which corresponded closely to nothing in the contemporary musical repertoire. Certainly there are some of his which match nothing in Aristides' collection, if only because there are seven of the former and only six of the latter, of which one, the Iastian, has no Eratoclean counterpart. Another of them, the Lydian, should probably also be excluded from any comparison, since there are reasons for thinking that Aristides' text does not preserve it in its original form.[24] Four of the cyclically generated systems, however, are sufficiently closely related to Aristidean *harmoniai* to stand as plausible rationalisations of them. All of the latter contain recognisable enharmonic substructures. At the same time they do not fit the rationalised patterns so exactly as to cast doubt on the musical authenticity and historical priority of the versions that Aristides records.

It appears, then, that someone in the later fifth century set themselves the task of analysing a set of attunements in contemporary use. The manner in which their analyses were later presented (probably by Aristoxenus) is likely to reflect their original expression in terms of linear distances between pitches. This points to the conclusion that their author was one of the *harmonikoi*, and that they are likely to have been conveyed in the form of a diagram. But the uses to which they were put may not have been merely descriptive, though in their historical context an attempt at an accurate, quantitative description would have been an original and ambitious project in itself. The evidence about Eratocles, and the relation between the descriptive analyses and his own, suggest that it was just these representations of *harmoniai* that gave him his starting point.

[24] See Winnington-Ingram 1936: 24–5, Anderson 1994: 155.

We even have a piece of evidence which licenses the guess that Eratocles did not devise the descriptive analyses himself, and which may give us the names of those who did. At *El. harm.* 36.15–37.7 Aristoxenus is complaining about the failure of the *harmonikoi* to identify the principles governing the ways in which notes and intervals can be combined to form *systēmata*, scales, or even to address the question whether such principles exist; and he comments sardonically on the very limited nature of the enquiries they did pursue. In this context he divides them into two groups. There were those who 'made no attempt to enumerate completely the differences between *systēmata*, but directed their investigations only to the seven octachords which they called *harmoniai*';[25] in the light of the earlier reference at 6.21–31 these must be Eratocleans. Secondly, however, there were 'those who did make the attempt, but conducted their enumeration quite haphazardly' (or perhaps 'by no means achieved a complete enumeration'), specifically 'the followers of Pythagoras of Zakynthos and Agenor of Mytilene' (36.28–37.1). Little is known of these individuals.[26] But what Aristoxenus denigrates as a 'haphazard' attempt at enumeration must be one that did not derive its results systematically from theoretical axioms. It must simply have been one which tried to specify the structures of such *systēmata* or *harmoniai* as came to these people's attention, that is, a collection of those that they found in current musical use. It would be rash to claim as a proven fact that the analysts who originally set out the Aristidean *harmoniai* (which clearly answer to this description) were, first, this non-Pythagorean Pythagoras, and secondly, several generations later, Agenor; but the evidence unquestionably fits.

I have spoken of Eratocles' cyclic system as a 'rationalisation', and as 'purely theoretical'. But his enterprise can also be envisaged in a more positive light. An inspection of Aristides' evidence reveals certain affinities between the *harmoniai* it describes, which would be all the more obvious if they were set side by side in a diagram. They all contain similar or identical components and substructures, variously placed within their ranges. A question that naturally occurs to a scientifically minded investigator is

25 On the text of this passage see n. 13 above.

26 Pythagoras of Zakynthos may belong to the early or mid-fifth century (Diog. Laert. 8.46). He is mentioned elsewhere only as the inventor and performer of a complex instrument called the *tripous* (Ath. 637b–f). Agenor of Mytilene is named by Isocrates (*Ep.* 8) as a notable musician and teacher who was still active in the middle of the fourth century, and he appears to be by some distance the most recent theorist to be mentioned in the *El. harm.* We have no other significant information about him. (Porphyry's reference to an Agenorian 'school' of harmonic theory, *In Ptol. Harm.* 3.5, is patently based on an entirely unreliable inference from his reading of Aristoxenus, as are his allusions to the 'schools' of Epigonus and Eratocles in the same passage.) Both Pythagoras and Agenor will be discussed further in Chapter 3.

whether they can all be understood as manifestations of the same fundamental system of order, a unique set of relations that constitutes the basis of melodic organisation as a whole. In the same period and in just the same way, Presocratic cosmologists were engaged in the task of interpreting all natural phenomena, in all their diversity, as arising by a few simple processes from the interplay of a few types of elementary constituent, and as manifestations of a single, intelligible system of cosmic order.

The idea that all musical attunements are generated from the same origin by a regular process of cyclic rearrangement has a strong appeal in this intellectual environment. The results it produces do not exactly fit the 'facts' that are to be explained. But they are close enough to encourage the thought that musicians' practices – which of course had not developed out of logical reflection on orderly abstractions – could be understood as approximations, more or less reliably guided by aesthetic intuition, to the structures on which an ideal art of melody would be based. The objects to which these notions refer differ in one important respect from those of the Presocratic scientists; musicians' works are not natural phenomena. They are artefacts. But in this period there was nothing new in the idea that works of human artistry should conform to principles that are not of human origin, and man-made music itself had long been conventionally represented as a distant echo of the music of Muses and gods. The thesis that 'art is an imitation of nature' is also one that was exploited in intellectual circles well before Plato made it famous and gave it his own unique interpretation.[27] Eratocles' project fits squarely into such patterns of thought. We do not know how he conceived the 'reality' inhabited by the perfectly organised structures he describes, as divine, for instance, in line with tradition, or as in some sense a part of the natural order, in the manner of fifth-century cosmologists. In either case his aim was apparently to look beyond the chaotic diversity of human musical practices, and to place the structures he described within a framework intelligible to the enquiring mind.

One further point needs to be made before we abandon Eratocles. Aristoxenus credits him and his associates with another insight, though (characteristically) he undercuts it with a bout of sharp criticism (*El. harm.* 5.9–23). He expresses it rather enigmatically. 'Eratocles and his followers said no more [on the issue under discussion] than that on either side of the interval of a fourth the melody splits into two' (5.9–12). Commentators have regarded this statement as referring to the fact that within the

[27] The idea is put to notable use, for example, in the Hippocratic treatise *De victu*, whose first book probably dates from about 400 BC. For its reflections on music see 1.18.

Unchanging Perfect System (see p. 17 above), when one has completed the fourth running upwards from *hypatē mesōn* to *mesē*, the series divides; one may proceed in conjunction into the tetrachord *synēmmenōn*, or through a tone of disjunction into the tetrachord *diezeugmenōn*. Aristoxenus' comments almost certainly reflect the same interpretation. But as a diagnosis of what Eratocles meant, it is, I think, only half-correct. There is no trace of this branching, two-octave scheme in sources earlier than the mid-fourth century, and Aristoxenus himself tells us that his predecessors (and specifically the Eratocleans) restricted their studies to the octave. Eratocles was alluding, I suspect, to a fact which is closely related – conceptually, musically and historically – to the one to which Aristoxenus and the commentators refer, but which emerges from an inspection of the Eratoclean cyclic system itself. Within that system, a regularly formed tetrachord (that is, in this context, an interval of a fourth divided in the pattern q,q,d) is sometimes immediately followed by the whole or the beginning of another structure of the same kind, while in other cases the sequence passes to the next such tetrachord (or the beginning of one) by way of the interval of a tone.[28] Whereas the Unchanging Perfect System, as standardly represented, assigns the division of the series to a fixed point, at the eighth note from the bottom, in the cyclic system the conjunctions and disjunctions may occur anywhere in the octave.[29]

What Eratocles made of this observation we cannot tell; Aristoxenus gives no further information, and the only other Greek writer who mentions him does no more than to record his name.[30] But we can draw three not insignificant conclusions: first, that Eratocles recognised the importance of the fourth as a structural unit, and may even have pioneered the idea, paradigmatic in all later theory, that the tetrachord is the fundamental unit of harmonic analysis; secondly that he not only set out the cyclic system in diagrammatic form but appended at least a few comments on details of the structures of the *harmoniai* it contained; and thirdly that since these comments survived until Aristoxenus' time, they must have been recorded in writing. It follows that in this one case at least, Aristoxenus was able

[28] Thus in Eratocles' Mixolydian, q,q,d; q,q,d; t, one tetrachord is followed immediately by another; in the sequence which he or later theorists called 'Hypolydian', q,d; t; q,q,d; q, the one regular tetrachord is followed by a single quarter-tone, as if setting off on another tetrachord of the same sort; whereas in his Dorian, q,q,d; t; q,q,d, the two tetrachords are separated by a tone.

[29] This is not to say that the structures deployed in Aristoxenian harmonics cannot handle the phenomena represented here as shifting conjunctions and disjunctions. But this will involve some quite elaborate and difficult manoeuvres connected with the theory of *tonoi*, and their relation to what are known as the 'species of the octave' (see pp. 215–28 below).

[30] See the reference to Porphyry in n. 26 above.

to draw on material about the *harmonikoi* over and above the diagrams themselves. However sketchy it may have been, there was an Eratoclean treatise.

(d) Empirical studies of the tonoi

Eratocles was a systematiser. Other *harmonikoi*, such as Pythagoras of Zakynthos and Agenor, and the person (if he is distinct from them) who described the Aristidean *harmoniai*, were apparently not. The project of direct, empirical analysis, in which these latter were engaged, can be recognised elsewhere in Aristoxenus' recurrent grumblings about his predecessors' failure to ground their findings in principles, and about the generally haphazard nature of their approach. He also makes it clear that these 'untheorised' investigations extended into at least one area that we have not so far discussed, that of the systems called *tonoi* and the relations between them. The topic of *tonoi* is probably the thorniest of all those involved in Greek harmonics, and I shall consider it more fully in a later chapter (pp. 215–28 below). For present purposes we can envisage them as roughly analogous to the 'keys' of modern musical discourse, that is, as a set of identically formed scales placed at different levels of pitch. They become important, in a scientific or theoretical context, when the structural basis of a sequence used by musicians cannot be located in a single, recognisable scale, but can be construed as shifting ('modulating') between differently pitched instances of the same scale-pattern. At a time when the basic scales themselves were only beginning to be definitively formulated, it cannot always have been easy to decide whether such a diagnosis was appropriate.

So far as this pre-Aristoxenian period is concerned, there is at least one significant difference between the ancient conception of *tonos* and the modern conception of key. Our keys are spaced uniformly at intervals of a semitone; but this arrangement, which seems so simple and obvious, is the fruit of several centuries of theoretical discussion and musical experimentation. It had an approximate Greek counterpart, as we shall see, though one which elided most of the difficulties that faced musicians and musicologists of the early modern era. But that comes later. In the fifth and early fourth centuries, conceptions of *tonoi* and of the ways in which they were related arose directly from the observation of current musical practice, rather than constituting a body of theory from which practices were derived. Extensive modulation was particularly characteristic of music performed on wind instruments (*auloi*). The relations between usable 'keys' depended, very largely, on the ways in which the instruments themselves were actually constructed, each being limited (with certain qualifications) to the patterns of attunement imposed by its dimensions and by the spacings of its finger-holes; and these depended, in part at least, on the demands made upon instrument-makers by composers and performers.[31]

What Aristoxenus tells us about his predecessors' accounts of relations between *tonoi* is, essentially, that they were both unsystematic and inconsistent with one another.

No one has said anything worthwhile about them, neither about how they are to be constructed nor on what basis their number is to be established. The expositions of the *tonoi* by the *harmonikoi* are just like the ways in which the days of the month are counted, as when the Corinthians call 'the tenth' what the Athenians call 'the fifth' and others again 'the eighth'. For in the same way some of the *harmonikoi* say that the Hypodorian is the lowest of the *tonoi*, that the Mixolydian is higher by a semitone, the Dorian a semitone above that, the Phrygian a tone higher than the Dorian, and the Lydian higher than the Phrygian by the same amount; whereas others, over and above those mentioned, add the Hypophrygian *aulos* below them, and others again, with an eye to the way in which the holes of *auloi* are drilled, separate the three lowest *tonoi*, the Hypophrygian, Hypodorian and Dorian, by intervals of three dieses, spacing the Phrygian at a tone from the Dorian, the Lydian from the Phrygian, once again, by an interval of three dieses, and the Mixolydian from the Lydian by the same amount. (37.10–34)

This contemptuous report makes it as clear as could be that the accounts of the relations between *tonoi* retailed by the *harmonikoi* were innocent of explicit theoretical presuppositions and axioms,[32] and that at least some of them were based on a study of the practices of performers on *auloi* and of the structures of the instruments they used.

This brings us to the last Aristoxenian passage we shall consider in connection with these *harmonikoi*. He devotes several pages (39.4–43.24) to the shortcomings of two general approaches to the science of harmonics, each of which had been adopted by some of his predecessors. The first of the two, as he presents them, conceives the objective of harmonics as the representation of melodic systems in a written notation. The second (not necessarily wholly distinct from the first) bases its account of 'the nature of

[31] See Pausanias 9.12.5, with West 1992a: 87, on the innovations attributed to Pronomus, around the end of the fifth century; [Plut.] *De mus.* 1138a on the special demands made on instrument-makers by Telephanes of Megara (fourth century); cf. Theophrastus' account of the processes by which the reeds for the instrument were prepared, and of changes introduced in response to new fashions in the style of performance, *Hist. plant.* IV.11.1–7, especially 4–5.

[32] This is the force of Aristoxenus' next remark: 'about the nature of the grounds on which they have decided to space the *tonoi* in these ways they have told us nothing'; and he adds a dismissive comment on the uselessness in this context of the procedure of *katapyknōsis* (37.34–38.5).

attunement' on observation of the physical characteristics of *auloi*. Since it connects directly with the topic we have just been discussing, we shall take the latter first.

(e) Harmonics and the study of instruments

The passage in question (41.24–43.24) makes no attempt to describe in detail the procedures of the exponents of this approach, or to explain why they adopted them; Aristoxenus, as usual, is concerned only with its defects. It emerges, however, that they tried to derive descriptions of the patterns of intervals underlying melodies from a study of the ways in which the finger-holes were spaced on the pipe of a wind instrument.[33] Aristoxenus is dismissive. 'No instrument tunes itself,' he points out (43.7). It is not because *auloi* are tuned as they are that attunements have their specific structures, but the other way round; the basis of the way in which *auloi* are constructed is in musicians' perception of harmonic order. The precise notes and intervals actually produced by these instruments are in any case not fully determined by the spacings of the finger-holes, but depend heavily on the technique of the performer.[34]

Aristoxenus' critical comments are clearly sound, as far as they go. But we should not be deceived; his predecessors need not have been guilty of all the errors of which he accuses them. In particular, they need not have supposed (and probably did not) that harmonic relations take the form that they do *because* the instruments are designed as they are. It was rather that they took the spacings of the finger-holes of the *auloi* as objective *evidence* about the systems of intervals underlying the melodies they were designed to produce. (The dependence of the exact nuances of the intervals on the performers' playing-techniques stands as a legitimate basis for criticism.) As Aristoxenus himself so often observes, they were not looking for the reasons why attunements fall into the patterns that they do, so much as for a clear-cut basis on which they could be identified and described. With the possible exception of the Eratocleans, the *harmonikoi* were engaged in

[33] Their choice of the *aulos* rather than a stringed instrument explains itself. The intervals between the notes of a wind instrument can be thought of – though misleadingly, as Aristoxenus points out – as rigidly determined by its permanent structure. The permanent features of a lyre's strings, on the other hand, like those of a modern violin or guitar, obviously fail to reveal anything about the pattern of intervals they will produce. Not only are they all the same length; they can in principle be tuned in whatever set of relations the performer decides.

[34] 42.30–43.24. Cf. Plato, *Phileb.* 56a, Ptol. *Harm.* 17.2–7.

something more like musical 'natural history' than in a fully explanatory mode of science.[35]

There is another feature of this procedure that needs to be disentangled from Aristoxenus' polemic. He repeatedly implies that the object on which its exponents focused was the instrument itself, with its 'holes and bores and other such things' (41.32–4), rather than the audible notes produced from it by an aulete (with which the supposedly objective data relied on by these *harmonikoi* are sharply contrasted, e.g. 42.7–22). Hence the scales and other sets of relations which they described were not simply transcriptions of what they heard, like those offered by other *harmonikoi*, and they did not measure intervals just by comparing them, by ear, with an audible unit of measurement. Their conclusions were based on inferences from what they could see, as features of the physical bodies of the instruments.

That cannot be quite the whole story, of course. They must have had some experiential basis for their inferences about the interval-patterns that a given arrangement of finger-holes would produce. But this creates a complication. Imagine a simplified wind-instrument with just three finger-holes, the first placed at a distance of eight units from the mouthpiece, the second at a distance of twelve units and the third at a distance of sixteen units. The distance between the first and the second is equal to that between the second and the third. One might naively suppose that the intervals between the notes they sound would be equal too. But that is not so. The intervals depend on the ratios between the relevant lengths of pipe, not directly on the sizes of the differences between those lengths. Given an 'ideal' pipe and an 'ideal' player, two intervals will be equal if the ratios between the lengths of pipe producing their bounding notes are the same, not if these lengths differ by the same amount. In the present case, the ratio of the second length to the first is 12:8 = 3:2, while that of the third to the second is 16:12 = 4:3. The first of the intervals corresponding to them (if we disregard variables introduced by the player's technique) is a perfect fifth, and the second a perfect fourth. We seem to be back in the territory of the Pythagoreans.

But Aristoxenus cannot be talking about Pythagorean or other mathematical theorists. He has large and general reasons, as I said earlier, for thinking their entire project irrelevant to his concerns, and has previously dismissed them in a phrase (32.20–8). He is most unlikely to be devoting extensive space to them here. Even if for some strange reason he had decided to do so, it would have been absurd to identify them as 'those

[35] Aristotle took the view that 'empirical' harmonics, as pursued by these *harmonikoi*, established only the 'facts', and that the explanations fell into the province of the mathematical branch of the subject (*An. post.* 78b34–79a6; see pp. 353–61 below).

who devote themselves to the study of *auloi*' (39.8–9); an interest in these instruments was hardly their most prominent peculiarity. These theorists must be among the *harmonikoi* whom Aristoxenus regarded as his legitimate if misguided precursors; and they must have represented intervals, like Aristoxenus, as linear distances, not as ratios, since it is precisely those who take the latter approach who are ruled irrelevant at 32.20–8.

If their conclusions bore any recognisable relation to the musical phenomena, their descriptions of the instruments' interval-patterns must have been linked in some way to ratios of lengths; but since Aristoxenus treats them as he does, they must have represented their results in terms of auditory 'distances'. They may even have conveyed them through diagrams of the same sort as those used by other *harmonikoi*. This kind of interplay between a ratio-based examination of the physical evidence and conclusions expressed by reference to distances in auditory space is rare in the technical traditions of harmonic science. It might be taken to suggest that the separation of approaches to harmonics, in its early phases, into two quite distinct schools of thought, totally insulated from one another, is an anachronism foisted upon musicological history by Plato, Aristoxenus and other tidy minded commentators. But that hypothesis would press the implications of this passage too hard. It does not prove that these *harmonikoi* were familiar with the work of the Pythagoreans and had one foot in their camp, though that is of course quite possible.[36] Knowledge of the relevance and the identity of at least the most important ratios was not restricted to Pythagoreans. There are reasons for thinking that it was widespread among instrument-makers themselves, though no doubt in the form of rules of thumb rather than as a set of abstract mathematical propositions.[37] The trade-mark of Pythagorean harmonics is not its reliance on ratios between the dimensions of parts of instruments, but its insistence that the intervals themselves must also be understood and represented as ratios (regardless of the fact that this way of portraying them generates nothing recognisable as a picture of the phenomena presented to the ear), coupled with the thesis, which we shall meet frequently in Part 3, that 'musicality' is determined by mathematical considerations.

These *harmonikoi*, then, need not have deliberately set out to mediate between rival approaches to harmonics. They may more plausibly be regarded as having adopted the same attitude to the representation of musical structures as all others whom Aristoxenus treats as his predecessors, but

[36] Hybrid approaches to harmonics are not entirely unknown in this period. In Chapter 10 I discuss a case where there seems to be slippage of a different sort between the two main traditions.

[37] See Landels 1963 and 1968, West 1992a: 94–101.

as having taken their evidence from a different set of empirical data. Their procedure did not require them to use any sophisticated, theoretically based method for 'translating' ratios into linear auditory distances. It was only the mathematical theorists who conceived the ratios as descriptions of musical intervals as such, which would demand such translation. The task facing them was rather a matter of absorbing the 'craft-knowledge' of instrument-makers, who knew, for example, how to make a pipe that would play in the Phrygian *harmonia*, and then of identifying, presumably by ear, the number of linear unit-intervals (quarter-tones) separating the notes sounded by an instrument built in that way. A modest amount of practical experience could have enabled them to formulate a collection of correlations analogous to the instrument-makers' rules of thumb, but in which the counterpart of a given ratio between lengths was not, for instance, 'the interval between the first and second notes of the Phrygian *harmonia*', but 'an interval of four dieses'.

Aristoxenus is probably right in insinuating that these theorists adopted their approach in order to pin their analyses to 'hard facts', objective and invariable data. Considered from this point of view the procedure has its flaws, as we have seen, but its motivation is by no means contemptible. An aspiration to provide exact descriptions of the patterns of attunement underlying real musical practice must face the difficulty that singers, for example, do not always produce the 'same' interval in precisely the same way, even in the same song, and that the nuances of a stringed instrument's attunement depend on the idiosyncrasies of the musician who tunes the strings, and will be different on different occasions. The intervals sounded by a wind instrument, by contrast, were guided (if not firmly fixed) by unchanging features of the instrument's construction; and these were established, at least in the best workshops, in response to the requirements of many different professional performers, rather than to the whims of one particular individual on a particular Saturday afternoon. They generalised, and so authenticated as a norm, the forms of the various species of attunement. In so far as a determinate set of data existed anywhere in the apparatus of Greek music-making, it was in the physical bodies of wind instruments. Aristoxenus' peevish criticisms are cogent, but the procedures of these *harmonikoi* nevertheless have much to recommend them.

(f) Harmonics and melodic notation

The second group of *harmonikoi* criticised in these pages of the *Elementa harmonica* are identified by their conception of 'the objective of the investigation called "harmonics"'. They hold, Aristoxenus says, that its 'limit', the ultimate achievement to which it aspires, is the notation (*parasēmainesthai*) of melodic sequences (39.4–8).[38] He finds a barrage of objections to this view, some of which are very abstract, based in broad preconceptions of an epistemological sort (41.3–24). The others, though they are elaborated at length and with plentiful detail, all revolve around a single thought: to present a melodic relation or group of relations in notated form requires the notator 'to grasp nothing but the sizes of intervals' (39.26–9). This, on Aristoxenus' view, makes it a hopelessly inadequate goal for a harmonic scientist, since there are a great many other features of such relations which a genuine expert in the field must understand. A capacity to grasp the sizes of intervals, he asserts,

> is not even part of [harmonic] understanding as a whole . . . for neither the functions of the tetrachords nor those of the notes, nor the differences between the genera, nor, to put it briefly, the differences between composite and incomposite intervals nor simple and modulating sequences nor the styles of melodic composition nor, one might say, anything else whatever, becomes known through the sizes [of intervals] themselves. (40.12–23)

There is no doubt some deliberate exaggeration here; the identification of the sizes of intervals plays a significant part in Aristoxenus' own science, though its role is not as central as it might at first appear (see pp. 175–92 below). But we need not concern ourselves with that issue in the present context, or with all the technicalities involved in his list of the topics which this conception of the science leaves unaddressed. What matters is that no part of Aristoxenus' cannonade of comments can have been relevant or plausible unless the system of notation he had in mind was in fact restricted, as he alleges, to expressing the sizes of intervals. It is worth noting that he does not suggest that it was inadequate even for that limited purpose, as surely he would have if he had thought it was; it seems reasonable to conclude that he reckoned it perfectly well adapted to its task, even if that task was trivial. Nor does he insinuate that those who used it did so incompetently. Since elsewhere he scatters accusations of observational incompetence freely, this points again to the charitable conclusion that he found their notational exercises tolerably accurate.

There has been some debate about the nature of the notation that these theorists were using.[39] Opinions divide, broadly speaking, into two camps.

[38] Aristoxenus says 'the notation of *melē*', which might be construed as 'melodies'. I shall discuss the term below.

[39] See e.g. Macran 1902: 270–2, Potiron 1964, West 1992a: 263–4.

On one view the notation must have been identical with the one we know from later treatises (especially that of Alypius) and from the surviving scores, or if it was not precisely identical with that one, it was at any rate an earlier and perhaps less developed version of the same scheme. This interpretation has one very strong point in its favour, the fact that with one exception which I shall consider below, other documents preserve no evidence at all to suggest the existence of any other form of melodic notation. The alternative reading holds that certain details of the present passage cannot be made to fit the 'Alypian' system; and its proponents usually argue that the individual symbols of its notation must have referred to intervals, not to notes as the Alypian notation does. Each symbol signified simply 'interval of such-and-such a size'.

The latter view is nowadays heretical, but I am convinced, at least, that the system in question cannot have been the one recorded by Alypius, or any near-relative of it.[40] Anyone who shares my view has to face an initial difficulty. Despite past uncertainties, most recent scholars accept that the Alypian notation had been devised, in its essentials, by the middle of the fifth century, and that it acquired approximately its final form not much later than 400 BC; and I think their arguments cogent.[41] In that case it is likely to have been already in use in the period when the *harmonikoi* were at work. Why, then, should they have taken the trouble to devise a new notational system when another was already to hand? But there is an obvious answer: that the existing system failed to serve the purpose which they had in mind, which was – if Aristoxenus is to be trusted – that of expressing exactly the sizes of intervals. As a comment on the Alypian system, that is a fair point. Not all the intervals between the notes indicated by its symbols are unambiguously fixed. We cannot tell, for instance, from the symbols it uses to notate a melody, whether the intervals inside its tetrachords are those of an enharmonic or a chromatic scale (though it does enable us to distinguish those classes of melody from diatonic ones); and it certainly does not allow us to determine which of the several variants of each genus was intended, or what the exact sizes of individual scalar steps would have been in performance.

It is also relevant that Aristoxenus does not seem to envisage these theorists as notating particular melodies, despite his reference to the notation of *melē* in the opening sentence of the passage. The word *melos*, in the *Elementa harmonica*, seems sometimes to mean 'melody', but elsewhere to

40 For a careful exposition and defence of the contrary view, see Pöhlmann 1988: 74–6.

41 For a valuable short discussion of the dating of the Alypian system see West 1992a: 259–63.

refer to a melodic structure, a scale or an attunement, rather than to an individual tune as such. In the present context the latter sense is guaranteed by the example Aristoxenus uses at 39.17–25, where he speaks of 'writing down the Phrygian *melos*', and compares it with 'writing down the iambic metre'. The expression 'the Phrygian *melos*' does not identify the tune of a particular composition, any more than 'the iambic metre' identifies the precise metrical scheme of a particular poem. It refers to a generalised structure on which many individual melodies can be based. Before the Roman period, by contrast, the Alypian notation was used, so far as we know, only to notate individual compositions; and it was used principally by musicians themselves for their own professional purposes. Further, the Alypian notation was a difficult and esoteric code, unintelligible to non-specialists. Yet Aristoxenus suggests that one motive for these theorists' project of notation may have been their desire to present 'lay persons' with a 'visible product' which would not only impress them, but would also be something on which they would feel qualified to pass judgement (40.29–41.6). A scale written out in Alypian form does not answer that description.

Two of the more specific details of the passage seem to point to the same conclusion. First, it is not true of the Alypian system that it fails to reveal all the features about which, according to Aristoxenus, the theorists' notation tells us nothing. A glance at any modern analysis of one of the surviving musical scores will show that scholars can confidently identify its individual notes, its tetrachords, its genus (with room for doubt in certain cases, as noted above), the points at which it modulates, and so on, none of which could be done on the basis of the notation to which Aristoxenus refers.[42] Secondly, as I have said, the symbols of the Alypian notation refer to notes, but in the present passage Aristoxenus nowhere speaks of the notation in these terms. Instead he refers to a person who 'sets down symbols of the intervals' (*sēmeia tōn diastēmatōn*, 39.30–1); and at 40.7–8 he says that the interval between *nētē hyperbolaiōn* and *nētē diezeugmenōn* and the interval between *mesē* and *hypatē mesōn* are written 'with the same symbol'. Here 'symbol' is in the singular, whereas with a system in which each symbol identifies a note, it takes two symbols to specify an interval.

[42] It is not good enough to evade these problems by dismissing Aristoxenus' criticisms as 'somewhat tendentious' (West 1992a: 264); they will be so only on the assumption that he is referring to the Alypian system. Nor is it relevant to remark that if that system was current at the time 'it would be surprising if Aristoxenus was unacquainted with it and regarded some other system as the only one available' (West *ibid.*). There is nothing in the passage to suggest that he knew of no other notation, only that it was not the one used by the *harmonikoi*; and I have indicated reasons why it would have been unsuitable for their special purposes.

Aristoxenus' grammar, as well as his turns of phrase and his list of the notation's inadequacies, apparently points to a non-Alypian scheme.

I mentioned above that it is not quite true that no notation other than the Alypian is known. A quite different system is presented, briefly and with little explanation, in a passage of Aristides Quintilianus,[43] who attributes it to 'the ancients'. If it has any claims to authenticity it must certainly be pre-Aristoxenian;[44] and it has features which encourage the hypothesis that it might be the one used by the *harmonikoi* whom Aristoxenus criticises. It does not discriminate between 'fixed' and 'moveable' notes or allow us to identify the tetrachord to which any given note belongs; it could in fact reveal nothing about a melodic or scalar sequence except the sizes of the intervals by which its notes are separated, as Aristoxenus says. It consists of a sequence of alphabetic symbols representing notes a semitone apart across the span of two octaves, between which reversed versions of the same symbols are inserted to indicate notes dividing each semitone in half. The whole system thus represents a two-octave space divided into steps of a quarter-tone each. A notation of this sort would have served the purposes of the *harmonikoi*, as Aristoxenus describes them, perfectly well; but there are two difficulties (over and above the questionable credentials of Aristides' report). One is that Aristoxenus tells us plainly that his predecessors considered no system larger than an octave, whereas this scheme runs across two. The other is that its symbols, like those of the Alypian system, refer to notes, or at least to sounds identified by their relative pitches, whereas Aristoxenus seems to say that the symbols in the notation of the *harmonikoi* did not represent notes, but that each represented an interval of a certain size.

These difficulties, however, are not enough to rule the hypothesis out of court. To deal with the second of them first, when Aristoxenus refers to 'symbols of the intervals' he may mean 'symbols which specify the sizes of intervals', as these do when taken in pairs; and when he speaks of the intervals between two different pairs of notes as being represented by 'the same symbol' in the singular, he may have been writing a little carelessly, or perhaps, as Pöhlmann has suggested (n. 40 above), the text should be emended to give a plural. But we should recognise that by abandoning the hypothesis of a notation whose symbols represented intervals rather than notes, we are not opening the door to an 'Alypian' interpretation; the other objections to that reading of the text still stand. Secondly, the two-octave

[43] The table of symbols appears at the end of *De mus.* 1.7.

[44] See West 1992a: 264–5 and 1992b, and cf. Winnington-Ingram 1973.

range of Aristides' 'ancient' system need not conflict with Aristoxenus' contention that the *harmonikoi* studied only one-octave scales, since it could have been used to display the relations between several one-octave scales set in different *tonoi* (see pp. 55–6 above). All in all, I think that this system is the one to which Aristoxenus is most probably referring. The Alypian notation simply does not fit his description, and though a purely intervallic notation could do so we have no independent evidence that such a thing ever existed.

In attributing this notation to the *harmonikoi* I am not crediting them with anything very sophisticated, or with anything (given our other evidence about them) that should cause surprise. Everything we know about them from Plato and Aristoxenus indicates that they focused heavily, if not exclusively, on the business of identifying the sizes of the intervals between the notes of musical systems. It would perhaps be more surprising if they had not devised a shorthand code through which sequences of these 'sizes' could be briefly and conveniently expressed. In effect, the notation would have been used as an alternative way of representing the schemata set out in their diagrams, conveying their results through the ordered and well-known sequence of the letters of the alphabet, rather than geometrically. Unlike the Alypian notation and perhaps the diagrams, a code of this sort could be easily read by intelligent lay persons, whose approval (so Aristoxenus insinuates) it was designed to elicit.

A passage in the Plutarchan *De musica*, almost certainly based on Aristoxenus, may preserve an echo of this notational scheme, though the people being criticised are not the same *harmonikoi*. The author is attacking people who claim that no intervals are melodically usable except those whose sizes can be determined through the manipulation of concords; this criterion excludes everything (in Aristoxenian terms) except the semitone and its multiples (see pp. 93–4 below). Hence they reject the use of the enharmonic diesis or quarter-tone.[45] 'They do not realise,' he continues, 'that in this way the third magnitude would be rejected too [as well as the first], and so would the fifth and the seventh, of which one contains three, one five and one seven dieses; and in general all those intervals which turn out to be odd-numbered would be dismissed as unusable' (*De mus.* 1145b–c). The mode of expression used here makes sense only on the assumption that

[45] Their rejection of the quarter-tone guarantees that they are not the *harmonikoi* discussed in the *El. harm.* If the source on which the writer is drawing is indeed Aristoxenus, they must belong, at least approximately, to his own period rather than to the fifth century; he refers to them explicitly as his contemporaries (*hoi nun*). In that case the passage will give us a rare glimpse of the views of late fourth-century theorists other than Aristoxenus.

all intervals greater than the quarter-tone are to be conceived as its multiples. This is flatly inconsistent with Aristoxenus' own analyses, in which the interval of one third of a tone, for example, can constitute a legitimate melodic or scalar step (e.g. *El. harm.* 25.12–22, 50.28–51.1). It makes perfect sense, however, in the context of analyses like those of the *harmonikoi*, in which the enharmonic diesis is the unit of measurement; and though it does not guarantee their use of any kind of notation, it would come naturally to anyone familiar with a notation of the sort with which I am crediting them. The 'first' interval is the one between a note represented by any given symbol and its immediate neighbour, the 'second' is the one between notes two places apart, and so on; the number of spaces between the points symbolised in the notation generates the titles by which the Plutarchan writer's source identifies the intervals.

The act of transcribing audible pitches into a notation, or representing them diagrammatically, inevitably imposes an artificial determinacy on the musical phenomena, especially in a culture such as that of fifth-century Greece, in which composition did not normally involve the use of written symbols, and musicians did not perform from a score. As Aristoxenus correctly pointed out, it is misleading to think of scale-patterns and forms of attunement as defined by sequences of intervallic 'sizes'; the perceptible characteristics which give each its recognisable identity may survive a good deal of variation in the measurable sizes of the intervals (see pp. 175–92 below). These acts of transcription also fix musical structures within the framework of the conventions embedded in the transcriptional code, in the present case within one defined by its grid of quarter-tones. Any intervals which in practice are nuanced in ways that the grid cannot capture must lose the fine details of their colouration.

At least to this extent the structures described by the *harmonikoi* were alien to the phenomena they sought to describe, but the formal rigidity which their analyses introduced was probably not an accidental and unwanted by-product of their procedure. In so far as they were looking for invariance and determinacy amid the flux of phenomena, their project sits squarely alongside that of fifth-century natural scientists, and also alongside those of intellectuals concerned more directly, like the *harmonikoi* themselves, with modes of experience internal to human culture. Rhetoricians tried to reduce the art of persuasive speech to a set of cut-and-dried rules; linguistic specialists, most notably Prodicus, tried to pin down precisely the meanings of words and to mark sharp boundaries between near-synonyms; Socrates, as depicted in Plato's earlier dialogues, sought to establish exact definitions of the concepts people used to organise and to reflect on their ethical and social lives. The quest for determinacy was endemic among the intellectual enquiries of the period, and it was plainly not always thought of as incompatible with an aspiration to provide an accurate account of the natural and cultural phenomena. (Socrates' question is 'What *is* this "virtue" that we talk about?', not 'What *ought* we to mean by the word "virtue"?') We have seen that with the exception of Eratocles and his followers (if indeed he had any), the *harmonikoi* made no deliberate attempt to reconstruct the musical data to fit preconceived notions of systematic order or to show us how it 'ought' to be. Their apparently humbler but in fact much more difficult task was that of describing, as accurately as possible, what the structures underlying real melodies actually were, in a form that made them accessible to the mind (often by way of images for the eye to survey) as well as to the ear. The questions that will concern us in the next chapter are who the people were who attempted these objective descriptions and depictions, why they undertook the task, and how their work was related to other cultural and intellectual enterprises of their era.

CHAPTER 3

The early empiricists in their cultural and intellectual contexts

Before the foundation of the great philosophical 'schools' in the fourth and third centuries, there were no institutions dedicated to scientific or other intellectual pursuits. Some thinkers recorded their ideas in writing, and these writings could travel widely, reaching other individuals at a long distance from their starting point. But no formally constituted community of scholars existed to provide a ready-made audience for the products of 'academic' endeavour. Who, then, were these pioneering harmonic scientists, and to whom did they address their work? Did they and their audiences fall into any clearly identifiable group or category, and if so, how were they related to other kinds of technical or intellectual specialist? Are they to be regarded as 'intellectuals' at all, comparable, for instance, to mathematicians, philosophers and astronomers? In what settings and for what purposes did they present their ideas? In this chapter I shall suggest two kinds of answer to these questions. They are different but perfectly compatible with one another; some at least of these theorists may very well have marched under both banners at once.

HARMONIC THEORISTS IN THE WORLD OF THE SOPHISTS

There is a small pointer towards my first suggestion in a passage of Aristoxenus which we glanced at in Chapter 2, where he speculates that those *harmonikoi* who set out their results in a notation did so in order to impress 'lay persons', people who were in no sense specialists in the musical sciences, and to give them something on which they could pass judgement (*El. harm.* 40.29–41.6). Aristoxenus is being deliberately cynical as well as speculative about their motives, but there would have been no sting in his insinuation if they had never talked of these things to anyone except fellow-experts. The passage of Plato's *Republic* from which we began has similar implications; the allegedly absurd activities of those who talk about *pyknōmata* and torture strings were evidently publicly visible, familiar to the musically uneducated Socrates as well as to Plato's readers, and open targets for witty caricatures.

On the face of it these hints are given a little more substance by a curious, fragmentary document preserved on papyrus, whose contents have been dated, with reasonable certainty, to the second decade of the fourth century. It has been much discussed by specialists in ancient musicology, though it has received less attention from scholars outside their ranks; and for that reason alone it needs to be considered here. I shall argue, however, that its relation to our concerns is a good deal less direct than at first appears. It is the opening section of a diatribe, presented as if it were a public speech (which very probably it was), attacking the pretensions of people who call themselves *harmonikoi*. Almost certainly it is the work of a sophist. Various attempts have been made to identify him, but none can be treated as certain.[1]

In the first sentence of the fragment, the author addresses his audience in the conventional manner of a speaker in a law-court or a public assembly: 'O men . . .'; but unfortunately the tattered papyrus leaves the phrase incomplete. 'O men of Athens' is as plausible a guess as any. He expresses astonishment at the way in which certain persons impose upon this same audience 'without your noticing', by presenting them with 'demonstrations' or 'displays' (*epideixeis*) on subjects in which they have no real expertise whatever. These people describe themselves as *harmonikoi*. They do not claim to be skilled performers; their specialism, so they say, is 'the theoretical branch' (*to theōrētikon meros*) of musical expertise. We are told, however, that their theorising is in fact haphazard and unintelligible, and they are forever dabbling in the art in which they disown any serious competence, wasting their time on the manipulation of stringed instruments, which they play very badly, and singing much worse than professional singers. 'As to what is called "harmonics" (*harmonikē*), of which they say they have a special understanding, they have nothing articulate to say, but are carried away with enthusiasm; and they beat the rhythm all wrong . . .'

If we stand back from the hostile rhetoric of the passage, a moderately clear picture emerges of these people's activities. They present their work *viva voce*, not in writing. It includes discussions of genuine musical compositions which they perform by way of demonstration, to the best of their

[1] *P. Hib.* I.13. For some recent discussions see Brancacci 1988, West 1992b, Avezzù 1994, Lapini 1994. There is a summary of views about the authorship of the fragment in Avezzù *ibid.* 116–17; currently the most favoured candidate is Alcidamas (active between approximately 390 and 365 BC). In previous publications I have expressed some scepticism about a fourth-century date, but am now persuaded that the piece was written in the region of 380 BC.

(admittedly limited) ability; they do not pretend to be skilled in the arts of performance. The author of the fragment treats their audience as identical with his own, one that can appropriately be addressed according to the conventions of a public meeting; and it is evidently a musically naïve and uninstructed audience, since he can represent it as having been repeatedly taken in by the pretensions of musicological charlatans. He himself plainly belongs, as I have said, to the world of the sophists, professional educators who offered instruction, for a fee, in a wide variety of subjects, but who advertised their wares and attracted paying pupils through public 'demonstrations' of the particular expertise they could provide.[2] The implied identity of his audience with that of the *harmonikoi*, together with the vitriolic character of his attack, indicates with something approaching certainty that they were rivals in the same line of business. These *harmonikoi* too were sophists, or were operating in a manner and a context closely akin to theirs.

The *harmonikoi* who are the targets of this polemic were not concerned only, if at all, with the abstract analysis of melodic structures. The principal focus of the speaker's assault is their contention that melodies of different sorts have different ethical effects on their hearers and performers.

> They say that some melodies make people self-controlled, some prudent, some just, some brave and some cowardly, failing to understand that the chromatic (*chrōma*) cannot make cowards of those who employ it, and the enharmonic (*harmonia*) cannot make them brave. For who does not know that the Aetolians and the Dolopes and all the [. . .], who use diatonic music (*diatonos mousikē*), are much braver than tragedy-singers, who always follow the practice of singing in the enharmonic? Hence it is obvious that the chromatic does not make people cowardly, and neither does the enharmonic make them brave. (lines 13–22)

Here we find our first surviving reference to the three melodic systems, enharmonic, chromatic and diatonic, which Aristoxenus calls *genē*, 'genera'. His contention that his predecessors considered enharmonic structures

[2] Modern scholars are virtually unanimous in placing the author among the sophists. It should be borne in mind that the term 'sophist' is not necessarily pejorative; it is used here, as it commonly was in the fifth and fourth centuries, merely as a label for persons who followed a particular profession. (Isocrates, for example, whose anxiety to present himself in a socially and morally respectable light amounts almost to an obsession, refers to himself as a sophist at *Antid.* 220. It is the appropriate designation for a professional teacher, especially a teacher of rhetoric, with the qualifications to do the job properly; and Isocrates' allusion in the next paragraph to ethically unreliable teachers as those who 'pretend to be sophists' shows that he regards it as an honourable title.) Its negative connotations have arisen, very largely, from the bad press given to these people in Plato's dialogues. For balanced accounts of the sophists and their environment see Kerferd 1981, Rankin 1983, de Romilly 1992. The nature of the fluid borderline between 'sophistic' and 'philosophy' is intriguingly and briefly examined in Cassin 2000, whose bibliography lists her earlier and fuller studies.

and no others may be reflected not only in the privileged ethical status assigned to them by the *harmonikoi* of this text, but also in its argumentative slippage between the other two categories, chromatic and diatonic. The Aetolians and others of that ilk are by common consent more courageous than tragedy-singers, even though the latter always sing music of the supposedly bracing enharmonic kind, while the former's music is diatonic. These are the 'facts' upon which the speaker bases his conclusion, and they contain no reference to chromatic music; yet the conclusion we are incited to draw from them is that cowardice cannot be induced by exposure to the chromatic. Diatonic and chromatic music, it seems, had not so far been clearly analysed and pigeon-holed as distinct types. The passage appears to imply, in fact, that diatonic music is just one kind of chromatic music. Chromatic or 'coloured' music was not originally thought of as a clearly definable musical form with its own determinate structure, but as a kind that abandons the rigidity of traditional structures in favour of flexible nuances of pitch and interval. It was music that had escaped from the disciplined world of the enharmonic, and was too slippery and variable to submit to the sharp-edged quantitative analyses of the *harmonikoi*.[3]

So far so good; but we should not assume too hastily that these self-styled *harmonikoi* belong to the same category as the students of 'empirical' harmonics discussed by Plato, Aristotle and Aristoxenus. There are at least two reasons for caution. First, the writer does not call them *harmonikoi* himself. On the contrary, his view seems to be that *harmonikē* is a reputable discipline whose genuine exponents would (presumably) deserve respect. He offers no criticism of the discipline as such, but contends only that the people he is attacking have nothing worth saying about it, and that when they venture into this area are merely talking at random (*schediazontes*). They are charlatans who pretend to an expertise which they do not possess, who deceive their audiences with 'displays' (*epideixeis*) of learning in fields outside their competence (*allotrias . . . tōn o[ikeiōn techn]ōn*) and do not deserve the title they claim. Hence even if we are right to locate them in the ambit of the sophists, it does not follow that the same is true of anyone who could properly be called a *harmonikos*, at least in this writer's opinion. If anything, we should draw the opposite conclusion, since the writer's opening sentence implies that the audience which both he and the victims of his onslaught address is entirely ignorant of harmonic theory, and for that reason is taken in by these people's false pretensions. This hardly

[3] On the development of the concept of 'genus' see Rocconi 1998 (on this issue in particular, 358–60).

encourages the hypothesis that purveyors of the genuine article were ready to hand in the intellectual market-place.

Secondly, as we have seen, there is nothing in this passage to suggest that these so-called *harmonikoi* had any interest in quantitative analyses of musical structures. The main theses attributed to them have to do with the psychological effects and ethical influences of various kinds of music, and although the writer seems implicitly to accept that they can distinguish enharmonic music from music which is diatonic or chromatic, he explicitly treats the distinction as familiar to everyone, no matter how technically ignorant they are. Let me quote the relevant passage again. 'Who does not know,' he asks rhetorically, 'that the Aetolians and the Dolopes and all the [. . .], who use diatonic music, are much braver than the tragedy-singers, who always follow the practice of singing in the enharmonic?' If everyone knows this, everyone must be familiar with the difference between the two types of music in question. Certainly he does not credit the people he criticises with a specialist's understanding of the differences between these kinds of music, or hint that they tried to justify their ethical distinctions by reference to the forms of scale upon which diatonic and enharmonic compositions were based. The theme of the (sadly mutilated) closing lines of the fragment seems in fact to be that they supported their ethical contentions in quite another and much more impressionistic way, by inviting their audiences to agree that certain melodies (which they apparently performed on the spot) have 'something of the laurel' about them and others 'something of the ivy', that is, that one kind is characteristically Apolline and the other Dionysiac.

Even when every possible allowance is made for the speaker's prejudices and polemical intentions, these people seem to have nothing in common with the *harmonikoi* we are concerned with here. They are apparently minor and perhaps incompetent exponents of a type of speculation which is familiar from comedy and reappears in a more sober guise in Plato's *Republic* and Aristotle's *Politics*; and we should note that in none of these sources are the writers' claims about the ethical attributes of melodies grounded in studies of the structures underlying them.[4] Even if (as I do not believe) Plato's account of the ethical affinities of the various *harmoniai* in the *Republic* are derived from the work of Damon, there is no acceptable evidence that he

[4] Eryximachus in Plato's *Symposium* (187c–d) and the Athenian in the *Laws* (670e) both claim, in their different ways, that a grasp on the structures of *harmoniai* is not enough to give one an understanding of a melody's ethical qualities. Theories propounded in the *Timaeus* approach the matter from an entirely different perspective and have no connection with harmonics in its 'empirical' guise. They will be examined in Ch. 12 below.

produced any analyses of scalar structures; and though negatives are notoriously hard to prove I think it overwhelmingly probable that he did not.[5]

I have argued that the comments preserved in this tantalisingly fragmentary text can tell us nothing about the *harmonikoi* with whom we are presently concerned; and though I have argued also that it makes it marginally less likely that our *harmonikoi* sold their wares in the same environment as the sophists, it certainly does not make that hypothesis untenable. It would indeed be no great surprise if it were true. Even if we leave aside the problematic case of Damon, the notion of a 'musical sophist' was familiar to Plato, who puts into the mouth of Protagoras the claim that Agathocles of Athens, Pythoclides of Ceos and many others were sophists who used discussions of music as a 'disguise' (we might say, less tendentiously, as a 'vehicle') for the 'wisdom' conveyed in their teaching.[6] At least one of the major sophists, the great polymath Hippias, also propounded theories on musical matters.[7]

Nor need this hypothesis set any very determinate boundaries around their purposes and their cultural roles. To call someone a 'sophist' was not to link him with any specific school of thought, or to attribute to him any specific interests, procedures or programmes. On the contrary, the sophists' intellectual activities were as various as the colourful individuals who pursued them. The title merely serves to pick out a profession which flourished in the culture of the fifth and fourth centuries, that of a person who earned his living by offering instruction in some form of 'wisdom', *sophia*, on topics or at an intellectual level beyond the range of the routines of a standard education. They are to be distinguished from people who passed on the secrets of their specialised craft or profession to others who planned to become specialists of the same sort themselves, sculptors or doctors or performing musicians, for example. Sophists found their clientele largely or exclusively among the intellectually curious or (in the case of rhetoricians) the politically ambitious elite, who had no need to train themselves for a money-earning occupation. In so far as they sold their *sophia* for a fee, their teaching was inevitably addressed to relatively small groups of students, more or less in private. But in order to attract fee-paying pupils they also needed a display-case for the wares they offered; and for this purpose the regular instrument was the 'demonstration-speech' or *epideixis*, presented

[5] See p. 47 above.

[6] Plato, *Prot.* 316e. Both of these people belong to the fifth century; they are said to have been teachers, respectively, of Damon and Pericles. Little more is known about them; for such other ancient references as survive see West 1992a: 350 nn. 101–2; cf also pp. 87–8 below.

[7] See Plato, *Hipp. Ma.* 285d, *Hipp. Mi.* 368d.

in a public place to anyone who cared to listen. It is from such *epideixeis* that Socrates, for example, can be imagined as having picked up the smattering of information he had acquired about Damon's analyses of rhythms (*Rep.* 400b–c).

These public 'demonstrations' constitute much the likeliest setting for the speech preserved in the papyrus fragment we have discussed, and for the 'lectures with musical illustrations' delivered by the *harmonikoi* it attacks. The writer of the fragment, as we have seen, describes his putative *harmonikoi* as presenting *epideixeis* on matters beyond their real comprehension. If I am right those people are irrelevant to our purposes; but the theorists who represented melodic structures in a notation designed (according to Aristoxenus) to impress gullible laymen may well have been operating in a similar context. (Some other products of the researches of the *harmonikoi*, such as their diagrams and the Eratoclean system of cyclically mutated octaves, perhaps figured only in the more advanced instruction given to fee-paying students.) *Epideixeis* were sometimes offered in more *recherché* surroundings, in the precincts of the houses of wealthy patrons of the arts and sciences, such as that of Callias in Plato's *Protagoras*; and we know that *harmonikoi* gave displays in such a setting in the later fourth century, since at *Characters* v.10 Theophrastus refers to the sort of person who lends his courtyard to philosophers, sophists, instructors in armed combat[8] and *harmonikoi* for their *epideixeis*.[9] These *harmonikoi* may well be theorists of the sort that Aristoxenus discusses, since in another passage Theophrastus unambiguously uses the term in this sense.[10] Two final examples will help to give the picture more colour and plausibility, and to convey something of the diversity of these activities and their exponents.

Pythagoras of Zakynthos, whom Aristoxenus mentions as an unsystematically minded *harmonikos* (p. 52 above), is known also from a passage in Athenaeus (637b–f, quoting from the third-century historian Artemon). Pythagoras, the passage tells us, invented an extraordinary instrument called the *tripous* ('tripod'). We need not tackle here the minutiae of Artemon's description. It was, in effect, a triple kithara,[11] uniting the essential elements of three differently tuned instruments in a single, three-sided structure, each

8 Cf. the opening speech of Plato's *Laches*.

9 In the MSS the passage is attached to the description of the Man who is Anxious to Please [*areskos*], but it is probably misplaced, and may belong rather to the portrait of the Man of Trivial Ambition [*mikrophilotimos*].

10 Theophrastus frag. 716.17–18 Fortenbaugh, where he refers to 'the *harmonikoi*, who judge by perception' and contrasts them with mathematical harmonic theorists.

11 The kithara was the stringed instrument used primarily by professionals in public performance, similar in its general structure to the familiar lyre, but with a wooden (instead of tortoise-shell) sound-box, and much larger and more resonant. For details and illustrations see Paquette 1984, Maas and Snyder 1989.

set of strings occupying one of its sides. The whole device could be rotated on its base by a movement of the player's foot.

> The ready movement on the base, propelled by his [Pythagoras'] foot, brought the different scales so swiftly under his hand, and he had practised the manual control of it to so high a degree of dexterity, that if one did not actually see what was happening, but judged by hearing alone, one would think that one was hearing three *kitharai*, all differently tuned. This instrument was greatly admired, but after Pythagoras' lifetime it quickly fell into disuse. (637e–f)

This Pythagoras, then, was simultaneously a theorist, an inventor and an accomplished performer. About the relations between his theorising and his invention we know nothing for certain, and our source tells us nothing about the context of his performances on the *tripous*. But they can hardly have fitted into the normal calendar of musical events in any Greek city; and one live possibility (I shall sketch out another shortly) is that they amounted to a free-marketeer's publicity stunt, calculated to capture the public's attention, to stimulate curiosity, and to attract fee-paying students eager to learn from so accomplished a master. In that case, whether or not it was accompanied by a spoken explanation, its function was closely related to that of a sophist's *epideixis*.

The only information we have about the date of Pythagoras of Zakynthos is vague; if it has any substance at all, he can hardly have lived much later than the middle of the fifth century.[12] My second example comes from the early decades of the fourth. The Athenian singer and kithara-player Stratonikos was a professional musician, one of the stars of his generation. Athenaeus records a whole medley of anecdotes about him, almost all of them, unfortunately, designed to illustrate his sarcastic wit rather than to explore his musical activities and ideas. There is one exception. Athenaeus tells us, on the authority of the late fourth-century scholar Phaenias, that Stratonikos was the first person to offer instruction in harmonic theory and to make use, in that context, of a diagram.[13] The thesis that he was the first to teach harmonics should not be taken at face value. The plethora of 'Stratonikos anecdotes' in Athenaeus shows that stories about him circulated in abundance, and Phaenias had evidently heard about this aspect of his teaching. Presumably he had not found similar reports about earlier musicians or theorists, the minutiae of whose doings were not the subject of such widespread gossip; but he cannot have known that no such predecessors existed. It was common for Greek scholars to look for individuals

[12] The information, such as it is, is at Diog. Laert. VIII.46, where this Pythagoras, along with two others of the same name, is said to have lived 'around the same time' as Pythagoras the philosopher, who 'taught' all three of them. But I would stake very little on the reliability of this evidence.

[13] Ath. 348d, Phaenias frag. 32 Wehrli.

in the past who could be treated as the 'inventors' of some art or practice or as 'the first' to do one thing or another. Few of their claims are reliable. The Plutarchan *De musica*, for instance, is awash with speculations of these sorts which the compiler found in the works of fifth and fourth-century writers; and certainly not all of them can be believed, if only because they often contradict one another. It is in any case wildly unlikely that harmonic theorists before Stratonikos had made no efforts at all to pass on their knowledge to anyone else, and I conclude that this element in Phaenias' report should be rejected. But we should accept his intriguing indication that there was sometimes direct interaction between the work of theorists (locatable in this case among Aristoxenus' *harmonikoi*, on the strength of the reference to a diagram) and the activities of musicians themselves. I shall say more about this matter shortly.

But there is another piece of evidence about Stratonikos which scholars have generally neglected. It appears in a dialogue called *Sisyphus* that has been transmitted under Plato's name; and though it is certainly not really from his pen, it is likely to be a fourth-century text. The indications it gives us are slight but important. Socrates, it tells us in the opening paragraph, has been attending an *epideixis* presented by Stratonikos, which 'put on display many excellent things, both in speech and in action' (387b). Though the text does not mention music, it is almost certain that the reference is to Stratonikos the musician. No one else of that name was sufficiently well known for the allusion to be taken in any other way; and music must be the strongest candidate for the role of being something presented 'in action' (*ergōi*) in an *epideixis*, as well as 'in speech' (*logōi*). The context makes it clear that the *epideixis* had been publicised in advance, since Sisyphus, the other character in the dialogue, had planned to be there too but was prevented by civic business. The scenario, of course, is fictional. It must be constructed, however, within the constraints of what its readers would have found plausible; and this suggests that Stratonikos was known to be a person who had taken his musical expertise into the market-place in the manner of the sophists. The passage's phrasing guarantees that what he presented on this imagined occasion was not a musical recital, though its demonstrations 'in action' were presumably live musical examples, like those of the *harmonikoi* mentioned in the Hibeh papyrus (but they would certainly have been performed a great deal more skilfully). It was a discursive exhibition of his *sophia*.

The dialogue unmistakably portrays Stratonikos' presentation both as an intellectual treat (even if there is a tinge of irony in Socrates' flattering remark) and as an 'attraction' capable of drawing a keen non-specialist audience; Socrates attended with unspecified companions, and Sisyphus is at pains to emphasise that he would surely not have missed it had it not been for the absolute imperative imposed on him by a summons to a consultation with the city council (387b–c). Much the same can be inferred, as we have seen, about the circumstances surrounding the speech of the writer of the papyrus fragment and the pronouncements of the *harmonikoi* he criticises, and though the latter have little or nothing to do with 'harmonics' in the strict sense, the example at any rate shows that musical speculations were not out of place in such a context. It is at least possible that it provided the setting for the demonstrations of Pythagoras on the *tripous*, the discussions of the empirical theorists mentioned in *Republic* 531, and the presentations of the 'notators' derided by Aristoxenus. By the later fourth century and the early years of the third the jargon of harmonic science was sufficiently familiar to be satirised, along with that of contemporary philosophers, on the comic stage.[14]

In the wake of the activities of the famous fifth-century sophists, Protagoras, Gorgias, Prodicus, Hippias and the rest, the *epideixis* seems to have become something of an institution, at least in Athens, not in the sense of being 'institutionalised', like the musical and dramatic festivals with their formal civic and religious roles, their battery of regulations and their established calendar, but simply in the sense that it became a common and popular feature of city life, providing intellectual stimulation combined with entertainment, and no doubt also the opportunity for vigorous argument. It was a prominent element in the free-wheeling, individualistic and entrepreneurial culture that flourished, in the major cities, alongside the regular repertoire of events for whose organisation the civic authorities were responsible. It would have been easy to represent an induction into the mysteries of harmonic analysis as a means by which musical enthusiasts, like the 'lovers of sights and sounds' whom Glaucon and Socrates gently tease in the *Republic* (475d–476b, cf. 476d–e, 479a, 480a), could deepen their understanding of the musical arts they admired; and there would be nothing surprising about the fact, if it is one, that exponents of harmonic theory, like enterprising specialists in other branches of learning, sometimes offered their educational goods for sale through the medium of the *epideixis*.

Well; there is nothing implausible about the hypothesis, but the direct evidence I have been able to cite in its favour is precariously thin, and

[14] See especially Damoxenus frag. 2 K–A, in which the art of cooking is represented in terms that ingeniously combine the technicalities of musical theory with those of Epicurean philosophy.

even if it is true it leaves important questions unanswered. In particular it does not explain how and why this style of harmonic theory came into existence in the first place, since it seems unlikely to have been devised with the sole purpose of giving its inventors something unusual to sell. Let us pass on to the second suggestion I promised to make. It can be introduced most straightforwardly by a list of the names of those who are known to have worked in this field or can reasonably be suspected of having done so, together with a brief review of our other information about them.

HARMONIC THEORISTS AS PRACTICAL MUSICIANS

In prime position on the list we must put the three *harmonikoi* whom Aristoxenus names in the *Elementa harmonica*, Eratocles, Pythagoras of Zakynthos and Agenor of Mytilene. To them we can add two others whom he mentions briefly and apparently regarded as having speculated on topics proper to harmonics, though he does not explicitly call them *harmonikoi*; they are Lasus of Hermione and Epigonus of Sicyon[15] (*El. harm.* 3.20–4). Information given in the Plutarchan *De musica* (1136d) about an analysis of the Mixolydian *harmonia* propounded by Lamprocles of Athens makes it almost certain that he too belongs among the *harmonikoi*; and there are some grounds for thinking that the musical historian Glaucus of Rhegium, on whose work this *De musica* draws from time to time (1132e and parts of 1132f–1133a, 1133f, 1134d–f) had at least very close affinities with them. We have already seen that the Athenian kitharist Stratonikos was conversant with their ideas and made use of them, even if he did not contribute to the subject's further development. A passage of Plato's *Protagoras* which I mentioned above (316e) suggests that we should consider, more tentatively, the case for including Pythoclides of Ceos (cf. [Plut.] *De mus.* 1136d) and another Athenian, Agathocles (cf. Plato, *Laches* 180d). I shall close the list there; various other names might also be proposed, but with progressively less confidence. So little is known about such people as Dracon of Athens and Metellus (or Megillus) of Acragas, for instance ([Plut.] *De mus.* 1136f), that even if they are entitled to a place on the list the fact would give us no worthwhile information.

Aristoxenus speaks of Eratocles as if his name was familiar to everyone, without even identifying him as 'Eratocles of such-and-such a city', and though he criticises him roundly he treats him as a prominent figure among

[15] According to Juba, cited at Athenaeus 183c–d, Epigonus was 'an Ambraciot by birth, a Sicyonian by adoption'.

the *harmonikoi*. But the only other passage of ancient literature where he is mentioned is patently an echo of the *Elementa harmonica* itself;[16] the writer had no independent information about him, and neither have we. We know nothing at all about his life, and nothing about his activities except what Aristoxenus tells us about his contributions to harmonics. His dates might fall anywhere within a span of about 150 years, and any attempt to set his investigations in some definite context would be a work of imaginative fiction.

Eratocles, fortunately, is an isolated case; what we know about the others does not amount to very much, but we are not completely in the dark about them. The two crucial points that emerge can be stated very briefly. First, no ancient writer treats any of them as a philosopher or a mathematician, or as involved in abstract intellectual speculations of any sort; and secondly, all of them are known to have been practical and in most cases professional musicians, either composers or performers or both. In these respects they are in absolute contrast to harmonic theorists in the mathematical tradition, of whom the reverse is true; none of them was a practical musician, and all of them were active in the fields of mathematics and philosophy. It would be a waste of time for me to examine all the evidence about each of the empiricists in detail here, in an attempt to prove that they were indeed musicians as well as theorists, since in most cases there is little doubt about it. But it seems appropriate to say a few words about each.

Lasus is well known as an innovative musical composer and poet, active in the late sixth century and the early fifth.[17] He seems to have been instrumental in the process whereby the dithyramb became a regular and important element in civic music-making.[18] He developed a more complex melodic style than his predecessors, and one writer represents his way of using 'more numerous notes, widely scattered about' as altogether revolutionary, transforming the music of the past and preparing the way for the allegedly outrageous novelties of Phrynis, Melanippides and other composers of the 'New Music' of the later fifth century.[19] As to his contributions to musical theory, the *Suda* tells us that he was the first person to write a book on music, and he is mentioned in a passage on harmonics and acoustics by Theon of Smyrna.[20] The passage, unfortunately, is corrupt and interrupted by a lacuna, and it is hard to be sure just what it attributes to Lasus, though

16 Porph. *In Ptol. Harm.* 3.5.

17 The evidence about him is conveniently assembled in Campbell 1991: 296–311. See also Privitera 1965; a new and thought-provoking study is Porter (2007).

18 Schol. to Pindar, *Ol.* 13.26 (test. 5 Campbell).

19 [Plut.] *De mus.* 1141C (test. 6 Campbell).

20 *Suda* s.v. Lasus (test. 1 Campbell), Theo Smyrn. 59 Hiller (test. 8 Campbell).

it appears to connect him with ideas about the physical basis of sound and pitch. Speculations of that sort are more characteristic of mathematical theorists, and (at that date) specifically of the Pythagoreans, as the text of Theon suggests, and they are alien to harmonic science as Aristoxenus conceived it. But Aristoxenus does not tar Lasus with the Pythagorean brush. He criticises him for the error of supposing that a note has 'breadth' (*platos*), and this odd conception, however exactly it is to be interpreted, fits much better into a picture in which notes in a scale are strung out along a line, as in the diagrams of the *harmonikoi*, than into a Pythagorean scheme of ratios. It may have been designed to explain why no note can be placed between adjacent notes of a scale, even though there are bound to be gaps between their pitches.[21]

Epigonus cannot be securely dated, but is probably a rough contemporary of Lasus. Aristoxenus mentions him, or rather, 'some of the followers of Epigonus', in the same breath as Lasus, and accuses them of the same mistake. If my interpretation of Lasus' thesis is on the right lines, they too must have been speculating about the reason why each note of a scale is musically 'next door' to its neighbours, a topic to which Aristoxenus himself devoted a great deal of attention. 'Followers of Epigonus' are mentioned by another author too, and here they are associated with activities of a quite different sort; they are said to have been the first performers on the kithara to produce on the instrument the effect called *enaulos kitharisis*, which is attributed also to another pioneering Sicyonian musician, a certain Lysander. We cannot be sure what this was, but it was evidently both novel and effective; the passage implies that it added resonance and dramatic colour to the instrument's tones.[22]

Epigonus himself is said to have been an 'expert musician', and to have invented an instrument with forty strings, named after its inventor as the *epigoneion*. Juba of Mauretania, who was a grammarian and historian as well as a king, reports that at a later date its construction was altered so that it could be played in an upright position, and we can infer that it was originally held horizontally, perhaps balanced on the seated player's knees.[23] Modern scholars have generally concluded that it was a board zither, with parallel strings running across the face of a wide, flat soundbox.

[21] Aristox. *El. harm.* 3.20–4 (test. 7 Campbell). It is worth emphasising that it is to a musical note (*phthongos*) that Lasus is said to have attributed breadth, and not merely to a 'sound', as Campbell translates it, though he gives 'note' as an alternative in a footnote. 'Note' is in fact not simply an alternative possibility; it is right and 'sound' is wrong. In Aristoxenus' writings the distinction is very important.

[22] Philochorus cited at Athenaeus 637f. For discussion see Barker 1982b, West 1992a: 341–2.

[23] Juba cited at Athenaeus 183c–d; the number of strings is given at Pollux IV.59.

Such instruments were uncommon in the Greek world, and West has suggested that it and a related instrument with thirty-five strings, called the *simikon* or *simikion* after its inventor Simos,[24] were originally designed 'for the academic study of intervals and scale-divisions' rather than for musical performance. He suggests indeed that their strings may have been tuned with the smallest possible interval between each and its successor, giving an extended sequence of quarter-tones and providing, as it were, a working model of the matrix of quarter-tones used by Aristoxenus' *harmonikoi*; and they would have been used, on that hypothesis, 'in order to experiment with and demonstrate' analyses of the sorts which those theorists offered.[25] It is an attractive idea, though plainly running well beyond anything our evidence can confirm; but whether it is true or not, our information about Epigonus and his circle will certainly allow us to conclude that they combined theoretical explorations with notable achievements in the realm of musical performance.

I have already mentioned all the substantial information we have about Pythagoras of Zakynthos, who was probably at work around 450 BC or a little earlier. Aristoxenus explicitly numbers him among his own predecessors, that is, among the empirically minded *harmonikoi*, and mentions him as someone who attempted (though inevitably, in Aristoxenus' view, he failed) to 'enumerate' all the distinctions between the various types of scale. He says the same of Agenor of Mytilene; and since Agenor lived about a century after this Pythagoras, we can infer that he was trying to improve on the latter's catalogue, and that he shared Aristoxenus' view that Pythagoras had not completed the project of 'enumerating' forms of scale and describing the ways in which they differed. No more did Agenor himself, according to Aristoxenus;[26] and it may be that the project was developed progressively and cumulatively by successive theorists over a long period (Aristoxenus mentions Pythagoras and Agenor only by way of example).

[24] The historian Duris describes Simos as a *harmonikos* in a passage quoted at Porph. *Vit. Pyth.* 3 (Duris *FGrH* 76F23, DK 56.2). He apparently belongs to the fifth century, and was accused by Pythagoras' son Arimnestus, according to Duris' anecdote, of claiming falsely to have invented the *kanōn* or monochord, and of mutilating the inscription which named Arimnestus as its real inventor. It is not clear that the story should be believed (on its credentials see Creese 2002: 22–7), but if it were true it would at least show, as Burkert says (Burkert 1972: 455 n. 40), that there were other harmonic theorists 'in competition' with the Pythagoreans in this early period.

[25] West 1992a: 78–9. The conjecture that the *epigoneion* was a board zither was first made by Sachs 1940: 137.

[26] Aristox. *El. harm.* 36.26–37.1. We should note that Aristoxenus actually speaks of 'those around' (*hoi peri*) Pythagoras and Agenor, that is, their followers or associates, as he does in the case of Epigonus. The phrase is notoriously vague, and it may be that Aristoxenus had information only about the individuals he names. But we know that Agenor taught music (Isocrates, *Epist.* 8), and it is more than likely that Pythagoras and Epigonus also had pupils who repeated or extended their conclusions.

We know from Isocrates (*Ep.* 8) that Agenor was also a distinguished musician, and to judge by the story of Pythagoras and his *tripous*, which I outlined above, the same is true of him; it shows that he was not only, like Epigonus, an inventor of instrumental novelties but also an accomplished performer. Artemon's account leaves little doubt that he played his remarkable instrument in public, or at least in front of audiences other than his own immediate circle of associates, and he must have been something of a virtuoso in order to inspire the admiration which his performances received. If the standard of his performances on the triple kithara had fallen noticeably below that of familiar players of the normal, single instrument, his audiences would merely have laughed at him. He evidently had a very well developed technique; we are told that the *tripous* fell into disuse soon afterwards because it was so difficult to play.

I speculated earlier that Pythagoras might have exhibited his skills in the course of *epideixeis* which combined performance with theoretical discussion (presumably about the three patterns of attunement represented on his instrument and the relations between them), and that these displays of practical and analytic expertise might have been designed to attract paying pupils. That remains a possibility, but Artemon says nothing to encourage the hypothesis, and it is equally possible that they were straightforward 'concert performances'. An anecdote about the singer and kithara-player Aristonikos of Olynthos shows that distinguished soloists sometimes gave public musical recitals outside the framework of the regular festivals and competitions (in which the *tripous* could have had no place),[27] and though Aristonikos was active towards the end of the fourth century the practice may well be much older. From the sixth century onwards there were also, of course, any number of musical performances, sometimes by eminent professionals such as Anacreon, at the symposia of wealthy citizens and at the courts of tyrants such as the Pisistratids in Athens and Hieron I in Syracuse. Pythagoras could have played in such contexts before relatively restricted audiences; or again, his performances might have taken place in the environment, though not on the official programme, of one or more of the great festivals. It appears that in the fifth and later centuries these festivals often had a flourishing 'fringe' in which speakers and performers of all sorts could exhibit their talents.[28] Both Pythagoras and Agenor,

[27] See Polyaenus, *Strat.* v.44.1.

[28] See e.g. Plato, *Hipp. Mi.* 363c–364a, 368b–d; fragments of Gorgias' speeches at Olympia, DK 82B7, B8, B8a, and evidence for his speeches at the Pythian festival, B9; cf. Dio Chrys. VIII.9–12, Philostratus, *Vit. soph.* I.11, Lucian, *Herod.* 1–4, 7–8, Aelian, *Vet. hist.* 12.32; for contests in painting at festivals in the fifth century, Pliny, *Hist. nat.* 35.58. On other 'unofficial' entertainments at the festivals, some of them musical, see West 1992a: 376 with the references in n. 81 to Sifakis 1967.

then, are clearly identifiable as *harmonikoi* who were also first-rate practical musicians; Agenor was certainly a professional and probably Pythagoras was so too.

Our information about Lamprocles is very limited. He should probably be dated around the middle of the fifth century, and he was certainly a poet and composer. Athenaeus says that he composed dithyrambs, and he seems to have been the author of a well-known song in honour of Athena whose first words are quoted by Aristophanes in the *Clouds*, as an example of the manly music taught to schoolboys in the 'good old days'.[29] The evidence for his activities as a theorist is slight but persuasive; all we have is a statement in the Plutarchan *De musica* (1136d), giving 'the *harmonikoi* in their historical writings' as its authority, which credits him with a new analysis of the structure of the Mixolydian *harmonia*. As we saw in Chapter 2, the form which Lamprocles' analysis assigns to it is (or is close to) the one found repeatedly in later accounts of the *harmoniai*, where it is represented as a regularly formed segment of the Greater Perfect System, spanning the octave between *hypatē hypatōn* and *paramesē*.[30] The passage of the *De musica* describes it in the same way, but since the GPS seems not to have been constructed until a century or so after Lamprocles' time,[31] his description must have been different. It is probably the one given in another phrase in the *De musica*; on Lamprocles' account, the Mixolydian 'does not have the disjunction where almost everyone supposed it to be, but at the top'. This formulation, identifying the *harmonia* by reference to the position of the 'disjunctive tone', fits quite naturally with the procedure which Aristoxenus attributes to Eratocles, in which each of the seven *harmoniai* (or 'species of the octave') is generated by removing the interval at the top of its predecessor and replacing it at the bottom (see pp. 43–4 above); the tone of disjunction, which in the Dorian *harmonia*, for instance, is the fourth interval from the bottom, will appear in a different position in each. I explained earlier how the opinion held previously by 'almost everyone' about where the disjunction falls in Mixolydian can be understood if the pattern of attunement they had in mind was the eccentrically formed system recorded by Aristides Quintilianus in his description of the 'very ancient scales' (pp. 49–50 above). Lamprocles' analysis gives it the regularised

29 Ath. 491c, Aristoph. *Clouds* 967 with the schol. *ad loc.*; *PMG* 735, Campbell (n. 17 above) 1991: 184–7. Opinions differed in antiquity about whether the song was by Lamprocles or Stesichorus; the fact may be that both composed songs which began with the same three words (see Campbell at 187 n. 3, with the additional references given there).

30 See pp. 49–50 above.

31 For an elaborate new attempt to push its date back to around 400 BC, see Hagel 2005, which also gives copious references to earlier studies. But even if Hagel is right, it cannot have been devised until four or five decades later than any plausible date for Lamprocles.

form which it has in the later sources, and which enabled it to be absorbed into Eratocles' cyclic scheme. But since Lamprocles is credited only with this isolated insight and not with comparable analyses of the other *harmoniai*, we should not regard him as having anticipated the latter's systematic approach, which must be a later development and appears to be motivated by more abstract and theoretical considerations.

Glaucus of Rhegium, who was active around the end of the fifth century, is a particularly interesting case. He is known to have been (among other things) a performing musician, and like Epigonus and Pythagoras of Zakynthos, he played a very unusual instrument. Our information comes from a scholium on Plato *Phaedo* 108d4, commenting on the phrase *Glaukou technē*, 'skill of Glaucus'.[32] It records that the early Pythagorean Hippasus made a set of four bronze discs, equal in diameter but differing in thickness; the ratio of the thickness of the first to that of the second was 4:3, to that of the third 3:2, and to that of the fourth 2:1. Hence they were tuned so as to give the pivotal notes of a scale, that is, a fundamental note and the notes at a perfect fourth, a perfect fifth and an octave above it.

It goes on to say that Glaucus noticed the discs' musical potential, and 'was the first person who set himself to making music on them'. The discs which Glaucus used are unlikely, of course, to have been those of Hippasus himself, who was at work about a century earlier; the sense is rather that it was his information about Hippasus that prompted him to tackle the project. No doubt he had a set of discs made to his own prescription and for his own use, and we may guess that there were more than four of them, creating a more nearly complete scale, if he was able to conjure out music that impressed its hearers. Evidently it struck them as remarkable; the scholiast says that as a consequence of his feats the phrase *Glaukou technē* became proverbial, and was used to refer 'either to things that are not accomplished easily, or to things that are made with great care and skill'. In the *Phaedo* it evokes the idea of exceptional expertise. The essential point to glean from this report is that although the apparatus was originally devised for the theoretical purposes of a student of acoustics and mathematical harmonics, Glaucus deployed it, or something based on its principles, in the quite different context of practical music-making. There is nothing here to suggest that he was interested in the relevant mathematics, or in harmonics of any sort for its own sake. Since his performances generated enough excitement to give birth to a catch-phrase, current perhaps some fifteen or twenty years later, they must have been presented in some sense

[32] The report is based on the reliable authority of Aristoxenus (fr. 90 Wehrli).

'in public', like those of Pythagoras of Zakynthos, and were admired by a good many people who were familiar with the polished virtuosity of other instrumentalists. Whether or not Glaucus was strictly a professional musician, he was certainly a very accomplished performer in his peculiar branch of the art.

The evidence for his interest in musical analysis comes from another direction. He wrote a book on the history of the art, entitled *On the Ancient Poets and Musicians*, snippets of whose contents are preserved in the Plutarchan *De musica*.[33] As the title suggests, all the people mentioned in the surviving excerpts were already 'ancient' in Glaucus' time; none is later than the sixth century. Two aspects of Glaucus' work stand out clearly. First, he was concerned to establish the chronological relations between the composers he discusses, asserting for instance that Terpander was earlier than Archilochus, and that Thaletas was later than Archilochus but earlier than Xenocritus (1132e, 1134d–f). Secondly, he was interested in the ways in which each composer's melodic and rhythmic styles were influenced by those of his predecessors. Sometimes his remarks on this topic seem simplistic; Terpander, he says, 'emulated' the epic rhythms of Homer and the melodies of Orpheus (1132f). Elsewhere they are rather more complex. The compiler cites Glaucus for the thesis that 'Stesichorus of Himera did not imitate Orpheus or Terpander or Archilochus or Thaletas, but Olympus, and made use of the *Harmatios nomos* and the dactylic form which some people say comes from the *Orthios nomos*' (1133f). Later he is reported as saying that Thaletas 'imitated' Archilochus' melodies but 'stretched them out' to greater length, and that he used two types of rhythm which Archilochus had not employed, and which Thaletas 'worked up' (*exeirgasthai*) on the basis of the aulos-music of Olympus (1134d–e).

We need not pause here over the details of these assertions.[34] The essential points can be stated quite briefly. First, when he says that one composer 'imitated' another's work he does not mean that he copied it slavishly; Thaletas, for instance, 'imitated' Archilochus' melodies only in the sense that he used some of their features and developed them in his own, apparently more elaborate way. (It seems to have been this conception of the

[33] Passages from it are summarised or paraphrased at *De mus.* 1132e–1133a (excluding an awkwardly placed citation of Alexander Polyhistor at the beginning of 1132f, which perhaps originated as a reader's note in the margin of a manuscript), 1133f, 1134d–f. The word in the title which I have translated 'poets' (*poiētai*) might equally be rendered as 'composers', since it was used freely in either sense. But this is unimportant; all the poets mentioned here were composers too, and their poems were the words of their songs.

[34] For some suggestions about Glaucus' comments on Stesichorus see Barker 2001b and (more briefly) 2002a: 44–55.

relation between them that allowed him to conclude that Thaletas was the later of the two, 1134d.) Secondly, he did not confine himself to linking sequences of musicians in a linear way as 'masters' and 'pupils' or 'followers'; the map he draws of the connections between the various composers is a good deal more sophisticated than that, with separate lines of 'influence' crossing and converging to form an intricate network. Thus Thaletas took the work of one predecessor as the model for his melodies and another as the source of his rhythms, and both Thaletas and Stesichorus mixed elements from two quite different kinds of repertoire, those of music for a solo aulete and of music for voice and kithara.[35]

Finally and crucially, if Glaucus' assertions have any foundation at all, they must have been based on the careful study of these various composers' works as they were presented in performance in his own time. They were certainly not preserved in written scores. Glaucus' conclusions about the relations between these long-dead composers must therefore have depended on the analysis of music perceived by the ear alone; and though our source preserves nothing but the bare bones of a few of those conclusions, the analyses which supported them must have been detailed and sophisticated enough to enable him to identify and classify a range of melodic and rhythmic structures or motifs, and to recognise them when they reappeared in quite different musical contexts. Whether these analyses were his own or taken from the work of others we do not know, though as a practical musician he was presumably familiar with a wide range of styles and compositions, and could have undertaken the scrutiny himself. Nor do we know for certain whether the method of analysis which he used was derived from the work of *harmonikoi* of the kinds discussed by Aristoxenus. We shall see later, however, that there were indeed people who applied techniques of harmonic analysis to the study of pieces by composers of an earlier period, and the most economical hypothesis about Glaucus would be that he either took over some of the harmonic theorists' conclusions for his own purposes, or else was actually, in Aristoxenus' sense, a *harmonikos* himself.

I need not say much more than I have already about Stratonikos, the widely travelled virtuoso kithara-player who is said to have been the first person to teach harmonics (but I have argued that Phaenias' statement that he was the first is unreliable), and to have made use of a 'diagram', almost certainly of the sort which Aristoxenus attributes to the *harmonikoi.* West may be right in thinking that the diagram depicted the structures of several

[35] Stesichorus was a *kitharōidos* (see especially West 1971) but took the music of the aulete Olympus as one of his models; Thaletas 'imitated' the melodies of Archilochus, who is treated in this text as a *kitharōidos*, but drew on Olympus for some of his rhythms.

'modal scales' at once and showed the relations between them (perhaps it amounted to a representation of the relations between *tonoi*; see pp. 55–6 above), since as he points out, the 'multiplicity of notes' which Stratonikos is said to have introduced into performance on the kithara is likely to have been connected with a 'modulatory style of music'.[36] Here, plainly, we have someone who was primarily a musician, but who found that analyses in the style of the *harmonikoi* were a valuable aid to musical understanding, at least in an educational context.

We have only a smattering of evidence about Agathocles and Pythoclides, and I am not convinced that either of them is relevant to our present concerns. Pythoclides is perhaps slightly the older of the two, but both can be placed in the early to mid-fifth century. The *harmonikoi* who discussed Lamprocles' study of the Mixolydian are said, in the same passage, to have named Pythoclides as the 'inventor' of this *harmonia* in the same 'historical writings'; and they specify that he was an aulete, which must mean that pipe-playing was his principal profession.[37] Whatever the truth about the Mixolydian's origins may be, the report indicates that Pythoclides had a reputation for musical innovation. One source says that he was among the 'clever people' (*sophoi*) who taught Pericles, and another that he was the teacher of Agathocles.[38] Agathocles himself is said to have taught Pindar, Lamprocles and Damon.[39] It seems clear that both of them were distinguished musicians whose names were still familiar in the fourth century.

The main indication that they had an interest in musical theory of some sort is in a passage of Plato's *Protagoras* which I mentioned earlier (316e). Protagoras has been listing various well-known figures from the past who, so he asserts, were really exponents of the 'sophist's art' (*sophistikē technē*), but who because of the hostility which this profession provokes disguised what they were really doing behind the screen of some other activity, presenting themselves, for instance, as poets or athletic trainers (316d). He mentions Pythoclides and Agathocles as examples of people who adopted the persona of a musician for this purpose, and describes Agathocles in particular as a 'great sophist'. But this scarcely proves that they were musical theorists of any kind, let alone *harmonikoi*, any more than the poets whom Protagoras mentions, Homer, Hesiod and Simonides,

36 West 1992a: 368, cf. 218 n. 2. But the hypothesis is hardly as secure as West's unqualified statements suggest.

37 [Plut.] *De mus.* 1136d. A few lines earlier Aristoxenus is reported as saying that Sappho was in fact the Mixolydian's inventor, and he is probably the compiler's source for the whole passage, including the record of the opinions of these *harmonikoi*.

38 Plato, *Alcibiades* 118c, and the schol. to that passage.

39 Schol. to Plato, *Alcibiades* 118c; Plato, *Laches* 180d.

were poetic theorists or scientifically minded students of metre. Protagoras is making a thoroughly tendentious case to support a position of his own, and the most we can infer is that these musicians, like the poets he names, had a reputation for generalised 'wisdom'; we might guess as much in any case from Pythoclides' association with the circle of Pericles. The statement by the scholiast on Plato *Alcibiades* 118c that Pythoclides was a Pythagorean may have a similarly slender foundation, since later writers commonly assumed that anyone known both as a musician and as a 'thinker' in this period must have been a follower of Pythagoras. Probably, then, we should expunge these people's names from our list.

Whether they are included or not, however, it turns out that with the possible exception of Eratocles, the only nameable individuals from this period who were *harmonikoi* in Aristoxenus' sense, or whose activities seem to have been closely connected with theirs, were accomplished musicians; most of them, indeed, are known primarily as composers or performers. This does not prove that the same is true of every exponent of empirical harmonics before Aristoxenus, or that they always thought of their harmonic investigations as bearing significantly on their musical activities; but it does make this a tenable hypothesis. Three passages where no names are given, two from Plato, the other probably based on a lost work of Aristoxenus, encourage the same conclusion.

In a speech in Plato's *Phaedrus* (268d–e) Socrates mentions someone who 'thinks himself to be a *harmonikos*', but does so merely on the grounds that he 'knows how to produce a very high and a very low note' (or perhaps 'the highest and the lowest note'). He imagines such a person being confronted by another, described as a *mousikos*, who points out that though anyone who is to be a *harmonikos* must indeed have this knowledge, its possession does not guarantee that they understand the least thing about the subject-matter of harmonics (Socrates says 'about *harmonia*'). Hence this person is not really a *harmonikos* at all. All he has grasped are some pieces of learning (*mathēmata*) which are necessary preliminaries to knowledge of *harmonia*, and he knows nothing of 'harmonic matters' (*ta harmonika*) themselves.

Two points need to be made about this passage. The first is purely negative, but it will be important; Socrates is not (as is sometimes said) drawing a strong contrast between two types of individual, the *harmonikos* and the *mousikos*. The person mildly chided by the *mousikos* is not a genuine *harmonikos* (any more than are the self-styled *harmonikoi* of the papyrus fragment we examined above), and the comments of the *mousikos* about his deficiencies therefore license no conclusions about the differences, if there are any, between *mousikoi* on the one hand and people who really deserve the title *harmonikoi* on the other. We cannot even tell from this excerpt alone whether, when Socrates calls the second person a *mousikos*, he means that he is a well-qualified practical musician or simply a person of refined general culture; in Plato's usage the term can have either sense. Socrates does imply that a *mousikos* will be civilised and well mannered, but he may be trading on the word's ambiguity and may still have a musician primarily in mind. I shall return to this issue shortly; all we can say for the present is that this *mousikos* knows something about harmonics, since he can say with authority that the other person is completely ignorant of the subject.

The second point arises from the context in which Socrates makes his remarks. He is arguing that the capacity to draw the various subtle distinctions and to produce the effects taught by rhetoricians does not amount to mastery of the art of oratory. Such things are merely 'pieces of learning which are necessary preliminaries to the art' (269b). The case of the putative *harmonikos* and the *mousikos* is just one of a series of examples which Socrates offers in order to make clear what he means; the others involve someone who purports to be a doctor on the strength of his ability to make a patient's body hotter or colder and the like (268a–c), and someone who supposes himself to be a skilful tragedian just because he can 'compose very lengthy speeches on small matters and very short ones on great matters' and do other things of that sort (268c–d). Their pretensions are patently absurd. But the crucial point is that these people imagine themselves to be respectively a doctor and a tragedian, that is, fully qualified *practitioners* of the arts in question, and not simply 'theorists' with an intellectual understanding of their principles. Correspondingly, the people to whom they make their ridiculous claims are genuine and indeed famous doctors (Eryximachus and Akoumenos) and tragedians (Sophocles and Euripides), not merely theorists or connoisseurs; and when Socrates comes to the case of oratory the criticisms are put into the mouths of eminent public speakers, the 'honey-tongued Adrastus' and the great statesman Pericles.

We are bound to conclude, then, that in the passage we are concerned with the person who claims (falsely) to be a *harmonikos* is claiming to be a skilful *practitioner* of the musical arts; and we can resolve the question about the meaning of the word *mousikos*. He must be a genuine practitioner of the art to which the other person pretends. He is the musical analogue of Eryximachus in medicine, Sophocles in tragedy and Pericles in oratory, and hence he is a musician in the full sense of that word, a Timotheus or a Stratonikos, and not merely a well-educated aesthete. It follows further that this passage, so far from marking a sharp distinction between the *harmonikos* and the *mousikos* in fact treats them as identical. If the picture it paints is on the same lines as the others, as it must be to play its part in the argument, the *mousikos* has precisely the qualifications which the other is wrongly representing as his own when he calls himself a *harmonikos*. We can go a step further. We noticed that the *mousikos* is evidently credited with some knowledge of harmonics; and we can now add that he must indeed be an expert in this field, since (if my reasoning is persuasive) he is the real possessor of the art or skill which the other man claims, and the title *harmonikos* belongs properly to him. On the evidence of this passage, then, the well-qualified musician and the genuine *harmonikos* are one and the same.[40]

The second Platonic passage comes from Book II of the *Laws*. The Athenian has been arguing that in order to be a competent judge of the merits of a piece of music, a person must know three things: he must know what it is designed to 'imitate' or 'represent'; he must know whether it is so constructed that it imitates its object 'correctly'; and finally he must know whether the resulting composition is 'good', that is, both aesthetically admirable and ethically edifying.[41] He points out that people whose education has equipped them to make these judgements reliably will have been much better trained than the ordinary members of contemporary choruses, who have merely been 'drilled in singing to an accompaniment and stepping in rhythm, but do not realise that they do these things without understanding any of them'. Unlike those choristers, he says, the fifty-year-old citizens who, in his imagined city, will need to decide what music is acceptable and what is not must have both an acute perception of rhythms and *harmoniai* and an intellectual understanding of them. If a person does not know what the 'constituents' (*hoti pot' echei*) of a melody are he cannot possibly know whether or not it is 'correct'; that is, he cannot fulfil the second of the conditions laid down (670b–c). These judges, then, must have a sound grasp on matters that are clearly in the province of a *harmonikos*. A little later the Athenian makes a further comment which has a clear bearing on our present concerns. People who fulfil all three of the relevant conditions, he says, will have had a 'more detailed and accurate education' (*akribesteran paideian*) not only than that of the mass of ordinary people, but even than that of the composers themselves. 'For though it is essential for a composer

[40] This may help to explain why Socrates, at *Philebus* 17b–e, refers to a person who has a full technical knowledge of intervals, notes and scales as a *mousikos* or an 'expert in *mousikē*'. We might have expected him to be called a *harmonikos*, but if every person skilled in empirical harmonics is in fact a musician, the choice of terms in this context is arbitrary.

[41] These conclusions are stated summarily at *Laws* 669a–b. For the thesis that all music is 'imitation' see 668a, and cf. 655d, where the pieces performed by choruses are described more specifically as 'imitations of characters'.

to understand (*gignōskein*) the facts about *harmonia* and rhythm, the third topic, whether a given imitation is fine and good (*kalon*) or not, is one of which he need have no knowledge' (670e).

Much might be said about this fascinating passage, but for the present the only conclusion we need draw from it is straightforward. When the Athenian describes the musical attainments of ordinary chorus-members and those of composers, he does not seem to be inventing a situation which will exist only in the 'best' kind of city, the one which he and his friends are imaginatively constructing. He is drawing on his own and his companions' knowledge of real-life choristers and composers. Typically, the former have no knowledge of the 'constituents' of melodies, the rhythmic and harmonic elements and structures which they contain; but composers must have a thorough understanding of these matters in order to pursue their craft. The Athenian tells us that they must grasp the details of *harmoniai* and rhythms both with acute perception (*euaisthētōs echein*) and intellectually (*gignōskein*, 670b), just like the well-qualified harmonic theorists of Aristoxenus' *Elementa harmonica* (particularly 33.1–10). It seems safe to conclude that at the time when he wrote this passage, Plato took it for granted that 'ordinary people' could not be expected to know anything about harmonics, but that every composer worthy of the name would be fully in command of the discipline. The qualifications of a harmonic theorist and those of a musician are once again wrapped up in the same person.

The third passage to be considered in this connection (mentioned briefly on pp. 65–6 above) runs through Chapters 38–9 of the Plutarchan *De musica* (1145a–d). Despite some difficulties which I shall mention in due course, its contents can confidently be attributed to Aristoxenus, and much of it, I suspect, has been preserved in his own words; its vocabulary and turns of phrase, as well as its themes and its waspishly sarcastic tone, are thoroughly characteristic of him. As a matter of convenience, at any rate, as well as conviction, I shall refer to the writer as Aristoxenus. It is a sustained attack on people who are described explicitly neither as theorists nor as musicians, but simply as 'people nowadays', *hoi nun* (1145a). The principal crime of which Aristoxenus accuses them in Chapter 38 is their rejection – on what he depicts as the flimsiest of grounds – of 'the finest of the genera, which was held in the highest esteem by the ancients on account of its majesty (*semnotēs*)', that is, the enharmonic. He argues, secondly, that if their reasoning were accepted a great many other familiar forms of scale would also have to be abandoned.

These people are said to offer two reasons for their thesis that the enharmonic genus has no place in melody. The first is that in their view one of its characteristic intervals, the enharmonic diesis or quarter-tone, makes 'no impression at all' on our perception, an assertion which makes Aristoxenus positively snort with rage. All it shows, he says, is that they are idle and lazy-minded, and imagine that the feebleness of their own powers of perception amounts to a proof of their position, 'as if anything whatever that escaped them must be completely non-existent and completely unusable' (1145a–b).

We should probably not take the phrase 'no impression (*emphasis*) at all' quite at face value. Unless the people criticised were altogether tone-deaf, they would certainly have been able to detect the fact that two strings which are in fact tuned a quarter-tone apart are not in unison. They are more likely to have meant that in such a situation it is impossible to determine by ear exactly how far apart the pitches are. Intervals as small as that cannot therefore form part of a recognisable scale, and they are out of place in the context of melody, since when we hear them we cannot tell what they are, and we have no way of deciding when they are tuned correctly and when they are not. This is evidently a continuation, at some distance in time, of the controversies lampooned in *Republic* 531a–b, where theorists dispute about the identity of the smallest identifiable interval. The people attacked in the present passage, who are Aristoxenus' contemporaries ('people nowadays'), had evidently broken ranks with the earlier *harmonikoi* discussed in the *Elementa harmonica*, whose prime focus was on the enharmonic and whose use of the quarter-tone as a unit of measurement guarantees that they believed it to be reliably identifiable by ear. But despite the apparent confidence of those *harmonikoi* and Aristoxenus' outraged contempt, the position which these 'lazy-minded' individuals adopt is defensible. There is no way of deciding objectively and 'scientifically', by ear alone, whether the tiny interval between two pitches is exactly one quarter of a tone or marginally more or less. Hence, contrary to the views of the older *harmonikoi*, the quarter-tone is useless as a unit of measurement; and contrary to the emphatic contentions of Aristoxenus, it is futile to assert that the 'finest' of all melodic systems is the one whose tetrachords are made up of one interval spanning exactly a ditone and two spanning exactly a quarter-tone each.[42]

[42] A ditone can be constructed with adequate confidence, by the method described in my next paragraph. One might argue that it is then possible to identify the two intervals remaining in the perfect fourth (where a fourth is defined in the Aristoxenian manner as spanning two and a half tones) as exact quarter-tones, on the grounds that they divide a half-tone into two equal parts. But the judgement that these parts are precisely equal is itself open to doubts which no available method could resolve.

This way of interpreting their position is supported by the second reason they are said to offer. It is that the quarter-tone 'cannot be constructed through concords, as can the semitone and the tone and other intervals of that sort' (1145b). The method of construction they have in mind is known from several sources including Aristoxenus,[43] and the principle underlying it is the basis of tuning-procedures still employed by string-players today. Aristoxenus states the principle clearly.

Since among the sizes of intervals, those of the concords have either no range of variation at all but are limited to a single size, or else have one that is completely negligible, whereas those of the discords have this attribute [that is, fixed and determinate size] to a much lesser degree, and since for these reasons perception judges the sizes of the concords much more confidently than those of the discords, the most accurate way of constructing a discordant interval is the one that proceeds through concords. (*El. harm.* 55.3–12)

What Aristoxenus says about concords and discords is true; it is much easier to reach agreement on when two strings are tuned at precisely an octave, a perfect fifth or a perfect fourth apart than it is to secure such agreement about intervals such as the tone and the semitone. When we are tuning a pair of strings, intervals of the former kind, the concords, seem to signal their presence when the interval reaches a certain fairly determinate size, and our ears will tolerate little or no 'range of variation', at least if they are reasonably well trained; whereas we may perceive both of two intervals as semitones, for example, even when in fact their sizes are substantially different. This is because when a concord is reached, the acoustic phenomenon known as 'interference' or 'beats' is minimised, and increases again quite detectably as soon as the relation between the pitches is altered to make the interval larger or smaller. The strategy for constructing discordant intervals 'through concords' is unproblematic, at least in principle. To construct (for instance) the interval of a tone downwards from a given note by ear alone, we first tune down a perfect fifth from the given note, and then up a perfect fourth from the note we have reached. Since the tone is defined as the difference between a fifth and a fourth, we will have done what we set out to do; we have arrived at the note a tone below our starting point, and we have done so by assessing intervals all of which are concords and can be reliably identified by ear.

The people under discussion evidently held that this method of constructing discords is reliable, and that no other is; that is, no other procedure will allow us to be sure that the discordant interval we have constructed

43 Aristox. *El. harm.* 55.3–56.12, [Eucl.] *Sect. can.* prop. 17.

is precisely the one we were aiming for. As the passage in the *Elementa harmonica* shows, Aristoxenus realises that there is a good deal of substance in this opinion. But he vigorously rejects the conclusions drawn from it by the theorists criticised in the *De musica*. As they themselves point out, the only intervals that can be constructed in this way are 'the semitone and the tone and other intervals of that sort'; more precisely, they are the semitone and its multiples. It follows that not only can the intervals of Aristoxenus' favourite system, the enharmonic, not be reliably constructed or identified or used; neither can 'the third size of interval or the fifth or the seventh', that is, an interval spanning three, five or seven quarter-tone dieses, and 'none of the divisions of the tetrachord is usable except one in which all the intervals employed are even [multiples of the diesis]' (1145b–c). This will in fact eliminate all but two of the divisions specified as 'familiar' in the *Elementa harmonica*, and the vehemence with which Aristoxenus dismisses his rivals' opinions may in part reflect his consciousness that his apparently confident descriptions of those divisions are uncomfortably vulnerable to objections of this sort.

I should like to comment briefly on the terms which Aristoxenus uses in this passage of the *De musica* to describe the intervals that cannot be constructed through concords. The point I want to make is important, though it is incidental to our immediate concerns. As he himself explains (1145b), 'the third size of interval' means 'an interval of three of the "smallest dieses" or quarter-tones', and so on for the rest; he talks in the same vein of 'odd' and 'even' intervals, by which he means those comprising odd or even numbers of dieses; and in summing up he says that on his opponents' hypothesis every tetrachord must be divided into intervals all of which are 'even'. He could have said 'intervals which are semitones or multiples of a semitone', but he does not; and he does not even mention the various small intervals involved in his divisions in the *Elementa harmonica* whose sizes cannot be expressed as some whole number of quarter-tones at all. His 'soft chromatic', for instance, uses intervals of one third of a tone, and intervals of three eighths of a tone are required by his 'hemiolic chromatic' (*El. harm.* 50.28–51.4). He is plainly using the quarter-tone as his unit of measurement throughout the diatribe in the *De musica*, and yet this is a practice which he rejects in the *Elementa harmonica* and attributes, as we have seen, to the wrong-headed *harmonikoi* who preceded him.

Were it not for the unmistakably Aristoxenian rhetoric of the passage, these facts might lead us to attribute it instead to one of the *harmonikoi* themselves, but I cannot believe that it is from any hand but his; his fingerprints are all over it. It is perhaps just possible that he found a record of the controversy in someone else's work and recast it in terms of their system of measurement but with his own evil-tempered embellishments, and it is true that in another passage which we shall consider later he may have added arguments of his own to a report which he attributes to others (n. 54 below). But there the case is quite different, since there is nothing in those other people's approach to which he himself would object. As an alternative we might suggest that the passage we are studying comes from a relatively youthful work; it was written at a time when he had not yet developed his own system and was relying on the methods of earlier *harmonikoi*, though his passion for the enharmonic and his penchant for intemperate polemic were already well established. If that were true it would add something to our knowledge of the *harmonikoi* and the controversies in which they were involved, and it might require us to assign an earlier date to the people under attack. But we have no independent evidence that Aristoxenus' intellectual career passed through a phase of that sort, and that he subsequently underwent something like a Pauline conversion. I find it an attractive hypothesis, but we cannot rely on it.

Let us return to our theme. So far, the people Aristoxenus is attacking would appear to be harmonic theorists. They are certainly to be regarded as 'empiricists', since their arguments depend wholly on their views about the intervals that can reliably be identified by the ear. But they differ markedly from the *harmonikoi* discussed by Aristoxenus elsewhere, since they completely reject the enharmonic systems to which those *harmonikoi* had devoted themselves, and claim that the interval which those theorists had treated as the fundamental unit of measurement is unidentifiable and unusable. Aristoxenus' critique continues, however, into the next chapter of the *De musica* (Chapter 39), and there they appear in a different light. These people, he says, are not only wrong but inconsistent; 'for they themselves make a great deal of use of divisions of the tetrachord in which most of the intervals are either odd-numbered or irrational, since they are constantly flattening the *lichanoi* and the *paranētai*' (1145c–d). Similar and even more outlandish practices are attributed to them in the remainder of the chapter, but we need not attempt to unravel the details. The point I want to make could not be simpler. It is that all these criticisms presuppose that these people are musical performers, and that when they play or sing they use intervals which are outlawed by their own theoretical position. Here again, then, we find that the theorist and the practical musician are one and the same, and Aristoxenus shows no sign that he thinks the fact strange. His view is rather that theory and practice should go hand in hand, and that there is something badly wrong with the approach of a person whose musical practices conflict with the results of his theorising.

THE PURPOSES OF EARLY EMPIRICAL HARMONICS

Every piece of evidence we have considered points to the conclusion that empirical harmonics before Aristoxenus was the preserve of musicians, and I know of none that casts doubt on it. The passages from Plato's *Phaedrus* and *Laws* suggest that the point might be put more strongly; every properly qualified musician must be equipped with a sound knowledge of harmonic theory, and no one but a musician (except the ideal and probably imaginary judge of musical excellence) is likely to have any grasp on the details of the science. That may be overstating the case; but it seems abundantly clear that if we are to understand the objectives of those who set out to analyse scales and attunements on an empirical basis, and to place their studies in an intelligible context, it is to the interests, activities and priorities of composers and performers that we should turn our attention.

Correspondingly, we are not likely to learn much in this connection from the writings of philosophers and other refined intellectuals. None of the major philosophers appears to have thought that the procedures and conclusions of the empirical harmonic theorists had any bearing on the issues with which they themselves were concerned. In the *Republic* Plato represents them as laughable and irrelevant. In the *Phaedrus* and the *Philebus* he treats them with more respect, but refers to harmonics only as one specialised body of knowledge among others, and uses it only by way of analogy with some other discipline or form of enquiry; and it is the latter, not harmonics and the other analogues for their own sake, that is the focus of his attention. His allusions are designed to bring out a feature shared by all established branches of knowledge, so as to encourage the idea that it will be found also in the discipline he is really examining. There is nothing here to suggest that harmonics as such is of general intellectual interest or has anything to offer a philosopher. We are told in the *Laws* that reliable judges of music must have a thorough grounding in the subject, and it is true that the discussion of the qualifications they need plays an important part in the conversation. But the Athenian does not mention any details of the discipline, nor does he imply that it has any significance outside this immediate context. It provides some (though by no means all) of the knowledge a person needs if he is to distinguish music that should be heard and performed in the city from music that should not, but its propositions have no bearing on anything else.

In Book VIII of the *Politics*, rather similarly, Aristotle occasionally refers to 'musical experts', though it is not clear that he is thinking of their harmonic analyses. They apparently have useful things to say, particularly about which of the *harmoniai* should be used in education, but Aristotle shows no more inclination to pursue their ideas in detail than does Plato. In the *Posterior Analytics* he represents empirical harmonics as a very humble discipline which can do no more than give factual descriptions of musical phenomena; it is left to its mathematical cousin to provide scientifically interesting explanations. Mathematical harmonics also provides Aristotle with ideas which he develops for his own purposes in other connections, and empirical harmonics does not.[44] Theophrastus, in a long and intricate passage, subjects some of the conceptual foundations of both mathematical and empirical harmonics to close critical examination; but he devotes far more of the discussion to the mathematical than to the empirical variety, and in any case concludes, in effect, that neither approach makes sense.[45]

If we are to find out anything about the contexts and purposes of the *harmonikoi*'s researches we must therefore look for information elsewhere; and it is thin on the ground. I have already reviewed the evidence that some of them tried to present harmonics as a marketable commodity in the manner of the sophists, but I have argued that it must originally have been developed in a different setting and for purposes more closely connected with its subject-matter. The passages from Plato and Aristoxenus which we have examined seem to show that it continued to be pursued by musicians, with serious and not merely mercenary intent, throughout the period that concerns us. We cannot expect much help from the composers or other musicians themselves, at any rate from the words of their songs, which

44 For allusions in the *Politics* see e.g. 1341b27–8, 1342b23–4; the principal passage in the *An. Post.* is at 78b–79a, cf. 75a–b. For Aristotle's use in non-musical contexts of ideas drawn from mathematical harmonics, see e.g. *De sensu* 439b–440a, 448a and Ch. 13 below.

45 Theophrastus frag. 716 Fortenbaugh, discussed further in Ch. 15 below. Only eighteen of its 126 lines are concerned with the empiricists' approach. We should notice the incidental comment at lines 17–18 that 'some people' regard the mathematical theorists, who ground their judgements on calculations with 'intelligible numbers', as 'more intellectually penetrating (*synetōteroi*) than the *harmonikoi*, who judge by perception'. Theophrastus does not endorse these people's view, but it seems to reflect Aristotle's opinion, and to have been widely shared in Peripatetic circles. Books XI and XIX of the Aristotelian *Problemata*, for example, make some use of ideas drawn from mathematical harmonics but none at all of the empiricists' work. It is debatable whether Theophrastus' target in the few lines he devotes to the empiricists is Aristoxenus or some of his predecessors or both; for my own wavering views on the question see Barker 1985 and 2004, and cf. pp. 421–8 below.

would be an improbable environment for revelations about the aims of harmonic theory. The best we can find is an elegiac fragment from the astonishingly versatile fifth-century poet and prose-writer Ion of Chios (he died before 421 BC[46]), which celebrates the introduction of an instrument with eleven strings in terms that imply a background of careful harmonic analysis.

> Eleven-stringed lyre, with your ten-step arrangement
> and concordant crossroads of attunement,
> in the past all Greeks plucked you seven-toned, by fours,
> arousing a meagre music.[47]

The technicalities underlying the 'ten-step arrangement (*taxis*)', the 'concordant crossroads', the 'seven tones' and the 'fours' are somewhat obscured by their re-coding in poetic language, but there is no doubt that they reflect conceptions proper to harmonics, and modern scholars have found persuasive ways of interpreting them.[48] The 'crossroads', for instance, are the points at which the system offers the alternative possibilities of conjunction and disjunction, and the 'fours' are the conjoined tetrachords of a seven-note scale; but for present purposes the details are unimportant. The passage shows at least that this composer was familiar with current ways of analysing attunements, and that he found it appropriate to draw on them in expressing the superior musical capacities of the newly elaborated instrument. The work of harmonic theorists, then, was a useful resource for musicians who wanted to talk or sing or write about their art, but the passage licenses us to draw no firmer conclusions.

Like so much other valuable though enigmatic information, the most helpful hints come from material preserved in the Plutarchan *De musica*. They seem to me to point in two different but probably related directions. In the first place, we have already seen from its citations of Glaucus of Rhegium that pieces by early composers were examined through some form of musical analysis in the service of musical history; and there is another passage, much discussed by specialists in ancient musicology, which shows more directly that the resources of harmonic theory were sometimes deployed for historical purposes. It is the discussion of the 'invention' of the enharmonic genus by the aulos-player and composer Olympus which

[46] Aristoph. *Peace* 834–7. Useful studies of Ion include von Blumenthal 1939, West 1985, Strasburger 1990: 341–51.

[47] Ion frag. 32 West, quoted at Cleonides 202.13–17 Jan.

[48] For discussion see Levin 1961, West 1974: 174, 1985, and 1992a: 227.

occupies Chapter 11 (1134f–1135b), and is introduced as Aristoxenus' report of an account given by 'the *mousikoi*'. He does not call them *harmonikoi*, but the content of the passage shows beyond doubt that they are well versed in the discipline, and his choice of the other designation is not to be construed as implying anything to the contrary. What it indicates, perhaps, is that they are experts not only in harmonics but in other aspects of the musical arts as well. At the beginning of the *Elementa harmonica* Aristoxenus draws a distinction between a *mousikos* and someone who is expert in harmonic theory alone, but makes it clear that in his view (which seems close to that of the Athenian in the *Laws*), a person needs a wide range of qualifications in order to merit the honourable title *mousikos*, and that these include a proficiency in harmonics.[49]

These *mousikoi*, according to Aristoxenus, assumed that all music before Olympus was diatonic or chromatic.[50] He continues:

> They suppose that the discovery happened in something like the following way. Olympus was moving around in the diatonic, and was making the melody travel across frequently to the diatonic *parhypatē*, sometimes from *paramesē* and sometimes from *mesē*, passing over [i.e. omitting] the diatonic *lichanos*; he appreciated the resulting beauty of character, and filled with admiration for the *systēma* constructed on this basis he adopted it, and composed pieces in it in the Dorian *tonos*. (1134f–1135a)

The part of the diatonic system from which they imagined that Olympus began covers a perfect fifth, and its notes, from the bottom upwards, are *hypatē*, *parhypatē*, *lichanos*, *mesē*, *paramesē*, separated by intervals of semitone, tone, tone and tone. Their historical reconstruction posits that Olympus was trying out the effect of moving downwards to *parhypatē*, both from *mesē* and from *paramesē*, without touching on *lichanos* along the way, thereby leaving a gap of two whole tones with no note inside it between *mesē* and *parhypatē*. As they go on to explain, the resulting system could be construed in several ways, since scales in all three genera can contain notes in these relations. That is, it might be interpreted either as an enharmonic scale lacking the *parhypatē* which would divide the semitone at the bottom into two quarter-tones, or as a diatonic or chromatic scale lacking its *lichanos*. But the undivided ditone is the trade-mark of the enharmonic, and they treat the pieces built on this system as 'the earliest enharmonic

[49] *El. harm.* 1.21–2.7. The passage will be further discussed in Chapter 9.

[50] They were almost certainly mistaken about this, but a similar view is reflected at *El. harm.* 19.23–9 and at [Plut.] *De mus.* 1137e.

compositions', of which the very first was Olympus' *spondeion* or 'libation-tune' (1135a). They assert further that the division of its semitone into two quarter-tones was not due to Olympus but to later composers, 'as one can easily tell if one listens to someone playing the aulos in the ancient manner' (1135b).

The *mousikoi* who devised this account were certainly not relying on any written record of what Olympus was doing, one day long ago in the seventh century. The little narrative they constructed is clearly a work of the imagination, and we need not pause to ask whether it is historically accurate.[51] What they had at their disposal were certain tunes which were still played in their own time, ones that were reckoned very ancient and were regularly attributed to Olympus, whose 'invention' of this form of the enharmonic earned him a reputation as 'the originator (*archēgos*) of music in the noble Greek style' (1135b).[52] The story they told was intended to explain how melodies of this special sort came into being, and to represent them as the direct ancestors of the more fully developed enharmonic music of the classical period which Aristoxenus so greatly admired. It was based wholly on their familiarity with these melodies and on their analyses of the scales underlying them, together with corresponding analyses of the 'classical' enharmonic, the commonest form of the diatonic, and (by implication) the commonest form of the chromatic. If it is true that the *harmonikoi* whom Aristoxenus regards in the *Elementa harmonica* as his predecessors studied only the enharmonic, these *mousikoi* must belong to a later period than them; perhaps they are not much earlier than Aristoxenus himself. Be that as it may, they are quite plainly directing the resources of harmonic analysis, of a kind akin to that of Aristoxenus, to the examination of ancient melodies accessible to them only through the ear; they are pressing their results into service in the interests of musical history; and they are using them also to explicate the relations between these 'inspirational' melodies[53] and the enharmonic music most revered by connoisseurs in their own time.

[51] For useful comments on this matter see West 1992a: 163–4.

[52] Evidence that melodies 'by Olympus' were known and admired in the fifth and fourth centuries is at e.g. Aristoph. *Knights* 8–10, Plato, *Ion* 533b, *Symp.* 215c, *Minos* 318b, Aristotle, *Pol.* 1340a. For the systematic omission of *lichanos* (and the corresponding note in the next tetrachord) in a diatonic context, generating a 'pentatonic' system and an effect of antique solemnity, see the opening section of the paean by Athenaeus composed for a ceremony at Delphi in 128/7 BC. The score and a transcription into modern notation are printed in Pöhlmann and West 2001: 62–9; the relevant passage is on pp. 62–3.

[53] For their inspirational qualities see the passages from Plato's *Symposium* and Aristotle's *Politics* cited in n. 52 above.

The passage has several other intriguing features,[54] as does its companion-piece later in the *De musica* (1137b–e),[55] but I do not think that further investigation of this stretch of text by itself would add much to our knowledge of the uses to which empirical harmonics was put. But even if it tells us only a little about the motives of these *mousikoi*, the mere fact that it shows the discipline being put to work in a discussion of actual melodies is itself important. No Greek treatise devoted to the science itself, in any period, does anything of the sort, and modern readers who restrict themselves to those works could be forgiven for assuming that students of harmonics in antiquity never envisaged the possibility that their analyses of scales and other such structures could be applied to the study of individual compositions, and certainly never intended them to be used in that way. It turns out that they were so used. Even Aristoxenus, as we learn elsewhere in the *De musica*, not only took an interest in conclusions which others had drawn from the exercise (as the present passage indicates), but also undertook similar investigations himself (1142f, 1143b–c).

We shall consider Aristoxenus' purposes later (Chapter 9), but for the present I want to press our enquiry about his predecessors one step further. We have found that some of their analytic work was used as a basis for reconstructions of musical history. But we have found also that most if not all of them were in the first place professional musicians, and it seems reasonable to ask whether such people are likely to have been devoted to historical research simply for its own sake. That is of course not impossible, but we may wonder whether they had aims more immediately relevant to

[54] From another part of ch. 11 (1135a–b) we learn more about the structure of the *spondeion*. (For discussion see Winnington-Ingram 1928, Barker 1984a: 255–6.) It also deploys a principle (that a sequence of two ditones is *ekmeles*, 'unmelodic') which is akin to one used by Aristoxenus in the *El. harm.* but interestingly different; whereas Aristoxenus' argument (63.34–64.10) outlaws only a sequence of two 'incomposite' ditones, ones between whose boundaries no note intervenes, the sequence condemned here involves one incomposite ditone and one that is composite, made up of two successive intervals of a tone. We could indeed find 'Aristoxenian' grounds for rejecting such a sequence, but Aristoxenus does not make the case himself. (A sequence of two ditones both of which are composite is perfectly acceptable, and can be found in a chromatic octave scale in one of its regular forms: semitone, semitone, tone-and-a half, tone, semitone, semitone, tone-and-a half. The second, third, fourth, fifth and sixth intervals make up such a sequence.) We may reasonably wonder whether the argument in the *De mus.* is an intervention by Aristoxenus himself, as is usually assumed, or is part of his report about the views of the *mousikoi*. If the latter were true, it would have the important implication that theorists prior to Aristoxenus, or at any rate theorists other than Aristoxenus himself, had already begun to formulate 'rules' to which melodically acceptable sequences of intervals had to conform.

[55] The passage gives information about a structure closely related to that of the *spondeion*. It also identifies notes used in the accompaniment of melodies based on this structure but not in the melodies themselves, and specifies their relations to the melodic notes which they accompany as concordant or discordant. It is one of the very few texts which tell us anything about instrumental accompaniments to melodies; for discussion see Barker 1995.

the practice of their craft, and were not studying compositions by eminent musicians of the past for their antiquarian interest alone. This brings me to the second suggestion which I said I would make.

None of the sources answers our question explicitly, but another report by Aristoxenus embedded in the Plutarchan *De musica* offers a promising clue.[56] It tells a story about a Theban musician called Telesias. We know nothing about him beyond what we are told here, but details of the story itself place him firmly in the early fourth century. It is a tale with a moral: that a composer's early musical training has a profound effect on what he can and cannot achieve in later life. Telesias, we are told, was trained when he was young in music of the finest sort, and learned the most admired works of earlier composers such as Pindar, Lamprus, Pratinas and others. He studied aulos-playing as well as compositions for voice and lyre (Theban auletes were reckoned the best of all in this period); and he 'worked very thoroughly on the other elements of a complete musical education'. Later in his career, however, he was seduced (the text says 'thoroughly deceived') by the flashy, 'theatrical and elaborate' music of modern composers like Timotheus and Philoxenus, and no longer valued the 'noble works on which he had been reared'; and he made an exhaustive study of these newfangled compositions, especially the most complicated and innovative of them. But when he tried his hand at composing first in one of these styles and then in the other, he found that he was quite unable to do anything successfully in the Philoxenian manner. 'The reason,' the report concludes smugly, 'was his excellent training in his youth.'

The passage makes it clear that the close study of pieces by existing composers, past or present or both, was a major constituent of a musician's training. We are not told what was involved in 'being brought up on' these pieces and 'learning them thoroughly'. No doubt it included learning to play and sing them, but if, as it seems, this instruction was designed to provide guidelines for one's own compositions, there was probably more to it than that. Aristoxenus does not specify here that students were taught techniques of harmonic analysis and applied them to the works they were examining. Neither harmonics nor rhythmics is mentioned; all we have is a generalised reference to 'the other elements of a complete musical education'. But we already know from a passage in the *Elementa harmonica* (n. 49 above) that in Aristoxenus' opinion, harmonics is one of the disciplines which a *mousikos* must master. This is plainly presupposed,

[56] *De mus.* 1142b–c. It is the beginning of a long Aristoxenian passage (1142b–1144e) which we shall study more closely in Chapter 9.

too, throughout the remainder of the discussion in the *De musica* of which this tale is the starting point. A person setting out on musical studies must be 'knowledgeable' (*epistēmōn*) about composition in all three of the genera which harmonics distinguishes (1142d). Harmonics is one of the relevant *epistēmai*, and can be used in the examination of individual compositions, though it cannot tell us everything we need to know about them (1142f–1143a, cf. 1143b–e). In order to 'follow' a piece of music intelligently and to grasp its merits and faults we must (among other things) understand what is involved in the 'continuity' (*synecheia*) of a sequence of notes, time-lengths or syllables (1144b); and so far as its application to notes is concerned, this is a topic high up on the agenda of harmonics, as Aristoxenus conceived the discipline.[57]

Summarily, then, we have the following facts. Empirical harmonics before Aristoxenus was very largely a musician's preserve. Its resources were at least sometimes deployed in the examination of individual pieces of music, notably ones by reputable 'ancient' composers. A close study of such compositions was an essential and major part of a musician's training; and this involved also the mastery of various other musical disciplines, of which (at least in Aristoxenus' opinion) harmonics was one. On this basis I propose the hypothesis that one context in which the techniques of harmonic analysis were brought to bear on specimens of real music was the professional training of musicians. Phaenias' report about Stratonikos, which we discussed above, shows that harmonics was used in this setting at least from the early fourth century, and I argued that the passage cannot guarantee that Stratonikos was anticipated by no previous maestro. Techniques which permitted close structural analysis would have been a great help to musicians during the period from about 450 BC onwards, when the pioneers of the 'New Music' were at work with their unfamiliar modulations and other innovations and complexities, both as a means of understanding what these iconoclasts were doing, and (especially for more conservative teachers) as a means by which the differences between their work and that of the older composers could be pinned down. My present hypothesis puts a little flesh on the bare bones of Phaenias' report by suggesting how the discipline could have actually been put to use in a practical musician's education.

I think it very probable, though certainly I can adduce no conclusive evidence, that this was the context in which empirical harmonics began, perhaps right back in the days of Lasus and Epigonus (whose innovations

[57] See especially *El. harm.* 27.15–29.1, 52.33–53.3.

may have seemed as strange to their contemporaries as did those of Phrynis and Timotheus to theirs), and that it remained very largely confined to that specialised environment, among the mysteries of a professional craft, through the fifth century and much of the fourth. It surfaced only occasionally in more public environments, in writings such as Glaucus' foray into musical history which may have been addressed to a wider audience, and in quasi-sophistic *epideixeis*, which certainly were. It had no larger intellectual pretensions. This would explain why Plato places it alongside other branches of practical expertise such as the doctor's or the tragedian's, and neither he nor Aristotle treats it as a subject of serious importance; and it would come as no surprise that when Aristoxenus wrenched it from its original setting and paraded it in new dress as a fully fledged natural science, he found its earlier manifestations wholly inadequate. There was no reason why the *harmonikoi* whom he criticises should have been interested in deriving conclusions from high-level principles and presenting harmonic theory as a systematically organised body of knowledge, if for them it was simply a tool of the trade designed to provide ways of describing structures actually present in compositions they knew and taught. For such purposes the elaborate paraphernalia of scientific demonstration and systematisation would have been useless and irrelevant at best. Given that the scales, attunements, schemes of modulation and the other structural underpinnings of compositional practice had developed haphazardly through custom and usage, and were never designed to fit some intellectually excogitated system or derived by reasoning from principles, no analysis which enclosed them within a theoretically well-organised scheme would have been a reliable guide to the realities of musical practice.[58]

[58] An 'anti-theoretical' position of this sort is adopted at the beginning of the fifth-century Hippocratic treatise *Ancient Medicine* (20.1–2). The writer denounces people who say that medical knowledge must be based on an understanding of the 'nature of man', of the kind sought by philosophers, and who insist that doctors will be unable to 'care for people correctly' unless they derive their practices from this abstract starting point. More generally, he continues, medicine has no need of 'empty postulates' of the sort used in enquiries into obscure matters about 'things above and below the earth', but should work on the basis of empirical observation and experience. I suggest that the *harmonikoi* of this early period, if they had considered the matter, would have taken a similar view about their own art.

CHAPTER 4

Interlude on Aristotle's account of a science and its methods

Aristotle tells us rather little about empirical harmonics. His most significant reflections on the subject are concerned with the relation in which it stands to the mathematical form of the discipline, which he finds a good deal more rewarding, and we shall examine his views on this relation in Chapter 13. But before we move on to the next major phase in the history of the empirical approach, we need to review briefly some aspects of his *Posterior Analytics*. This pocket-sized treatise condenses into a mere seventy pages of text a meticulous study of the structure of scientific knowledge. We can only scratch the surface here, and in any case not all of it is relevant for present purposes; some other parts of it will be discussed when we come to consider Aristotle's comments on harmonics in their own right. But a sketch of some central theses is essential at this point, along with a rather closer account of certain details, because of the influence it exerted, a generation later, on Aristoxenus' conception of his project in the *Elementa harmonica*. I shall refer back to it repeatedly in Chapters 6–8. One of Aristoxenus' main aims was to improve radically on the manner in which his predecessors approached their research and set out its results, which he regarded as methodologically haphazard and inadequate, and to transform the discipline into a properly constituted science. There is little doubt that in tackling this task he took the prescriptions of the *Posterior Analytics* as his principal guide.

Something is known scientifically, Aristotle says, if we know not only that it is true, but that it *must* be true, and why; scientific knowledge of a truth is grounded in its explanation, which is, simultaneously, its proof. Knowledge of this sort is established and expressed through 'demonstration' (*apodeixis*) or 'demonstrative reasoning', and the principal focus of attention in the *Posterior Analytics* is on the conditions that any such demonstration must meet. It is important to recognise at the outset that Aristotle is not offering a 'logic of scientific discovery', though certain issues bearing on the process of investigation and discovery are discussed in the closing pages of the treatise. What he describes is the goal of enquiry rather than the procedures through which it is to be reached. It is only by being presented as a complex of demonstrations that a completed science, if there were such a thing, could display its credentials as a body of firmly established scientific knowledge.[1]

An *apodeixis* is a type of argument, and like any argument it derives its conclusions from other propositions, its premises. If it is to serve its purpose its reasoning must be logically valid, which means, for Aristotle, that it must follow the pattern of one or other of the valid forms of syllogism examined in the companion-work, the *Prior Analytics*.[2] Equally obviously, however, an argument cannot count as having demonstrated and explained its conclusion's truth merely because it is valid. It will do so only if its premises match up to certain special and rigorous standards. They must, says Aristotle, be

> true, primary and immediate, better known than the conclusion, prior to it and explanatory of it; for in this way they will also be the proper principles of the conclusion. There will be a valid argument even if these conditions are not met, but there will be no demonstration, since it will not produce scientific knowledge. (71b19–25)

The condition that the premises must be true is unproblematic; one cannot establish the conclusion as something known by deriving it from false premises. The claim that they must be 'primary and immediate' means that they must not themselves call for demonstration, but must in some way be known without being proved on the basis of additional premises. They must in fact be such that they *cannot* be demonstrated, since nothing that can be demonstrated can be known without demonstration (71b26–9); if something can be demonstrated and we lack the demonstration, we lack the grounds on which its truth depends. There is of course nothing wrong with an argument whose premises do need to be demonstrated, so long as the chains of reasoning involved in their demonstration can ultimately be traced back to premises that are primary and immediate. In such cases a full demonstration of the conclusion would include all the links in those chains (though in practice this will often be done in a shorthand form, by noting that one or more of the demonstration's

[1] See e.g. Bolton 1991, the introduction to Barnes 1993, Lloyd 1996: 7–37.
[2] In fact only a limited number of the *Prior Analytics*' syllogism-types can figure in the *apodeixeis* of the *Posterior Analytics*, but neither that issue nor other problems about the relation between the two works need be pursued here. See e.g. Barnes 1981, Smith 1982.

premises has already been demonstrated elsewhere[3]). The notion of truths that are known 'immediately'[4] and without demonstration raises questions to which we shall return; but Aristotle insists that they are essential if scientific knowledge is to be possible. If no such truths exist, and if every premise calls for demonstration, the chain of proofs will stretch back to infinity, and no part of it will be firmly established (72b5–15).

The last three conditions go closely together. The premises must be 'explanatory of' the conclusion because, we are told, 'it is when we know the explanation for something that we know it scientifically' (71b30–1); we cannot claim to have understood the necessity that something must be so if we do not know why it is so. They must be 'prior to' the conclusion because the truths explaining why something is so are prior to it, more fundamental in the order of nature; it is *because* they are true that it is true. A fact can indeed be proved from premises that are not prior to it and do not explain it, but on that basis it is not fully understood; it remains a bare fact, unaccounted for within the propositions of the science.[5]

The expression 'better known than' the conclusion is at first sight puzzling. Aristotle explains what he means by drawing a distinction between two ways in which something can be 'better known', which correspond also to two ways in which it can be 'prior'. It may be prior and better known 'by nature' or prior and better known 'to us', and these are different. In using the phrase 'better known to us' he is not speaking of scientific knowledge; he means, approximately, 'more readily accessible to our awareness'. The items that fall into this category are those 'closer to perception', derived more directly from the evidence of our senses. Those closest of all to perception are facts about particular objects we encounter, expressible in statements like 'this object here is a black and white dog'; at a slightly greater distance will lie crude empirical generalisations, 'dogs chew bones', or 'dogs bark'. Those 'better known by nature' are further from perception, dealing not with particular objects but with universals at high levels of abstraction: 'a dog is an animal with such-and-such essential attributes' – the attributes that constitute every dog *as* a dog. When

3 Thus in the Euclidean *Sectio canonis*, for instance, later arguments frequently refer back to propositions proved earlier in the treatise, and some take as premises propositions that have been proved in earlier work, outside the treatise itself (specifically, in Euclid's *Elements*), alluding to them as propositions that are already established and known, but without repeating the proofs. There are explicit allusions of this sort in Propositions 2 and 9.

4 In an argument of the form 'Every A is B; every B is C; therefore every A is C', the terms A and C in the conclusion are linked by the mediation of another term (the 'middle' term), B. If the proposition 'Every A is C' is known immediately, this means that the connection between its terms is known without such mediation.

5 Compare Aristotle's examples of arguments to do with the planets at 78b22–79a4.

we reach the point at which we know something scientifically, we must, Aristotle asserts, have come to know such universals *better* than we know those that are closer to perception. We have scientific knowledge of the latter, rather than mere acquaintance with them, only as it were at second hand, derivatively, through the grasp we have on the former (71b33–72a5, 72a25–b4).

What the opening pages of the *Posterior Analytics* envisage, then, is a fully axiomatised science, every one of whose propositions is either a principle of the privileged, 'immediate' sort, or is demonstrated by being derived deductively from such principles. Much of the remainder is devoted to more detailed elucidations of the conditions so far specified, and of their consequences (Aristotle postpones until the end, as we shall too, the question how the non-demonstrable principles can themselves be known). We can by-pass most of these minutiae here, but there is one small thicket that has to be explored. Without some account of it we shall make no sense either of Aristotle's own comments on harmonic science or of the approach adopted by Aristoxenus.

If something is known absolutely (*haplōs*), Aristotle says, it must be impossible for it to be otherwise. What is known through demonstration must therefore be necessary (*anangkaion*), and as a consequence the principles from which the demonstration derives it must be necessary too (73a21–4; the point is argued more fully at 74b5–75a17). This condition can be fulfilled only if what is said of a subject holds of the subject 'in itself', *kath' hauton*, or 'as such' (75a18–37).

Aristotle explains what he means by 'in itself' at 73a34–b24. The passage bristles with difficulties which I shall not attempt to unravel;[6] a few musical examples will be enough, I hope, to convey the central points. There are two kinds of case. Suppose, as a preliminary, that the perfect fourth can properly be defined as the interval by which an octave exceeds a perfect fifth, and consider first the statement 'The perfect fourth is an interval.' Here 'interval' holds of 'perfect fourth' in itself or as such, because being a perfect fourth necessarily involves being an interval; 'interval' enters into the definition of what it is to be a perfect fourth. Such a statement, plainly, is necessary in the sense that Aristotle intends; it is impossible for it not to be true.

Secondly, consider the statement 'The interval of a perfect fourth is concordant.' The situation here is different, since although there still seems to be a particularly intimate relation between the subject of the statement and the attribute with which the statement credits it, 'concordant' does

[6] See Barnes 1993 *ad loc.*

not figure in the perfect fourth's definition. The definition of 'concordant', on the other hand, must include the term 'interval', since everything that is (in the literal sense) concordant must be an interval, and the definition must make this clear. Not all intervals are concordant; but concordance is, in itself or as such, a property that attaches essentially to intervals, just as 'straight' attaches essentially to lines, though not all lines are straight. Hence if it is true that the interval of a perfect fourth is concordant, 'concordant' holds of it as such, in the guise in which it has been specified, that is, as an interval. If this proposition expresses a necessary truth, it does not wear the fact on its face; it is not known 'immediately'. But it is a *prima facie* candidate for being demonstrated scientifically; it is the sort of proposition which *might* be shown to be necessary if we were able to derive it from suitable premises. No proposition, Aristotle insists, can be necessary unless its subject is related to its predicate 'in itself', in one or other of these ways. Since only necessary propositions can figure in demonstrations, either as premise or as conclusion, the 'in itself' relation must hold between the subject and the predicate of every proposition in every demonstration.

Aristotle now claims that from the conclusions so far reached, it follows that 'it is not possible to demonstrate while crossing over from one kind (*genos*) to another' (75a38). As he goes on to explain (75a39–b2), the 'kind' envisaged here is that to which the subject of the demonstration's conclusion belongs. The predicate attributed to it in the conclusion must hold of it in itself. In that case it must belong to the same kind as the subject, either because it is part of the subject's definition, or because the subject is incorporated in the predicate's definition. By the same sort of reasoning, every term in the demonstration must belong to the same kind. Then if the demonstration runs:

(i) Every A is B;
(ii) Every B is C;
Therefore
(iii) Every A is C,

A and C must be in the same kind, for the reasons given above, and since the two premises are 'in itself' predications, B must belong to the same kind as A and C. What Aristotle means, then, is that no demonstration can contain terms designating items which belong to different kinds, and thereby 'cross over' in its reasoning from one kind to another. He does not pause to explain what he means, in this context, by a 'kind', or by 'belonging to a kind', but his intention is clear enough. Items belong to the same kind if and only if the terms designating them can be linked together by an 'in itself' proposition or by a sequence of them. Thus 'egg-laying' and 'feathered' are linked to one another through 'in itself' propositions involving 'bird', and though no giraffe or rabbit is egg-laying or feathered they belong to the same kind as those in the previous group through their shared relation to 'animal'. The same will be true of number, ratio, factor, two, triple and the like; and if a term from one of these groups were inserted into the other it would find itself in alien company.

But there is more to it than that. Any object studied by a scientist can be conceived in different ways and referred to by different terms. A falling conker, for instance, may be conceived as an object with a certain mass, and as such its behaviour will be explained by reference to the laws of physics. The terms used to construct a demonstration explaining why it must fall as it does must all belong to the same kind, and it will be a kind that belongs within the domain of that science. But it may also be conceived as the seed of a type of plant, the horse-chestnut tree, and under that description its behaviour in detaching itself from the parent plant will be explicable only in terms proper to botany. Aristotle evidently assumes that the various sciences occupy quite separate territories, such that no kind specified by a term in one scientific domain can be also part of another's subject-matter. Thus a physicist can demonstrate nothing about the behaviour of the conker as such (that is, conceived botanically, as the seed of a certain kind of tree), and the term 'conker' cannot figure in any of his arguments, since the physicist (as such) knows nothing about such things. What his demonstration must refer to in this context is not a conker but a material object with a certain mass.

By showing that one cannot cross over between kinds in the course of a demonstration, Aristotle therefore takes himself to have shown that it is impossible to demonstrate a conclusion belonging to one science from premises any of which, or any of whose terms, belongs to a second. 'One cannot demonstrate what belongs to one science by means of another' (75b14); every component of a demonstration must fall within a single domain. In previous publications I have called this thesis the 'same domain' rule, and I shall stay with that title here. We need pursue these thorny discussions no further in this abstract form. In particular, the question whether Aristotle's reasoning holds water is beyond my scope. But we shall find that the 'same domain' rule is pivotal to Aristoxenus' conception of his project, and it will also turn out to be the springboard from which we can reach Aristotle's own most interesting and provocative remarks about harmonic science.

We have not quite finished with the *Posterior Analytics*. Any scientist who aims to present his results as the conclusions of Aristotelian demonstrations needs access to a set of principles which provide them with their foundations. The premises of a demonstration must be 'immediate', in the sense that they are known without themselves being demonstrated; or, if we relax that condition slightly, they must at least have been logically derived from principles of that sort. Every fully demonstrative science, then, must possess a set of these fundamental principles, and every other proposition it contains must be shown to follow from them.

The principles are known without being demonstrated, but they are not simply 'given' to an enquirer as part of the pre-existing furniture of his mind.[7] Where, then, do they come from and how can they be established? Aristotle faces these questions in the final chapter of the *Posterior Analytics*. He finds it unbelievable that we somehow possess these principles all along without realising it, and he is simultaneously convinced that knowledge cannot be built out of nothing at all; there must be some basis out of which it grows.[8] His preliminary conclusion is that we must possess some capacity (*dynamis*) whose exercise can lead to knowledge of principles, but which is not, as he puts it, 'more exact' than such knowledge itself. This is an abbreviated way of saying that the truths to which it gives access are not 'superior' to the principles in the way in which the premises of a demonstration are superior to its conclusion. Demonstrations work from above downwards, establishing their conclusions from higher, 'prior' principles, that is, from principles that are wider in scope and more fundamental in the natural order. We reach the principles, by contrast, as the culmination of a process that works upwards from below, and sets off from the operations of some relatively humble cognitive faculty.

This faculty, Aristotle goes on, is an inborn power of judgement that is shared by all animals, the power of sense-perception. Some animals, though not all, are capable of retaining what they perceive – that is, they can remember – and when many similar items have been perceived and retained, some animals endowed with memory, though again not all, are capable of isolating the feature which binds all these items together, and of representing them in a single general 'account' (*logos*, 99b34–100a3). Aristotle calls the condition in which we find ourselves at this stage 'experience' (*empeiria*), and describes it as the foundation of both knowledge and skill (100a5–8).

[7] That they are in a sense 'given' in this way is a thesis embodied in the Platonic doctrine of *anamnēsis* or 'recollection', first introduced in the dialogues at *Meno* 81a ff. as a solution to a puzzle propounded at 80d–e. Aristotle mentions the puzzle explicitly at *An. post.* 71a29; cf. 99b22–7.

[8] This is a major theme of the first chapter of *An. post.* Book 1; it reappears in the closing pages of the work at 99b28–32.

The remainder of his discussion recapitulates and clarifies these ideas with the help of a graphic metaphor, and adds one crucial new point. So far he has given no explanation of the way in which several distinct items, perceived and stored in the memory, can come to be grasped as one, that is, as instances of a single kind. This gap is filled by the thesis that even when just one perceived item is retained in the mind, we possess a 'primitive universal' (in Barnes' translation), since although what is perceived is a particular individual, 'perception is of the universal' (100a15–b1). Aristotle means, I think, that when we perceive anything, we perceive it not just as an individual, but as something of a certain kind, as a human being, for instance. In perceiving it we register the existence of a type to which other things may also belong. The accumulation of instances which are perceived and remembered as instances of that kind can then lead us to grasp what it is about them that constitutes them as members of that kind, and hence to an account of what such-and-such a kind of thing is. Once we have achieved that level of understanding we are in possession of a principle capable of standing as a premise in a demonstration. We are in a position to show that a human being *must* have characteristics of a certain sort because *this* is what a human being is.

Aristotle's account in this passage is brief and dense. The exact meanings of many details are debatable, as are their relations to other passages in his writings which cover similar ground (especially the first chapter of the *Physics*). It is far from clear, in particular, how the principles reached through the process he describes can acquire the degree of certainty they need if they are to play their allotted role in demonstrations. But the general thesis is clear. We reach the principles of a science from a starting point in perception, empirical observation, and we proceed from there by a route very similar to what is nowadays called 'induction' (Aristotle's term is *epagōgē*). It is not just a matter of generalising features found in particular cases, but involves the refinement of 'primitive universals' (such as 'human being', or perhaps 'concord' or even 'well-formed musical scale') to the point at which we have abstracted what is essential to them, what makes them what they are. Sketchy as it is, this account provides a basis on which an empirical science can go beyond the collection of facts to knowledge of the principles by which the facts can be explained, and might eventually become capable of presenting itself axiomatically. All facts within its domain, apart from the principles themselves, will be derived from the principles as the conclusions of Aristotelian demonstrations. These notions are fundamental to the conception of harmonic science developed by Aristoxenus in the *Elementa harmonica*, to which we shall now turn.

CHAPTER 5

Aristoxenus: the composition of the Elementa harmonica

We have now reviewed virtually all the significant data we have about Aristoxenus' predecessors in the empirical tradition, and have put in place the Aristotelian ideas about scientific method which form an essential background to his own work in harmonics. They will figure extensively in Chapters 6–8. Most of his surviving reflections on the subject are contained in the text we know as the *Elementa harmonica*. The present chapter is concerned mainly with the structure of this work as we now have it, and we shall tackle the much-debated question whether the whole of the surviving text originally belonged to the same treatise, and if it did not, how the relations between its parts are to be understood. These are issues which any serious student of Aristoxenus must address; but they are quite intricate and involved, and I cannot pretend that this chapter is easy reading. Some readers may prefer to cut to the chase, and after glancing at my preliminary paragraphs on Aristoxenus' life and writings, to jump to the discussion of the substance of his theories which begins in Chapter 6. If so I am happy to forgive them; they may perhaps be motivated to come back to the present chapter at a later stage.

ARISTOXENUS' LIFE AND WRITINGS

Aristoxenus was born in Taras, a Greek city in south-east Italy (Tarentum to the Romans, modern Taranto); the date of his birth is uncertain, but can be no later than about 365 BC. His father was a professional musician, but though he trained him in the art from an early age, Aristoxenus never embarked, so far as we know, on a career as a composer or performer. Instead he turned to philosophy, initially, and perhaps for a good many years, in the company of well-known Pythagoreans. There is nothing surprising in that. The cities of southern Italy were this tradition's main stronghold in the fifth century and the early fourth, and during Aristoxenus' childhood the dominant statesman in Taras was the Pythagorean Archytas; a report implying that he was an acquaintance of Aristoxenus' father probably originates with Aristoxenus himself (Aristox. frag. 30 Wehrli). But by the time he was a young man most Pythagoreans had left Italy (probably for political reasons) to settle in mainland Greece (frag. 18); and Aristoxenus travelled to various cities where groups of them had come together. There he pursued his studies with a Pythagorean from Thrace named Xenophilus,[1] spent some time in Mantinea, visited Corinth, where he met the exiled Dionysius II of Syracuse (frag. 31), and became familiar with the Pythagorean circle based at Phleious; our report (frag. 19, cf. frag. 18) names Phanton, Echecrates, Diocles and Polymnastus, former students of Philolaus and Eurytus. It describes them also as 'the last of the Pythagoreans', an expression which the writer, Diogenes Laertius, apparently took from one of Aristoxenus' own works. Its sense is probably that they belonged to the last generation in the continuous Pythagorean intellectual tradition of mathematicians, philosophers and scientists, by contrast with those whose claim to the title 'Pythagorean' depended less on their philosophical commitments than on their adoption of an 'alternative' lifestyle as other-worldly mendicants.

The next thing we know of Aristoxenus' life is that he came to work with Aristotle in Athens. Since Aristotle was elsewhere from 347 to 335 BC, and it was only after 335 that he founded his own school, the Lyceum, Aristoxenus must have been thirty or more when he joined it, and he therefore came under Aristotle's influence only after a long engagement with Pythagorean thought. By the time of Aristotle's death in 322 BC he was a prominent enough figure in the Lyceum to anticipate being appointed as its new head, but Aristotle in fact bequeathed the position to Theophrastus.[2] No more of his life-history is known, not even whether or not he stayed in Athens; all we have is a remark in the *Elementa harmonica* which shows that some of his writings must belong to this later period.[3]

The *Suda*'s statement (frag. 1) that he wrote 453 books need not be taken at face value, but his output was certainly vast. Beside the musical writings for which he is most often remembered, we hear of extensive works on Pythagoras, Pythagoreans and Pythagoreanism, treatises on educational

[1] See frag. 1 and frags. 18–20b. In frag. 20a Xenophilus is described as a *mousikos*, and is said to have lived for many years in Athens; but the source (Lucian) is not necessarily reliable.

[2] See frag. 1. My estimate of his date of birth and my inference about his age when he joined the Lyceum depend mainly on the report that he expected to succeed Aristotle; if the expectation was in the least realistic he must have been at least forty in 322 BC, and was probably older. Frag. 1 (an entry in the *Suda*) also states that he was active during and after the reign of Alexander (336–323); given the convention that a person reaches his *floruit* at age forty, this tells a similar story.

[3] *El. harm.* 30.16–19. The tense and the form of expression, 'as Aristotle used constantly to relate . . .', unmistakably indicate that Aristotle was dead when it was written.

and political institutions, biographies of Archytas, Socrates and Plato (the last two of them thoroughly gossipy, not to say scurrilous), an essay on the soul and various miscellanea. The musical works themselves included several treatises on harmonics and one or more on rhythmics, an *On Music* which combined history and analysis, a biography of the composer Telestes, essays on listening to music, on composition, on instruments,[4] on choruses, on dancing in tragedy, on auletes and on tragedians, along with certain other works (*Praxidamantia*, *Comparisons*, *Sympotic Miscellanies*) which had musical content but whose principal focus is not known.

None of these works survives intact. The writings on Pythagoras and his school were used extensively by later authors, more extensively, in all probability, than those authors' explicit citations reveal. We have enough fragments from the biographies of philosophers to recapture their general flavour; and we can reconstruct a tolerable outline and some of the details of the work *On Music*. Of the other musical treatises we have only scattered remnants, with two notable exceptions. A substantial fragment survives from the second book of his *Rhythmics*, along with a good deal of evidence about the contents of the remainder, incorporated into other writers' work.[5] Most importantly, we have the writings that have been transmitted as the *Elementa harmonica*, and which are our primary concern here. They too are incomplete, but can to some extent be supplemented from later treatises that were based on them.

THE STRUCTURE OF THE *ELEMENTA HARMONICA*

(a) A preliminary survey

In its surviving form, the *Elementa harmonica* is in three books, of which the third breaks off in mid-flow. At that point the programmes announced in Books I and II are some way from being completed, but we can be reasonably confident, on the basis of later reports, that Aristoxenus did not abandon the project or die before finishing it; his account of the remaining topics did at one time exist, and it survived (perhaps only in summary form) until at least the second century AD. A good deal, then, has been lost. But the situation is more complex than that, since there are signs that the three surviving books did not all originally belong to the same work. In particular, Book II does not read like a continuation of Book I. A preliminary list of

[4] It is not clear whether the works cited as *On Auloi and Instruments* and *On the Piercing of Auloi* (frags. 100–101) were separate essays or parts of the compendious *On Instruments* (named in frag. 102).

[5] See Pearson 1990.

the more obvious obstacles to such a reading would contain at least the following items.

(i) The opening passage of Book II is apparently designed to introduce the subject as a whole, and would stand most naturally at the beginning of a treatise. Placed where they are, thirty pages into the work, its reflections on the importance of offering an informative introduction at the outset could only be construed as ironical; and Aristoxenus is not given to self-ironisation.

(ii) Book II includes a list of the 'parts' of harmonic science which differs substantially from the corresponding list in Book I, and it sets off on a study of them as if none of the science's parts had so far been addressed.

(iii) Certain of its theses seem incompatible with those of the first book, or at the very least can be accommodated to them only with difficulty.

(iv) Book II is noticeably different from Book I in pace and style, and in its more leisurely and more sophisticated forays into questions of methodology.

(v) A concept which Book II makes central to an understanding of harmonic structure, that of melodic 'function' (*dynamis*), is completely missing from Book I.

(vi) Finally, a long passage of Book II covering several topics (44.21–55.2) is so nearly identical in both content and arrangement to a passage of Book I (19.16–29.34) that one must be a revised version of the other; they cannot have been intended to stand cheek by jowl in a single work.

Book III, on the face of it, might be a continuation of either Book I or Book II, but on closer inspection the important part played in it by the concept of *dynamis* must associate it with the latter. In view of these and similar considerations I have argued elsewhere that the *Elementa harmonica*, as we have it, is the remains of two different treatises, written some years apart, of which the earlier contained Book I, the later Books II–III. This view is quite widely though not unanimously shared by other scholars,[6] and I still think it is essentially correct. But in one respect the position I have adopted in the past is misleading. Book II is not simply a second attempt at precisely the same project as that undertaken in Book I. If it were, in so far as it was successful it would make Book I redundant, and it does not. On the

[6] There is a survey of views on the matter in Bélis 1986, ch. 1; she argues there that the surviving text is the remains of a single work, divided into two parts, *Principles* and *Elements*. For the hypothesis that it is the remains of two works, one represented by Book I and the other by Books II and III (with which I agree), and that Book I is the later (with which I do not), see Yamamoto 2001 (in Japanese but with an English summary).

contrary, though the introduction to Book II plainly marks the beginning of a new work, and though the treatise represented by Books II–III was certainly based on revisions of its predecessor, as we shall see, it does not stand alone. It presupposes much of the material in the first two thirds of Book I, and occasionally refers to it directly. Though Book I was a segment of a project that had been superseded, that part of it, at any rate, continued to be 'required reading' (or, more probably, listening) for students of its successor.

So much by way of a preliminary verdict. But though the issue may seem arcane, it is really too important to be left there. If parts of the *Elementa harmonica* were composed at different periods of Aristoxenus' life, his approach to harmonic science may have shifted during the intervening years; and in that case it would be misleading to try to force them into conformity with one another, or to try to read the first consistently in the light of the second. Conversely, we might go badly wrong if we assumed without further ado that a hypothesis assigning Book I and Books II–III to separate and differently conceived works is correct, and consequently refused to allow either to inform our interpretation of the other. A good deal may hang, in fact, on the conclusions we reach about the relations between the work's three books. I make no apology, then, for devoting the rest of this chapter to a more detailed – though certainly not a complete – examination of the relevant evidence. I shall not pretend to be doing so entirely without prejudice; my project, at each stage, will be to see how closely the evidence can be fitted to the hypothesis outlined above, and what adjustments are needed to the latter in order to sustain it. As an initial step, I shall draw some comparisons between the lists of topics offered as programmes for harmonic science in Books I and II. This will highlight some of the problems and put some flesh on the bones of the suggestions I have made so far. It will also be important to decide how faithfully the programme announced and summarised early in each book is pursued in practice in the sequel.

(b) *The programmes announced in Books I and II*

The agenda laid down for the science in Book I, at 3.5–8.13, is not set out with perfect clarity, and there are a few uncertainties about the divisions between its topics. But the following summary, which agrees in most respects with that of Bélis,[7] will not be far from the mark.

[7] Bélis 1986: 40–1.

A(i) The nature of the movement of the singing voice,[8] and the 'space' or 'place' within which it moves (3.5–26).

A(ii) Five subsidiary topics needed in order to gain proper understanding of the first: relaxation and intensification of pitch (*anesis* and *epitasis*), depth and height of pitch (*barytēs* and *oxytēs*), and pitch itself (*tasis*) (3.27–33).

A(iii) The extent of the distance there can be between low pitch and high (3.33–4.3).

A(iv) The interval (*diastēma*): first an account of it in general (*katholou*), then a survey of the various types of distinction that can be made between one interval and another (4.3–5).

A(v) The system or scale (*systēma*) treated in a parallel way, first *katholou*, then through the distinctions that can be made between one type and another (4.6–8).

A(vi) The nature of musical *melos* (roughly, 'melody'), first sketched in outline and distinguished from non-musical types of *melos*, then dividing the broad conception (*to katholou*) into its kinds or 'genera' (*genē*) (4.8–22).

A(vii) Continuity (*synecheia*) and succession (*to hexēs*) in *systēmata* (4.22–5).

A(viii) The differences between the genera (*genē*) of melody which depend on the shifting positions of the moveable notes, and the ranges of the latter's movements (4.25–33).

A(ix) Incomposite intervals (4.34–5.1).

A(x) Composite intervals, and the principles governing the orderly combination (*synthesis*) of intervals into *systēmata* (5.1–31).

A(xi) The *systēmata* constructed out of orderly combinations of intervals, distinguished according to size (*megethos*), form (*schēma*), composition (*synthesis*) and position (*thesis*) (5.32–7.1).

A(xii) *Systēmata* involving mixtures of genera (7.1–6).

A(xiii) Notes (7.7–10).

A(xiv) 'Regions of the voice' (i.e. pitch-ranges) and the *systēmata* that belong to them; the *tonoi* or 'keys', considered in connection with modulation (7.10–8.3; 8.4–13 rounds off the passage as a whole).

The discussions pursued in the main course of Book 1 follow precisely this arrangement with little deviation at least as far as topic A(vi). Thus A(i) is discussed at 8.13–10.20; A(ii) at 10.21–13.30; A(iii) at 13.30–15.12. The

[8] The expression 'the voice', in this and many other Aristoxenian contexts, will include the sounds of musical instruments considered as bearers of melody.

presentation of A(iv) and A(v) is a little more involved. A(iv) is prefaced by a very brief definition of 'note' (*phthongos*).[9] The part of A(iv) that deals with the interval *katholou*, in general, is at 15.13–33, followed immediately by the corresponding part of A(v) on the *systēma* (15.34–16.15); then come the parts of the discussions which consider distinctions between types of interval (16.16–34) and of *systēma* (17.1–18.4). Topic A(vi) is examined at 18.5–19.29.

As for the remainder of Book I, it is less difficult to accommodate to the projected programme than commentators have sometimes suggested. The passage that follows directly after topic A(vi) (19.30–21.19), admittedly, does not seem to fall squarely into the sequence. It picks up one of the distinctions between intervals listed under A(iv), that between concords and discords, and addresses the task of identifying the largest and smallest concords that fall within the scope of music-making. Why it appears here is uncertain, but it seems not to be an accident, since there is a parallel discussion at exactly the corresponding point in Book II, directly after a cursory list of the genera and shortly before a full examination of the differences between them. In both books the passage on concords is followed by a brief list of ways in which the tone can be divided (into halves, thirds and quarters, 21.20–31, 45.34–46.18), which is presumably designed to prepare the way for the uses to which these fractions will be put when the genera are studied in detail. Possibly the passage on concords is placed where it is because most of the remainder of the work, as envisaged in the programme announced in Book I (and also, as we shall see, in Book II) concerns *systēmata*. The smallest *systēmata* that can be treated in their own right as musically meaningful wholes are tetrachords,[10] structures contained by the smallest concord, the fourth. They are also the main building-blocks for more extensive *systēmata*, and the principal variations between tetrachords will shortly be examined in the context of the genera, under topic A(viii). Since all significant *systēmata* are bounded by concords,[11] the largest concord that can be used in music (given the limitations of any actual voice or instrument, 20.23–21.19) will determine the range of the largest *systēma* with which harmonics concerns itself.

[9] This gives only the minimum needed to make possible a definition of *diastēma*, interval, and is not to be confused with the discussion of notes promised as topic (xiii), and whose strangely late position in the list will be discussed in a later chapter (p. 187 below). Bélis is mistaken in representing the fourth item on the list in the introductory passage as 'De l'intervalle et du son' (Bélis 1986: 40).

[10] Aristoxenus sometimes uses the term *systēma* to refer to a smaller structure, the *pyknon* or its diatonic counterpart (e.g. *El. harm.* 24.30, 29.2), but these are not musically significant, or even definable, except in relation to the tetrachords of which they form parts.

[11] *Systēmata* can indeed be bounded by discords (17.1–7), but those that are will always be understood as incomplete fragments of systems bounded by concords.

Those passages' places in the programme can only be hypothetically identified, but they are brief, and most of the rest of Book I falls into place straightforwardly. Topics A(vii) and A(viii) are transposed, so that the long section on the genera, A(viii), appears at 21.32–27.14, and A(vii), on continuity and succession, at 27.15–29.1. They are followed by a list of propositions presented as principles to be assumed (29.1–34), but whose function is not explained. In the parallel passage of Book II the role of such principles is made clear; they are principles governing the ways in which incomposite intervals, topic A(ix), can be combined to form composite intervals and *systēmata*, topic A(x). But there are difficulties in construing those listed in Book I in quite this way (see pp. 131–3 below). With these principles the Book ends, leaving topics A(ix)–A(xiv) unaddressed.

Book II offers its own list of the 'parts' of harmonics at 35.1–38.26. It is shorter and simpler than that of Book I, and it announces before it starts that there are only seven of them (34.33–4). The parts are these.[12]

B(i) The genera, and the ways in which the differences between them are determined (35.1–25).

B(ii) Intervals, including all the distinctions by which they are differentiated (35.26–36.1).

B(iii) Notes, and specifically whether they are to be understood as pitches or as 'functions', *dynameis* (36.2–14).

B(iv) *Systēmata*, enumerating and distinguishing all the types, and explaining how they are put together out of notes and intervals (36.15–37.7).

B(v) The *tonoi* and the relations between them (37.8–38.5).

B(vi) Modulation (38.6–17).

B(vii) Melodic composition (38.17–26), a topic apparently excluded from harmonics in Book I (1.24–2.7, 8.4–8; but see pp. 138–40 below).

This list of topics has a good deal in common with the latter part of the programme outlined in Book I. If we ignore melodic composition, it corresponds broadly to A(viii)–A(xiv) in the earlier catalogue, though the order is not quite the same. In fact, however, Book II provides a direct discussion of only one item on its list, the first (46.19–52.33). Apart from the general introduction (30.9–32.9) and the enumeration of the science's parts, all the earlier part of Book II (that is, up to 44.20) is concerned with conceptual and methodological issues, in several cases contrasting Aristoxenus' approach with those of others. Between them these reflections occupy about nine pages of text. As I mentioned above, the passage dealing with topic B(i),

[12] Cf. Bélis 1986: 42–3.

like its counterpart in Book I, is prefaced by a brief list of the genera and by discussions of the concords and the divisions of the tone (44.21–46.18). It is followed, again as in Book I, by remarks on the nature of melodic continuity and succession (52.33–53.32), and then by a statement of principles (53.33–55.2) corresponding, though more economically, to those enunciated at 29.1–34. The whole of 44.21–55.2 seems therefore to be a revised version of 19.17–29.34. We shall return to the details of their relationship later (pp. 124–34 below). The contents of the remaining three pages of Book II cannot be mapped directly onto any part of either of the lists of agenda, though they do reflect the subject of a discussion mentioned but not undertaken in Book I (24.3–8), and must also, on my hypothesis, be revisions of material that appeared in the earlier treatise. They do not set out further definitions, facts or principles, but explore two practical procedures, one by which certain discordant intervals can be constructed through manipulations of concords (55.3–56.12), and another through which investigators can satisfy themselves whether the interval of a fourth amounts to exactly two and a half tones (56.13–58.5).

The most interesting sections of Book II are those devoted to methodology, and we shall examine them in Chapter 7. Its contributions to the substance, by contrast with the methods and conceptual basis of the science, are restricted almost entirely to the passage (44.21–55.2) which corresponds to 19.17–29.34 in Book I. Its very close links with the earlier treatment are underlined by the way in which the topic for which the introductory list prepared us, B(i), is preceded and followed by unannounced discussions in exactly the same sequence as before. In revising the earlier treatment, if the hypothesis of revision is correct, Aristoxenus has evidently allowed the previously established arrangement to impose itself on his new version, despite the absence from the contents list in Book II of anything corresponding to 44.21–46.19, or, more surprisingly, to 52.33–53.32, the counterpart of A(vii) in Book I.

(c) Relations between Book III *and Books* I *and* II

Something rather similar seems to be the case in Book III. I take it to be part of the same treatise as Book II, partly (as I have said already) because of the use it makes of Book II's new concept of melodic *dynamis*.[13] There are other

[13] Book III contains only one explicit reference to *dynamis* (69.9), but I shall argue (pp. 213–16 below) that it nevertheless plays a crucial role in the book's reasoning.

reasons too. The list of principles stated in Book I, for instance, unlike that in Book II, contains several that are redundant in the arguments of Book III (see pp. 131–4 below). Another (29.25–8), which stands oddly as a 'principle', is repeated in Book III, apparently as something new, in the more appropriate guise of a definition (60.10–17). Remarks found only among the principles of Book II (at 54.22–31) help to explicate an otherwise mysterious passage of Book III (59.33–60.9). Other details which I shall not pursue seem to point in the same direction.

But when we consider the actual sequence of discussion in Book III, it matches the second half of Book I's list of topics much more closely than it does that of Book II. An introductory passage uses the principles stated in Book II to elaborate the notion of succession, *to hexēs*; this was topic A(vii) in Book I (though it was discussed after topic A(viii)). Book III next defines and discusses incomposite intervals, topic A(ix). There follows a long series of theorems, occupying most of the rest of the book, which specify in detail the ways in which incomposite intervals can and cannot be combined in sequence, topic A(x). The book breaks off part-way through an enumeration of the forms (*schēmata*) of the concords, which will evidently contribute to topic A(xi); both this passage and the theorems themselves, as I shall argue in Chapter 8, must also have made important contributions to Aristoxenus' discussion of A(xiv), on *tonoi* and modulation. It is also worth noting that it is in connection with the ways in which intervals can be combined to produce *systēmata*, A(x), that Book I introduces the notion of *apodeixis*, demonstration,[14] and that the theorems of Book III, unlike anything else in the *Elementa harmonica*, plainly aspire to the condition of demonstrations in the Aristotelian sense.

Book III fits less comfortably with the programme announced in Book II. The main part of Book III, the theorems, corresponds clearly enough to B(iv). But according to Book II's list of the science's parts, this should be preceded by a study of intervals, including all the distinguishing features of the different types, B(ii), and by a study of notes, B(iii), settling, among other things, the question whether a note is a pitch or a *dynamis*. Nothing in Book III fulfils these promises. Here again, then, it seems that in revising his first treatise to accommodate the new ideas of the second, Aristoxenus allowed the sequence of topics to be dictated by their arrangement in the first, rather than reorganising them to match the programme that the second sets out.

[14] 5.21–2, 6.4–14. A word cognate with *apodeixis* appears only once elsewhere in Book I, at 26.15.

(d) The first treatise and its revisions in the second

We have not yet finished with the evidence bearing on the relations between the parts of the surviving text,[15] but it will be helpful at this stage to restate, rather more elaborately and with some adjustments, the hypothesis which I think best fits the facts. There are good reasons for thinking, to begin with, that these works were not originally published as writings but as lectures, whose contents were discussed, after delivery, by Aristoxenus and his students.[16] At some point in the post-Pythagorean phase of his career, probably during the period when he was working with Aristotle in the Lyceum, Aristoxenus composed a series of lectures covering the whole of harmonic science, as he then conceived it, beginning with our Book I. The series was completed in a treatment we no longer have, but whose footprints can be traced confidently enough, as we have seen, in the sequence of topics addressed in what we know as Book III. The first lecture-series was itself preceded by another discussion, referred to at 2.25–30, in which Aristoxenus examined and criticised the work of his predecessors. Aside from that, it presupposed no specialised knowledge of the subject.

At a later date, certainly after Aristotle's death, Aristoxenus decided that his first account of the science needed revision, particularly in its treatment of the more advanced and intricate topics. It called, in fact, for a new perspective to which the concept of *dynamis*, not mentioned in Book I, is fundamental. Discussions with his students had made him aware that he also needed to address methodological and conceptual issues more fully and explicitly, and there may have been some real developments in his own understanding of the methods appropriate to the discipline.[17] We shall need to consider whether his pronouncements on these matters in Books II–III are merely expansions of ideas already implicit in Book I, or whether they signal a significantly different view about the way in which the science's investigations should be pursued and its results expounded.

The first part of Book II (30.9–44.20) is entirely new, and we have seen that the programme it announces differs in several respects from that of

[15] The evidence will go on accumulating, in fact, until we have completed our study of Aristoxenus in its entirety. After the end of the present, rather laborious chapter I shall not return explicitly to the issue, but it may be worth a reader's while to continue to bear it in mind throughout the rest of our study of the *Elementa harmonica*.

[16] The principal evidence is at 30.9–32.9; cf. e.g. 59.5, 60.17, 68.13, 73.33, where Aristoxenus seems to be responding to problems raised by his audiences.

[17] Several of the long and thoughtful passages of methodological reflection in Books II–III are prompted by questions that his audiences are said to have raised, or by their misunderstandings of statements he had previously made; see n. 16 above.

Book I. We have also noted, however, that after this introductory passage the programme of the original treatment re-imposes itself, and revision takes place within its framework. But the material contained in the revised version does not include all the topics of the earlier work; from that earlier perspective it begins in the middle, taking as the first item on its agenda the topic that came eighth in the programme of Book I (though it was actually examined in seventh place). Yet an exposition that begins with the genera, without setting out the fundamental theses and concepts of the first six topics in Book I, scarcely seems to make sense;[18] and in fact the content of A(i)–A(vi) is plainly presupposed throughout Books II–III.

Despite the indications given in its introduction, then, the series of lectures to which these books belong, with its unexpected starting point and its repeated references to earlier discussions, was not conceived as an examination of the entire subject from the ground up. Unlike the audience of the original series, students of the second had heard Aristoxenus holding forth about harmonics before; and these later lectures constituted an advanced course of study, building on previous knowledge. They dealt almost exclusively with topics from the second half of the original programme, and were mainly constructed through detailed revisions of its treatment; the earlier version of this material was now abandoned. But he seems to have been satisfied that the earlier account of the preliminary topics was still adequate as an introduction to harmonics as a whole, and retained our Book I, largely or completely unrevised, to serve for the instruction of beginners.

In doing so he left in place a long passage overlapping with part of the advanced course in Book II. Scholars have found these extensive repetitions puzzling, though one might argue, on general grounds, that if my hypothesis is correct they need cause no great surprise. No Aristoxenian course on harmonics, however preliminary, would be complete without an account of the differences between the genera, which forms the core of the passage; by the same token it needs to be included in the advanced programme too, as part of the study of *systēmata* with which that programme is concerned, and also, more mundanely, to remind students of what they had previously been taught. All this seems reasonable enough. But we need to see whether it will survive exposure to a closer comparison between the overlapping passages, one that picks out in more detail the ways in which they are both similar and different. The material falls straightforwardly into six sections, each of which is addressed twice, once in Book I and once in Book II. We shall consider these pairs of passages one by one; for convenience I shall again refer to Book I's discussion as **A** and Book II's as **B**.

[18] Criticisms of this sort were already current in antiquity; see Porph. *In Ptol. Harm.* 81.23–5.

(i) 19.16–29 and 44.21–7. The main function of these short sections is to list the three melodic genera (in Book I this is the final part of topic A(vi)). **A** embroiders the list with a few rather speculative comments on their relative antiquity and sophistication; these are absent from **B**, which confines itself to a brisk catalogue. In another way, however, it adds to the information given by its predecessor. According to **A**, 'every melody, among those that are attuned on the same basis, is either diatonic or chromatic or enharmonic' (19.20–3). The corresponding statement in **B** is this: 'Let this be laid down, that every melody will be diatonic or chromatic or enharmonic or a mixture of these or common to these' (44.24–7). The references to generically 'mixed' melodies, and to those whose interval-patterns are common to the genera, indicate the ways in which a melody may not be 'attuned on the same basis' throughout, as **A** puts it. Thus **B** not only pares the material down to its essentials, excising the intriguing but scientifically unnecessary elaborations of **A**, but also inserts information appropriate to more detailed study.

(ii) 19.30–21.19 and 44.28–45.33. These passages consider the maximum and minimum sizes of the concords, and how many of them there are. **A** begins with a reference to 'the second difference between intervals that we have mentioned', that between concordance and discordance (the list of these differences, numbered 'first', 'second' and so on is at 16.16–34). It announces concordance as its subject, and adds that among the various ways in which concords differ it will focus only on difference in size, *megethos* (19.30–20.6). **B** also sets off by referring to 'the second distinction between intervals', that between discords and concords (44.28–30); but the phrase directs us to nothing in Book II itself. It must be either a deliberate reference back to the list of distinctions in Book I, or an accidental residue of the earlier version of the passage, overlooked when it was revised and inserted into its new context in **B**. Next, where **A** merely selects one of the distinctions between intervals from its list, and then one of the distinctions between concords, without discussing the reasons for its choice, **B** attempts a more integrated preface to the subject. Difference in size and difference in respect of concordance and discordance, Aristoxenus says, are the most familiar and significant (*gnōrimōtatai*) kinds of difference between intervals; and the latter is 'contained in' the former, since every concord differs from every discord in size. Furthermore, difference in size is itself the most familiar and significant kind of distinction between concords (44.30–45.6). We are no longer just presented with the topic 'the sizes of concords' as an item taken arbitrarily from a list, but are offered a sequence of rather abstract considerations designed to persuade us of its importance.

There is no disagreement between the two books about the sizes of the concords or about how many there are. But they present the matter in quite different ways. Both tell us that the size of the smallest concord is determined by the 'nature of melody' itself, since it is simply a fact – a 'fact of nature', as we might put it – that all intervals smaller than the fourth are discordant. **A** then goes on to explain that the largest concord is not determined in the same way; since the addition of an octave to any concord produces another concord, there is in principle no upper limit to their magnitude. But according to our usage, that is, in our actual music-making with the voice or with an instrument, the limit is reached at two octaves and a fifth, 'since our range does not extend as far as three octaves'. This limit, Aristoxenus continues, must be defined by the range of a single instrument, not by that of several instruments taken together.[19] The distance between the highest note of the smallest aulos and the lowest note of the largest, or between those of a boy's voice and a man's, exceeds two octaves and a fifth. The same is true even of a single aulos, if one considers the interval between its lowest note when played normally, and its highest when the device called the *syrinx* is employed (this was a device comparable to the speaker-hole on a modern wind-instrument, giving access to higher harmonics). It is from such cases, says Aristoxenus, that we have discovered that the triple octave, the quadruple octave and even greater intervals are concords. But in the present context they are irrelevant, and as a consequence there are, in practice, just eight concords.[20]

Here, then, the proposition that there are just eight concords is presented as the conclusion of an argument based on musical experience, and it is also elaborately qualified. It is true only of human usage, and even there only if we consider just one voice or one unmodified instrument at a time; and Aristoxenus allows himself a fourteen-line digression (20.32–21.11) on cases where this condition is not met, including a little parade of organological learning. **B** begins, by contrast, by stating **A**'s conclusion as an unargued assumption: 'Let there be eight sizes of concords' (45.7). It identifies the fourth as the smallest, on the same basis as **A**; next comes the fifth, since all intervening intervals are discords, and then, for the same reason, the interval compounded from the fourth and the fifth, that is, the octave.

[19] This is presumably, though he does not say so, because although several instruments were sometimes played together in Greek musical performances, the melodic line was taken by just one of them, or possibly by more than one in unison. It was not shared out between them across a wider range than a single instrument could encompass.

[20] Aristoxenus does not list them all, but they are evidently the following: fourth, fifth, octave, octave plus fourth, octave plus fifth, double octave, double octave plus fourth, double octave plus fifth.

These facts, Aristoxenus says, have been handed down by our predecessors, but we must establish the rest for ourselves.[21] Yet he does not establish them, or not here. The remainder of the passage is confined to the proposition that the addition of an octave to any concord makes another concord, as **A** also states, and that concords are not produced when what is added to a concord (other than the octave) is a fourth or a fifth, or an interval compounded from either of them and an octave. He does not set out **A**'s contrast between the limitlessness, in principle, of the sizes and the number of concords, and the limits imposed by human usage, nor does he revisit its reflections on the latter. As a result he leaves the injunction at 45.7, 'Let there be eight sizes of concords', with no interpretation and no grounds. Plainly he expects his audience to remember their previous discussions of the matter. Here, in a manner characteristic of Book II, he is more interested in elaborating the abstract principle about octaves (which **A** merely states and uses) than in relations between the propositions of harmonics and the facts of practical music-making, or the capacities of voices and instruments.

(iii) 21.20–31 and 45.34–46.18. Here **B** repeats, with little change, the content of **A** on the tone and its melodic division into halves, thirds and quarters. It makes two significant additions. Unlike **A**, it includes a comment about the size of the fourth, stating that it amounts to two and a half tones, which **A** postpones until part-way through its discussion of the genera, at 24.8–10. This seems to be a sensible if unremarkable piece of revision; we need this information before we begin to consider how the fourth is divided in each genus. Secondly, in an explicit reference to discussions held on earlier occasions, Aristoxenus comments that 'many people have previously misunderstood, supposing that I was saying that a tone divided into three equal parts can be performed melodically', that is, that one can introduce into a melody three successive intervals of one third of a tone. These people have failed to grasp 'that it is one thing to construct a third part of a tone, and quite another to perform, melodically, a tone divided into three' (46.8–16). He is not denying that intervals of one third of a tone occur in melody; that, indeed, is part of what this passage asserts, and two such intervals occur in succession in one form of the chromatic. Where people go wrong is in assuming that if one third of a tone is a legitimate melodic interval, then a melody can proceed from one boundary of a tone to the other through three equal steps. But these, Aristoxenus insists,

[21] This corresponds to the claim made in Book I, e.g. at 2.15–25, that earlier theorists restricted themselves to systems spanning an octave. It also points to the conclusion that the two-octave structure known as the Greater Perfect System had not previously been identified as the framework within which harmonics should conduct its analyses.

are two entirely different propositions. The first is true; the second (as will be shown in Book III) is false.

(iv) 21.32–27.14 and 46.19–52.32. These are the longest and most detailed parts of the overlapping passages, and I shall examine only a few of the issues they raise (for other details see pp. 155–9 below).

A begins with a careful exposition of the basis of distinctions between genera, which depend on movements of the two moveable notes within the space demarcated by two fixed notes a perfect fourth apart.[22] It goes on to show how these relations can be exemplified in just one tetrachord, whose notes (from the top down) are *mesē, lichanos, parhypatē* and *hypatē* (21.32–22.24). In **B**, this measured introduction is abandoned, and the essential points are summarised in a single short sentence (46.19–23).

Both passages next discuss the total size of the 'space' or 'range' (*topos*) within which each of the moveable notes travels, considering first *lichanos*, then *parhypatē* (22.24–24.1, 46.24–47.8). Their conclusions about these *topoi* are the same, and so, with one important exception, are the reasons they use to support them. The exception takes the form of an extensive and discursive passage included in **A**, designed to defend the thesis that *lichanos* can lie as much as two whole tones below *mesē*. It does so on aesthetic and historical grounds, arguing that though modern performers, with their tasteless preference for the 'sweetness' of chromatic systems, always place *lichanos* less than a ditone below *mesē*, a complete ditone in that position is characteristic of certain 'ancient' styles, and the forms of composition that use it are the noblest and most beautiful of all.[23]

None of this is in **B**; but **B** now goes on to a much longer digression of its own (47.8–50.14). It is prompted by a problem that some of Aristoxenus' students have raised, about how it can make sense to treat notes at different distances from a given point as the same note, and to give them the same name. How can both *this* note and *that* note be *lichanos*, for example, when one is a tone below *mesē*, the other a ditone? Aristoxenus' elaborate response takes us to the heart of his conception (as represented in Book II) of the way in which melodic phenomena are to be intellectually understood; we shall examine it in Chapter 7 (pp. 178–92 below). What is of immediate interest here is the contrast between the scope and character of this digression and those of the one found in the same context in **A**, at 22.30–23.22. Where **A** confronts the aesthetic obtuseness and cultural ignorance of contemporary musicians

[22] Here I follow Da Rios in accepting the MSS text at 22.5–12, which is excised by Westphal and Macran. Like Da Rios, I also think unnecessary most of the editors' emendations in the preceding lines, and Macran's indication of a lacuna after 22.4.

[23] 22.30–23.22. We shall return to this passage in Ch. 9.

and audiences, **B** responds to the intellectual uncertainties of Aristoxenus' academic students. His argument in **A** requires his listeners to expand their musical experience beyond the currently fashionable repertoire; in **B** it requires them to think, and specifically to think about the criteria according to which notes (and other melodic entities such as the *pyknon*, the chromatic genus, and so on) are assigned their identities. **A** is concerned to convince us that Aristoxenus' statements match the realities of music itself; **B** reflects, in a very abstract manner, on the conceptual resources we need in order to understand the nature and the dynamics of musical relations in general.

The next major project, in both treatments, is to identify the positions of the moveable notes, *lichanos* and *parhypatē*, in six different divisions of the tetrachord, each of which is assigned to a genus and given its own name. The two versions preface this with very similar definitions of the *pyknon* (24.11–14, 50.15–19), preceded in **A** by a statement postponing the question whether the fourth is exactly two and a half tones, and resolving to assume, for the present, that it is; **B** had introduced this equation earlier (46.1–2), without suggesting that it is debatable.[24] The passages dealing directly with the structures of the six tetrachordal divisions (24.15–27.14, 50.19–52.32) give broadly equivalent results, but again their presentations are quite different.

A devotes all but the last twenty lines of its discussion to the positions of the *lichanoi*, first locating them by reference to the size of the (composite) interval between them and the bottom of the tetrachord (24.15–25.11), then laboriously calculating, arithmetically, the fractions of a tone by which their positions differ (25.11–26.7). It goes on to state that every *lichanos* below the lowest chromatic is enharmonic, and that every *lichanos* above the lowest chromatic and below the lowest diatonic is chromatic; and it adds the important thesis that there are, in principle, an infinite number of *lichanoi*, 'for wherever you arrest the voice in the space that has been shown to belong to *lichanos*, there is a *lichanos*' (26.13–16), contrasting this view with that of other theorists (26.8–28). Its remarks on the positions of *parhypatē* are relatively sketchy. It does not identify their precise positions, but contents itself with the statement that the ranges of diatonic and chromatic *parhypatē* are the same, and that every *parhypatē* below this range is enharmonic (26.29–27.2). Since nothing has been said to justify the (rather unexpected) proposition about this note's diatonic and chromatic ranges, Aristoxenus adds an argument, based on principles not previously mentioned, in its

24 But in fact the issue is discussed at length later in Book II, at 56.13–58.5.

support (27.2–14). It raises issues we shall not consider in this chapter (see p. 293 below); it is also compressed almost to the point of unintelligibility.

B's treatment is less confusing and more economical, and gives more complete accounts of the divisions. Instead of discussing the *lichanoi* and *parhypatai* separately, it anatomises each division one by one, identifying all its intervals and so locating both of its moveable notes. It omits **A**'s awkward calculations of the sizes of the intervals separating the *lichanoi* of the various divisions, presumably because they tell us nothing of any importance to the science. It may seem more surprising that it does not repeat **A**'s thesis about the infinite number of *lichanoi*, about which Aristoxenus had certainly not changed his mind when he wrote Book II. But at this point in **B** it does not need repeating, since it is part of a wider thesis that has been carefully expounded in the digression at 47.8–50.14 (see especially 49.8–30); and we are reminded of it in the immediate context by the statement at 50.19–22 that the divisions to be considered are not all those that are possible, but only the most 'notable and familiar'. Finally, **B** offers comments principally concerned with the positions of *parhypatē* (51.34–52.8), roughly parallel to those in **A** at 26.29–27.2 but more detailed and precise; and like **A** it also adds arguments designed to explain and justify their surprising assertion about the ranges of *parhypatē* in diatonic and chromatic (52.8–32). The main line of argument is the same as before, but here Aristoxenus expounds it more fully and clearly, and provides more illustrative examples. Difficulties remain, but he has evidently made some attempt to improve on the almost impenetrable version in **A**.[25]

(v) 27.15–29.1 and 52.33–53.32. Little need be said here about these passages on melodic continuity and succession. They make the same points, rather differently expressed but without any detectable shift in perspective; **B** presents them rather more economically than **A**. Both emphasise that their treatments are no more than preliminary sketches. In **A** we are told (28.33–29.1) that details will be demonstrated 'in the elements', or perhaps 'in *The Elements*', treating *ta stoicheia* as a title (see pp. 134–5 below); and in **B** they will not be expounded 'until the combinations of the intervals are set out' (53.15–17). Both references point clearly to material of the kind found in Book III.

[25] The substance of 24.15–25.11 and 50.22–51.33, outlining the structures of the six tetrachordal divisions, reappears in a quotation from Aristoxenus at Porph. *In Ptol. Harm.* 138.14–29. The language is unmistakably Aristoxenian, and there is no reason to doubt that Porphyry is quoting accurately; but it is expressed quite differently from either of the passages in the *El. harm.* Aristoxenus seems to have revisited this material on at least three different occasions.

(vi) 29.1–34 and 53.32–55.2. We come finally to the passages that set down propositions to be treated as fundamental principles. In **A** there are seven such principles, in **B** only two; and whereas **A** merely lists them, in the form 'Let it be laid down that . . .', without introduction or explanation, **B** offers some discussion of their status and the roles they are to perform. The relations between the two passages will be the more easily clarified if we begin, this time, with **B**.

B begins by announcing that what it will specify is 'the first and most essential of the conditions that bear upon the melodious combination of intervals' (53.32–54.2). It is the rule that 'in every genus, the melody proceeding from any note through successive notes, both downwards and upwards, must make either the concord of a fourth with the fourth note in succession or the concord of a fifth with the fifth' (54.2–7). I shall call this the 'law of fourths and fifths'. Aristoxenus goes on to emphasise that though this law is the most fundamental principle of all, and though if it is broken musical attunement (*to hērmosmenon*) is destroyed, it is not by itself sufficient to guarantee that intervals are 'melodically' combined into *systēmata* (54.7–21). The same applies to the second principle, which concerns relations between tetrachords belonging to any one *systēma*; we need not examine its content here (54.21–55.2).

These principles, then, govern the ways in which intervals can be combined in sequence to form *systēmata*. The combinations are examined in meticulous detail in Book III, whose theorems demonstrate which are melodically acceptable and which are not; and as **B** leads us to expect, their proofs depend almost wholly on the first principle, the law of fourths and fifths. The second principle, on relations between tetrachords, has some relevance to discussions early in Book III, but none to the main body of theorems, which do not consider constructions involving several tetrachords arranged in sequence. Such constructions must certainly have been addressed later, in the missing sequel to Book III, and the principle may have been exploited there. The crucial point at present, however, is a simple one. Aristoxenus' treatment of these principles points forward to the work they will do as the basis of demonstrative arguments in Book III, and perhaps beyond.

In **A**, the list of principles comes right at the end of the book. Aristoxenus does not explain what they are for,[26] and it is hard to construe all seven

[26] The fact that they follow directly after the passage on continuity and succession perhaps implies that they are related to this topic, and no doubt in some sense they are. But Aristoxenus gives no hint of what the relation might be; there is not even a connecting particle to link them with what has gone before.

of them in the same light as those in **B**. The law of fourths and fifths is included, as the second principle in **A**'s catalogue, but receives no special emphasis. By the standards of **B** and Book III, the first, third and fourth propositions laid down as principles in **A** are not principles at all. The first amounts to no more than a summary of statements made a little earlier in **A**, at 28.6–17; and all three of them can be derived, in the manner of Book III, from the law of fourths and fifths.[27] The fifth 'principle' is in fact a definition, specifying what is meant by 'incomposite interval'; the sixth, which states that no concord is divided into magnitudes all of which are incomposite, seems loosely related to it. But this latter principle has no known role elsewhere in Aristoxenus' work; and **A** does not alert us to the crucial distinction between 'incomposite interval' (*diastēma*) in the fifth principle and 'incomposite magnitude' (*megethos*) in the sixth, which is carefully explored in Book III at 60.10–61.4. (The same passage also provides and explicates a sharper definition of 'incomposite interval' itself.) The seventh principle in **A** is again a definition; it defines *agōgē*, 'consecution', a term which Aristoxenus might have found helpful in the context of Book III, but which as it turns out he does not use.

A's list of principles is a very odd assortment. If I am right in believing that Book I was originally followed by discussions analogous to those of Book III, there seem to be two ways in which its peculiarities might be explained. On one hypothesis, the list was included in the original treatise or lecture-course in much the same form as we have it, and its task was the same as that of the principles in **B**, to underpin the argumentative demonstrations of its sequel. In that case we must conclude that Aristoxenus later came to see that most of them are unnecessary for that purpose. This hypothesis cannot be disproved; but if all **A**'s principles were really put to work in the arguments that followed, these arguments must have been revised from the ground up when Aristoxenus wrote the surviving Book III. For reasons I have given already (p. 122 above), I am not convinced that his revisions were as radical as that would suggest.

The second possibility is that **A**'s list of principles did not appear in the earlier work, or not in that form, and that it contained instead something

[27] The sense of the first is that after the two lowest intervals in a tetrachord, a melodic sequence moving upwards must pass next through an interval no smaller than the remainder of the perfect fourth, and moving downwards must pass through one no smaller than the tone (29.1–6). The third requires that when a sequence proceeds through a perfect fifth, starting from the two intervals at the bottom of the tetrachord, it travels through the remaining intervals in the fifth (i.e. the interval completing the tetrachord and the tone of disjunction) in the opposite order upwards and downwards (29.14–22). The fourth states that two notes which make the same concordant interval with two successive notes are themselves successive (29.23–5).

closer to what we find in **B**. On that hypothesis, Aristoxenus introduced them at the end of Book I when he gave that book its new role as a self-contained preliminary course for beginners. They would have two purposes. In the first place they would round off his introductory treatment by summarising, very compactly, conclusions that have either already been sketched or are needed to complete a survey of the science, but which could only be established firmly through discussions at a higher level. That might account for the presence in the list of the first, third and fourth principles, which are interesting in themselves, but which can be shown, with some intellectual effort, to follow from the second. Secondly, they would provide his audience with a taste of what was to come if they pursued their studies further, outlining a selection of propositions and indicating the new mode of thought and discussion that these more advanced investigations would demand. Aristoxenus would be allowing them a glimpse of what higher-level harmonics would offer, and perhaps also issuing an implicit warning about its difficulty, much as a modern lecturer might do at the end of a first-year course. An interpretation of this sort may be moderately plausible, but it is disappointingly vague and no more verifiable than my first hypothesis. Neither can it altogether dispel the impression that **A**'s catalogue is something of a rag-bag. They are merely the best strategies I can suggest for making some sense of the jumble.

We can be reasonably satisfied, however, that the evidence gleaned from these comparisons between the overlapping passages is consistent with my interpretation of the relations between Book I and Books II–III. It even does a little to support it. What I take to be the later and higher-level treatment does not wholly supersede the earlier, since certain propositions in Book I, in its new role as an introductory survey, are assumed to be familiar to students of Book II. At several points **B** abbreviates parallel passages in **A**, whittling them down to their essentials. Elsewhere it adds clarification; on the central matter, the differences between the genera, it adds detail and eliminates a tedious and unnecessary set of calculations included in **A**. Here and there **B** presents **A**'s theses in a different and more helpful order. In **A**, remarks and digressions that are tangential to the main themes relate to historical and aesthetic issues, and to the vocal and instrumental resources of musical practice; their counterparts in **B** are more abstract, 'philosophical' in a recognisable sense of that elastic expression, calling for intellectual reflection rather than for observation and experience. In **B**, but not in **A**, these digressions are cast as responses to problems raised previously by Aristoxenus' audiences, so adding to the impression that Book II is the sequel to an earlier account. Finally, the two principles enunciated in **B** provide a much smoother and more economical transition to the theorems of Book III than does the untidy collection of seven in **A**, which can only with difficulty be given any coherence at all. The case is not proved, but I think it has much in its favour.

THE EVIDENCE OF PORPHYRY

Thoughts prompted by allusions in a later writer will make an appropriate appendix to the main business of this chapter. Porphyry's commentary on Ptolemy's *Harmonics*, written some six centuries after Aristoxenus, is peppered with quotations from his work, along with looser paraphrases and other references, over twenty of which reflect passages in the *Elementa harmonica* as we now have it. Four are of interest here, since they give the titles, as Porphyry knew them, of the writings from which they were taken. At *In Ptol. Harm.* 80.22–7 he quotes a passage of our Book I (*El. harm.* 14.18–28), not quite 'mot pour mot', as Bélis puts it (1986: 29), but with only minor verbal variations; and he tells us that it comes from the first book of the treatise *Peri archōn* (*On Principles*). At 81.23–5 he comments on the fact that Aristoxenus puts the genera first in the sequence of topics which harmonics should consider – which he does only in Book II – and assigns this to a work called *Harmonika stoicheia* (*Harmonic Elements*). A passage from the middle of Book II, *El. harm.* 45.3–46.1, is quoted fairly accurately at 124.15–125.8, and comes, we are told, from the first book of the *Harmonika stoicheia*; and Porphyry's reference (quoting Didymus) at 28.22–3 to 'the introduction to the first book of the *Harmonika stoicheia*' seems to allude to a discussion near the beginning of *El. harm.* Book II, starting at 32.10.

Porphyry therefore thought of the contents of our *Elementa harmonica* as belonging to two separate works, the *Peri archōn* and the *Harmonika stoicheia*, each of which was divided into at least two 'books'. The first book of the *Peri archōn* included at least part of *El. harm.* Book I, and much if not all of *El. harm.* Book II was in the first book of the *Harmonika stoicheia*. Our texts of Aristoxenus refer often enough to *archai*, sometimes in the sense 'principles' and sometimes simply 'beginnings', but none of these references makes it probable that *Peri archōn* was Aristoxenus' own name for the treatise including Book I, or that he thought of it in the way this title implies.[28] A better case can be made for assigning the title

[28] Relevant passages are cited in Bélis 1986, 30–2. I do not think that they justify the conclusions she draws from them on pp. 32–4.

Harmonika stoicheia to the work that included Books II–III. I mentioned earlier a remark near the end of Book I, rounding off its preliminary sketch of the concepts of melodic continuity and succession. 'How they come about, and which interval is placed after which and which is not, will be shown in the elements [*stoicheia*]' (28.33–29.1). In our surviving texts this promise is fulfilled in Book III, and Book III, I have argued, is a revision of the original continuation to Book I. We cannot tell whether the statement I have quoted was present in Book I from the outset and pointed to the original continuation, or whether it was inserted later as a signpost to Book III itself. But in either case the expression 'the elements' could be read as the title of a work, or a section of a work, devoted to the detailed, argumentative demonstration of propositions in theorematic style.

A similar expression occurs in Book II, in a context that generates an intriguing contrast. It comes right at the end of the book's preliminary discussion, just before Aristoxenus begins his exposition of factual data in the passage that overlaps with Book I. 'These, more or less, are the things one might say by way of preface to the study of harmonics (*tēs harmonikēs pragmateias*); but those who are setting off to address the study of the elements (*tēi peri ta stoicheia pragmateiāi*) must understand in advance things of the following sort' (43.25–30). These latter 'things' turn out to be logical and methodological principles governing the procedure of demonstration, and are heavily indebted to Aristotle.[29] Once again the study of 'elements' must at least include the theorems of Book III, though it might also extend to the contents of Book II, as Porphyry's use of the title suggests, in so far as the latter can be thought of as establishing the premises on which those theorems will depend. What is particularly interesting is that Aristoxenus' antithesis in the passage I have quoted is not, like Porphyry's, between 'elements' and 'principles', but between the study (*pragmateia*) of the elements and the *pragmateia* of harmonics as a whole. These are apparently two quite different projects, calling for different kinds of preface. Aristoxenus' students are about to be introduced to the second. His remarks in the preceding pages of Book II, reviewing the 'things one might say by way of preface' to the first, serve partly to remind them of the wider context within which their specialised investigations will take place. As we shall shortly discover, however, they are designed also to provide a new and much more sophisticated perspective on the science as a whole.

[29] They are stated at 43.30–44.20; see pp. 193–6 below.

CHAPTER 6

Aristoxenus: concepts and methods in Elementa harmonica *Book* I

In his personal life, for all we know, Aristoxenus may have been a charming companion, generous and open-hearted with his friends, modest among his intellectual peers, unfailingly gentle with animals and kind to children and old ladies; but this is not the persona he constructs for himself in the *Elementa harmonica.* He mentions earlier exponents of the science repeatedly, but always to criticise them and often to dismiss them with contempt. Their main function in his writings is to point up, by contrast, his own immeasurable superiority. His references to discussions he has had with his students or other contemporaries are invariably patronising or bad-tempered or both, and display nothing but their errors and confusions, from which, so his remarks suggest, only he is immune. He is the solitary master of the field of harmonic science, surrounded by ignorant incompetents whose existence he will acknowledge only to expose their follies. One may hope, without much confidence, that he was not really as arrogant as he seems. But a serious point emerges from his unattractive self-presentation. Very few of his withering outbursts are concerned with points of musicological detail. Their regular focus is on deficiencies of a larger sort, on other people's failure to understand the conditions which harmonics, as a genuine science, must meet, which topics do and do not fall within its scope, which methods and approaches are proper to it and which are not, how its propositions are to be established and connected with one another, through which conceptual filters the phenomena it considers are to be understood and described, and much more in the same vein. The unique achievement that raises him (in his own opinion) above all his predecessors and competitors is that he has grasped completely, as no one else has done, what the pursuit of the science involves, what its goals should be and how they are to be reached.

In examining the fundamental concepts and procedures of harmonics, as the *El. harm.* portrays them, I shall begin by discussing Book I by itself, allowing myself only a few comparisons with Books II–III, in line with the hypothesis about their composition adopted in Chapter 5. When that task is finished and we turn (in Chapter 7) to the later treatise in its own right, we shall be better placed to raise questions about the relation between the earlier and later pictures, if not always to answer them. Neither work includes a complete, systematic exposition of the relevant issues (the nearest approximation to such a survey is in the opening phase of Book II, at 30.9–44.20). In order to tackle the project we need to collect remarks scattered throughout the texts, and to look for an integrated way of interpreting them. Book I poses the more intricate problems in this respect, since its reflections, on the whole, are much shorter, more fragmented and less elaborately developed than their counterparts in Books II–III.

I shall assemble my comments on Book I into four rather loose groups; they will inevitably overlap at the edges. We shall consider first its brief account of the scope and limits of the science; secondly its picture of the singing voice (or the 'voice' of an instrument) as a traveller in a special kind of 'space', a picture that provides the background against which all Aristoxenus' discussions of melodic phenomena are to be interpreted; thirdly a set of formal conditions which the science must try to fulfil, and the methods by which it does so; and finally Aristoxenus' crucial thesis that there is such a thing as the 'nature' of melody, inherent in which is an orderliness governing the form of all melodic structures, and whose characteristics and consequences the scientist must conceive, discover and demonstrate in appropriate ways.

THE SCOPE AND LIMITS OF HARMONICS

Harmonics, says Aristoxenus in the first sentences of Book I, is one of many varieties of *epistēmē* ('knowledge' or 'science') that are concerned with *melos*. But it has a special status among them. It is 'first in order' and 'foundational', since it conveys understanding of the 'first things' – that is, of things that are in some sense primary.[1] The gist of this is clear enough, but its precise sense depends on what is meant by the word *melos*. There are passages in Greek literature where it refers to music or song considered as a complex whole, including words and rhythms as well as melody, and others where it singles out melody in particular, in abstraction from the rest.[2] If Aristoxenus has the broader sense in mind here, there will be no difficulty in specifying

[1] In this passage, 1.11–19, I read *tōn prōtōn theōrētikē* in line 19 with Macran (whose conjecture is supported by Porphyry, *In Ptol. Harm.* 5.26–7, a clear echo of the present text), in place of the MSS *prōtē tōn theōrētikōn*.

[2] The contrast is conveniently illustrated by two nearly adjacent passages of Plato's *Republic*, where the former sense is explicit at 398c11–d2, the latter at 399e11–400a2.

branches of knowledge which deal with other aspects of the subject. But on this interpretation the priority assigned to harmonics (over rhythmics and poetics, for instance) is hard to understand; and in its frequent appearances elsewhere in the *El. harm.* the word always has its second, more restricted meaning. If then we read the passage in the light of Aristoxenus' regular usage, it may no longer seem strange or arbitrary to assign a foundational status to harmonics, but it becomes much more difficult to guess what the 'many other sciences' concerned with *melos* might be. I do not know how this problem should be resolved.

Aristoxenus now specifies the 'primary things' which form the subject-matter of harmonics. They are 'everything that bears upon the study of *systēmata* and *tonoi*' (1.19–22). It is an intriguing formulation. As is clear from Book I's list of the parts of harmonics, which I summarised in the previous chapter (p. 118 above), the topics which it examines or proposes for later examination include many others as well as these two, even if we include the analysis of the genera within the study of *systēmata*, as we reasonably might. The implication is that those others, most of which fall into the earlier part of the programme, are relevant to harmonics only in so far as they 'bear upon' the investigation of *systēmata* and *tonoi*. If I may relax for a moment my self-imposed ban on premature comparisons with Book II, we can see that the two books' perspectives are in this respect consistent. Book II represents harmonics as concerned with the ways in which the singing voice can arrange intervals, as it moves upwards and downwards, while respecting the 'nature of *melos*', that is, with the ways in which intervals can be strung together to form *systēmata* (32.10–14); and its programme, as we have seen, begins with the genera (see p. 120 above). Much of the content of Book I consists, therefore, of necessary preliminaries to harmonics proper as Book II conceives it, and from that perspective forms an appropriate introduction to the project of Books II–III. The central business of the science is not with the elementary components of melody – notes and intervals – or with the nature of distinctions between high and low pitch, concord and discord, and so forth, or not for their own sake. It is with the patterns formed by the orderly arrangement of these components into melodic structures, and with the principles that govern their construction.

When it has dealt adequately with these issues, Aristoxenus continues, harmonics has completed its task, and nothing beyond them falls within its scope. 'Matters studied at a higher level, when *poiētikē* is already making use of *systēmata* and *tonoi*, no longer belong to this science but to the one that embraces both this science and the others through which everything to do with music is studied; and this is the accomplishment of a *mousikos*'

(1.22–2.7). These declarations, along with an echo of them at 8.4–8, have often been taken as an explicit denial that matters to do with the composition of melodies fall within the province of harmonics. Certainly 'melodic composition' (*melopoiia*) is not included in Book I's list of topics, as it is in Book II's. But so far as the present passage is concerned, Aristoxenus' position is not as clear-cut as that interpretation suggests.

In the first place, the word he uses is not *melopoiia*, as one would expect if he was thinking of melodic composition, but *poiētikē*. Taken literally, with a noun such as *technē* understood, it means 'the skill of making something'; and this broad usage is common in Plato, Aristotle and many other authors. In suitable settings it has a more specialised meaning, to do with poetry; it is 'the poetic art', 'the skill characteristic of a poet'. Occasionally the context allows us to infer that the art to which it refers includes the use of strictly musical elements, melody and rhythm as well as words and metres, and may even extend to the art of performance as well as of composition (see e.g. Plato *Gorgias* 502c–d). Similarly, and more pertinently still, there are passages derived from Aristoxenus himself in the Plutarchan *De musica* in which *poiētikē* is, quite straightforwardly, the art of musical composition, the *poiētēs* is the composer, *poiein* is 'to compose', a *poiēma* is a composition and *poiēsis* is the process of composing.[3] But *poiētikē* in those passages includes all aspects of composition – rhythmic, melodic, verbal, instrumental and so on – taken together. The word never has a sense that is straightforwardly equivalent to *melopoiia*, 'the art of composing melodies'.

Secondly, though Aristoxenus would certainly agree that *poiētikē*, conceived in this broad sense, falls outside the province of harmonics, that is not the point he is making here. He says that harmonics does not extend to matters that are studied 'when *poiētikē* is already making use of *systēmata* and *tonoi*', and he describes the person equipped to engage in such studies as a *mousikos*. As we have seen in earlier chapters, this word has a wide range of applications, and can sometimes refer to a professional musician, whether a composer or a performing artist. Quite possibly Aristoxenus means to include such people under the description here,[4] but he is unlikely to be

[3] There are at least four instances of this sort, for instance, in the passage at [Plut.] *De mus.* 1142b–f, and three more at 1144e.

[4] The Plutarchan *De musica* includes a long passage on issues of this sort which is plainly derived from Aristoxenus. It will be discussed in Chapter 9; but it is worth noting here that the kind of understanding which goes beyond harmonics, rhythmics and other special disciplines, and which is essential for anyone who is to pass reliable evaluative judgements on musical compositions, is there attributed to the practical musician, the *technitēs*. Authoritative judgement of this sort is his *ergon*, his proper function (1143a). It is evidently implied that his training, if it has done what it should, will have equipped him for the task, but it does not follow that the necessary kind of understanding is inaccessible to everyone else.

thinking of them alone. When used without qualification and in the context of an all-embracing understanding of music, the term *mousikos* is more naturally read in a broader sense, indicating a person of broad and sophisticated musical culture, whether or not they are also a specialist in some branch of the art. This person, the passage tells us, has the capacity to consider what is made out of *systēmata* and *tonoi* when the art of *poiētikē* puts them to use, that is, actual compositions and performances, and this is beyond the competence of someone who is an expert in harmonics alone.

Aristoxenus' explicit views on the relation between specialised knowledge and the capacity to form sound critical judgements about pieces of music will be discussed more fully in Chapter 9, where we shall find that they are entirely consistent with the position I am attributing to him here. For the present, if he is thinking of that large capacity and not specifically about the study of melodic composition, the significant consequence is that his comments do not, after all, contradict his contention in Book II that *melopoiia* is one of the topics included in harmonics. Let us agree that a harmonic scientist's knowledge of *systēmata* and *tonoi* does not by itself give him an adequate basis for passing judgement on compositions. It might nevertheless equip him to analyse and classify, for instance, the ways in which such systems and their constituent substructures can be deployed, varied and inter-related melodically when *melopoiia* puts them to work, or to establish principles which limit the repertoire of manipulations to which a composer can subject *systēmata* while constructing his melodies. There might, in fact, be a good deal of work on *melopoiia* which a specialist in harmonics can legitimately undertake. We still have to face the fact, of course, that *melopoiia* is not named in Book I, as it is in Book II, as one of the topics with which harmonics is concerned. But the two books do not flatly contradict one another.

THE MOVEMENT OF THE SINGING VOICE

The length of Aristoxenus' discussion of this topic and of a collection of closely related concepts gives some indication of its importance; it runs from 8.13 to 13.30, to which we may add his comments in the preliminary list of the parts of the science, at 3.5–33. The 'voice' (*phōnē*) to which he is referring is primarily that of a singing human being, and its mode of movement will be contrasted with that of the same voice when it speaks; but his remarks apply equally to the 'voice' of an instrument, when it is used to play a melody (and not merely to emphasise a rhythm, for example). The voice, as it sounds one pitch after another, is depicted in this passage as a traveller in a dimension which Aristoxenus describes simply as 'space' or 'place', *topos*, in which the pitches are locations and the intervals separating them are the stretches of territory between those locations, to be measured and compared as distances.

The conception of pitch as inhabiting a dimension analogous to geometrical space was implicit, long before Aristoxenus, in the approaches taken by the *harmonikoi*, and was graphically represented in their diagrams (see Ch. 2 above), where pitches were set out as points marked on a line, and the intervals were represented by the gaps between them. By Aristoxenus' time it must have been moderately familiar; the practice of describing pitch-relations in spatial terms was not a novelty that called for extensive exposition, and he does not claim the credit for having pioneered it. The first thing to notice about this passage, then, is that it does nothing whatever to elucidate the nature of this spatiality, and does not purport to do so, nor does it undertake to persuade us that the imagery derived from it provides an appropriate framework for the representation of melodic phenomena. All that is taken for granted; it is not this space itself that is the focus of Aristoxenus' attention. His concern is with the nature of the movement of the voice within the space, and with the differences between the modes of movement involved in song and in speech. This is a topic, he tells us earlier, that has not previously been examined with sufficient care. If it is not clarified, it will be very hard to say what a note (*phthongos*) is; but once the nature of the voice's melodic movement has been defined, 'many later matters will become clear' (3.5–26).

Aristoxenus enunciates, and underlines heavily, what amounts to an instruction to his audience about the way they are to construe his description of the movement of the singing voice. He is describing it as it appears to the perception of a listener, 'according to the representation of perception' (*kata tēn tēs aisthēseōs phantasian*, 8.23, 9.2–3). He distinguishes his account sharply from descriptions given by people who represent notes themselves as movements and assert that sound in general is movement (12.4–7), those, that is, who discuss these matters from a physicist's point of view, as they 'really are', by contrast with the way in which they are perceived. Aristoxenus does not mean that these people's view is mistaken. It is merely irrelevant, and issues arising from it 'belong to a different enquiry, unnecessary in the context of the present investigation' (9.5–7, compare 12.4–32). He plainly finds it extremely important that his audience should be alerted to the point he is making, and be left in no doubt that his business is with the phenomena or 'appearances' presented within the perceptual field, and with nothing else. He reminds them time after time of this exclusive perspective, not only through explicit contrasts with the approach of a physicist, but through insistent repetitions of words conveying the notions of 'appearance', 'seeming' and so on, scattered liberally throughout the discussion; expressions of this sort occur twelve times in the first fifty lines of the passage alone.

Nowhere in Book I does he explain why he adopts this approach, why the alternative perspective is irrelevant or why the issue is so crucial. Book II unravels the issues much more explicitly (see pp. 166–8 below). In Book I the present passage provides guidelines for interpreting Aristoxenus' subsequent statements, all of which are to be related to the 'representations' given in our auditory experience, and not to unperceived physical processes underlying them; but only a few scattered remarks cast further light on these guidelines' significance.[5] Perhaps the most revealing is one that appears early in the book, at 6.23–9. 'Although the constitution of melody displays an astonishing orderliness, those who have mis-handled the discipline we are discussing have caused many people to accuse music of extreme disorderliness. Yet none of the objects of perception (*ta aisthēta*) possesses so great and so excellent an orderliness.' The point is not only that melody is an object of perception, an *aisthēton*, but that the 'astonishing orderliness' which Aristoxenus attributes to it belongs to it *in its character as* an object of perception, and that this orderliness places it on a higher level than other *aisthēta*. He does not say that it is superior in this respect to other ordered complexes of any sort whatever, the abstract systems of relations constructed within pure mathematics, for example. If melody is essentially an *aisthēton*, then its patterns must be discovered within the perceptual data themselves, not in relations between underlying but unperceived physical events; and in that case they must be identified empirically, not through speculations about their hidden causes, let alone through mathematical reasoning.[6] Thus at 22.30–23.23 two propositions about the distance between the notes *lichanos*

[5] Several other passages, for instance, dismiss as irrelevant to the present enquiry issues that are to be decided by reasoning to which the evidence of musical hearing does not contribute; see especially 15.6–12 and cf. 20.12–27 (the 'unnecessary distinctions' mentioned at 16.32–4 are irrelevant for more obvious and straightforward reasons).

[6] One might usefully compare a passage of Plato's *Meno*, in which Socrates expresses dissatisfaction with a definition of colour based on Empedocles' theory about the processes by which colours and other phenomena affect our senses (*Meno* 76c7–e9). It is not, I think, because in his view the Empedoclean theory is false, but because the putative definition is of the wrong sort. We cannot say what colour is by identifying the causal processes underlying it; its nature as a phenomenon presented in our visual experience will be exactly the same no matter what causal explanation turns out to be correct. The type of definition Socrates prefers ('shape is the limit of a solid', 76a6–7, cf. 76e7) is independent of such explanations. Similarly, for Aristoxenus, the defining characteristics of the perceptible phenomenon named *melos* are independent of the processes through which such phenomena are produced.

and *mesē* (one of them clearly controversial) are to be established by presenting people with the experience of listening to melodies in certain styles; they are to become 'accustomed' to them, and to draw the appropriate conclusions on the basis of 'induction', *epagōgē* (23.6–11, cf. 22.30–23.1). The truths that a harmonic scientist expounds (such as 'the total range in which *lichanos* moves is a tone', 22.27–8) are universals abstracted directly from musical experience, and are applicable to the contents of auditory experiences and to nothing else.

We may infer, then, that a procedure which determines a relation such as that between *lichanos* and *mesē* on the basis of mathematical calculation[7] is going about its business in the wrong way, and that in any case it fails to represent the facts under appropriate descriptions. '*Mesē* stands to *lichanos* in the ratio 5:4', for example, conveys nothing that is true, or could in principle be true, of notes in a melody as they are perceived by the ear. We may also think it likely that if there are principles governing the 'astonishing orderliness' of melody, they too, in Aristoxenus' opinion, will be established by abstraction from auditory experience, not excogitated by pure reason or borrowed from some other scientific domain, and that the same will hold of the 'nature of melody' by reference to which we shall come to understand 'which interval the voice places naturally and melodically after which' (28.20–3). But these are not issues that are pursued in Book I.

Aristoxenus presents his account of the movement of the singing voice by contrasting it with another form of movement within the space or place of pitch. He labels the former as *diastēmatikē*, 'intervallic', the latter as *synechēs*, 'continuous'; and he explains a little later that movement of the continuous kind is proper to speech. In singing it must be strictly avoided (9.21–10.10). The contrast is best conveyed in his own words.

> For any voice capable of moving in the manner I have mentioned [that is, in 'place'] there are two species of movement, the continuous and the intervallic. In the continuous kind of movement the voice appears to perception to cross some space in such a way as to stand still nowhere, not even at the limits themselves, at least according to the representation of perception, but travels continuously to the point of silence. In the other kind of movement, which we call 'intervallic', it appears to move in the opposite way. For as it crosses the space it stands still on one pitch and then on another; and when it does this continuously – I mean, continuously in time – passing over the spaces bounded by the pitches, and standing still on the pitches themselves and uttering them alone, it is said to be making melody [*melōidein*] and to be moving in intervallic motion. Both of these descriptions are to be understood according to the representation of perception . . . (8.15–9.3)

7 Such procedures are fundamental to the analyses of Archytas, and of Plato in the *Timaeus*; see Chs. 11–12 below.

The summary Aristoxenus gives as he sets off on his next, closely related topic provides some additional clarification. In singing, the voice 'makes its tensions and relaxations of pitch imperceptible, but utters the pitches themselves and makes them perceptibly evident'. When it crosses the space of an interval in either direction, its passage across this space must be undetectable (*lanthanein*), whereas it must give out the notes bounding the intervals in such a way that they are 'clear and stationary' (10.11–20).

These descriptions are lucid and need no elaborate analysis. A minor point to be noted is that they give no hint that the directions in which the voice travels in this space are to be imagined as 'upwards' and 'downwards' in a vertical dimension. The words corresponding to our 'high' and 'low' are as usual *oxys* and *barys*, 'sharp' and 'heavy', and those indicating shifts in pitch-level draw on the conception of greater or lesser tension, *epitasis* ('tightening') and *anesis* ('relaxation'). Aristoxenus does nothing to obstruct the natural assumption that movement in place, unless otherwise qualified, is more or less horizontal movement, as when one walks along a path.[8]

It is also worth making explicit a feature of Aristoxenus' position which the passage does not express or very obviously imply: the pitches on which the singing voice comes to rest exist only as boundaries between the intervals it traverses. They themselves have no extension in the space, and are in this respect closely analogous to geometrical points. This becomes clear partly from a caustic, off-hand comment he has made earlier (3.20–4) about people who suppose that notes have 'breadth' (*platos*), and partly from his treatment of such matters as the ranges within which the moveable notes move inside the tetrachord. 'The spaces (*topoi*) [belonging to *lichanos* and *parhypatē*] do not overlap, but their limit is their conjunction; for when *parhypatē* and *lichanos* arrive at the same pitch, as the one's tension is increased and the other's relaxed, there the spaces reach their limit. The space below this limit is that of *parhypatē*, the one above it that of *lichanos*' (23.28–24.1). Since the spaces or ranges of the two notes do not overlap, even when, at the limit, their pitches coincide, that limit cannot occupy any amount of the space shared out between their ranges. Correspondingly, the distance of two and a half tones between the boundaries of the tetrachord is completely used up by the intervals between the notes, none being left over for the notes to occupy. In an enharmonic tetrachord, for instance, there

[8] Cf. Porph. *In Ptol. Harm.* 95.13–17. 'The Aristoxenians posit that an interval is spatial (*topikon*). For they explain the thesis that there is an unmoved space, inhabited by the voice, in which we move the voice through some distance, by placing their feet in different positions within the space in which they walk. By stretching their strides more widely they mark out a larger interval in the space, and a small one by taking small steps.'

are two intervals of a quarter-tone each and one spanning a ditone. The notes stand at positions in the space without using up any of it themselves.[9]

Those details should be borne in mind. But the most striking feature of Aristoxenus' discussion is its treatment of the voice itself. What we are faced with, when we hear a singer in full cry, is not just a series of sounds coming one after the other as disconnected events, first one note and then another, even though the notes are separated from one another by empty 'spatial' gaps.[10] It is the movement of a single, persistent entity, the voice, travelling in its characteristically melodic way from place to place. We might compare it to the movement of a weasel, for instance, running through long grass. It pauses and sits, upright and visible, first in one place, then in another; but as it moves between these points it is invisible, running low to the ground, hidden by the leaves. We see it here, we see it there, but not in between; nevertheless it remains the same self-identical weasel, and we perceive it as such. The analogy is of course inexact; the weasel 'continues to be' while it moves, we suppose, in a sense acceptable to a zoologist or a physicist, not just in our reading of the contents of our perceptual field.[11] Aristoxenus' depiction of the persisting vocal traveller, on the other hand, must be construed *kata tēn tēs aisthēseōs phantasian*; it describes only the way in which the phenomena present themselves to us, and 'whether it is actually possible or impossible for a voice to move and then stand still on a single pitch is a matter for a different enquiry, unnecessary to the present investigation' (9.1–7).

But this conception of the voice as an enduring, moving subject is of fundamental importance.[12] When Aristoxenus states propositions about the positions of notes or the sizes of intervals, or discusses the principles at work in the orderly formation of *systēmata*, he is not thinking in terms of an abstract blue-print or recipe which can be used to assemble diverse components into coherent complexes, but about an order inherent in the

[9] This point is occasionally (but rarely) made explicit in definitions of 'note' (*phthongos*) given by later authors. The clearest example is Nicomachus, *Harm.* 261.6–7, where the definition of a note as 'a sound with no breadth, having no extension in space' is the last of three definitions listed; as to its origin, Nicomachus says merely that it is the definition given by 'some people'. The first sentence of an earlier (but clearly interpolated) passage in the same text offers the same idea: 'a note is the breadthless pitch of a melodic sound' (242.21–243.1).

[10] They are not perceived as separated by empty spaces of time. Aristoxenus says at 8.27–9.1 that when the voice is 'making melody' (*melōidein*), in the course of its movement it stands still first on one pitch and then on another, and does so 'continuously in time'. This can only mean, I think, that a necessary condition of its movement taking a melodic form is that there should be no discernible temporal gap between one note and the next. I do not know what Aristoxenus would have said about staccato performance, or about melodic phrases in which this continuity is broken by rests.

[11] The analogy breaks down in another way too, since the weasel's appearances are separated in time.

[12] I discuss this matter and its 'metaphysical' or ontological implications in Barker 2005a.

behaviour of a single enduring subject, the melodically moving voice. The compositions and performances of musicians and the constructions of theorists are melodically well formed if and only if they do no violence to the regular patterns of behaviour in which this subject's nature is expressed.[13] These patterns manifest themselves in sequences of movements unfolding one after another over a period of time; and Aristoxenus' central question, as we have seen, is 'Which interval can the voice, of its own nature, place melodically after which?' (e.g. 28.22–3). It is not 'Which patterns of relations, abstracted from any temporal sequence, will provide the framework upon which a melody can be built?', a question to which the Pythagoreans' integrated systems of ratios, grounded in principles involving no reference to succession in time, might be conceived as offering answers.

The remainder of this passage (10.21–13.30) may seem to call for no comment at all. It defines and distinguishes five concepts: 'intensification' (*epitasis*) and 'relaxation' (*anesis*) of pitch; the attributes which we would call 'height' and 'depth' of pitch, *oxytēs* and *barytēs*; and pitch (*tasis*) itself.[14] Aristoxenus' accounts of these items seem clear and appropriately tailored to the conception of vocal movement we have been considering. But there are curious features that should not be overlooked.

Three and a half pages of text, to begin with, seems much more than is needed to mark out five straightforward and familiar notions. Nearly a page of it (12.4–34) is easily accounted for; it reflects, once again, Aristoxenus' anxiety to distinguish his way of conceiving and talking about these matters from that adopted by students of physical acoustics. We should not be confused, he says, by their thesis that notes and sounds in general consist of movements, where what we are calling 'stability' or 'motionlessness' of pitch is represented as movement maintaining a steady speed. Their way of speaking is of no concern to us. 'We shall still go on describing the voice as "standing still" when perception displays it to us as setting off neither towards high pitch nor towards low' (12.4–16).

Even after this passage is subtracted, however, there remains a good deal of apparently unnecessary discussion. Most of it is devoted to ridding people of a confusion to which Aristoxenus says they are prone (10.30–11.3), since they mistakenly identify intensification of pitch (i.e. the process of raising it, *epitasis*) with height of pitch (*oxytēs*), and relaxation of pitch (*anesis*) with depth of pitch (*barytēs*). He attributes this error to 'most people', not

[13] In most contexts in Aristoxenus, expressions such as 'the nature of the voice', 'the nature of *melos*', 'the nature of *melōidia*' and 'the nature of that which is attuned (*to hērmosmenon*)' are virtually interchangeable.

[14] The interpretation of *tasis* as 'pitch' will be qualified below.

to theorists, and it hardly seems likely to be theoretically motivated; it is so elementary, in fact, that it seems unlikely that anyone had really confused the phenomena which Aristoxenus labels with these terms. Perhaps their 'mistakes' were merely terminological, and consisted in their using the words *epitasis* and *anesis* where *oxytēs* and *barytēs* would have been linguistically correct. In that case Aristoxenus' laboured correction of their errors (11.3–32) seems more pedantic than enlightening. There might be good reasons for it if the distinctions in question played a significant part in his later deliberations, but they do not; there are only three places in Book 1, after this passage, where the nouns *epitasis* and *anesis* or the corresponding verbs occur, and they play only minor roles in their contexts (18.15, 22.23, 23.32).

I shall return briefly to this passage below, in the light of some more obviously interesting issues raised by Aristoxenus' account of *tasis*. The literal sense of the word is 'tension'; in writings on harmonics it often means 'pitch', or 'a pitch', and most translators give it that rendering here. The passage runs from 12.1 to 13.30, but this includes his discussion of the distinctions between his treatment of *tasis* and that of the physical theorists, and another in which he again insists on identifying, at apparently unnecessary length, the ways in which this concept differs from the other four on his list. We can leave all that aside. It is the brief definition itself that demands attention. 'What we want to call *tasis* is something like a steady motionlessness of the voice' (12.1–3).[15] Elsewhere in the discussion he calls it 'rest' *(ēremia)* as contrasted with 'motion' of the voice (12.27, 31, 13.3), and identifies it with the situation in which the voice 'stands still', specifically when it 'stands still on a note' (12.14–15, 22).

In the context of his overall account of intervallic vocal movement there is nothing problematic about these characterisations. What seems curious is that they do not quite correspond to what we mean by 'pitch' and most Greek writers mean by *tasis*. As Aristoxenus defines it here, *tasis* is not 'pitch in general', the dimension in which high and low notes are located (his word for that is *topos*, 'space' or 'place'), and neither is it 'a pitch', a position in this dimension on which a note may fall. It is the 'resting of the voice' *at* such a position. He seems to be trying to replace the more abstract concept of a *tasis*, as a locus in a continuum, with one in which it is a phase of the singing voice's observable behaviour. The word names an activity in which we perceive the voice engaging, that of 'standing still'; a *tasis*, we might say, is a 'pitching' rather than a pitch, and as such it is

[15] The phrase I translate as 'a steady motionlessness' is *monē kai stasis*, which I interpret as a hendiadys. It is almost impossible to render literally in English; 'a remaining and standstill' is about the best we can do.

an immediate object of our musical experience. It is not simply part of the formal background against which the objects of experience are envisaged, or a 'place' in which they appear.

A glance back at his discussion of 'height of pitch' or 'sharpness', *oxytēs*, shows that in one passage at least he treats it (and its opposite, *barytēs*) in a similarly concrete way. It is not represented as a region to which the voice may move within the dimension of pitch, but as a state of affairs that comes into existence only when a string, for example, has been appropriately tuned and sounded (11.12–15). It is a specific variety of the voice's 'standing still', existing only when such a standstill actually occurs (11.15–21). Aristoxenus, then, is apparently attempting to reinterpret a whole set of familiar musical terms, *tasis*, *oxytēs* and so on, in such a way that they refer to perceptible features, states or activities of the singing voice, rather than to aspects of the 'space' within which it is perceived as moving, but which is not itself perceived. The novelty of this approach and the stresses it imposes on the Greek language may do something to explain the length to which the discussions are drawn out. In the end, however, he is unable to maintain consistently a mode of expression which reflects his revisionary posture. The more usual picture insinuates itself at least once even into the present passage, when at 13.15–16 he speaks of the voice as 'standing still on one *tasis*'; this is nonsensical if we take *tasis* strictly in the sense he has specified. Elsewhere in the *El. harm.* he regularly reverts to the familiar, pre-existing pattern of usage (e.g. 15.15–16, 23.30–1, 47.4–7, 71.30–3).

These lapses, if such they are, are in my view due more to the recalcitrance of the language than to any loss of faith by Aristoxenus in the ideas that the earlier passage conveys. The whole stretch of text we have been examining can be read as a manifesto, setting out what Aristoxenus took to be the basis of a revolutionary approach to harmonic science; and it is worth spending a moment in a summary review of its themes and some of its implications. Musical melody exists only as a phenomenon presented to our hearing. Descriptions of its elements and structures must therefore refer to the 'appearances' and to nothing else; they are all to be construed *kata tēn tēs aisthēseōs phantasian*. The phenomena present themselves within a purely auditory dimension which Aristoxenus calls 'space', *topos*. When a melody is heard, it is not perceived as a series of discontinuous events, but as a single subject in motion, a voice travelling melodically from one point of rest to another within this space. Its actual movements between its resting places are imperceptible (and apparently instantaneous), but perception still grasps it as a continuous subject, whose development over time is an expression of its nature.

The space within which it moves is featureless in itself. It is not calibrated in advance, as it were, marked out with pitches placed at pre-determined distances from one another for the voice to use as its stepping-stones. The landmarks from whose locations we can infer general truths about melodic structure are created by the voice itself, in its moments of 'steady motionlessness'; they are not features of the topography of auditory space in its own right. In that sense it is the 'pitchings' of the voice that constitute the 'pitches' with which harmonics is concerned, and no intervals are relevant to the science except those which separate the resting-places natural to the voice itself. The science's subject-matter is the behaviour of the singing voice, and it cannot be too heavily emphasised that all the structures it detects and describes, and all the principles governing them, have their origin in the nature of 'melodic voice' or of *melos*. No rules are imposed on *melos* externally, from the repertoires of mathematics or physics; correspondingly, it imposes its nature on nothing else. Its behaviour, and the principles underlying it, have no implications of a metaphysical or cosmological sort; there can, for instance, be no authentically Aristoxenian theory of the 'harmony of the heavens'.

Since the subject whose behaviour harmonics investigates inhabits only the perceptible realm, and since its nature is autonomous, the evidence on which the science can legitimately draw is in the strictest sense empirical. It does not even infer from the perceptual phenomena the existence of unperceived physical events which are their causes; such events no doubt exist and are discussed by appropriate experts, but they have no relevance to harmonics. Nor does it say anything about the physiological or psychological processes involved in the act of perceiving. Aristoxenus nowhere suggests, however, that the 'appearances' which he examines are 'mere appearances' in a sense that would brand them as objectively unreal. The idea that they might be regarded in that way does not seem even to have occurred to him. Melodically moving voices are as real, in the domain accessible to our ears, as are dandelions and crocodiles in the visual realm. Precisely as in botany or zoology or in any natural science, facts about the behaviour of the singing voice in general are discovered by abstraction or induction from observation; and the 'astonishing orderliness' that its behaviour displays alerts us, as in the other sciences, to the fact that it is the manifestation of a 'nature', a consistent and unified kind of being, unfolding itself in time like a plant from a seed. The ultimate task of the science is to discern what that nature is, and to show how it is responsible for the regularities inherent in melodic movement. This requires that the scientist's understanding of the nature of *melos* be translated into words, as an integrated set of principles which jointly define it, and all relevant propositions about its behaviour will be shown to follow logically from these principles, through demonstrations in the Aristotelian style. These prescriptions will be more fully explored when we reach Books II and III.

FORMAL OBJECTIVES AND PROCEDURES

In the course of Book I, and especially in the introductory passages (up to 8.13), Aristoxenus repeatedly makes points about the proper conduct of harmonic investigations by criticising the failings of his predecessors. His first move of all, after the initial paragraph on the scope of the science, is to comment on the blinkered perspective of the *harmonikoi*, who restricted their studies to the enharmonic genus, and within that genus only to eight-note *systēmata* spanning an octave. 'About the other magnitudes and arrangements in that genus itself, and in the others, no one has tried to learn anything' (2.18–21). Similar comments reappear in connection with many of the topics in his list of those which harmonics should address. No one has carefully defined the difference between the modes of movement involved in singing and in speech (3.16–17), or between *anesis*, *epitasis*, *barytēs*, *oxytēs* and *tasis* (3.31–3). No one has ever had even the slightest understanding of the differences between the genera, or about the ranges within which the moveable notes move (4.28–30). Most of the *harmonikoi* did not even realise that one must investigate the ways in which incomposite intervals are combined into *systēmata*, and the one who did, Eratocles, failed to answer or even to ask a whole series of essential questions (5.6–23). Only Eratocles, again, made any attempt to enumerate different forms of *systēma*, and his account, besides its other faults, left out the great majority of them (6.11–31).

There is more in the same vein, but it would be pointless to continue the catalogue. What Aristoxenus is hammering home is that the science should aim at complete coverage of its subject-matter, and is defective to the extent that it falls short of that ideal. His predecessors' work is found to fall short of completeness at two distinct points in the enquiry. First, the science must address all topics that 'bear upon the study of *systēmata* and *tonoi*' (1.19–21). Secondly, wherever items of a kind that it considers can be divided into distinct varieties, all these must be enumerated and described. These two stages and the difference between them are drawn to our attention repeatedly. One must first discuss the interval in general, *katholou*, and then 'divide it in as many ways as is possible', that is, identify all its distinct varieties (4.3–5). Exactly the same must be done with *systēmata* (4.6–8) and with musical *melos* (4.17–22). The mode of exposition that these statements indicate is followed systematically throughout Aristoxenus' treatment of the topics later in the book.

Aristoxenus does not explicitly discuss the criteria by which a scientist should decide whether completeness has been achieved at each of these stages; and he does not explain what satisfies him either that his own list of the parts of harmonics is complete, or that he has enumerated every relevant variety of interval, *systēma* and so on. On these issues we can make only rather general comments. A topic will earn its place in the list of parts if it contributes something essential to the study of *systēmata*. The most obvious way of identifying these essential contributors is by working backwards to them from an attempt to analyse what *systēmata* in general are, and how their varieties differ. They will be all those musically significant items to which such an analysis must refer; and most of the topics on Aristoxenus' list can be picked out by such a method. This approach presupposes, of course, that we understand in advance or can discover in the course of our enquiry the terms in which the analysis will appropriately be conducted, and our criteria of appropriateness will depend, in turn, on broad, pre-existing assumptions about the nature of the subject under investigation. A Pythagorean, for instance, will use different criteria from an Aristotelian. Aristoxenus' discussion of the voice's movement and of the five concepts associated with it is in part an expression of his own commitments at this general level.

When he sets off to enumerate the ways in which intervals differ from one another, he specifies that the distinctions to be included are only those which are *chrēsimoi*, 'useful' (16.21); and when he has listed five kinds of distinction, he comments that all others can be set aside, since they are not 'useful to the present enquiry' (16.32–4). These remarks can be interpreted in the light of the picture I have just been drawing. These five distinctions between intervals, and no others, contribute to the analysis of *systēmata*. It makes sense, too, that he does add a similar qualification to his list of seven ways in which *systēmata* themselves may differ from one another. They are relevant not because they are 'useful' in the analysis of something else, but because they belong directly to the subject whose analysis is the goal of the science. They mark the distinctions between all the varieties of *systēma* that differ in melodically significant ways. Hence the task of identifying them calls for a different approach, and nothing can serve the purpose except the meticulous empirical observations of a musically discriminating listener. The decision as to which distinctions between *systēmata* mark off significantly different melodic structures, and which do not, depends on the judgement of the ear. We might wish to push these issues a step further, and to ask what is involved in perceiving a distinction as 'melodically significant' and by what criteria others are excluded; but such questions are pursued only in Books II–III (see pp. 175–92 below).

Aristoxenus' criticisms of his predecessors in Book I focus most often on their failure to pursue essential topics, distinctions and other issues; and they are plainly designed to advertise his own achievement in reaching a comprehensive grasp on the subject as a whole and in all its parts. But there are other criticisms too, highlighting other qualities which a rigorous scientific treatment must possess, and which are lacking in everyone's work but his own. Thus we are told that the *harmonikoi* studied 'in an unsatisfactory way', or perhaps 'unmethodically',[16] even the subjects they did not overlook, and discussed them 'inadequately' (2.25–3.2). At 3.31–3, similarly, people are said to have addressed certain topics either not at all or confusedly. These comments and others like them seem to refer to faulty conceptions of method as much as to merely factual mistakes, but they are too vague to reveal the respects in which the methods of the *harmonikoi* were inadequate, and in which those of Aristoxenus, presumably, are not.

Elsewhere, however, his methodological criticisms are more specific; the clearest examples appear in his denunciation of Eratocles at 6.11–31. Its principal role is to emphasise, through the contrast between Eratoclean failings and Aristoxenian virtues, key points he has made in the preceding sentences about the study of *systēmata* (5.33–6.11). Once the ways in which incomposite intervals can be combined in sequence have been demonstrated, he says, we must discuss the *systēmata* formed out of them. From what we know about the combination of intervals, we must demonstrate 'how many *systēmata* there are and what they are like, setting out the distinctions between them in respect of size, and for each of their sizes the distinctions in respect of arrangement and combination and position, in order that no size, arrangement, combination or position proper to the *melōidoumena* [the "things that are sung" or "melodically performed"] may be undemonstrated'. This is where he shifts into critical mode.

No one else has ever touched upon this part of the enquiry. Eratocles did attempt, without demonstration, an enumeration over one segment of it, but we saw earlier, when we examined his work in its own right, that what he said amounted to nothing; it was all false, and as an account of what appears to perception his account was completely mistaken. (6.11–19)

[16] The phrase used here, *oudena tropon*, literally 'in no manner', carries in itself no very specific meaning, and the context offers nothing that would dispel its vagueness. We need not suppose that Aristoxenus intended any sharply defined criticism; he is offering only the general observation that their approach is flawed.

More specifically,

> Eratocles took one *systēma* in one genus, and tried to enumerate the arrangements of the octave, displaying them without demonstration by moving the intervals around in a circle, not understanding that unless the arrangements of the fifth and the fourth have been demonstrated in advance,[17] and along with them the nature of the way in which they can be melodically combined, it will turn out that many times the seven arrangements can be produced. (6.19–31)[18]

The passage identifies at least four kinds of flaw in Eratocles' procedure. The first is familiar; he addresses only one small part of the subject, considering *systēmata* in only one genus instead of three, and *systēmata* of only one size, when really there are many. Completeness of coverage is one of the aims most heavily underscored in the preceding lines. Secondly, Eratocles' results are false, in the sense that they do not reflect what is presented to perception. He is convicted, therefore, either of careless observation or of neglecting the task of comparing his results with what observation reveals, and in either case this makes his account of the matter worthless. Thirdly, he presents his theses 'without demonstration', *anapodeiktōs*, merely 'displaying' them (*deiknus*) through the cyclic rearrangement of intervals (see pp. 43–4 above). This probably means that he literally 'displayed' these rearrangements in diagrams, in which his audience could see the seven *systēmata* being generated as cyclic transformations of one another. Aristoxenus seems to imply that he failed to confirm that this procedure gives melodically acceptable results; and he explicitly asserts that Eratocles did nothing to show that arrangements of the intervals outside this group of seven are melodically unacceptable.

Expressions evoking the concept of demonstration, *apodeixis*, reappear several times in both the positive and the critical sections of this passage. The noun itself is not used here or anywhere else in Book 1; but between 5.21 and 6.31 the cognate verb *apodeiknumi* occurs twice and the compound form *proapodeiknumi* once, the adjective *apodeiktikos* once and its negative counterpart *anapodeiktos* four times (twice in adverbial form). These eight instances form a significant cluster, since no words belonging to this family are found elsewhere in Book 1, apart from an isolated example at 26.15–16; and there it seems to be used non-technically, without reference to 'demonstration' of the sort described in the *Posterior Analytics*.

In the present passage they must certainly designate a specific and strict mode of demonstration or proof, whose absence from the work of Eratocles

[17] Here I accept Monro's emendation, reading *proapodeichthentōn* ('demonstrated in advance') for the *prosapodeichthentōn* ('demonstrated in addition') of the MSS.

[18] On Eratocles and his investigations see pp. 43–55 above.

guarantees its inadequacy; and if the discussions and procedures of Books II–III are a reliable guide, the model is without doubt Aristotelian (though with some qualifications; see p. 168 below). Their high profile here and their absence from other contexts in this book points to the conclusion that it is only in connection with the topics currently under review that *apodeixis* is appropriate and necessary. There must be demonstrations, first, of the ways in which incomposite intervals can be combined in sequence (5.33–6.1); and when these have been established they must be used as premises in demonstrations showing how many kinds of *systēma* there are and what forms they can take (6.3–11). One of the reasons why Eratocles' procedure was unsatisfactory was that he tried to enumerate (*exarithmēsai*) the forms of an octave *systēma* 'without demonstration', *anapodeiktōs* (6.22–3). Another, which brings us to Aristoxenus' fourth criticism, was his failure to understand that the number of octave *systēmata*, and their characteristics, cannot be established unless a series of other propositions (about the arrangements of intervals within a fifth and a fourth, and the ways in which they can be combined) have been 'demonstrated in advance' (6.25–31). The scientist, in this phase of his project, must find the right way of organising the thicket of propositions to which his empirical researches have led him, so as to produce a systematic hierarchy of demonstrations in which the prior and the subordinate are correctly assigned their roles.[19]

The thesis that the forms which *systēmata* can take must be established demonstratively from prior principles and propositions may seem to conflict with a suggestion I made earlier, that they can be discovered only through direct empirical investigation (pp. 151–2 above). In fact there is no conflict; in Aristotelian theory, as we saw in Chapter 4, observation and demonstration are complementary phases of an integrated scientific procedure. From repeated and meticulous attention to the melodies presented to our ears we abstract descriptions of the various types of *systēma* underlying their construction. Analysis of these *systēmata* uncovers their essential, distinguishing features; and from this information we discover which kinds of distinction between the intervals from which they are built contribute to the differences between them. We consider also the ways in which these intervals are combined in the formation of the *systēmata*, and by abstraction from these we arrive at principles which govern all such combinations. Only after all this is done can the task of demonstration begin. It will show first that the principles legitimise all the ways of combining intervals that are perceived as melodic and exclude all others; and it will go on to prove

[19] Compare 43.34–44.3 (in Book II), and pp. 193–6 below.

that from these combinations and the principles governing them all and only the observed varieties of *systēma* can be derived.

The procedure might be accused of circularity. Logically speaking, the accusation is mistaken, since while the observed regularities follow from the principles, the principles do not follow deductively from the observations, but are abstracted from them by a kind of intuitive or inductive insight.[20] If they are true, they are true independently of any of the observations, and could legitimately be used to prove the truth of what the latter report. It would however be a weak kind of proof, since despite the independent standing of the principles, our faith in them rests on the observations on the basis of which we reached them. But all this misses the point. The principal task of the demonstrations is to show that the *systēmata* display an 'astonishing orderliness', which arises from their shared status as manifestations of the single nature that the principles define. It is not so much a matter of proving propositions not previously known to be true, as of organising and presenting them in such a way as to 'demonstrate' their integration under the aegis of the principles. To the extent that all melodically acceptable *systēmata* can be shown to conform to the principles, and that the principles exclude all arrangements of intervals which strike the ear as unmelodic, that task has been done.

One more issue needs to be tackled under the present heading, 'formal objectives and procedures'. When the three melodic genera have been identified, they and their variants must be analysed in a way that brings out the differences between them (4.25–8, 21.32–4). In the course of this analysis (21.32–27.14), Aristoxenus identifies the sizes of the intervals between notes in tetrachords of various types, and of the ranges within which the moveable notes can change their positions. Such a project can obviously not be pursued without the help of a system of measurement, in two senses of that phrase. Aristoxenus must be equipped with a metrical terminology through which the sizes of intervals can be conveyed, and also with a procedure that allows him to assess or 'measure', in terms of that metric, the sizes of the actual intervals he encounters in his musical experience.

When the *harmonikoi* quantified intervals, they did so by reference to a unit of measurement that was treated as 'minimal', in the sense that the ear can identify no smaller ones. Larger intervals were expressed as its multiples, and because it was the smallest identifiable interval, they had, at least in principle, a procedure by which the larger intervals could

[20] On this matter see Barnes 1993: 267–71, a careful study of Aristotle, *An. post.* 100b5ff.; see also the studies by Irwin, Bolton and Kahn to which he refers.

actually be measured. According to this method, one discovers the size of an interval presented to the ear by determining how many of the smallest identifiable steps are needed to move from one of its boundaries to the other. Aristoxenus, too, accepts the thesis that there is an interval of determinate size which is the smallest identifiable by ear, and it is also the smallest that the voice can produce melodically.[21] But in other respects Aristoxenus' approach to measurement is not like that of the *harmonikoi*. When he explains the sense in which intervals smaller than the quarter-tone are unrecognisable, he says that such an interval is one which 'the voice cannot utter clearly, and the ear cannot perceive what part it is either of a diesis or of any other of the familiar (*gnōrimōn*, perhaps 'recognisable') intervals' (14.21–5). This implies that an acceptable way of measuring small intervals, one which in the cases he is discussing is not available, is by reference to a 'familiar' interval which is not a minimal unit, but of which the interval being measured is an identifiable fraction. This view is reflected in Aristoxenus' actual practice. The point of reference for all his measurements is not a minimal unit but the interval of a tone. The sizes of other intervals are specified as multiples of the tone or as fractions of it, or in some cases as the sum of a multiple and a fraction (the fourth, for instance, is two and a half tones).

Like the dieses of the *harmonikoi*, the tone is a 'size' or a 'distance' in an exclusively auditory dimension, and can be identified only by the ear. It fulfils one of the crucial conditions that a reference-point for measurement must meet, that of being accurately recognisable. This is because it is, by definition, the difference between a fourth and a fifth, which – because they are concords – can themselves be constructed and recognised with very little margin of error.[22] They also figure, uncontroversially, as fundamental elements in most of the familiar patterns of attunement. Hence reliable instances of the tone can readily be found and displayed, and an audience will have no difficulty in grasping what is meant by the statement that an interval spans such-and-such a number of tones or such-and-such a fraction of a tone.

So much is unproblematic, and gives a clear framework for the exposition of the results of the scientist's researches. But complications are bound to arise at an earlier stage, in the context of practical measurement, where the task is to assess the sizes of intervals in relation to the tone, in practice and by ear. We can understand the statement that the lowest interval in an enharmonic tetrachord is a quarter-tone, or that the smallest chromatic

[21] See 14.6–25. This 'least *melōidoumenon*' is the quarter-tone; see 24.15–19, 25.11–25, cf. 28.10–15.

[22] See 55.3–10, and its detailed application in the sequel (to 56.12).

diesis is one third of a tone; we can even agree, on the basis of simple arithmetic rather than perceptual judgement, that the difference between them is one twelfth of a tone (25.12–25). But although the quarter-tone, unlike one twelfth of a tone, is a melodic interval, 'capable of being placed in a *systēma* in its own right' (25.24–5), it is very far from clear how we are to establish in practice that the interval which we hear, and recognise as a familiar enharmonic relation, is indeed one quarter of a tone.

This is something that Aristoxenus does not explain. The case is different with the semitone, since he shows in detail how a semitone, as well as a tone or a ditone, can be constructed reliably through manipulations of concords, so long as we accept that the fourth spans exactly two and a half tones; and the same passage goes on to explain how we can satisfy ourselves whether this latter proposition is true.[23] But intervals smaller than the semitone, and larger ones which are not exact multiples of it, cannot be constructed by this or any comparable method. In an Aristoxenian passage of the Plutarchan *De musica* which we glanced at in Chapter 3, this is given as one reason why 'people nowadays' exclude the enharmonic diesis from the class of melodic intervals; a consequence of their position (which Aristoxenus points out, but which, he alleges, they have failed to notice) is that no division of the tetrachord can be used except the very few in which all intervals are even-numbered multiples of a quarter-tone.[24]

The other contention on which these people base their view is that the enharmonic diesis makes no impression (*emphasis*) on our perception.[25] Aristoxenus dismisses this as merely a sign of their laziness and their lack of perceptual acuity (*anaisthēsia*, 1145b), but the point cannot really be waved aside so easily. Let us grant that these tiny intervals are 'melodic', in Aristoxenus' sense. They can be sung, and recognised by the ear; both the enharmonic and the chromatic diesis can take their places as individual steps in *systēmata* and in melodies, and the ear can tell which is which. It does not follow that it can tell what their sizes are, that the former is a quarter of a tone and the latter a third. That could only be done through laborious and arguably unreliable procedures analogous to those of the *harmonikoi*, whose example Aristoxenus shows no sign of following. So far as our evidence goes,

[23] These passages are in Book II, at 55.3–58.5. But a remark in Book I (24.4–8) plainly points forward to a treatment of the kind found there; and Aristoxenus must therefore have included, or at any rate planned to include, a similar discussion in the treatise to which Book I belongs.

[24] [Plut.] *De mus.* 1145a–c; see pp. 93–4 above.

[25] The word *emphasis* generally denotes the 'appearance' which something presents to the senses, on the basis of which we take it, rightly or wrongly, to be a thing of a certain kind. Though the enharmonic diesis, according to the formulation in this text, is something that falls into the province of perception, it does not present itself *as* anything we can discriminate.

he has no reliable system of practical measurement, and I can see no way of avoiding the conclusion that the appearance of precision which he gives, when specifying the sizes of intervals (other than the semitone and its multiples) in the various generic tetrachords, is illusory.[26]

This conclusion, however, is less damaging than it might seem. Certainly it undermines the details of several of his analyses of tetrachords. But Aristoxenus gives less emphasis to them than to the broader task of specifying the ranges traversed by each of the moveable notes, whose shifts of position are responsible for the differences between genera. He tells us explicitly that the note *lichanos* can stand at any point whatever in its range (26.13–28). In that case the divisions he gives can be no more than exemplary,[27] and the fact that his descriptions of them cannot be confirmed in all their detail through the judgement of the ear is of no great importance. What matters more is that the ranges of *lichanos* and *parhypatē* should be accurately determined. According to Aristoxenus, the highest point of the range of *lichanos* is a tone below the fixed note *mesē*, and the lowest is a ditone below *mesē* (22.27–30); and both of these estimates can be checked by the procedure I have called the 'method of concordance'. *Parhypatē*, at its highest, is again a ditone below *mesē*, or, equivalently, a semitone above *hypatē*; at its lowest it is a quarter-tone above *hypatē* (23.25–30). Only the first of these claims can be confirmed through the manipulation of concords. The second depends on the assumption that the 'least *melōidoumenon*' is a quarter-tone, so that *parhypatē* can lie at a quarter-tone from *hypatē*, but can come no closer. Hence it cannot be regarded as completely secure, though it has at least the merit (though Aristoxenus might not have regarded it in that light) of agreeing with the views of the *harmonikoi*, and can be further supported in the way I have suggested in n. 26. Aristoxenus is at any rate on firmer ground when measuring these ranges than when he is attempting to quantify the intervals of specific tetrachordal divisions. No more can be said about these issues on the basis of the material in Book I, but the relative unimportance of exact quantification becomes an explicit theme in Book II,

[26] I commented earlier that this criticism can be mitigated slightly in the case of the enharmonic. Since the upper interval of its tetrachord, according to Aristoxenus, is a ditone, and since this fact can be established by the 'method of concordance', we can be confident (given that the tetrachord spans two and a half tones) that the two lower intervals add up to a semitone. If then they 'appear to perception' to be equal in size, they must be quarter-tones. But even if we reckon this procedure reliable (as perhaps we should not), no such strategy is available for the 'soft' chromatic, in which the *pyknon* comprising the two lowest intervals is said to amount to two thirds of a tone, or for the 'hemiolic' chromatic, where it is three quarters of a tone, or for the 'soft' diatonic, where the highest interval is a tone and a quarter and the second interval is three quarters of a tone.

[27] Cf. 50.19–22, and the allusions to divisions other than the six that have been analysed, at 27.9–11, 52.13–18, 30–2.

one that brings into sharp focus the new ideas about harmonic analysis which Aristoxenus develops there (see pp. 175–92 below).

THE NATURE OF *MELOS*

I have mentioned several times Aristoxenus' thesis that melody has a 'nature' (*physis*), and that it is the task of a harmonic scientist both to discover what it is and to show how it is responsible for the 'astonishing orderliness' displayed in the behaviour of actual melodies. These are themes to which we shall return in Book II. For the present we shall put most of the issues aside and concentrate, principally, on just one. Given that we must reach an understanding of the *physis* of melody through reflection on our experience of melodic 'behaviour', that is, on what we hear, in what kinds of relation between the audible elements of melody does that *physis* manifest itself, and what conception of the phenomena will best guide us towards this understanding?

Before we tackle that question, however, I should say a little more about the general notion of a *physis*, as Aristoxenus employs it. At the beginning of Book II of the *Physics*, Aristotle explains in outline what is meant by saying that something is what it is 'by nature', and what a 'nature' is. As examples of natural things he lists animals, plants and the 'simple bodies' (i.e. the elements, earth, fire, air and water). What distinguishes them from things that are not 'put together naturally', he says, is that 'each of them has in itself a source (*archē*) of change and stability'. This is not true, for instance, of artefacts such as beds or items of clothing, or at least not as such; if they do possess any inherent tendency to change in specific ways, it is not because they are beds or clothes, but because of the nature of the materials from which they are made. The source of a natural thing's impetus towards characteristic forms of change (in place, in size or in its qualities), and towards the maintenance of certain states of stability, is its own inherent nature (*Phys.* 192b8–33). When Aristotle says, then, that something exists 'by nature', he does not mean merely that it occurs as an item in what we call the 'natural world', but that its characteristic behaviour arises from an origin in itself, its nature, and is not imposed on it by some external agency, as is the case with an artefact.

Aristoxenus' conception of a nature inherent in *melos* is closely related to the Aristotelian model. It too is an origin or principle intrinsic to that in which it exists, and is responsible for melody's regular and characteristic patterns of change; they are not imposed on it externally by human agents. This seems paradoxical, and at odds with the sharp distinction Aristotle draws between natural things and artefacts, since it can hardly be disputed that melodies fall into the latter category. But Aristoxenus does not mean, of course, that melodies compose themselves or 'grow' of their own accord. The manifestations of the nature of *melos* which are studied by harmonic scientists are not melodies, but *systēmata* and their inter-relations. Melodies themselves are constructions that we create while 'using' *systēmata* in ways that do not conflict with their natural orderliness (1.24–2.3). The patterns formed by the *systēmata* are not artificial but natural, and we discover them rather than creating them.[28] This is a striking and perhaps surprising doctrine, but it is unquestionably what Aristoxenus meant. Only on that condition can harmonics of his sort be regarded as a science, and without it his 'demonstrations' that only those scalar progressions which conform to certain principles are genuinely melodic would be no more than expressions of personal prejudice. From our own perspective we might be tempted to regard them in precisely that light, but the perspective is not shared by Aristoxenus. As we shall see in Chapter 9, he distinguishes objective truths about the conformations proper to the nature of *melos* very sharply from judgements of what we would call 'taste', which figure very rarely in the *Elementa harmonica*. If a composer fails to respect the modes of organisation in which the nature of *melos* expresses itself, and by which its manifestations are distinguished from what is *ekmelēs*, 'unmelodic', the constructions he is creating are not bad melodies but non-melodies. They do not fall under the category of *melos*, and have no status or value as music.

An Aristotelian nature displays itself primarily though not exclusively in orderly patterns of change specific to itself, and one of the forms which such changes can take is movement, change of place. According to Aristoxenus, as we have seen, the voice, when making melody, is a persisting subject that moves from one 'place' to another across gaps or intervals, in which it does not appear. This 'intervallic' style of movement is natural to melody, an aspect of the way in which its nature is manifested, in that it distinguishes the voice's melodic behaviour from its movement in speech. But it does not distinguish it from movement which, while being intervallic, is 'non-harmonic', 'unmelodic' or 'faulty', where the difference depends on the ways in which individual, incomposite intervals are combined into sequences

[28] Compare 19.23–9, where the diatonic is the first of the genera which human nature 'hits upon', and enharmonic is the last, since it takes much effort for one's perception to become 'accustomed' to it, and to recognise its musical credentials. Cf. also [Plut.] *De mus.* 1137e, perhaps derived from Aristoxenus: 'When we speak of one genus as "older", we must do so with respect to its discovery and employment by mankind; for so far as the nature of the genera themselves is concerned, none is older than another.'

(18.16–19.1). The principle which will give us the most significant insight into the nature of *melos* is the one which governs, and so unifies, all such sequences of melodic movement without exception. 'While there are many distinctions displayed by that which is attuned (*to hērmosmenon*) in the ways its intervals are combined, nevertheless there is something which will be found to be one and the same throughout everything that is attuned, whose power is such that if it is destroyed it destroys attunement' (19.4–10). Though Aristoxenus does not explain in this passage what this crucial 'something' is, all the indications are that it is the principle I have called the 'law of fourths and fifths' (p. 131 above). Its identity, however, is less important here than the fact that what it governs are the sequences of movements across intervals that will qualify as expressions of the nature of musically attuned melody.

Aristoxenus returns to these matters late in Book I, in a discussion of melodic continuity and succession, *synecheia* and *to hexēs*, which he himself describes as a preliminary sketch and no more (27.15–29.1). An understanding of these topics would allow us to answer questions of the following sort. When a voice, in its melodic movement, has reached a given note after running through some sequence of intervals, how far away is the nearest note to which it can next move in either direction? Where are the notes that are 'successive' with the one at which it has arrived, and which will form with it a segment of a 'continuous' sequence? Between any two notes there is always a gap, an interval; by what criteria, then, can we conclude that notes separated by such-and-such an interval are next-door neighbours, and that although it is always logically (and may be physically) possible to insert an intermediate pitch in the gap between them, such a pitch has no place in a melody or a scale? (Let us ignore, for the present, complications introduced by modulations between different *systēmata*.)

The passage begins with an analogy between 'the nature of continuity' in the sphere of melody and the ways in which letters (or rather, the sounds symbolised by them) can be arranged in sequence to form the syllables of speech. In neither case can the voice place the relevant items, letter-sounds or notes and intervals, in just any random order. In speech there is 'a kind of natural growth in the process of combination', and similarly in singing 'the voice appears to maintain a natural process of combination in placing intervals and notes successively, and does not sing just any interval after any other' (27.17–33). The status of certain sequences as 'continuous' and of their constituent steps as 'successive' thus belongs to them 'naturally'; it is intrinsic to melodic movement, not established by human tradition or convention but determined by the nature of *melos* itself.

How, then, are we to discover which intervals are by nature successive with one another, and where does the next-door neighbour of a given note lie? As so often in Book I, Aristoxenus clarifies his position negatively, through criticisms of previous approaches.

Continuity should not be sought in the way in which the *harmonikoi* tried to set it out in their 'densifications' (*katapyknōseis*) of the diagrams, where they displayed as lying successively with one another those notes which happened to be separated by the smallest interval. For the voice is so far from being able to utter melodically twenty-eight dieses in succession that it does not even add a third diesis, but in its upwards progress it sings at least the remainder of a perfect fourth – it is incapable of anything less – . . . and in its progress downwards from two dieses it cannot utter melodically anything less than a tone. (27.34–28.17)[29]

Aristoxenus is not disagreeing with the *harmonikoi* about the size of the smallest interval that can be sung; for both of them it is the enharmonic diesis or quarter-tone. Nor is he suggesting that there is some feature of our physiological apparatus which makes us incapable of singing three very small intervals in a row. It is a musical impossibility, not a biological or physical one; after a sequence of two dieses, a third one cannot be 'uttered melodically' (*melōideisthai*). His negative thesis, then, is that we cannot determine the sequence of intervals which forms a melodically continuous series simply by mapping out, one after another, a sequence of steps each of which spans the smallest interval that any melody can employ.

He also dismisses, briefly and enigmatically, a second approach to the issue. 'We should not press the question whether continuity arises sometimes from equal and sometimes from unequal intervals' (28.18–19). What I suppose Aristoxenus to mean is that we should not look for the principles governing melodic continuity in formulae which depend in any way on the distinction between equal and unequal intervals, such as (perhaps) 'A sequence of two equal intervals is always followed by an interval unequal to them,' or 'After two unequal intervals, the next interval must be equal to one or other of them.' The point is not that all such propositions are false (though in Aristoxenus' view very few will in fact be true), but that their proponents are looking in the wrong place for the key to the problem. Continuity does not consist in the preservation of some ordering of equal and unequal elements.

In what, then, does it consist, and how are we to discover what it is? We should not rely on either of the rejected approaches, Aristoxenus says,

[29] On the diagrams of the *harmonikoi* and the procedure of *katapyknōsis* see pp. 41–3 above.

but we should try to focus on the nature of *melōidia*, and concentrate on understanding which interval the voice naturally places after which in accordance with *melos*. For if after *parhypatē* and *lichanos* it cannot utter melodically (*melōidēsai*) any note closer than *mesē*, it will be *mesē* that comes immediately after *lichanos*, no matter whether it marks off an interval twice or many times the size of the interval between *parhypatē* and *lichanos*. What we have said, then, shows reasonably clearly the way in which we should look for continuity and succession; but how they occur, and which interval is placed after which and which is not, will be shown in the *Elements*. (28.20–29.1)[30]

By asserting, at the end of this passage, that his remarks bring out his main point 'reasonably clearly', Aristoxenus is claiming that though they have been sketchy they are not uninformative; and in fact they are neither as vague nor as naïve as they might seem. We should try to discover what constitutes melodic continuity by examining what melodies actually do, and do consistently, so revealing aspects of the *physis* that determines the patterns into which their movements fall. The procedure of *katapyknōsis* has nothing to tell us about this; and careful inspection of melodic behaviour will show that its regularities are independent of the equality or inequality of the intervals through which a sequence passes. What will emerge is that in any sequence that makes melodic 'sense', every note will be perceived as playing a role which is somehow associated with its name. Some note will be perceived as playing the role of *lichanos*, for example, and this will carry the implication that the next note in succession above it will always play that of *mesē*. Any note that is perceived as *mesē* stands, as a fixed note, at the top of the tetrachord which includes *lichanos*, and it is melodically successive with *lichanos* no matter how close to the bottom of the tetrachord this particular *lichanos* may lie.

There are deep mysteries involved in this approach, which Aristoxenus does not address in Book I. At a more superficial level it might be criticised for relying too simplistically on established but arguably mutable conventions by which the Greeks identified and named the notes of their scales (cf. also Book II, 53.18–32). Aristoxenus' point, however, is that unlike other procedures it is built on firm empirical foundations. Melodies always do presuppose continuous scalar sequences in which notes acquire the characters of *lichanos*, *mesē*, and so on, whatever exactly those characters may be; and if we disturb the note-series by inserting others between them, the

[30] On the question whether the phrase 'in the *stoicheia*' refers to a work by its title, *The Elements*, see pp. 134–5 above.

result is no longer recognisable as melodic.[31] It is therefore in this series' manner of progression that the nature of *melos* should be sought.

Book I hardly hints at the radical consequences of these ideas for the methodology of the science, which are faced more squarely in Book II; perhaps Aristoxenus had not thought them through when he wrote the earlier treatise. But one such consequence seems evident enough. Given that the sizes of the intervals between the named notes of the scale were in practice so variable, and that Aristoxenus goes so far as to accept the melodic legitimacy of an unlimited number of these variations (26.13–18), the role of quantification in elucidating the structure of melodically continuous sequences will be severely restricted. It does not follow, however, that the conditions they must meet – 'which interval is placed after which and which is not' – cannot be specified in sharp-edged detail. Aristoxenus promises to expound these intricacies in the *Elements*; our surviving texts preserve such expositions in Book III, and there is nothing vague or over generalised about them. I shall argue, none the less, that this is not because Aristoxenus has smuggled back into his treatment a quantitative approach which, by that stage, he has explicitly renounced. Many of his propositions are dressed in language that gives them a quantitative air, and one can be misled into thinking that they can apply only to sequences of intervals whose sizes are precisely specified. If that were so, they would be inappropriate and inadequate as representations of the patterns of melodic movement which are dictated or excluded by the nature of *melos*. But their quantitative costume is borrowed finery, whose function in the business of exposition is not difficult to discern; and Aristoxenus himself provides the means by which we can penetrate the disguise. We shall return to the issue in due course (pp. 208–15 below).

[31] I continue to ignore issues related to the practice of modulation; see pp. 215–28 below.

CHAPTER 7

Elementa harmonica *Books II–III: the science reconsidered*

In Meibom's edition of 1652, whose pagination modern scholars use as their standard reference-point, Book II of the *El. harm.* occupies a little less than twenty-eight pages of thirty-four lines each. No more than about six pages are taken up with statements and elucidations of facts about musical structures, or of principles governing them, and all those pages fall within the 'overlapping passage' discussed on pp. 124–34 above. Virtually the whole of their contents, by contrast with their manner of presentation, is already familiar from Book I (see 44.21–47.7, 50.14–52.32, 53.32–55.2). Of the remainder, there are two pages of preface (on the value of prefaces in clearing away potential misunderstandings, 30.9–32.9); the revised list of the science's parts occupies nearly four pages, the bulk of which is devoted, as in Book I, to comments on the procedural failings of Aristoxenus' predecessors (35.1–38.26); one page of the 'overlapping passage' is its study of the ways in which the topic of melodic continuity should be addressed (52.32–53.32); three pages describe procedures for constructing discords by the 'method of concordance' and for assessing the size of the perfect fourth (55.3–58.5). All the rest, amounting to nearly twelve pages, takes the form of a series of long digressions, concerned with the concepts that must be brought to bear on the subject if it is to be properly understood, and with the methods by which its theses are to be established and expounded. (These are at 32.10–34.34, 38.27–44.20 and 47.8–50.14; the first two are subdivided into two and three separate 'digressions' respectively.)

Around twenty-two of the book's twenty-eight pages, then, are devoted, broadly speaking, to methodological and conceptual issues. In the light of these crude statistics it seems clear that it is these issues, and not substantive points of doctrine, that form its main agenda, and that the passages I have called 'digressions' (some of which are indeed presented as such) are really nothing of the sort. They contain, in fact, most of the ideas that distinguish Book II from Book I, and together with one comparable passage in Book III (68.13–69.28) they will occupy most of our attention in this chapter. I shall not revisit Aristoxenus' comments on his predecessors, except in the first section below, and at one or two other points where they add something significant to the complaints he makes about them in Book I. Most of them do not.

The chapter is in four sections. In the first I examine Aristoxenus' brief introductory remarks on the contrast between earlier theorists' approaches and his own. The second considers the intellectual and perceptual resources that the harmonic scientist should bring to bear on his subject, their roles and the relations between them. The third discusses the role of non-quantitative conceptions in Aristoxenus' harmonics, and attempts, in particular, to elucidate the crucial concept of melodic 'function', *dynamis*, and the way the science puts it to work. Finally, we shall examine the statements Aristoxenus makes in Book II about the procedure of *apodeixis*, 'demonstration', which he will use extensively in Book III.

THREE APPROACHES CONTRASTED

Aristoxenus follows his two-page preface to Book II (30.9–32.9, which we shall leave untouched until Chapter 9) with a paragraph setting out the general aims of the science and contrasting his own procedures with those of other theorists. It is succinct, carefully structured and packed with detail, and is worth quoting in full.

> One should realise that our enquiry, taken as a whole, is concerned with the question how the voice, in every *melos*, naturally places the intervals as it rises and falls.[1] For we assert that the voice has a movement that is natural to it, and does not place intervals at random. We shall try to provide for these matters demonstrations (*apodeixeis*) which are in agreement with the data of perception (*tois phainomenois*), not in the manner of our predecessors, some of whom uttered irrelevances, pushing perception aside on the grounds that it is inaccurate, while devising theoretical explanations and claiming that it is in certain ratios of numbers and relative speeds that high pitch and low pitch arise, making statements that are completely irrelevant and wholly at odds with the perceptual data; others, by contrast, made oracular pronouncements about isolated topics, without giving explanations or demonstrations, and without even enumerating the perceptual data correctly. We, however, shall try to adopt principles that are all perceptually evident (*phainomenas*) to those experienced in music, and to demonstrate what follows from them. (32.10–33.1)

We have touched earlier on most of the important issues raised by this passage, but a little recapitulation will be helpful before we move on. Just

[1] Literally, 'when it is tensed and relaxed', *epiteinomenē kai aneimenē*.

like Plato in *Republic* 531, Aristoxenus distinguishes two existing approaches to the subject, and contrasts both of them with his own, superior procedure. His line of demarcation between the two groups of previous theorists falls in much the same place as Plato's. His comments on the second group, who make 'oracular', unsubstantiated statements on specific topics, whose observations are faulty and incomplete and who neglect the business of explanation and demonstration, are unmistakable echoes of his criticisms of the *harmonikoi* in Book 1. The first group deploy ratios of numbers as representations of relations between pitches, basing their approach on the supposition that high and low pitches depend on differences in 'speeds'; and the considerations to which they appeal are theoretical or intellectual instead of being grounded in perception, whose evidence they are quite content to 'push aside' and dismiss as inaccurate.[2] The objects of this critique are plainly the Pythagoreans and those others who followed in their footsteps, including Plato in the *Timaeus* and the unnamed proponents of 'mathematical harmonics' mentioned by Aristotle in the *Posterior Analytics* and elsewhere.

What Aristoxenus does not say here is as revealing as what he does. He does not allege that the *harmonikoi*, like the mathematical theorists, dismissed the evidence of their ears on principle. They were aiming to describe what they heard, and to that extent were addressing the subject correctly; they were merely incompetent observers. Nor does he accuse those who spoke of ratios of neglecting the whole business of demonstration, as the *harmonikoi* did. Their mistake was to look in the wrong place for the principles on which their arguments depended, drawing them from mathematics and theoretical physics rather than seeking principles 'perceptually evident to those experienced in music', as Aristoxenus will do. As a consequence, their arguments are irrelevant.

The thought underlying this accusation of irrelevance is similar to one which we encountered in Book 1's discussion of the movement of the voice (9.3–11, 12.4–34), where Aristoxenus urges his audience not to concern themselves with the thesis that sound, from a physicist's perspective, is a form of movement (pp. 141–2 above). A harmonic scientist should focus exclusively on what *melos* is 'according to the representation of perception', *kata tēn tēs aisthēseōs phantasian*. In the present context, where the issue is to do with explanations and demonstrations, the point can be sharpened. Aristoxenus uses a striking expression to convey the irrelevance of the

[2] The participle I translate as 'pushing aside' is *ekklinontes*. The verb from which it comes can mean 'bend out of the true', 'distort', but here the sense 'push out of the way', as at Plato, *Laws* 746c2, seems more apt.

mathematical approach, *allotriologountes* (32.20), literally 'saying things that belong to something else' (the verb may be a novel coinage of his own); a few lines later he expands it into *allotriōtatous logous legontes*, 'uttering other-belonging statements' (32.27). These remarks fall into place against the background of Aristotle's theory of demonstration, which I sketched in Chapter 4, as reflections of his thesis that 'it is not possible to demonstrate while shifting from one class of things to another' (*An. post.* 75a38). I have called this the 'same domain rule'. Aristotle, as we shall see later, treats harmonics as an exception to this rule (or as falling under it only in a qualified sense), with empirical harmonics providing the facts, and mathematical harmonics – which works in a different 'domain' – providing the principles from which the facts are explained and demonstrated (see pp. 353–61 below). Aristotle's approach is in this respect closer to that of the people Aristoxenus is criticising than to that of Aristoxenus himself. Aristoxenus insists on the 'same domain rule' even more strictly than its original advocate. Demonstrations in harmonics can be based on no principles other than those accessible through reflection on the 'representations of perception' given in our auditory musical experience.

But this is not quite how he puts the point here. His formulation is blunter: 'We, however, shall try to adopt principles which are all perceptually evident to those experienced in music' (32.31–4). The word I have translated 'perceptually evident' is *phainomenas*, which in many contexts might mean 'apparent', 'apparently true', or 'evident' without any special reference to perception. In Aristoxenus, however, the verb *phainomai* repeatedly marks his insistence on the authority of perceptual, auditory 'appearances', and has occurred in that role three times already in this short passage (32.19, 28, 30). If we construe his statement literally, then, the principles he will adopt are part of the content of what those 'experienced in music' actually hear. But principles can obviously not be heard. His mode of expression is designed for emphasis rather than precision, and what I have called 'reflection' must enter the matter somewhere. Aristoxenus owes us some account of the way in which perceptual and intellectual activities are related in the proper conduct of his science.

PERCEPTION, THOUGHT AND MEMORY

A few lines later he seems to be laying the foundations of such an account. 'The enterprise depends on two things, hearing and thought (*dianoia*); for we judge the sizes of the intervals through hearing, and we discern their functions (*dynameis*) through thought' (33.4–9). Aristoxenus is apparently trying to distinguish the roles of the two faculties by carving up the territory of harmonic science between them, assigning to hearing the task of assessing the sizes of intervals and to thought that of identifying their *dynameis*.[3] He seems to imply that their ranges do not overlap; hearing tells us nothing about 'functions', and thought cannot judge the sizes of the intervals.

This is puzzling. Certainly it is Aristoxenus' view that we cannot assess the sizes of intervals by thinking about them; the ear is the only arbiter. But the converse proposition, that the *dynameis* do not fall within the scope of hearing, seems both paradoxical and at odds with plain statements that he makes elsewhere. It is paradoxical, one might argue, because if the *dynameis* are inaccessible to hearing, they can play no part in what a *melos* is 'according to the representation of perception', and in giving them a pivotal role, as Aristoxenus does, we will be convicted of *allotriologein*, 'uttering irrelevances', as surely as the mathematical theorists he has criticised. We shall consider some of his plain statements later, noting only, for the present, that if hearing is restricted to judging the sizes of intervals, the information it gives us will, in Aristoxenus' own estimation, contribute nothing of any importance to the science (see especially 40.12–23, with pp. 175–92 below). This seems quite alien to his general position, and might drive us to the conclusion that his preliminary statement about the tasks of hearing and thought is seriously misleading, if not (by his criteria) quite false. We shall revisit the issue at the end of this section.

Hearing and *dianoia* reappear in tandem at the end of the present passage (34.27–8); in the intervening discussion Aristoxenus says nothing about 'thought', only about *aesthēsis*, 'perception'.[4] The discussion falls into two distinct but closely related parts. The first insists on the importance of 'becoming trained to judge particulars accurately' (33.9–10). One cannot adopt, in harmonics, the strategy used in geometry, where one can say 'let this be a straight line' without worrying about whether the visible line drawn

[3] Editors have been dissatisfied with the expression 'their *dynameis*', where 'their' (*toutōn*) must refer to the intervals just mentioned. Macran argues that it is only to notes, not to intervals, that Aristoxenus assigns *dynameis*, and therefore emends *toutōn*, 'their', to *tōn phthoggōn*, 'of the notes'. In fact Macran is mistaken; the range of items to which *dynameis* are assigned is wider than he supposes (see p. 186 below). But a difficulty remains, since even if intervals have *dynameis* it is certainly not they *alone* that have them, as the MSS text seems to imply. If emendation is needed, the simplest solution would be to delete *toutōn* and put nothing in its place. This would leave the unembroidered sense 'and we discern the *dynameis* through thought'.

[4] We shall consider later whether the shift in terminology from *akoē*, hearing, to *aesthēsis*, perception, is significant. *Aisthēsis*, rather than *akoē*, is Aristoxenus' usual term for the faculty through which we perceive melodies and melodic relations, and in the immediate context the transition facilitates the comparison he draws at 33.10–26 between harmonics, where accurate (auditory) perception is essential, and geometry, to which accurate (visual) perception contributes nothing.

in the diagram is really straight or not. A geometer does not need to decide whether what he sees is straight or circular or anything of that sort, and has no need for training in visual discrimination. But a harmonic scientist cannot say 'let this be a perfect fourth' regardless of what he is hearing. Although the conclusions he will aim to reach are about things of certain sorts, not individual specimens as such (about enharmonic *systēmata* in general, for instance, not about the melody now being played), they must all be grounded in observations of particular instances, and they will be worthless if those observations are careless or imprecise. For any student of music, perceptual accuracy is of the highest importance (33.9–26).

The thesis that this accuracy demands training, specifically training through practice (*ethisthēnai*, 33.9, cf. 34.32), carries a significant implication about the status of the musical 'object' that we perceive. Although a *melos* exists only in perceptual representations, it is not identical with the subjective content of any and every hearer's perception. That content will vary in its outlines from one perceiver to another, reflecting differences in their physiological make-up, their powers of musical discrimination or the degrees of attention they happen to be exercising. The audible object possesses its properties objectively, and it is entirely possible for a hearer's impression of it to be inaccurate or mistaken. If that were not so, Aristoxenus' demand that we should learn to perceive particulars accurately would make no sense, and neither would his allegations that his predecessors were careless or incompetent observers. The impression we receive of the characteristics of a melody may be correct or incorrect. Aristoxenus would apparently have no truck with the Epicurean doctrine that all sensory impressions are true, and that falsehood only enters the scene with the opinions we form about them and their causes.[5] Our thought-processes too must be trained (34.27–30), but all the emphasis in this passage is on the need to instil accuracy into our perceptions themselves.

In the second part of the discussion (33.28–34.30), Aristoxenus draws attention to one central and recurrent feature of the phenomena that we are to perceive accurately and seek to understand. 'Understanding (*xynesis*) of music is concerned simultaneously with something that remains constant and something that changes, and this applies – to put it straightforwardly – throughout virtually the whole of music in all of its parts' (33.28–32). He clarifies and illustrates this statement with a string of examples (33.32–34.25). We perceive (*aisthanometha*) differences between genera

[5] For translations of relevant Epicurean texts and a lucid account of this theory see Long and Sedley 1987: 78–86.

when the boundaries of an interval remain in place and the notes between them change (or 'move'). While the size of an interval stays the same, we identify it sometimes (for instance) as that between *hypatē* and *mesē*, sometimes as that between *paramesē* and *nētē*, 'for while the size of the interval remains constant the *dynameis* of the notes change'. An interval placed in one position creates a modulation, while the same interval in another position does not. Further examples include several to do with rhythm rather than melody, supporting Aristoxenus' claim that his thesis applies to music 'in all of its parts', not just in connection with the subject-matter of harmonics.

Music, and specifically melody, thus conducts its business by keeping some elements constant while others change. It is through our grasp on these interconnected constancies and changes that we can attain understanding of music; and this grasp is both perceptual and intellectual. 'Since the nature of music is of that sort, it is essential in matters to do with attunement too [that is, as well as in the study of rhythms and other musical phenomena] for thought (*dianoia*) and perception to be trained jointly (*synethisthēnai*) to assess well that which remains constant and that which changes' (34.25–30). In sharp contrast to the problematic statement at 33.6–9, this comment treats thought and perception as collaborative faculties both of which are concerned with the same objects; and those objects include both the sizes of intervals and *dynameis*, of which the former had previously been assigned to hearing, the latter to thought. It now seems clear, then, that we encounter *dynameis* in our perceptual experience of melodies as well as in intellectual reflection on them.

In all the melodic examples that Aristoxenus gives here, the element that remains constant is the size of an interval, but this need not be so. We 'perceive' differences of genus, as the present passage tells us, when the boundaries of a tetrachord remain fixed (so that the size of the interval between them is constant), while the intermediate notes move. But Aristoxenus also insists in very similar language, later in the book, that the genus can remain constant while the sizes of intervals inside the tetrachord change, within determinate limits. 'Each of the genera presents its characteristic form of movement to perception while employing not just one division of the tetrachord but many. It is therefore clear that while the sizes of the intervals change the genus remains constant, since when the sizes of the intervals change, up to a certain point, it does not change with them but continues to remain constant' (48.33–49.6). In listening to melodies as a scientist should, we must thus attune our powers of perception to constancies and changes in both quantitative and non-quantitative aspects of them. Both fall within the scope of musical perception, and an aspect of either sort may change while one of the other kind remains unaltered.

The 'changes' of which Aristoxenus writes are not merely 'differences', such as the differences between scales in the enharmonic genus and the diatonic, or between the *dynameis* of the notes *mesē* and *nētē*. He is thinking of them as changes that take place in time, as a melody runs its course. This becomes explicit in a short paragraph at 38.27–39.3, which begins by recapitulating, with an important addition, statements made in the passage we have been reviewing.

It is clear that to understand (*xynienai*) melodies is to follow, with both hearing and thought (*dianoia*), the things that are coming into being in respect of all their distinctions. For melody consists in coming-into-being (*genesis*), just as do the other parts of music. Understanding (*xynesis*) of music arises from two sources, perception and memory; for one must perceive what is coming into being, and remember what has come into being. It is not possible to follow the contents of music in any other way.

This is a clear statement of Aristoxenus' view that all musical understanding, including that of a harmonic scientist, is concerned with processes taking place over time. Harmonics is not the study of static or abstract structures. What matters from a musical perspective is not even the pattern of attunement existing, for instance, in the strings of a well-tuned instrument, or not as such, but the ways in which notes and intervals follow one another in melodic sequences, and the relations which these temporal successions set up.[6] Memory, then, is an essential part of the student's equipment, since one can only understand what is happening in the present moment in the light of its relationships with what has happened earlier in the melody, and again by looking back to it from the perspective of what 'comes to be' later. No note or interval can be assigned its melodic character and identity (as e.g. *mesē* or *hypatē*, as the lowest note of a diatonic tetrachord, as introducing a modulation), or even assessed as melodic or unmelodic, *emmelēs* or *ekmelēs*, except in its context in the process of a melodic *genesis*.

Aristoxenus' assertion that musical understanding arises from perception and memory should not be taken to cancel his previous claim about its dependence on hearing and thought (*dianoia*), which is reaffirmed in

[6] We might expect him to argue, though he nowhere does so explicitly, that it is from an understanding of melodic sequences that a musician can work out appropriate patterns for the attunement of a lyre or the positioning of finger-holes on an aulos. What makes an attunement (in this 'practical' sense) appropriate is that it provides for performances of melodically acceptable sequences. It is not the structures of the attunements that are fundamental, and determine which sequences make musical sense (though this is a position which mathematical theorists may possibly have espoused).

the paragraph's opening sentence. In a closely related discussion in the Plutarchan *De musica*, certainly Aristoxenian in origin, perception and *dianoia* are inserted explicitly into the enterprise of 'following' music as it unrolls in time.[7] We are told that perception and *dianoia* must 'run alongside one another' (*homodromein*) when we are assessing any aspect of music; neither should be allowed to rush ahead of the other or lag behind (1143f–1144a). Aristoxenus (as paraphrased by the Plutarchan compiler) does not explain the point further, contenting himself with some sardonic allusions to people whose perceptions are too hasty or too sluggish, or who somehow manage to succumb to both faults at once. We may suppose that when thought outstrips perception it is jumping to premature conclusions about the patterns of movement exhibited by the melody, and when we perceive new notes and intervals while our mind is still occupied with the past, these new elements will be heard uncomprehendingly.

However that may be, the suggestion that perception and *dianoia* must not only work on the same aspects of a *melos*, but must do so simultaneously, requires us to construe *dianoia*, 'thought', in an unexpected way. It cannot be 'thinking' of a discursive or argumentative sort, despite the fact that philosophers very commonly use it in precisely that sense.[8] It is perhaps more like the cognitive interpretation of what we perceive, as when we interpret a set of visual representations not merely as a bundle of colours and shapes but as a dog or a cow.[9] If this is correct, Aristoxenus is apparently at odds with Aristotle's position at *An. post.* 100a14–b5, where the classifying conception of the object perceived is given in perception itself (see p. 112 above). In separating the perception from its interpretation Aristoxenus seems closer to the Stoics, except that in their view, while our 'conception' of something, once established, is housed, as it were, more or less permanently in our mind, and is distinct from the perceptual impression made by such

7 The passage is at [Plut.] *De mus.* 1143f–1144c. Fascinating though it is, and despite its strong resonances with the passage in the *El. harm.*, not all of it is relevant here, since it is concerned with musical judgement extending beyond harmonics to all other facets of music at once. It focuses particularly on the complexities involved in perceiving several kinds of element in a performance simultaneously (a note, a rhythmic duration, a word or syllable), separating out from one another the strands (the sequences of movements) to which each belongs, and 'surveying what is faulty in each of them and what is not'. See further p. 237 below.

8 For an unambiguous example, see Plato, *Soph.* 263e3–5, where *dianoia* is described as the soul's internal 'conversation' (*dialogos*) with itself.

9 It is hard to find pre-Aristoxenian cases where *dianoia* is clearly given this sense. On the other hand it is often used to designate the 'meaning' of a word or an utterance (e.g. Plato, *Lysis* 205b, Aristotle, *De an.* 404a17); possibly Aristoxenus has transferred it here to the faculty through which we discern such 'meanings'. A usage akin to this can be found, a little later, in Stoic sources; see e.g. the report of Diocles at Diog. Laert. VII.49 (this is passage 39A in Long and Sedley 1987; cf. their discussion in vol. I: 239–41, especially 240).

a thing on our senses, nevertheless in human adults the latter is always accompanied by the former, and is interpreted through it. According to Aristoxenus, it seems, the perception may or may not be grasped in the light of an accompanying interpretation; and when it is, it is because our *dianoia* is being brought into active engagement with the object perceived, simultaneously with our perception of it.

Let us return, finally, to the puzzling statement at 33.6–9: 'we judge the sizes of the intervals through hearing, and discern their *dynameis* through *dianoia*'. Everything in the sequel seems inconsistent with this remark. *Dynameis*, and other non-quantitative features of melodies, fall into the province of perception (*aesthēsis*) just as surely as do the sizes of intervals; and *dianoia*, when properly exercised, is concerned with precisely the same objects as those of perception, and at the same time. If there is a solution to this problem, it looks as if it must depend on a distinction between *akoē*, hearing, and the wider notion of *aisthēsis*, perception. Aristoxenus' statements become more nearly reconcilable if *akoē*, taken by itself, is conceived as lacking any capacity for discrimination of a specifically musical kind. All it perceives are pitched sounds, and all it can do with them is to judge the distances between them. The noun *akoē* occurs only nine times in the whole of the *El. harm.*, and in all of them it can comfortably be understood in this restricted sense. *Aisthēsis*, by contrast, is perception of a sort that includes sensitivity to the musical 'meanings' of notes and intervals in their various contexts, and incorporates within itself the operations of *akoē* and interpretative *dianoia* together. Aristoxenus uses the noun *aisthēsis* and its cognate verb *aisthanesthai* repeatedly, many more times than he uses *akoē*; and it seems to be his chosen term of art for the mode of perceptual sensitivity on which an understanding of music fundamentally depends. On this reading, then, *akoē* and *dianoia* are distinct, but when trained and employed as harmonic investigation demands, they are fused into a single, complex activity which Aristoxenus calls *aisthēsis*.

This hypothesis offers a reasonably secure anchorage to the statement at 33.6–9, and has the minor advantage of allowing Aristoxenus' conception of *aisthēsis* to chime more harmoniously with Aristotle's. One might wish it were sustainable, but it comes at some cost. In two of the three crucial passages in the *El. harm.* where *dianoia* is distinguished from some other faculty, the faculty in question is *akoē* (33.6–9 and 38.27–30). They can therefore be understood in the way I have suggested. In the other, however (34.28), it is *aisthēsis* that must be trained together with *dianoia*, and this resists my interpretation. It is hard to avoid the inference that *dianoia* is not after all an element of *aisthēsis* itself, however closely related their functions may be. In the Plutarchan paraphrase, similarly, it is *aisthēsis* and not *akoē* that must 'run alongside' *dianoia*, and not get ahead of it or lag behind. We may perhaps discount the paraphrase, which might have substituted *aisthēsis* for the *akoē* of its Aristoxenian original. But *El. harm.* 34.28 cannot be disposed of in the same way. It seems to me preferable to accuse Aristoxenus of terminological carelessness here than of a wholly unacceptable assertion at 33.6–9. But I cannot claim that the case has been proved.

THE ROLE OF NON-QUANTITATIVE DISCRIMINATIONS, AND THE CONCEPT OF *DYNAMIS*

Some people, says Aristoxenus, think that they will have reached the limit or goal of harmonic understanding when they are able to represent any *melos* in a written notation (39.4–8). We discussed their position in Chapter 2. Aristoxenus devotes nearly three pages (39.4–41.24) to a critique of this view, in the course of which several of his own most interesting ideas about the science emerge. In the first part of his commentary he argues that the information conveyed in a notated score is limited to just one aspect of melody out of many, and that even this one aspect is of minimal importance. We need not reopen the question whether his complaints are justified; the point is rather that his catalogue of these people's alleged omissions is a list of items to which, in his view, a harmonic scientist must pay serious attention, and which must be grasped if the goal of understanding (*xynesis*) in this field is to be reached.

The only ability a person needs in order to notate a *melos*, he asserts, is that of 'perceiving accurately' (*diaisthanesthai*) the sizes of its intervals (39.27–9). This was of course the task for which we were told that hearing, *akoē*, was to be trained (33.6–10). Here, however, we learn that the knowledge it gives contributes nothing of any importance to harmonic understanding.

> The fact that accurate perception of the sizes [of the intervals] themselves is no part of an overall understanding (*xynesis*) was stated at the start, and can easily be grasped from what we shall say now. For neither the *dynameis* of the tetrachords nor those of the notes nor the differences between the genera nor, to put it succinctly, the differences between composite and incomposite [intervals] nor the simple and the modulating nor the styles of melodic compositions nor, one might say, anything else whatever becomes known through the sizes [of intervals] themselves. (40.12–23)

It is easy to be blown away by the rumbustious rhetoric of this passage, and to suppose it to be telling us that there is no purpose at all to be served by training ourselves to assess accurately the sizes of the intervals we hear. That would evidently contradict Aristoxenus' earlier statements and make nonsense of prominent phases of his own expositions. But it is not what he says. For all the sound and fury of his denunciations he has taken care to express his point with some precision, and in such a way that it does not carry that alarming implication.

Accurate perception of 'the sizes themselves', he says, is 'no part' of the understanding that the science seeks. None of the items in his list becomes known 'through the sizes themselves'. Earlier in the passage he has asserted that skill in notating melodies is 'not even a part' of scientific knowledge in harmonics; and he draws a comparison with the science of metrics. In that field, if a person knows how to 'write down' the iambic metre in an appropriate notation, it does not follow that he is equipped to understand 'what the iambic is'; and just so, in harmonics, a person who can 'write down' the Phrygian *melos* need not for that reason understand 'what the Phrygian *melos* is' (39.13–23). None of these remarks implies that accurate perception of the sizes of intervals – which can be displayed in the ability to translate the melodies we hear into notated form – is out of place among the tools of a harmonic scientist's trade. What they tell us is, first, that it is not coextensive with harmonic understanding, and secondly and more provocatively, that it is not, as such, even a 'part' of such understanding. The understanding which is the goal of the enquiry is not constituted even partially by this perceptual acuity, and none of the propositions through which its contents are enunciated concerns itself with the sizes of intervals as such. But these contentions are perfectly compatible with the thesis that the ability to assess accurately the sizes of intervals has an essential role to play in the process through which we move towards the science's goal. The characteristics of sequences in the various genera, for example, are not in the end to be specified by reference to the sizes of their constituent intervals (this claim will be discussed further below); but we shall not be able to establish what those characteristics are without first determining, by ear, the sizes of the intervals included in a substantial number of their instances. The results of Aristoxenus' own empirical studies of these 'sizes' are set out at 50.19–52.33.

Let us recapitulate Aristoxenus' list of the melodic features of which mere perception of intervallic magnitudes leaves us ignorant, and which, unlike those magnitudes, are to be considered 'parts' of the contents of harmonic understanding. It includes (i) the *dynameis* of the tetrachords, (ii) the *dynameis* of the notes, (iii) the distinctions between the genera, (iv) the distinctions between composite and incomposite intervals, (v) the simple and the modulating, and (vi) the styles (*tropoi*) of melodic compositions. Aristoxenus' throw-away phrase at the end, 'nor anything else whatever', shows that the list is not intended to be exhaustive.

We can deal fairly briskly with the last three items. An incomposite interval is one 'bounded by successive notes' (60.10–11). No note falls melodically between its boundaries, and it is not divided into smaller intervals; the opposite is true of a composite interval. In the sequel, responding to some people's puzzlement about this proposition, Aristoxenus explains that one cannot specify which intervals are composite and which are not by reference to their sizes; a ditone, a tone or a semitone, for instance, will sometimes be composite and sometimes not. Thus the ditone between *mesē* and *lichanos* in an enharmonic tetrachord is incomposite – no note can be placed melodically within it – whereas the ditone between *mesē* and *parhypatē* in a diatonic tetrachord is composite, since it is divided by *lichanos* into two intervals of a tone each. 'For this reason,' Aristoxenus concludes, 'we say that the property of being incomposite does not depend on the sizes of the intervals but on the notes that bound them' (60.17–61.4). The size of an interval will therefore give us no help in discovering the principle which determines whether or not a note can be placed between its boundaries. Knowledge of whether it is incomposite or not depends wholly on knowledge of the notes surrounding it, in a sense that requires these notes, in turn, to be specified according to their roles in some *systēma*; and these considerations are not quantitative.

If we turn now to item (v), we find that similar criteria apply. Aristoxenus says little about *metabolē*, modulation, in the surviving parts of his work, but later sources dependent on his lost writings allow us at least to grasp what sort of phenomenon he has in mind. We shall pursue the topic more closely in Chapter 8. For the present it is enough to say that a melodic sequence is 'simple' if it stays within the structure of a single, continuous *systēma* throughout, and 'modulating' if it shifts, part way through, to a *systēma* of another sort. One type of modulation (but by no means the only one) involves a change of genus, so that the sequence q, q, d, t, s, t, t, for example, plainly modulates from enharmonic to diatonic in the upper part of its range. Clearly this judgement is based on the 'sizes of intervals themselves', which are all that my rudimentary 'notation' reveals, though it also involves prior information about the genera, which is not all of that kind. But consider the sequence d, t, s, taken in isolation. It might be part of the one given in the previous example, and thus would be 'modulating'. Equally, however, it might be consistently enharmonic, merely omitting the note that would lie a quarter-tone above the tone. Or it might be purely diatonic, omitting the note a tone above its starting point. From the sizes of the intervals alone we cannot tell. We can answer the question whether a sequence modulates, like the question whether an interval is incomposite, only if we grasp the positions and roles of the intervals and their bounding notes in a *systēma* (or several *systēmata*).

I shall say nothing here about item (vi), the styles of melodic composition; some remarks on the subject will appear in Chapter 9. Our text of the *El. harm.* gives virtually no clue about the ways in which Aristoxenus identified, distinguished and classified them. But it seems obvious enough that distinctions between 'styles' of composition, in any recognisable sense of that word, cannot be made to turn on the sizes of the intervals they use. Even if that consideration has some part to play in the matter, it will carry no great weight.

Item (iii) in Aristoxenus' catalogue is 'the distinctions between the genera'. It has emerged in our earlier discussions (pp. 158 and 171 above) that one cannot define a melodic genus by attributing specific sizes to the intervals contained in its tetrachords, since these sizes can vary, within determinate limits, while the genus remains unchanged. In a later passage (48.15–20) we are told that if we insist that two structures are of the same kind and deserve the same name only if their intervals are of equal sizes, our misguided focus on equalities and inequalities will make us 'throw away' our capacity to discern what is genuinely alike and unlike. We would be stipulating that such terms as '*pyknon*', 'chromatic' and 'enharmonic' always designate groups of intervals of the same size:

> And it is obvious that none of these correspond to the representation of perception. For perception looks to the similarity of some single form (*eidos*) in identifying the chromatic and the enharmonic, and not to the size of some single interval.[10] I mean that it identifies the form of a *pyknon* in all cases, just so long as the two intervals occupy a smaller space than the one,[11] for the sound (*phōnē*) of something compressed (*pyknon*) is perceptually evident in all *pykna*, even though they are unequal; and it identifies the form of the chromatic wherever the chromatic character (*ēthos*) is perceptually evident. For each of the genera moves with its own characteristic kind of movement (*idian kinēsin*) in respect of perception, while employing not just one division of the tetrachord but many. (48.21–49.2)

The thesis that there is a *pyknon* 'just so long as the two intervals occupy a smaller space than the one' clearly presupposes an accurate discrimination

[10] One might expect 'some single size of an interval' here, but that is not what our texts offer, and the intended sense is clear enough.

[11] That is, the two lowest intervals in a tetrachord constitute a *pyknon* so long as they are smaller, taken together, than the one interval above them.

of the sizes of intervals. But the phrase quoted does not define what a *pyknon* is; if it did, there would be no reason for supposing that the class of small structures labelled *pykna* has any particular melodic significance. It would be an entirely arbitrary category. What makes it a structure whose characteristics a harmonic scientist must grasp is the fact that its presence or absence transforms the qualities of a *melos*, as perception experiences them, giving to a melody whose structure contains a *pyknon* a perceptually different 'character' from one that lacks it.[12] The specification of its quantitative limit is the result of scientific observation, which presupposes but is not identical with musical perception, and it tells us, one might say, where a *pyknon* is to be found, rather than what it is.

There are quantitative limits, similarly, within which the positions of the moveable notes will give a tetrachord a chromatic character, but their identification gives no insight into what 'chromatic character' is; that is something we must in fact already be able to distinguish, before setting out to discover the extent of the range of cases in which it presents itself to perception. In grouping structures together as 'chromatic', for instance, and distinguishing them sharply from others that are 'diatonic', we are said to be respecting musically significant 'similarities and dissimilarities', as we are not if we try to group them through purely quantitative considerations of 'equality and inequality' (48.15–17). The 'form' or structure of the chromatic will be present just so long as the 'chromatic character', *ēthos*, or the 'characteristic movement' of this genus presents itself to perception. It is this 'character' that constitutes what the chromatic is.[13] Aristoxenus, wisely perhaps, makes no attempt to define this *ēthos*. One might as well try to define the 'character' of the major mode in the music of the nineteenth century. Certainly one can say a little about it, as the Greeks did about their genera, but such descriptions are invariably elusive and vague.[14] The distinctions between genera are grasped directly in *aisthēsis*, perception, once the perceiver has become saturated with experiences of listening closely to melodies of every kind; and this seems to be the *only* way in which they can be thoroughly grasped. Knowledge of the aesthetic qualities of a genus

[12] The absence of a *pyknon* from the tetrachord is the distinctive mark of a diatonic system, 51.19–22.

[13] *Ēthos* should not be understood here in an 'ethical' sense, implying that the essence of the chromatic lies in its effects upon the moral dispositions of the human soul. It is a purely musical or 'aesthetic' category, designating a quality directly perceived in a melody by the listener. For further discussion see Ch. 9 below.

[14] Thus according to one author, the diatonic 'displays a character that is somewhat solemn and powerful and well-tensioned', and the chromatic one that is 'more mournful and emotional' than the diatonic (Theo Smyrn. 54.14–15, 55.6–7).

can apparently not be transferred from one person to another in speech or writing.

The point can perhaps be made more clearly in connection with the *pyknon*. In this case Aristoxenus does offer a brief characterisation of its musical effect; wherever there is a *pyknon* we perceive 'the sound of something compressed' (48.29–30), where the word translated 'compressed' is again *pyknon*. The fact that in this description the property assigned to the *pyknon* is that of being (or striking perception as) *pyknon* does not make it vacuous; its role is to draw attention to the sound-quality metaphorically evoked by the structure's technical name. What makes this example helpful to a modern reader is that it identifies something that corresponds broadly to a feature of our own musical experience. Any pair of intervals spanning less than half the tetrachord, that is, less than a tone and a quarter, is a *pyknon*. By our standards, any interval in the region of a tone and a half will be perceived as an approximation to a minor third, a relation that strikes us as 'concordant' or 'smooth'. Any interval in the region of a tone, or less, strikes our ears more harshly, as a 'discord'. The point at which the harsher quality replaces the smoother is arguably less determinate than Aristoxenus' formulation suggests, but around a tone and a quarter seems a plausible estimate; and to me, at least, a description of the harsher relation as sounding 'compressed' seems to convey its quality quite well. But we can make sense of the description only because we can associate it with something in our own musical experience. It would, I think, be meaningless to someone who had never encountered such a relation in their own perceptual 'representations'. The description can serve to identify and sketchily characterise the phenomenon only *after* we have heard it and become aware of its aesthetic effect.

Where then does this leave harmonic science? Aristoxenus seems to have argued it into a corner where it is seeking understanding of items which cannot be defined and about which little or nothing can usefully be said. All that the scientist can specify with any precision are the conditions under which these melodic characters or forms will appear, and the specifications will (at least in these cases) be quantitative, whether they pin down the conditions to specific sizes of intervals or allow for variation within a determinate range. Such accounts have their uses, but will provide no insight into the character of the phenomenon which presents itself under these conditions.

I do not think that Aristoxenus has overlooked this problem. Quite the contrary, in fact; it is not so much a problem as a point that contributes substantially to his conception of the discipline. He insists, as we have seen, that students of the subject must become familiar with melodies of many sorts, and must train their hearing, their memory and their interpretative sensitivity to grasp what is going on in any melody as it unfolds. What he calls the *pragmateia*, the 'enterprise' or 'activity' of studying harmonics depends essentially on the use of these faculties. Analogous statements would be unsurprising in a modern handbook for students of the natural sciences, who might very reasonably be told that they should train themselves in careful observation, in interpreting what they see through a microscope, and so on. But the two situations are entirely different. We think nowadays of science as a continuing research project; a scientist is someone who will contest received theories and doctrines, and will seek constantly to extend the range of what is known and to deepen our understanding of it. Students of physics or biochemistry are – at least in theory – being taught to think and behave as enquirers and potential researchers, and are not merely being familiarised with established 'facts'. Aristoxenus, in sharp contrast, presents himself as the authoritative fountain-head of truth in harmonic science. His instruction aims to give a student access to the same knowledge that he possesses, and there is nothing in his writing to suggest that he expects or hopes that his students will be able to contest his views cogently and improve on his insights. In this respect he keeps company with the great majority of Greek scientists and other intellectuals.

The only reason, then, why they must become experts in musical observation is that without such skills and experience they will not understand what Aristoxenus is trying to teach them. One cannot achieve understanding in this discipline just by reading books or listening to lectures, even if they express, meticulously and exhaustively, everything that can truly be said about its subject-matter. Students can grasp the meaning and relevance of what Aristoxenus says about the *pyknon*, for instance, only if they have already acquired, through experience, an acute perceptual sensitivity to its character and its melodic roles. The point is put forcefully in the engagingly sententious introduction to a short treatise from a much later period, not firmly datable but perhaps written in the fourth century AD. Far removed in time though it is, its sentiments are thoroughly Aristoxenian.

It is essential for anyone who will listen to the accounts given of these things[15] to have trained and exercised his hearing in advance,[16] through experience, so as to hear notes and recognise intervals accurately, both the concordant and the

[15] In the preceding lines 'these things' have been specified as notes, intervals, *systēmata*, *tonoi*, modulations and forms of melodic composition in all the musical genera.

[16] 'To have trained and exercised in advance' translates *progegymnasthai*, a metaphor from the gymnasia and wrestling schools.

discordant, so that by matching appropriately what is said to his perception of the characteristic properties (*idiōmata*) associated with the notes, he may achieve complete understanding through personal experience now augmented by speech.[17] But if anyone has come to listen to these accounts without being able to hear a note accurately and without having trained and exercised his hearing, let him depart, shutting the doors on his ears.[18] For even if he is here he will be blocking his ears by not recognising in advance, through perception, the things with which the accounts are concerned. (Gaudentius, *Intr. harm.* 327.8–18)

These Aristoxenian themes and their echoes in Gaudentius' vigorous overture point towards a significant distinction between widely held ancient and modern conceptions of the purpose of scientific activity. In so far as they are not motivated merely by the desire to further their careers, to improve the status of their institutions or to attract financial support, modern researchers are inclined to think of themselves as seeking to increase the corpus of human knowledge. That, at any rate, is a mantra very commonly recited in this connection. They are agents in an enormous, loosely defined, global and trans-generational enterprise in which the work of any one individual is significant only in so far as it is made public and becomes available for others to build on. A scientist's 'product', typically, takes the form of published research papers or books. In such a context, one could hardly accept with equanimity the thesis that the most important characteristics of items in the domain under scrutiny cannot be conveyed in words or in mathematical or other symbolism, and can be grasped for what they are, from a scientific perspective, only within the private experience of each individual observer.

Aristoxenus, however, has nothing but scorn for people who suppose that what a scientist should aim at is 'some visible product' (*ophthalmoeides ti ergon*, 40.33). To say that the achievement which crowns understanding is the production of something publicly visible is to turn the truth on its head; faced with something visible or otherwise perceptible the scientist's real task is to understand it. It is in understanding, not in writing and publication, that his work reaches its fulfilment; and understanding (*xynesis*) is not publicly perceptible or directly communicable, but is 'hidden somewhere in

[17] In my translation, 'what is said' and 'speech' are renderings of *logos*, and 'accounts' represents its plural form, *logoi*. I am therefore interpreting the passage differently from Zanoncelli, for whom *logoi* in the plural are indeed 'accounts' or 'lectures' ('la trattazione', 'queste lezioni'), as they must be, but the singular instances refer not to speech but to reason ('ragione'). On her view (Zanoncelli 1990: 352–3), Gaudentius' remarks engage with the familiar issue of the relation between perception and reason in harmonics; but the supposed transition to that topic dislocates quite unnecessarily the passage's train of thought.

[18] This carries an allusion to a line from an Orphic poem quoted by Gaudentius at the beginning of his introduction. 'I sing for initiates; shut the doors, ye profane!' (327.3 Jan; see West 1983: 34, 82–4).

the soul' (41.6–17). It is something that can be achieved only by individuals, one at a time and inside themselves, in the privacy of their own minds. There is no place here for the notion of the 'sum of human knowledge'.[19]

Modern scientists – or scholars in the humanities, for that matter – are usually also teachers, concerned with the intellectual progress of individual students. But this fact has much more to do with institutional and economic arrangements than with contemporary conceptions of the scientist or the scholar as such. If Caroline or David or Mandip or Konstantinos were guided by my publications towards an understanding hidden somewhere in their souls, it would please me enormously; but that is not, fundamentally, what efforts in research are nowadays supposed to be for. In the eyes of the intellectual community, they are there to be banked, as it were, partly just to be 'there', and partly so that others can use them, critically or constructively, in generating other 'visible products' of a similar sort. (In the natural sciences, of course, there is also the possibility that one's published ideas may find more obviously practical applications and uses.) For Aristoxenus, by contrast, 'scientific knowledge' is not something stored in libraries or data banks, but a condition of an individual's mind. Hence, if we leave self-aggrandisement aside (large though it seems to have bulked in Aristoxenus' motivational repertoire), the only worthwhile purpose that can be served by 'research publications' is that of helping individuals in the improvement of their own, personal understanding. 'Research' and 'teaching' come together in a far more intimate way than is envisaged in the pious declarations of any modern university. In the Greek scenario there is nothing in the least anomalous in a scientific treatise which exhorts its readers to grasp, through their own efforts, features of experience for which neither this nor any other concatenation of words or symbols can stand proxy, and which directs them to ways in which this mode of experience can be attained. To put the point as starkly as possible, the goal is *epistēmē*, scientific knowledge, or *xynesis*, understanding, and the *El. harm.* itself does not contain them. These are conditions that exist in the mind of Aristoxenus, and which his writings may perhaps help to stimulate in the minds of Caroline, David, Mandip and Konstantinos. If they do not, they are worthless.

Since the first two items on Aristoxenus' list of non-quantitative objects of harmonic understanding are (i) the *dynameis* of the tetrachords and (ii) the *dynameis* of the notes, I cannot postpone any longer my discussion of

[19] In Aristoxenus' time there was in any case no large-scale community of like-minded scholars to be addressed. Compare Netz 2002: 201–8, on the number of mathematicians at work in the Greek world; Netz 1999: 271–312, and his catalogue in Netz 1997. The number of specialists in harmonics was certainly a good deal smaller.

the concept of *dynamis*. The account of it which Aristoxenus apparently promises at 36.11–12 is postponed indefinitely, since if it was written it has not survived. So far as we know, Aristoxenus was the first theorist to give the term *dynamis* a special application in harmonics. No earlier source uses it in a comparable way; and later writers seldom seem to attach much importance to the concept. They refer to it infrequently, and their treatments are of limited value in the task of interpreting Aristoxenus. Passages of Ptolemy, Cleonides and Aristides Quintilianus provide some clues, but though their ways of handling the term are not inconsistent with Aristoxenus', they are, I think, a good deal more restricted. We are left to make the best sense we can of Aristoxenus' usage in the passages that survive, with rather little help from other musicological texts.

On the other hand, Aristoxenus presumably did not choose the term arbitrarily, and we would expect its sense in his writing to be based fairly directly on usages available in the non-technical or the philosophical and scientific Greek of his time. He could evidently assume that his audience would understand his way of using it at least reasonably adequately in advance of the full explanation he promises. In general, then, the commonest meanings of the noun *dynamis* fall within the range of our words 'power' and 'ability'. In scientific contexts it often refers to the intrinsic 'capacities' of the elements or other bodies (the capacity of fire to heat things, for example), or to those 'faculties' of organisms which enable them to carry out such activities as seeing, breathing, moving and self-nourishment. *Dynamis* is also one of the most important words in Aristotle's philosophical vocabulary, where (putting aside special metaphysical subtleties) it is usually a 'potentiality' for becoming something or acquiring certain attributes, or else a 'power' to do or produce something or to be affected in some specific way. Things that Aristotle describes as 'having' such *dynameis* are sometimes described, instead, as 'being' *dynameis*; a seed, for example, 'has' the potential to grow into an animal or a plant, but also 'is' a potentiality, in the sense that it *is* a potential animal or plant.[20]

Two of the word's other uses are the most likely to be relevant here.[21] First, in Platonic and later usage the *dynamis* of a word or a statement is its significance, what it means.[22] Secondly, though much less commonly,

[20] Compare e.g. Ar. *De gen. an.* 726b11 with 725b14.

[21] It also has important uses in mathematics, but Aristoxenus' conception is plainly not of a mathematical sort.

[22] In this respect its functions overlap with those of *dianoia*, whose use in connection with meanings we noted earlier (n. 9 above). The two words envisage meaning from rather different perspectives; *dianoia* represents it as the 'thought' in or behind the expression, *dynamis* as its 'power' to signify something, what it 'can do'.

it may be used of items which depend for what they are on their relation to other things, or on the context in which they are set, and specifically to attributes such as 'heavier' or 'health-giving'.[23] The reason for classifying these attributes as *dynameis* is perhaps that they are not attributes which any object or substance (a rock or a medical potion, for instance) possesses in its own right, simply in virtue of what it is, but ones which such a thing *can* possess, and will do under suitable circumstances.

With this broad and flexible repertoire of usages in mind, let us now begin from the other end, and inspect the passages in which Aristoxenus uses the word in the way that concerns us here.[24] Since the issues are important and the passages are relatively few (there are seven in Book II and one in Book III), I have thought it worth presenting them all as a collection. Some are already familiar.

(i) 'We judge the sizes of the intervals through hearing, and discern their *dynameis* through thought' (33.6–9; see pp. 168–9 above).

(ii) 'While the size [of the interval] remains constant, we call one instance [the interval between] *hypatē* and *mesē*, and another that between *paramesē* and *nētē*; for while the size remains constant the *dynameis* of the notes change' (34.1–5).

(iii) 'Since the intervals are not sufficient by themselves to enable us to discriminate the notes (for every size of interval, to put it briefly, is common to several *dynameis*), the third part of our enterprise as a whole deals with notes, saying how many there are and by what criteria they are recognised, and whether they are pitches (*taseis*), as most people suppose, or *dynameis*, and explaining exactly what a *dynamis* is' (36.2–12).

(iv) 'We shall say the same about the *dynameis* which are made by the natures of the tetrachords, for the [interval between] *nētē hyperbolaiōn* and *nētē* and that between *mesē* and *hypatē* are written with the same symbol, and the symbols do not distinguish the difference between

[23] This usage appears, with these and other examples, in the work of the third-century Epicurean philosopher Polystratus. 'It is surely obvious to everyone that greater and smaller are not seen to be the same everywhere and in relation to all magnitudes . . . So too with heavier and lighter things, and the same applies to all other *dynameis* without exception. For the same things are not health-giving or nourishing or destructive to everyone . . .' For the excerpt and its context (Polystratus, *De contemptu* 23.26–26.33) see passage 7D in Long and Sedley 1987.

[24] Like any other Greek author, he sometimes uses it in more commonplace ways, not specific to harmonics. At 33.15, for instance, he refers to the 'power' or 'faculty' of perception, and at 19.9 to the 'power' of a musicological principle. I shall ignore these and similar cases, which are easily identified. (It is puzzling that some of them seem not to have been identified by Da Rios, whose index entry for *dynamis* is misleading; not all the occurrences listed as carrying the meaning 'potentia, vis soni vel intervalli . . .' really belong in this group.)

the *dynameis*, since they correspond only to the sizes themselves, and to nothing beyond that' (40.4–11).[25]

(v) 'For neither the *dynameis* of the tetrachords nor those of the notes, nor the distinctions between the genera nor . . . becomes known through the sizes [of the intervals] themselves' (40.16–24; see p. 175 above).

(vi) 'For we see that *nētē* and *mesē* differ from *paranētē* and *lichanos* in respect of *dynamis*, and so do the latter from *paramesē* and *hypatē*, which is why a special name is assigned to each of them individually, while the interval between these notes is in every case the same, a fifth. Thus it is evident that a difference between notes cannot always coincide with a difference between the sizes of intervals' (47.29–48.6).

(vii) 'Thus it is clear that while the sizes [of intervals] change, the genus can remain constant . . . and while it remains constant it is only to be expected that the *dynameis* of the notes will remain constant too' (49.2–7; see p. 171 above).

(viii) 'One must specify everything in music, and assign it to the sciences, in so far as it is determinate, and in so far as it is indeterminate one must abandon it. Thus in respect of the sizes of the intervals and the pitches of the notes, the facts about melody seem to be somehow indeterminate, but in respect of *dynameis*, forms and positions they appear to be determinate and consistently ordered. The progressions [literally 'routes'] downwards from the *pyknon*, for instance, are determinate in *dynamis* and in their forms, and are only two in number; the one through a tone leads the form of the *systēma* into disjunction, while that through the other interval, whatever size it has, leads it into conjunction' (69.3–18).

The items which are most often said to have *dynameis* or to be *dynameis* in these passages are notes, but there are a few exceptions. The apparent attribution of *dynameis* to intervals in (i) is questionable (n. 3 above), but (v) assigns them to tetrachords; (iv) speaks, more enigmatically, of *dynameis* 'made by the natures of the tetrachords'; and in (viii) the progressions from the *pyknon* are said to be 'determinate in *dynamis*'. It seems clear, however, that their prime possessors are notes. It is in that connection, in passage (iii), that Aristoxenus promises a full-dress account of the topic; and in later authors, as we shall see, only notes are credited with *dynameis*.

[25] There are textual difficulties here, but for our purposes they are unimportant. My translation follows the text printed by Da Rios.

Although passage (iii), having raised the question whether notes are pitches or *dynameis*, does not explicitly answer it, it leaves us in no doubt as to what the eventual answer will be. That they are pitches is what 'most people' suppose, and the expression is a sure indication that Aristoxenus disagrees. Further, passage (viii) asserts that so far as the pitches of notes are concerned, the facts are 'indeterminate' and so fall outside the scope of a science. Presumably this reflects the thesis that there is no limit to the number of positions available to the moveable notes in a tetrachord, so that they cannot be determinately enumerated. This thesis, however, was already asserted in Book I (26.13–18); and yet there Aristoxenus seems quite content with the view that what makes a sound a note is its 'incidence on a single pitch' (15.15–16).

I think we can do something to reconcile the two positions expressed in Book I, while recognising that Aristoxenus had not yet thought the issue through, and had not yet arrived at the concept of *dynamis*, which figures nowhere in that book. After stating that a note is 'the incidence of the voice on a single pitch' he continues: 'for it is when the voice appears to stand still on a single pitch[26] that there appears to be a note such as is capable of being placed, in an orderly way, in an attuned *melos*' (15.17–20). What he specifies at 15.15–16, then, are the conditions that something must meet if it is to be *capable* of being made an element in a melodic structure. It must be a sound with a steady pitch. But what he calls a 'note' in Book II is not this, but something that actually *has* a place in such a structure. A pitched sound becomes a note only when it is appropriately placed among other notes, and can be assigned one of the names by which notes are identified.[27]

The *dynamis* of a note is plainly not identical with its name; but all Aristoxenus' references to a note's *dynamis* can be understood as allusions to those of its attributes, whatever they may be, that entitle it to one of the accepted Greek note-names, *lichanos*, *mesē* and the rest.[28] These names are assigned on the basis of a note's relation to others – its place in the structure of the *systēma*, not the 'distance' at which it stands from those others – so that *lichanos*, for instance, could provisionally be defined as the note

[26] Here I accept the supplement to the text offered originally by Meibom, followed by Marquard, Westphal and Macran.

[27] See passage (vi), and notice Aristoxenus' insistence on the relevance of the notes' names, both in the immediate sequel to that passage (48.7–13), and in the more extensive argument that follows passage (vii), at 49.8–30. Compare also his contention that 'successive' notes are (or perhaps 'include') 'those which we have used since the distant past, such as *nētē* and *paranētē* and those continuous with them' (53.29–31).

[28] The names implied in passage (vii) may include a specification of genus, 'chromatic *lichanos*', 'enharmonic *lichanos*', and so on. Such expressions are common in the later sources.

immediately above *parhypatē* and immediately below *mesē* in a continuous *systēma* (cf. e.g. 28.24–8, 68.6–12). It is in this connection that the concept of *dynamis* reappears in post-Aristoxenian writings. Ptolemy, for instance, distinguishes two methods of naming notes. One is by 'position', *thesis*, where a name tells us only whether a note comes first, second or third, and so on, in order from the bottom of the series in question. The other is by *dynamis*, where the name relates it to its role in the system's structure. *Mesē*, for instance, is 'the lower note of the higher disjunction', and *hypatē mesōn* is 'the note common to the two lower conjoined tetrachords above the lower disjunction'.[29] Aristides Quintilianus does not distinguish the two methods of naming notes, but his use of the term *dynamis* is the same; so is that of a (perhaps interpolated) passage in Cleonides.[30]

Here, perhaps, we have our first clue to Aristoxenus' reasons for calling the notes' identifying attributes their *dynameis*. Like the attributes described as *dynameis* by Polystratus (n. 23 above), they are relational, and they are ones which sounds of determinate pitch *can* possess, but do not actually possess until they are placed in an appropriate context. But Polystratus' usage is uncommon, and I do not think that this is the whole explanation. Aristoxenus' conception of melodic structure is rooted in the notion of the voice's movement, its ways of progressing, melodically, through a series of intervals from one steady pitch to another. If it has arrived in the course of its travels at a note whose role is that of *lichanos*, for example, the fact has consequences for its further progression. To be a *lichanos* is to have a kind of 'power', which determines features of the route that the voice can take next, and the pattern of relations into which its subsequent movements can fall. It is not just a fixed point, a pitch, but something with its own dynamic properties, which (for example) impel the voice to move next, in its melodious trajectory, through no distance upwards less than that which separates *lichanos* from the highest note of its tetrachord, *mesē*.

This may seem no more than a fanciful way of putting a simple point: any pitch that is to function melodically as a note must take the role of one or another of those that appear in the Perfect Systems.[31] From *lichanos* a melody can move upwards to no point closer than *mesē* because the systems assign no melodic role to an intermediate pitch; and it will therefore not amount to a 'note', in the sense given to this word in Books II–III. But from Aristoxenus' perspective there is more to it than that, and not only because of his 'dynamic' conception of melodic form. The Greater

[29] Ptol. *Harm.* II.5. [30] Arist. Quint. *De mus.* 15.15–19, 101.5–6, Cleonides, *Intr. harm.* 207.10–13.

[31] For the Greater and Lesser Perfect Systems and their combination as the Unchanging System see pp. 13–17.

and Lesser Perfect Systems, after all, are not themselves 'given' musical phenomena, but constructions worked out by theorists (perhaps, in their canonical form, by Aristoxenus himself[32]) to form a framework within which musicians' practices could be intelligibly coordinated. It therefore becomes reasonable to ask why these systems, and no others, capture all the melodic 'roles' taken by notes in musical melodies. Why are there just so many notes, with just such constraints on their relationships with one another, in the space between *hypatē* and *mesē*? What precisely is the obstacle that prevents us from inserting other notes between them? Aristoxenus construes such questions, as we noted earlier, as scientific questions about the manifestations of a *physis*, the 'nature of *melos*', not as questions about the development of conventional expectations within a culture.

Since the structure combining the two Perfect Systems includes all the notes, placed in their proper relations to one another, successive notes in any continuous, non-modulating progression will be precisely those that are represented as successive in some part of that structure. Now one of the tasks of Book II, with its extensive discussions of conceptual and methodological matters and its introduction of the concept of *dynamis*, is to prepare the way for the propositions and arguments of Book III. These spell out rules to do with melodic progressions ('which interval can be placed melodically after which, and which cannot') in minute detail, and demonstrate their truth. Aristoxenus seeks to show that they follow inescapably from a few very simple principles which express essential and unalterable aspects of the *physis* of melody. In Chapter 8 I shall argue that these 'rules of progression' cannot be satisfactorily interpreted as statements about intervals and sequences of intervals as such, identified by their sizes. What they set out are the imperatives imposed on melodic sequences by the *dynameis* of the notes; the routes that a melody can take after passing through an incomposite ditone, for example, are determinate (in a sense to be explored) not, fundamentally, because the interval spans two tones, but because its lower boundary is a *lichanos*. It is these rules that determine the structure of the Perfect Systems, not the other way round. The rules reflect the *dynameis* of the notes, which are prior both to the rules and to the systems whose form they govern, and these *dynameis* arise in turn from the *physis* of melody itself, as expressed in the governing principles.

It follows from these points that the structure of the Greater Perfect System, for example, is not essentially to be conceived in quantitative terms. This may seem a perverse conclusion, since the system's principal

[32] For an intricate and ingenious development of a case for a different conclusion, see Hagel 2005.

landmarks, the 'fixed' notes, are apparently separated from one another by quite determinate distances; *hypatē mesōn* stands at exactly a perfect fourth below *mesē* and a perfect fifth below *paramesē*, for example. But the point is that the concordant relations of a fourth and a fifth themselves are not to be defined as intervals of certain sizes. Like the *pyknon*, they are items that present specific 'characters' to the musical ear. They exist wherever that character is perceived, and we can recognise them prior to, and independently of, any attempt to identify the sizes of the intervals in which this character occurs. They differ in this respect from the *pyknon* only because, as it turns out, the sizes of the intervals which present themselves as instances of these concords admit of very little variation.

Aristoxenus' way of putting this point is instructive. 'Among the sizes of the intervals, those of the concords appear to have either no range of variation at all, being restricted to a single size, or a range that is only a hair's breadth wide (*akariaios*), whereas this is much less true of the sizes of discords' (55.3–7). Intervals of several different sizes, then, can create what is musically the 'same' discord; and although in the case of a concord the range of variation is either zero or vanishingly small, in principle the same could be true. If it could not, and the essential nature of each concord were constituted by its size, Aristoxenus could not even envisage the possibility that it had any range of variation, however minute. He himself works on the assumption that the perfect fourth, for instance, can be reckoned as spanning two and a half tones; but he is not dogmatic about it. The method of construction he offers at 56.13–58.5 is not represented as a proof, but as a procedure through which we can form our own judgement about whether that estimate is accurate or not.

The framework of the Perfect Systems, then, is not determined quantitatively, but through a pattern of relations which present the characters of the various concords. If a harmonic scientist can assign more or less determinate sizes to them, that may be convenient in certain phases of his work; but any quantitative determinacy that they have is incidental to the system's musical nature and status, and to the primary concerns of a scientist investigating the characteristics of *melos*. The determinacy of the system's structure is not fundamentally a quantitative determinacy, and it would survive even if the sizes of the concords were more variable than in fact they are. More generally, as passage (viii) puts it, it is the *dynameis* of the notes and not the sizes of the intervals that are determinate, and can be specified in such a way as to be proper objects of a science's attention. Any constraints that turn out to be placed on the sizes of the intervals themselves will be consequences of the 'powers' exerted by the notes; and since, as we learn in passages (ii), (iii) and (v)–(vii), the *dynameis* of the notes are not adequately definable in terms of the distances between them, the nature of these constraints cannot be captured in rules specifying which sizes of intervals can succeed one another and which cannot.

We saw earlier that Aristoxenus occasionally attributes *dynameis* to things other than notes, specifically to tetrachords, which possess them according to passage (v) or whose *physeis* 'make' them according to passage (iv), and to the melodic progressions from the *pyknon*, in passage (viii). These cases, however, are arguably secondary, in the sense that the *dynameis* of these items are derived from, and perhaps reducible to, those of the notes involved in them. Aristoxenus commonly identifies a tetrachord by naming the notes that mark its boundaries: one tetrachord is that 'of *mesē* and *hypatē*' and another that 'of *nētē hyperbolaiōn* and *nētē*'.[33] Since every tetrachord in a regularly formed *systēma* has the same internal interval-pattern as every other, their distinctive *dynameis* must display themselves in their relations to other elements of the system, perhaps in particular to other tetrachords. The point is not merely that one tetrachord will lie in a higher or lower region of pitch than another; it is that the routes by which the voice can progress from one tetrachord into those above and below it, through a disjunction or a conjunction, are different in each case, as a glance at the map of the Unchanging Perfect System on p. 17 will show. Thus from the tetrachord *diezeugmenōn*, for example, the voice can progress upwards into *hyperbolaiōn* through a conjunction and downwards into *mesōn* through a disjunction; from *mesōn* it can move downwards through a conjunction into *hypatōn* and upwards either through a conjunction into *synēmmenōn* or through a disjunction into *diezeugmenōn*, and so on. In no two cases are the available patterns of movement the same, and in this sense each tetrachord has its own distinctive *dynamis*, expressed in the 'power' to impose a particular form of order on the melody.

But it is clear that there is no need to attribute *dynameis* to tetrachords independently of those of their constituent notes. In describing the consequences of the former we are merely isolating and bringing together one special group of the consequences of the latter. The *dynameis* of the two downwards progressions from the *pyknon* reflect *dynameis* inherent in the relevant notes in a similar way. The lowest note of a *pyknon* has the *dynamis* of the fixed note at the bottom of one or other of the tetrachords, and from such a note the *melos* can proceed downwards only into another tetrachord, through a disjunction or a conjunction. This is tantamount to saying that when it moves to the next note in succession, it must arrive at a note whose

[33] See passage (iv). The latter *nētē* is *nētē diezeugmenōn*. Cf. also passage (ii), where the notes by which intervals are identified are again the boundaries of tetrachords. For another example see 46.19–21.

dynamis is either that of the fixed note at the top of the next tetrachord, or that of its second note, its *lichanos* or *paranētē*.

One final question needs to be raised here, though I am not sure that it can be definitively answered. Aristoxenus has repeatedly contrasted knowledge of the sizes of intervals with knowledge about non-quantifiable features of notes, structures or relations, some of which he has called *dynameis*. Others have been described as 'characters' or 'forms', or in phrases such as 'the differences between the genera', 'the characteristic movement of a genus', and the like.[34] What he leaves unclear are the relations between these items. The 'form' of a melodic progression may perhaps be determined by the *dynameis* of its constituent notes, and statements about the former may even be reducible to statements about the latter. But no analysis of the *dynameis* of notes seems likely to be able to express or to account for the 'character' of a *pyknon* or a concord, or the aesthetic 'form' of a chromatic sequence.

The concept of *dynamis* is unquestionably central to Aristoxenus' thinking in Books II–III, and *dynameis* are prominent among the non-quantitative melodic phenomena that the harmonic scientist must grasp. But other such phenomena also contribute in significant ways to our perceptual appreciation of *melos*, and need to be integrated into the scientist's understanding of its nature. If the *dynameis* of the notes and the characters and forms of various other items are as independent of one another as they seem, it is not clear how such integration is to be achieved. If it is not achieved, doubts will arise about Aristoxenus' thesis that what we experience as intrinsically 'melodic' qualities are all genuine manifestations of a single *physis*. Perhaps, so far as their 'nature' is concerned, they are essentially unrelated aspects of sound-sequences, which mere custom and tradition have led us to group together in constructing the concept of *melos*. If that is true, *melos* will have no 'nature' of its own, in the Aristotelian sense. Aristoxenus would certainly reject this view, but he does nothing, in the present context, to explain how such doubts can be put to rest. I shall return briefly to the issue at the end of this chapter.

THE REFLECTIONS OF BOOK II ON *APODEIXIS*, 'DEMONSTRATION'

Aristoxenus repeatedly denounces earlier theorists for neglecting the task of 'demonstrating' the theses they propound; and in Book III he generates an elaborate set of *apodeixeis* of his own, which we shall explore in Chapter 8.

[34] See especially 40.16–23, 48.21–49.21, 69.6–22.

Though Book II contains no such demonstrations, it offers one short but important paragraph of general comment on the subject. It comes immediately after the introductory discussion of approaches to harmonics reviewed in the first section of this chapter. It identifies six points which a student must keep firmly in mind when embarking on more advanced and detailed work in this discipline; I have numbered them in the translation.

That is what one might say, more or less, as a preface to the enterprise called harmonics. But people who are going to set out on a study of the elements need to know such things as the following: that it is not possible to go through the subject thoroughly unless the three conditions I am about to specify are satisfied in advance. (i) First, the phenomena themselves must be properly grasped; (ii) secondly, the prior and the posterior among them must be correctly distinguished; and (iii) thirdly, what goes together and agrees must be duly recognised.[35] And since it is necessary, in every science that consists of more than one proposition, to adopt principles from which the things dependent on the principles will be demonstrated, we must adopt them with the following two points in mind: (iv) first, that each of the foundational propositions should be true and perceptually evident (*phainomenon*), and (v) secondly, that each should be such as to be recognised by perception as one of the primary parts of harmonic science; for anything that somehow demands demonstration is not foundational (*archoeides*, 'having the form of a principle'). (vi) In general, we must take care, as we begin, not to set off into alien territory[36] by starting from a conception of sound as a movement of air, nor to turn back too soon,[37] leaving out many things that belong to the subject. (43.25–44.20)

Some of these six points, notably the first and fourth, with their heavy emphasis on perception, have a special resonance in their Aristoxenian setting. Almost all of them, however, also echo requirements laid down for scientific demonstrations by Aristotle in the *Posterior Analytics*. I shall not

[35] The phrase I have translated, fairly literally, as 'what goes together and agrees' is *tou symbainontos te kai homologoumenou*. The diversity of uses to which the verb *symbainein* is put in philosophical Greek makes the phrase hard to interpret with confidence; I shall explain below what I think it means here. I cannot accept the view of Laloy and Da Rios, that it can be translated as 'their essential attributes' (cf. Aristotle's expression, *ta kath' hauta symbebēkota* at e.g. *An. post.* 75b1), referring to attributes necessarily attached to the subject but not expressly mentioned in its definition (cf. Ar. *Metaph.* 1025a30–2). That would imply that the items with which they 'agree' are subjects capable of bearing attributes, notes or intervals, for example. But here, it seems, the subjects whose attributes they would be are the 'phenomena', which in Aristoxenus' regular usage are not items of that sort. They are not 'things', but facts evident to perception, such as the fact that a fifth plus a fourth makes up an octave. It makes little sense to speak of such facts as possessing attributes, essential or otherwise.

[36] The noun is *hyperoria*, 'a place beyond the boundary or frontier'.

[37] This is a metaphor from racing; the phrase means literally 'to turn inside', i.e. to turn inside the post marking the point round which the runners must go before racing back towards the place where they started. To 'turn inside' is to 'cut corners'.

harp on their Aristotelian pedigree again here, but it is at any rate plain that they locate the work of an advanced student of harmonics in the context of demonstrative reasoning. Meticulous observation is the starting point, of course, and its authority remains paramount; but the principal task will be to show how the observed facts hang together in a logically concatenated structure.

This structure is hierarchical. At the highest level are the principles on which other things depend and from which they will be demonstrated. Like Aristotle, Aristoxenus explicitly denies that the principles of his science can themselves be demonstrated, but he says rather little about the basis on which their truth and their foundational status can nevertheless be known. We are told only that they must be 'perceptually evident', and that their primacy must be such as to be 'recognised by perception'. We may reasonably guess that this 'perception' is that of an experienced student who has deployed his observational faculties widely and well, and that it amounts to an inductive summary of all his direct encounters with melody. One might indeed introduce here, quite consistently with Aristoxenus' general line of thought, a complete recapitulation of the final chapter of the *Posterior Analytics* (see pp. 111–12 above), tailored to the specific topics and techniques of harmonic analysis. It would be an interesting exercise and might bring us quite close to an accurate reconstruction of Aristoxenus' own position. But I suppose one should resist the temptation to conjure up an entirely fabricated 'ancient text' of our own. Aristoxenus drops only a couple of hints, and says no more on the subject.

Secondly, there is a hierarchy among the phenomena themselves (point (ii)). Some are prior, others posterior, and these must be correctly distinguished. Those described here as prior need not all be non-demonstrable principles, which are introduced as a separate topic in the second half of Aristoxenus' list. A part of his critique of Eratocles in Book I (see pp. 153–4 above) helps to clarify the relation he has in mind. Eratocles, he says, tried to establish that there are just seven ways in which the intervals of an (enharmonic) octave can be arranged,

> failing to understand that unless the arrangements of the fifth and the fourth have been demonstrated in advance, and in addition to them the manner of combination according to which they can be put together melodiously, many times the seven arrangements will be shown to arise. (6.25–31)

Aristoxenus is not disputing Eratocles' conclusion. There are indeed only seven ways in which the intervals of an octave can be rearranged melodically.

Though many other permutations are theoretically available,[38] they are not melodically acceptable. But the reason why the acceptable permutations are limited to seven lies in another set of facts or 'phenomena', facts about the restricted number of ways in which the fifth and the fourth can be melodically arranged and combined with one another. The facts about the fifth, the fourth and their combinations are thus 'prior' to the fact about the octave, in the sense that the latter is true only because the former are true. The latter can be established only if the former are established first, and are used as the basis of the latter's demonstration.

Finally, we must consider the phrase 'what goes together and agrees', in point (iii) on Aristoxenus' list. I have already commented on its obscurity (n. 35 above), and I shall be dogmatic in my interpretation. I do not think that it identifies items at yet another level in the logical hierarchy, below principles, prior phenomena and posterior phenomena. The only relevant items that would fall below the phenomena at the third level in that sequence would be other phenomena to which the level-three phenomena are prior in their turn. The hierarchy might indeed be extended in that way, so that phenomenon Q is posterior to phenomenon P and prior to phenomenon R. But Aristoxenus' shift of terminology indicates that he is making a point of a different sort, not remarking on the possibility of iterating the previous relation.

His intention, I suggest, is not altogether vague, but it is large and general. He is alluding to no particular group of items, but to the relations of 'going together' and 'agreement' as such. What he is exhorting us to notice is how superficially unrelated facts, regardless of their level in the hierarchy, fit together harmoniously with one another to form a rationally integrated whole. There are complex but genuine lines of implication between – for example – the routes by which the voice can proceed melodically from the *pyknon*, the ways in which tetrachords can be linked in sequence, the acceptable arrangements of the octave and the rules governing modulation between *tonoi* or 'keys' (some of these relations will be explored in Chapter 8). None of these 'phenomena' will have been adequately understood until the respects in which they 'go together' and 'agree' have been duly recognised. As our intellectual grasp on these intricate patterns of logical connection increases, so too will our appreciation of the 'astonishing orderliness' of the constitution of *melos* (5.23–4), and of the reality and unity of the 'nature' of which all individual melodies, in all their enchanting diversity, are partial expressions.

[38] For further comments and some details, see Ch. 8 n. 19.

This brings us back to certain issues that were raised in the second and third sections of this chapter. No doubt the impressions we receive of the 'characters' of the various genera, of *pykna* and non-*pykna* and other such items of musical experience cannot be directly linked to one another in a logical nexus, if only because they cannot adequately be specified in words. But a person in whose experience they are already embedded might be brought to see that they arise, regularly and consistently, from patterns of melodic movement which can be so specified, all of which are governed by the same principles and are in other respects logically interconnected, and that they are musically meaningful only within the context of this tightly integrated system. If the demonstrative theorems of advanced harmonic science can achieve this much, they will have done their work.

CHAPTER 8

Elementa harmonica *Book* III *and its missing sequel*

The third book of the *El. harm.* consists almost entirely of a set of twenty-three demonstrative proofs of propositions about melodic sequences (62.34–74.8). Before them come four preliminary arguments establishing points on which the proofs will rely (58.14–62.33); and they are followed by the beginning of an argument designed to show that there are three and only three melodically acceptable ways of arranging the constituent intervals of a perfect fourth (74.9–25). At this point our manuscripts break off.

Here and there in the course of his exposition Aristoxenus pauses to examine a methodological issue or to clarify theses which his hearers, so he says, have previously found obscure. The bulk of the text, however, is devoted to the logical derivation of rules about melodic sequences, one at a time, from axiomatic principles. Though the arguments are set out with various degrees of formality, all are presented within the conventions familiar to us from Greek mathematical works, notably from Euclid's *Elements*. A proposition is stated; there follows an argument deriving it directly or indirectly from the agreed axioms; finally, in many cases but not all, the proposition is restated as an established conclusion. The unadorned rigour of the mathematical treatises is reflected also in Aristoxenus' style of writing in this book; it is severe and concentrated, admitting few of the rhetorical flourishes and none of the barbed allusions to other theorists which enliven Books I and II. These remarks apply to the four preliminary arguments as well as to the main sequence of twenty-three theorems or 'demonstrations'; and though the reasoning is missing from our texts, the remnant of the discussion of the arrangements of the fourth is plainly the beginning of an argument cast in the same mould.[1]

[1] An argument is evidently required. Any fourth of the sort which Aristoxenus has in mind (a tetrachord) is occupied by three intervals, and from a purely mathematical perspective there are six ways in which they can be arranged, not three. The argument which is about to begin at the point where the text runs out must have been designed to prove that half of these theoretically possible arrangements are melodically improper. It could readily have been constructed on the basis of propositions established earlier in the book.

The whole of this phase of Aristoxenus' enterprise, then, is conceived and presented as a sequence of theorems. (At the end of this chapter I shall raise the question whether this form of presentation is likely to have been maintained in the missing portion of the work.) His adoption of a *modus operandi* so clearly copied from the mathematicians may seem surprising, in view of the dismissive comments he makes about quantitative analysis in the preceding books. But mathematics is also treated as the paradigmatic science in the first book of Aristotle's *Posterior Analytics*, from which Aristoxenus' picture of scientific demonstration was drawn; and its attraction for Aristoxenus lies in the clarity, precision and axiomatic orderliness of its reasonings, rather than in quantification as such. It is true that some of the propositions are expressed in quantitative terms. I shall argue later that he has nevertheless not abandoned the hostile attitude to quantitative analysis which he expresses so vigorously in earlier parts of the work.

I shall say little about the first four propositions. The role of the first, third and fourth, to put it in general terms, is to set clear limits on the agenda of the theorems that follow. The first (58.14–59.5) shows that tetrachords of the same form, succeeding one another in a continuous sequence, must be linked either in conjunction (where the highest note of one tetrachord is the lowest of the next), or in disjunction (where there is an interval, but no intervening note, between the highest note of the first and the lowest of the second); and that where there is a disjunction, the interval separating the tetrachords must always be a tone.[2] The third (61.5–34) shows, on the basis of the first, that in changes of genus it is only the constituent intervals of the tetrachord that are altered, not the interval of disjunction. The fourth (62.1–17) shows that the number of different-sized intervals in any one genus is, at the most, the same as the number of intervals into which a perfect fifth is subdivided, that is, four.[3] This proposition too is established by reference to the first. The principle from which the first proposition is derived is the corner-stone of the entire enterprise, and underlies most of the theorems that follow. It is the principle I have called the 'law of fourths and fifths', which states that in any continuous progression from any given note, either the fourth note in order stands to the first at the interval of a fourth, or the fifth note stands to it at the interval of a fifth, or

[2] Aristoxenus goes on, at 59.5–60.9, to make various other statements about tetrachords that belong to a single continuous sequence. Their interpretation raises thorny problems, but I shall by-pass them here; these statements have no detectable bearing on the surviving discussions that follow.

[3] Aristoxenus appends a brief response to people who have found aspects of this proposition problematic, 62.18–33.

both (54.2–10; see p. 131 above). From this point onwards I shall refer to it simply as **L**.

These three propositions, as I have said, establish basic, general facts about the melodic sequences that are to be discussed. First, these sequences occur within a structure whose tetrachords are either conjoined, or disjoined by the interval of a tone.[4] Secondly, no matter how radically the intervals contained in melodic sequences differ between one genus and another, the differences are all contained in the intervals occupying the span of a fourth. Finally, when we are considering which sequences of intervals are and are not melodically acceptable, the number of different-sized intervals whose possible permutations need to be reviewed is limited, in any one genus, to four.

There remains the second of Aristoxenus' preliminary propositions. It is different in kind from the rest, since it is in effect a definition: 'an incomposite interval is one bounded by successive notes' (60.10–11; discussion of this proposition continues to 61.4). The definition is relevant because the melodic sequences discussed in the subsequent theorems are conceived, precisely, as sequences of incomposite intervals; the propositions they establish deal with the routes that can and cannot be taken, from a given starting point, through melodically indivisible steps. The crux of Aristoxenus' definition is that the status of an interval as 'incomposite' or 'melodically indivisible' does not depend on its size but on the identity of the notes that form its boundaries. I commented on this point earlier (p. 177 above); its significance for the interpretation of the principal arguments of Book III will emerge in due course.

We turn now to the twenty-three theorems about melodic sequences. I shall not try the reader's patience by examining them all individually. I shall start by looking in some detail at just two reasonably straightforward specimens; and I shall go on to review the pattern into which the whole collection of theorems falls. After that I shall focus on three substantial difficulties. In the course of our attempts to solve them we shall find that they are intimately related, and the solutions I shall offer will tell us a good deal, I believe, about the way in which Book III should be interpreted.

Briefly, the three puzzles are these. The first arises from the fact that a significant group of Aristoxenus' arguments seem, on the face of it, logically absurd; and their absurdity is compounded by the fact that at least some of

[4] At no point has Aristoxenus shown, argumentatively, that every melodically acceptable series must be analysable as a sequence of tetrachords. This is an assumption, presumably a perceptual 'given', whose credentials are nowhere examined in the *El. harm.*, or indeed by any other Greek author.

the relevant propositions could apparently have been established straightforwardly by other means at his disposal. Aristoxenus was no fool, and it is beyond belief that he intended the arguments to be read in a manner which entails these obvious mistakes. We must find another way of understanding them. Secondly, there is the broader problem to which I have already alluded, of how the quantitative mode of expression used in many of the theorems can be reconciled with Aristoxenus' earlier statements about the worthlessness of quantitative analyses. If propositions about the sizes of intervals form no part of harmonic understanding (40.11–14), why does he devote so much tedious logical effort to the task of demonstrating them? Thirdly, at a still more general level, I want to debate the difficult issue of what these theorems are for. It has regularly been assumed that from a methodological perspective they are the pinnacle of the work, marking Aristoxenus' achievement, in his own estimation, of the goal of a 'completed science', as envisaged in the *Posterior Analytics*. I have previously taken that view myself. But in saying this we are eliding the plain fact that the work as we have it is incomplete. The programme that Aristoxenus announced in Book II, together with the summary accounts given by later compilers, gives some clues about the contents of the rest. We should at least consider the possibility that these theorems are not, or not only, a triumphant dénouement, presenting Aristoxenus' results in the axiomatic form proper to a fully accomplished scientific enterprise, but are staging-points on the way to something else. I shall suggest that this hypothesis goes a long way towards making sense of his most obtrusively peculiar arguments.

TWO SPECIMEN THEOREMS

Before we set off to examine Aristoxenus' theorems, it may be helpful to remind ourselves of the ways in which tetrachords in each of the three genera are divided. In enharmonic, the division which Aristoxenus treats as correct places a pair of quarter-tones at the bottom, above which lies a ditone, completing the interval of a fourth. He identifies several divisions of the chromatic; as examples we may mention one in which the two small intervals at the bottom are semitones and the remainder is an interval of a tone and a half, and another in which the small intervals are one third of a tone each, while the interval above them amounts to eleven sixths of a tone. The most straightforward version of the diatonic which he discusses has a semitone at the bottom with two successive tones above it. We need also to keep in mind the meaning of the term *pyknon*, which plays a large role in the theorems. It designates the pair of intervals at the bottom of a tetrachord, in those cases where they add up to less than half the span of a perfect fourth, that is, on Aristoxenus' estimate, to less than a tone and a quarter. Every enharmonic and chromatic tetrachord contains a *pyknon*, and no diatonic tetrachord does so.

By way of an initial example of Aristoxenus' theorems, let us take the first. It is very brief.

> No *pyknon*, neither the whole nor a part of one, is uttered melodically if it is placed next to another *pyknon*. For [if it is so placed] the consequence will be that the fourth note in order will not form the concord of a fourth with the first, and nor will the fifth note form the concord of a fifth; and notes that are placed like this are unmelodic, as has been said. (62.34–63.5)[5]

Like a good number of others in this collection, the proposition proved is negative; it identifies a sequence that is not melodic. This reflects a general feature of Aristoxenus' strategy. Having established, in his preliminary arguments, that the only different-sized incomposite intervals there can be in any one genus are those inside any given tetrachord, along with the tone of disjunction, he is setting off to examine the melodic credentials of every conceivable way in which two of them could be placed side by side. A significant proportion of these sequences turn out to be melodically improper, 'unmelodic'. The argument for his negative proposition, again characteristically, takes the form of a *reductio ad absurdum* (or perhaps more properly, a *reductio ad immodulatum*): x cannot be placed next to y; for suppose that it is; it will turn out that an agreed axiom (of harmonics, not of logic) has been breached. Such arguments do not fall into the syllogistic pattern privileged in Aristotle's *Analytics*, but they were widely used and regarded as probative by philosophers, orators, mathematicians and others. There is no mystery about Aristoxenus' reliance on them.

The axiom or principle that will allegedly have been breached is **L**. It is easy to see that a sequence of two complete *pykna* is incompatible with it. Any *pyknon* occupies, by definition, less than half the span of a perfect fourth (50.15–19), and two successive *pykna* will contain four intervals and five notes. Hence the fifth note in order will be less than a perfect fourth away from the first, and neither of the conditions stated in **L** can be satisfied. A sequence in which a *pyknon* is followed by just one 'part' (that is, one interval) of another, by contrast, will necessarily conflict with **L** only if another assumption is granted. There will be only four notes

[5] This translation is in some respects rather loose. A more ingenious translator than I might find ways of transposing the structure and the pithiness of the Greek more directly into English without loss of clarity, but there would be no dispute between us about its sense.

in this sequence; and if the next note after these forms the concord of a fifth with the first, **L** will not have been broken. But the resulting sequence cannot exist in a system constructed out of tetrachords. No matter what sizes the intervals in the *pyknon* and the part-*pyknon* may be, neither the sequence I have posited nor any formed by repetitions of it can contain groups of three successive intervals jointly spanning a perfect fourth. I noted above that Aristoxenus takes it for granted that the tetrachord is the basic unit of melodic structure, and that all extended melodic sequences are concatenations of tetrachords. His theorems repeatedly presuppose that assumption, but nowhere explicitly state it or allude to it, and it does not figure among the principles listed in either Book I or Book II. It reflects, apparently, a fact so fundamental in musical experience that the theorist does not even notice that it is an assumption, and that he is relying on it.

This assumption has a bearing on an apparent oddity in Aristoxenus' first theorem, one which in fact recurs throughout the book. He tells us several times (for instance at 65.19–20) that his concern is with the ways in which incomposite intervals can be placed in sequence. But here the basic melodic 'step' to which others are hypothetically appended is the *pyknon*, and since the *pyknon* always contains two intervals it is not in any obvious sense incomposite. Aristoxenus nevertheless treats it, almost invariably, as an indissoluble unit. He specifies, for instance, the relations in which it can stand to the ditone and the tone, but nowhere discusses the sequences that can follow just one interval of the sorts into which a *pyknon* can be divided, a quarter-tone, for example, considered simply as such. Even when he reviews, at 71.23–72.11, the progressions that can begin from the middle note of a *pyknon* (and which therefore begin with one of these small intervals), he is assuming that such elements as the quarter-tones of enharmonic and their counterparts in chromatic always hunt in couples. At least once (66.29–34) he makes a remark which implies quite directly that the *pyknon* is an incomposite interval, though of course he knows it is not.

Part of what lies behind this approach is, once again, the assumption that melodic structure is based on the tetrachord. But it involves more than that. It will be clear to any reader of Book III that Aristoxenus is also assuming that every tetrachord is either enharmonic, chromatic or diatonic. This has of course been asserted in Books I and II, but no attempt is made to prove it, and it is not deliberately laid down as a 'principle'. It is simply a fact that is patent to everyone, and calls for no special treatment or argument. The first two books also record that enharmonic and chromatic tetrachords are distinguished from those in diatonic by containing a *pyknon*; this too is simply a fact of observation for which no argument is offered. It follows that in two of the three genera the two small intervals invariably hang together. They form, in effect, a single though composite unit, and it is only in diatonic that it becomes necessary to consider the three intervals of the tetrachord, one by one, as starting points for melodic sequences. In the light of these reflections it becomes puzzling, however, that in his first theorem Aristoxenus is prepared to envisage (though he also rejects) a situation in which the opening move of a sequence is followed by a step of just one part of a *pyknon*. According to the assumptions I have outlined, such a step could never occur independently, and need not even be considered. This is a small fore-taste of larger problems we shall face later.

Our discussion of Aristoxenus' innocent-looking opening theorem has brought out at least one important fact. On the face of it, he is demonstrating rules of melodic sequence axiomatically, on the basis of explicitly enunciated principles (in this and many other cases, **L**), as a free-standing exercise in scientific deduction. The fact is that this impression is false. The demonstrations, of which in this respect the first is typical, depend heavily on unstated assumptions, for which the student has been prepared in earlier parts of the work, but which have not been explicitly integrated into the axiomatic framework of Book III. The impression of logical autonomy which it initially conveys is apparently an illusion; by the standards of Euclidean geometry, for example, it seems to be a resounding failure.

But perhaps these are the wrong standards to apply. We have noted Aristoxenus' insistence that students of harmonics must be equipped with well-trained ears, and with wide and considered experience of melodies; and he himself contrasts the harmonic scientist's dependence on perceptual acuity with its irrelevance to the work of a geometer (33.10–26). To appreciate the force of demonstrations in harmonics, one must come to the subject already primed, through personal experience, with certain primitive insights into what is a melody and what is not. The tetrachordal structure of all melodies, and their exhaustive classification into the diatonic, chromatic and enharmonic genera, are apparently among the data derived directly from experience that a student must bring with him to the higher regions of the science. One must also be good at reasoning; but if that is all that is needed in geometry, it is insufficient in harmonics.

The theorem we have been considering calls for one further comment, which is simply that its reasoning is exceedingly compressed. We have seen that the argument – given certain additional assumptions – is essentially sound. But Aristoxenus has not taken us through all the steps that would be needed to secure a logically water-tight case. Several other theorems (for instance at 65.3–7 and 66.25–30) are presented in a similarly abbreviated form, whereas others are argued out in minute detail. Aristoxenus' manner of exposition is in this respect inconsistent, again unlike that of the best mathematicians. He is perfectly prepared, on occasion, to leave his readers to fill in the logical gaps.

I need to extract only one point from my second example. We have considered a negative proposition; here is a positive one, the argument in whose favour is even more abbreviated than the first.

> A semitone can be placed melodically on either side of a group of two or three tones. For either the fourth notes in order will form the concord of a fourth, or the fifth notes will form the concord of a fifth. (66.22–5)

Once again the underlying principle is **L**, and again Aristoxenus' readers are left to work out essential details for themselves. In fact what Aristoxenus says is true; the sequences in question will not fall foul of **L**.

The point to be noted is that this is all that the argument can show. The fact that these progressions are consistent with Aristoxenus' fundamental axiom is treated as sufficient evidence that they are melodically acceptable. This is entirely characteristic of the positive propositions of Book III. Despite the fact that when stating principle **L** in Book II, he comments explicitly that conformity to it is not sufficient to secure melodic legitimacy (54.11–19), in Book III no proofs of acceptability beyond conformity to **L** are offered at any stage. This is no doubt problematic, but it is unsurprising. Aristoxenus can state principles to which melodic sequences must conform, and he can adopt, explicitly or otherwise, various other assumptions about the form they must take. But his procedure provides no purchase to the claim that any sequence of a certain sort *must* be melodically acceptable. All he can do is to show that it is consistent with some principle or principles. We can apparently formulate on Aristoxenus' behalf a rather generous meta-rule: if a sequence is not in breach of an identified principle, then it is melodic.[6]

HOW THE COLLECTION OF THEOREMS IS ARRANGED

To place these theorems and those we shall study later in the settings to which they belong, the next task is to consider how the sequence of theorems

[6] I call it generous because he does not assert, and does nothing to show, that he has identified *all* the relevant principles. In most of the theorems at issue, in any case, he only refers explicitly to a sequence's conformity with one of them.

is divided into distinct groups, and to map the relations between them. With a handful of real or apparent exceptions, the theorems fall into five such groups.

The first (62.34–65.24) deals with successions of equal incomposite intervals, in each case demonstrating the melodic acceptability or unacceptability of the sequence in question. It begins with the first theorem we discussed: (i) neither a part nor the whole of a *pyknon* can be placed next to another *pyknon*. Then, after a pair of theorems of a different sort, which we shall come back to shortly, it is shown (iv) that a ditone cannot stand next to a ditone.[7] The remaining demonstrations show (v) that two tones cannot be placed successively in enharmonic or chromatic; (vi) that in diatonic there can be sequences of two or three tones, but no more; and (vii) that a sequence of two semitones cannot occur in diatonic, though it can in chromatic. The task of the two theorems inserted immediately after the first (63.5–33) is to identify the relations in which the bounding notes of (ii) the ditone and (iii) the (disjunctive) tone stand to the *pyknon*. The arguments supporting them pose considerable problems which we shall examine later, but there is no mystery about why they stand where they do, near the beginning of the series. Their conclusions will play an essential role in Aristoxenus' argumentative strategy, and are drawn on repeatedly in later proofs. The first of them, for instance, is the whole basis of the proof about a succession of ditones, proposition (iv).

The five propositions in the second group (65.25–66.25) are concerned with successions of unequal intervals. (viii) A *pyknon* can be placed both above a ditone and below. (ix) A tone can be placed next to a ditone only above it; and (x) it can be placed next to a *pyknon* only below it. (xi) In diatonic there cannot be a semitone on both sides of a tone; but (xii) there can be a semitone on either side of a sequence of two tones or of three. The survey is more exhaustive than it might appear, given that the *pyknon* is being considered as an indissoluble unit, and that when Aristoxenus refers to the ditone, this is shorthand for 'whatever interval, over and above the *pyknon*, completes an enharmonic or chromatic tetrachord' (see n. 7). Semitones can occur in chromatic as well as in diatonic contexts; the fact that they are not considered independently in chromatic reflects their regular role, in that environment, as parts of a *pyknon*. Finally, the thesis that in diatonic a

[7] In most cases where he is not referring specifically to diatonic sequences, Aristoxenus states his propositions in language appropriate to the enharmonic, with its ditone and its two-quarter-tone *pyknon*. But his intention is more general, and in all cases his reasoning will apply equally to any chromatic sequence, regardless of the sizes of its *pyknon* and of the remaining interval in its tetrachord. See especially 68.2–12. This is a point to which I shall return (see pp. 210–13 below).

semitone can stand either above or below a tone, provided that there is no semitone on the tone's other side, is not formally stated; but it is plainly implied by the conjunction of propositions (xi) and (xii).

Aristoxenus' third group of theorems (66.28–68.12) identifies the number and nature of the 'routes' (*hodoi*) that can be taken, in either direction, from a starting point in each of the incomposite intervals. Their reasoning is based squarely on conclusions already reached in propositions (i)–(xii); from one point of view they merely reorganise those conclusions into a different pattern. They are as follows: (xiii) From a ditone there are two routes upwards, to a *pyknon* and to a tone, and one downwards, to a *pyknon*. (xiv) From a *pyknon* there are two routes downwards, to a ditone and to a tone, and one upwards, to a ditone. (xv) From the (disjunctive) tone there is one route in each direction, downwards to a ditone and upwards to a *pyknon*. It is at this point that Aristoxenus adds that the same proposition holds of chromatic sequences, if instead of mentioning the ditone we refer to the interval, whatever its size may be, that lies between *mesē* and *lichanos* in the variety or 'shade' of chromatic that is in play. He adds also that a corresponding proposition will hold of the diatonic. Here too there is one route in each direction from 'the tone common to the genera', that is, the tone of disjunction, downwards to the interval (regardless of its size) that lies between *mesē* and *lichanos* in the relevant shade of the diatonic, and upwards to the interval between *paramesē* and *tritē* (the two lowest notes of the next tetrachord above). We should notice that he can only specify these intervals clearly by reference to the notes that bound them; they cannot be identified by their sizes. This raises an important issue, and we shall come back to it.

After a discursive digression (68.13–69.28), prompted by some people's 'confusion' about aspects of proposition (xv), Aristoxenus moves to his fourth group of theorems (69.29–72.11). The first two are preliminary, preparing the ground; they argue (xvi) that in chromatic and enharmonic every note 'participates' (*metechei*) in a *pyknon*, that is, that at least one of the two intervals immediately adjacent to any such note is an element in a *pyknon*; and (xvii) that there are, in this sense, just three places in a *pyknon* where a note can occur (at its upper or its lower boundary, or between its two constituent intervals). Armed with these points (of which the former raises difficulties which we shall postpone), Aristoxenus proceeds, once again, to examine melodic 'routes', and again his arguments depend wholly on conclusions already established. But this time he specifies each of these routes as beginning, not from some interval but from a note, identified by its position in the *pyknon*.

Since there are only three positions in the *pyknon*, there are only three propositions in this group, after the two preliminaries. (xviii) From the lowest note of a *pyknon* there are two routes in each direction, downwards to a tone or to a ditone and upwards to a *pyknon* or to a tone.[8] (xix) From the highest note there is one route in each direction. Aristoxenus does not explicitly specify them, but his argument leaves no room for doubt about what he has in mind; the route downwards will be to the upper interval of the *pyknon*, whatever its size may be, and the route upwards will be to the ditone, or to one of its chromatic counterparts. (xx) From the middle note there is one route in each direction. This seems obvious; these routes will lead to the upper and lower intervals of the *pyknon*. Aristoxenus reaches this result, but through an extraordinarily convoluted piece of reasoning which shares its most obvious peculiarities with several others that I have already noted as problematic. They will be studied as an ensemble in the next section of this chapter.

It is perhaps misleading to treat the last three theorems in the series (72.12–74.9) as a fifth 'group'. The second and third of them plainly belong together, but the first seems unrelated to them, and has obvious affinities with those in the group that precedes them. But parcelling it up with them and separating off the second and third would conceal one important point. The three final propositions do have something in common. It is that they seem to introduce new issues, and to suggest a movement towards an agenda extending beyond the simple progressions examined in propositions (i)–(xx). Summarily, they are these. (xxi) Two notes that differ in their positions in relation to the *pyknon* cannot melodically be placed on the same pitch. (xxii) A diatonic sequence can include incomposite intervals of two, three or four different sizes. (xxiii) An enharmonic or a chromatic sequence can include intervals of either three or four different sizes; the formal argument supporting this thesis is followed by another short digression, disposing of a problem raised about it by Aristoxenus' uncomprehending audience. The reasoning of the last two theorems is in fact easy enough to follow. They are puzzling only in the sense that we cannot immediately tell why Aristoxenus enunciates them, and why he does so here. That is true also of the first; but in this case the argument itself is tortuous and peculiar, and in the context of the earlier theorems the proposition itself seems at the very least strange. We have so far identified nothing in those theorems which could make sense of the notion that two distinct notes, whatever

[8] The contention that a tone may lie immediately above the lowest note of a *pyknon* seems bizarre, to put it mildly. This too is a difficulty we shall address below.

their relations to the *pyknon*, can coincide on a single pitch. We shall find, however, that an interpretation treating proposition (xxi) as a preface to a part of the *El. harm.* which no longer survives will also help to resolve some of the most taxing difficulties presented by its own reasoning and by that of several earlier theorems. In this light it may be construed as offering some of the best clues we have about the way in which the theorems of Book III were related to their missing sequel.

This concludes our survey of the way in which the twenty-three theorems that form the core of the book are organised. Only seventeen further lines of text survive. They introduce the notion of 'arrangement' or 'form', which refers to the order in which the constituent intervals of a composite structure are placed; and they proceed, as I have said, to the beginning of a demonstration that there are just three melodically acceptable arrangements of the intervals in a perfect fourth. We shall return to this fragmentary passage in the next section, during our discussion of the principal problems that Aristoxenus' theorems pose.

THREE MAJOR PROBLEMS

As I said at the start, there are problems of three main sorts: about the reasoning of some of the theorems, about the uses they make of quantification, and about the role of the set of theorems in the project of the *El. harm.* as a whole. They clearly raise difficulties of different sorts, but I shall try to show that they are connected, and that the best solution to any of them is one that brings out the relation between them.

The difficulties about the logic of Aristoxenus' argumentation hang together in a more straightforward way, amounting, in fact, to a single problem. In its first appearance it is scarcely detectable. Proposition (ii) (63.6–20) states that the lower of the notes bounding the ditone is the highest note of a *pyknon*, and that the higher of the notes bounding the ditone is the lowest note of a *pyknon*. The proof is based on the fact that when tetrachords are linked in conjunction, ditones and *pykna* must always occur alternately; and that is true. It is also true, on the other hand, that when tetrachords are linked in disjunction, separated by the interval of a tone, the upper bounding note of the ditone is no longer the lowest note of a *pyknon*; it is the lower boundary of the disjunctive tone.

At this stage we might suppose that Aristoxenus is guilty of nothing worse than careless formulation, and that he means only that the ditone's upper boundary *can* be the lowest note of a *pyknon*, not that it invariably is so. But the next and quite similar proposition should raise the suspicion that something more complex is afoot. Proposition (iii) (63.20–33) asserts that each of the notes bounding the disjunctive tone is the lowest note of a *pyknon*. Yet while the note immediately above the disjunctive tone is indeed the lowest note of the next tetrachord and so, in enharmonic and chromatic, of a *pyknon*, the note immediately below it cannot possibly be the boundary of a *pyknon*, since it must lie between the tone and (in enharmonic) the ditone. Aristoxenus' contention entails that it is simultaneously the lower boundary of a *pyknon* and the lower boundary of the disjunctive tone, which seems absurd. The grounds he offers for this extraordinary thesis are that 'the tone, in a disjunction, is placed between tetrachords of such a kind that their bounding notes are the lowest notes of a *pyknon*, and it is by these that the tone is bounded' (63.22–6). Now the two bounding notes of an enharmonic tetrachord of the relevant sort will indeed both be the lowest notes of *pykna* when the tetrachords are put together in conjunction; but the higher of them can never abut on a *pyknon* in the case envisaged here, where the tone is introduced to disjoin the tetrachords. This case is more extreme than that of proposition (ii), which appears to be tenable in the sense that it is sometimes true; one part of proposition (iii), on the face of it, is *never* true.

There would be little point in dwelling on these strange contentions if they were isolated oddities, but they are not. They underlie the reasoning of several subsequent theorems. The first case in point is in fact the next in the series, proposition (iv), the proposition that two ditones cannot be placed next to one another (63.33–64.10). The proposition itself is sound, and could be proved straightforwardly by reference to **L**; a sequence of two ditones amounts to more than a perfect fifth, making it impossible for it to satisfy either of the conditions that **L** specifies. But this is not how Aristoxenus goes about demonstrating it. His argument is this.

> Let them be so placed. Then a *pyknon* will follow the higher ditone on its lower side, for we saw that the note bounding the ditone on its lower side is the highest note of a *pyknon*; and a *pyknon* will follow the lower ditone on its upper side, for we saw that the note bounding the ditone on its upper side is the lowest note of a *pyknon*. In that case two *pykna* will be placed in succession; and since that is unmelodic [by proposition (i)] it will also be unmelodic to place two ditones in succession. (64.1–10)

The whole basis of this argument, apart from the final allusion to proposition (i), is patently proposition (ii); and it is now clear that the charitable interpretation that might be offered for that proposition, taken in isolation, is untenable. The present argument depends on the thesis that the upper boundary of the ditone is *always* the lowest note of a *pyknon*, not that it can sometimes be so. It also shares the peculiarity of proposition (iii), in that it seems to envisage two entirely different structures as coexisting simultaneously in the same place. In the (admittedly hypothetical and rejected) situation that it conjures up, the note lying between two ditones lies, simultaneously, between two *pykna*, as though two quite distinct sequences were superimposed on one another. We can now see, in fact, that proposition (ii) must also imply a scenario of the same sort, in which, whenever tetrachords are linked by a disjunction, the upper note of the ditone is at the same time the lowest note of a *pyknon* and the lowest note of a disjunctive tone.

Arguments of a very similar form to that of proposition (iv), based on the same premises (that is, on propositions (ii) and (iii), and also on the less problematic proposition (i)) and entailing peculiarities of just the same sort reappear in propositions (ix) (65.30–66.8), (x) (66.9–17), (xvi) (69.29–70.14) and (xx) (71.23–72.11). I shall not examine them here, since no significantly new points would emerge. The contagion spreads, however, to one further proposition, proposition (xxi) (72.12–28), and this deserves special attention.

It states that two notes which differ in their positions in the *pyknon* cannot melodically be placed on the same pitch. The argument supporting this thesis is straightforward, depending wholly and uncontentiously on proposition (i). It is the proposition itself that is puzzling. As I said earlier, in my review of the whole sequence of theorems, we have been given no clue about the circumstances in which two different notes, whatever their relations to the *pyknon*, could possibly coincide on one pitch. At the same time, Aristoxenus' formulation is helpful, to the extent that it goes some way towards confirming that our interpretations of the other propositions in this group are not too far from the mark. He is indeed prepared to envisage situations in which two different melodic structures are somehow superimposed. The fact that he is concerned to identify one kind of case in which such a superimposition is melodically unacceptable strongly suggests that there are others in which it is not. They are, presumably, cases in which the notes whose pitches coincide are related to the *pyknon* in identical ways. This, as we shall eventually discover, provides an important clue to the way in which the conundrums presented by this group of theorems can be resolved.

There is, however, another step to be taken before we can approach that grand resolution. All the propositions in the first three groups, apart from propositions (ii) and (iii), are concerned with sequences of successive intervals (if we may for the present continue to treat the *pyknon* as an 'interval' for these purposes, as Aristoxenus does). These intervals are identified as ditones, tones, semitones or *pykna*, and the first three of these designations, taken at face value, are plainly quantitative. So too, by implication, is the last, given that the *pyknon* is repeatedly treated as the complement to the ditone in an enharmonic tetrachord, and must therefore span a semitone. On a casual reading, Aristoxenus seems to be offering proofs of the melodic acceptability or impropriety of sequences of intervals of certain sizes, simply as such, largely on the basis of **L** and some simple arithmetic, as if he had never uttered his strictures against people who try to define melodic sequences by the sizes of the intervals they contain.

But this cannot be so. For one thing, if it were, Aristoxenus' set of theorems would be radically incomplete, since he says nothing about intervals of the various other sizes involved in his earlier analyses of tetrachords in the three genera. We would be left in the dark, for instance, about the sequences in which the intervals of one third of a tone and eleven sixths of a tone can and cannot appear (these intervals are involved in the 'soft' chromatic). The collection of theorems, in fact, would be in principle uncompletable, given that the note called *lichanos* can be placed at any of indefinitely many points in its range (26.13–18), and there is therefore no limit on the number of different sizes available for the interval between it and *mesē*. Aristoxenus puts the point clearly towards the end of Book III. 'In respect of the sizes of the intervals and the pitches of the notes, the facts about *melos* seem somehow to be indeterminate' (69.6–8).

In fact, as I said earlier, Aristoxenus' expressions 'ditone' and '*pyknon*' do not serve here to identify intervals by their sizes. 'Ditone' stands for any interval which is the complement of a tetrachord after a *pyknon*, and *pykna*, correspondingly, may be of any size that the definition of a *pyknon* permits (48.27–31, cf. 50.15–19).[9] No doubt it would have been clearer if he had explained this point in set terms at an early stage, instead of reserving such explanation as he offers for a special context at 68.1–13 (see n. 7 above).

[9] The status of 'tone' and 'semitone' in the theorems seems rather different. Aristoxenus' exemplary tone is the tone of disjunction, and though he has not said quite unambiguously that it is invariable in size (61.5–7 says only that it is unchanged in changes of genus), he seems to treat it as such. When he writes of sequences of several tones (e.g. proposition (vi)), he evidently has in mind intervals all of which are the same size as the disjunctive tone. One might argue that since the tone is by definition the difference between a fifth and a fourth (45.34–46.1), and since Aristoxenus seems to allow that those concords may vary in size 'by a hair's breadth' (55.3–6), the tone, too, may be minutely variable; but this is a quibble. The semitone plays a relatively small part in the theorems, always in connection either with the form of the diatonic whose tetrachord contains a semitone and two tones, or – very much in passing – with the 'tonic' chromatic (two semitones and an interval of a tone and a half). In no case, I think, can the term be conceived, like 'ditone', as a shorthand reference to intervals of some wider class. This is especially clear in proposition (xii), 66.18–21.

His intention to formulate propositions which apply to enharmonic and chromatic sequences in all their forms is already evident, however, in the laborious argument he offers for proposition (v) (64.11–65.2); and it is a fact that the reasoning supporting all the theorems dealing with ditones and *pykna* will apply, unchanged, to the intervals of any 'shade' of these genera.

Proposition (iv) is particularly revealing in this context. This is the theorem which rules out a sequence of two ditones, the peculiar character of whose reasoning we noted above (pp. 209–10). No matter how that argument is to be interpreted, it seems strange that Aristoxenus should have entangled himself and the reader in its serpentine coils when the sequence conflicts so obviously with **L**. But it does so only if 'ditone' means precisely what it says. If the argument is to apply equally to any other interval that can complement the *pyknon* in a tetrachord, an appeal to **L** will not by itself be adequate. In the tonic chromatic, whose *pyknon* is composed of two semitones, the remaining interval in the tetrachord is a tone and a half (a 'trihemitone'). If two such intervals are placed in sequence they do not, like two ditones, exceed the compass of a fifth, making it flatly impossible for **L** to be respected. To make the first and fifth notes of the sequence concordant at the interval of a fifth would require the two trihemitones to be followed by two quarter-tones, giving a pattern that is in other respects outrageous; but it does conform to **L**. Straightforward reliance on **L**, then, would not have given Aristoxenus a compendious argument capable of being applied to sequences of all the intervals for which, as a class, the term 'ditone' does duty.

This does something, at any rate, to explain why Aristoxenus does not take the superficially obvious route to a proof that a sequence of two ditones is unmelodic. But it gives no help in explaining why his argument follows the particular and very peculiar course that it does, or how the premise on which it is founded, proposition (ii), is to be understood. To move a little closer to a solution to this and other significant problems, we must shift from the negative to the positive mode. If the intervals with which the theorems are concerned are not to be defined by their sizes, how are they to be correctly identified? To put the question more broadly, what are Aristoxenus' theorems really about, if they are not about sequences of steps covering specific distances in melodic space?

The stretch of text that does most to help answer this question runs from 68.2 to 69.28. It is in two parts, the first (68.2–12) being an immediate appendix to proposition (xv) (67.25–68.1), the second a long explanatory digression addressing objections that some people have raised against that proposition (68.13–69.28). Proposition (xv) states that there is one route in each direction from the tone, downwards to the ditone and upwards to the *pyknon*. The 'immediate appendix', which we touched on above, explains how the scope of this thesis can be extended to cover chromatic and diatonic sequences as well as those in enharmonic. All we have to do is to substitute for 'ditone' a reference to the interval, whatever its size may be, that lies between *mesē* and *lichanos*; in the case of the chromatic the reference to the *pyknon* can stand, though the size of this structure will be different and variable, while in diatonic (which has no *pyknon*) we should refer instead to 'the interval between *paramesē* and *tritē*'.

The crucial implication of these remarks seems perfectly clear. If we are to rid ourselves of quantitative conceptions, and to identify intervals in such a way that Aristoxenus' rules of sequence will apply to them in their full generality, they must be identified by reference to the notes which form their boundaries.[10] The overall plan of Aristoxenus' series of theorems itself suggests that there is something to be gained by revising our conception of the starting point of a sequence, treating it no longer as an interval but as a note. As we have seen, after the three groups of theorems which adopt the former approach, the fourth group shifts to the latter. Its propositions restate consequences already reached, in different and – as will appear – more illuminating terms, by taking the factor which makes some sequences melodically possible and others impossible to be an aspect of the note from which they begin.

This way of considering the matter fits well with what we learned earlier about notes. They are not points of pitch, mere inert abstractions, but *dynameis*, elements of melody that have the 'power' to impel the melodic traveller, the voice, along certain pathways and to obstruct others; or to put it another way, they have the 'potential' for further melodic development in a determinate number of directions. In the second part of the present passage, the long digression which I mentioned above (and shall not explore in full), Aristoxenus also emphasises the way in which regularities in melodic sequences can be brought within the ambit of a science if they are specified by reference to the *dynameis* of the notes they contain, that is, for immediate purposes, by identifying those notes by name; whereas an attempt to deal with them scientifically while they are conceived as sequences of intervallic magnitudes must fail, since these, unlike the *dynameis*, are 'indeterminate', *apeira* (69.2–12). This does not mean only that they are indefinitely many, but also that for any scientifically identifiable regularity in melodic

10 Compare the definition of 'incomposite interval' by reference to its bounding notes rather than its size, and the subsequent discussion (60.10–61.4).

sequences, there are no particular sizes of intervals to which it applies. Its range of application cannot be pinned down in quantitative terms.

In the light of these points it becomes a matter of some interest that all the theorems whose arguments pose serious interpretative problems involve direct allusions to notes, not merely to intervals like most of the propositions in the early parts of the series. These notes are not identified by name, but by reference to their positions in the *pyknon*, and I shall return to that point shortly. But the thesis that there is something about the character of each note that determines which courses the singing voice can and cannot next pursue suggests a way in which the puzzles surrounding those propositions can – at least partially – be resolved. Take, for instance, the thesis of the third theorem (63.20–33), that 'each of the notes bounding the tone is the lowest note of a *pyknon*'. This seemed paradoxical, because its lower bounding note is, *ex hypothesi*, the lower boundary of the tone, and the tone cannot form part of a *pyknon*. Aristoxenus argues for the thesis on the grounds that the notes in question are the upper and lower boundaries of tetrachords of a sort to which he has repeatedly referred before (those lying between fixed notes), and that the bounding notes of these tetrachords 'are the lowest notes of *pykna*'. Yet the upper boundary of a tetrachord is no longer the lowest note of a *pyknon* when it is succeeded by a disjunction. Specifically, when *mesē*, the highest note of the tetrachord *mesōn*, is succeeded by the tetrachord *synēmmenōn*, it is the lowest note of a *pyknon*; when it is succeeded by the tone of disjunction, leading to *paramesē* and the tetrachord *diezeugmenōn*, it is not. But as we can now see, the point is that it is still the same note, *mesē*; its character and 'powers' have not changed. If, then, one aspect of its *dynamis* is that which characterises the lowest note of a *pyknon*, inferences may still continue to be drawn from that fact even when it is not made evident by the pattern of intervals currently in play.

All the other theorems in this problematic group can to some extent be elucidated in the same way. In previous publications, however, I have pressed this interpretation harder than I now think is warranted. It does not, after all, fully resolve the difficulties of these passages, and the part of the story that is missing is in some ways the most interesting. There are, in particular, two features of the passages that it does not satisfactorily explain.

The first is straightforward. In specifying the notes with which they are concerned, these theorems do not identify them by their names, which point directly to *dynameis*, but only by their relations to the *pyknon*. The same is true of the theorems in Aristoxenus' fourth group, and of the theorem immediately following them (proposition (xxi), which I assigned to the rather miscellaneous group at the end). No doubt a note's position in the *pyknon* constitutes part of its *dynamis*, and tells us a good deal about the sequences that can follow from it, as the fourth group of theorems shows. But it certainly does not tell us everything about it. *Parhypatē mesōn* and *tritē diezeugmenōn* are identical in their relations to the *pyknon*, but they are not the same note, and they differ in *dynamis*; some progressions that follow a simple scalar sequence from the latter do not do so from the former.

Secondly, the most puzzling feature of these theorems, at first sight, was their apparent implication that two different melodic sequences can be superimposed on one another, so that *mesē*, for instance, lies simultaneously both immediately below a tone and immediately below a *pyknon*. The interpretation I offered above, in terms of the *dynameis* of notes, eliminates the need to think of the two sequences as coexisting at the same time; it is rather that the note retains the whole *dynamis* that equips it for both sequences even when only one of them is in fact activated. This interpretation can be made to work, perhaps with some strain, for most of the theorems at issue; but it falls at the final fence, in proposition (xxi) (72.13–28). Here, it will be recalled, Aristoxenus argues that 'two notes which differ in respect of their positions in a *pyknon* cannot melodically be placed on the same pitch'; and there is no doubt at all that in this case at least, he is envisaging the actual, simultaneous cohabitation of two notes in the same place. His argument leaves no room for us to wriggle out of this uncomfortable conclusion. He is not talking about a single note whose *dynamis* allows it to participate in sequences of two sorts, but about two distinct notes inhabiting the same location in pitch.

TONOI, MODULATION, AND THE MISSING CONTINUATION TO BOOK III

Under what circumstances can such a coincidence of notes occur? Plainly Aristoxenus is not referring to what we call 'chords' (apart from anything else, the two notes of which he speaks have the same pitch), or about the relations between two semi-independent strands in a composition, those of melody and accompaniment, for example. His topic is rigorously restricted to linear melody. Only one answer to the question is possible. Two notes can coincide on one pitch in the context of certain forms of modulation.

Consider what we would call a modulation of key, for instance the modulation from C major to G major. Given a suitable harmonic context, a modulation of that sort might occur while a melody passed through the sequence c-e-g-b-d′, and we might say that the modulation takes place at the note g. It forms a kind of junction between the two keys. Those keys lie in different pitch-ranges, but are identically structured, and we have a terminology by which we can specify the 'function' of each note in the structure to which it belongs. Here g is both the 'dominant' of C major and the 'tonic' of G major.

In the Aristoxenian scheme, the note-names are comparable (though by no means precisely parallel) to functional designations such as 'tonic' and 'dominant'. As in modern music, a complete system of notes characterised in this way can be placed at any of several different levels of pitch; and it is possible for a melody which sets out in a sequence belonging to the system when it is placed at one pitch-level to modulate into a sequence which occurs in the system only when it is transposed through some interval, upwards or downwards. Just as in our example, the note g is both the dominant of C major and the tonic of G major, so, in the Greek context, a pitch which is presented to our ears, by the sequence leading up to it, as (for instance) *hypatē mesōn*, might simultaneously form the beginning of a sequence in which its function or *dynamis* is that of *mesē*. A modulation has taken place, and at the junction between the two systems, items which by Aristoxenus' criteria are two different notes coincide on the same pitch.

In Greek musical theory, the systems which correspond approximately to our 'keys' are called *tonoi*, and modulation of the sort I have sketched is called 'modulation of *tonos*'.[11] *Tonoi* and modulations are announced, in the programme of Book II (37.8–38.17), as topics that Aristoxenus will address towards the end of his work. The remarks he makes at 37.8 ff. about his predecessors' treatment of the *tonoi* show that there was no agreement about the distances at which these 'keys' should be separated from one another, and no accepted principle on the basis of which agreement might be reached (see p. 56 above). The intervals through which modulations could take place, it appears, were fixed only by 'custom and practice', and the experts were far from unanimous about what they were, or should be.

Aristoxenus' discussions of *tonoi* and modulation have not survived, but later writers give us the gist of his conclusions. We know that he postulated thirteen *tonoi*, spaced uniformly a semitone apart, so that the highest is an octave above the lowest; and we know that he formulated various rules governing the ways in which a melody could modulate between them.[12] Our sources do not reveal, however, the grounds on which he justified the

[11] There are other sorts of modulation, for instance from one genus to another, but these will not concern us.

[12] The principal sources are Cleonides, Ptolemy and Aristides Quintilianus. Their evidence will be discussed below.

conclusions he reached. What I shall suggest is that they are based squarely on the theorems of Book III. Most of those theorems are concerned, at least explicitly, with non-modulating sequences only; but it is clear from proposition (xxi) – and in the light of that proposition, from the others in the argumentatively problematic group – that he is prepared to apply those theorems' conclusions universally, to cover modulating sequences as well. The theorems are not just a self-contained episode in Aristoxenus' exposition, rounding off his examination of simple melodic sequences in a properly demonstrative manner; they are the foundation on which he will build his account of more complex structures, relations and patterns of melodic movement.

The 'problematic' theorems alert us to the likelihood that Aristoxenus has issues about *tonoi* and modulation between them already firmly in his mind. The feature of Book III that allows us to make some headway towards a more detailed understanding of its connection with the missing material is the importance it assigns to the positions of notes in the *pyknon*.[13] Later authors often record a classification of notes on this basis, generally referring to them as *barypyknoi, mesopyknoi, oxypyknoi* and *apyknoi*, that is, respectively, as notes lying at the bottom, in the middle or at the top of a *pyknon*, or being such that they have no place in a *pyknon* at all.[14] But they do little to explain why the classification is worth making, or what might follow from the fact that a note is *oxypyknos*, for example.[15] The distinctions, as they appear in those later texts, carry with them a strong odour of tedious and essentially pointless scholastic exercises. But Aristoxenus does not waste his time on irrelevances, and must have thought that his distinctions could be put to work for some useful purpose. He even goes out of his way to insist, in the teeth of superficial appearances, that *every* note in enharmonic and chromatic has a place in the *pyknon*, coping with the apparent exceptions with the help of his deviously argued theorem about the bounding notes of the tone (proposition (iii)).[16]

One of the later Aristoxenian writers gives us the crucial clue. The *Harmonic Introduction* usually attributed to an otherwise unknown Cleonides,

[13] The notion of 'position in the *pyknon*' is fundamental to all the theorems in Aristoxenus' fourth group, and to all those I have been calling 'problematic'.

[14] See e.g. Cleonides, 186.1–187.2, 195.8–198.13, Arist. Quint. 9.13–24; cf. Bacchius 302.7–13, 307.15–16.

[15] One theorist who may have put the classification to constructive use is a certain Archestratus; see Porph. *In Ptol. Harm.* 26.26–27.16. But the passage (discussed in Düring 1934: 145–9) is exceedingly obscure. Nothing more is known about Archestratus' theory, and little about Archestratus himself.

[16] Not all the later theorists agree with this thesis, as their use of the category *apyknos* shows. Aristoxenus says nothing explicitly about the way in which the statements referring to the *pyknon* in propositions (xvi)–(xxi) could be recast so as to apply to the diatonic, in which no *pyknon* occurs. But plainly they could; the passage at 68.6–12, which I touched on above (p. 206), shows how it could be done.

and written, perhaps, around AD 100, is probably the most faithful of all such works to the views of Aristoxenus himself, though it is little more than a summary of 'doctrines', and by the standards of Aristoxenus' own quarrelsome and colourfully digressive presentation it is as dry as dust.[17] In his twelfth chapter he records, with little embroidery, that there are thirteen *tonoi*, attributing this thesis explicitly to Aristoxenus; he lists their names and specifies that they are spaced a semitone apart (203.4–204.15 Jan). In the thirteenth chapter he turns to the subject of modulation, distinguishing four types, of which one is modulation of *tonos*. His remarks about it need to be quoted in full.

> There is modulation of *tonos* whenever a modulation takes place from Dorian into Phrygian, from Phrygian into Lydian or Hypermixolydian or Hypodorian, or in general from one of the thirteen *tonoi* into another. Modulations begin from the semitone and continue up to the octave; and some of them are made through concordant intervals, others through discordant intervals. Of these, those made through concordant intervals and through the interval of a tone are melodic. Of the remainder, the closer ones are less melodic or unmelodic, and those that are further away are more melodic; or rather, in those cases where there is more in common they are more melodic, and in those where there is less they are more unmelodic, for it is necessary in every modulation that there is something in common, either a note or an interval or a *systēma*. 'Having something in common' is understood through a similarity between notes. For whenever, in modulations, notes that are alike in respect of their positions in the *pyknon* coincide on the same pitch [lit. 'fall upon one another'], the modulation is melodic, and whenever unlike ones do so, it is unmelodic. (Cleonides 205.6–206.2)

Even if Cleonides' treatise was not at every point designed as a résumé of Aristoxenus' ideas, we could be confident that this passage is so, in view of its close relationship with the section on *tonoi* where Aristoxenus is directly named. For the most part it is plain sailing, and we need not pause to unscramble minor obscurities. The part which most obviously concerns us comes near the end.

When a modulation takes place, the two *tonoi* involved must have something in common, and the more they have in common the smoother or 'more melodic' the modulation is. They count as having something in common if, at the points where notes from the two *tonoi* coincide in pitch, the coinciding notes have the same position in the *pyknon* (within the sequences proper to their respective *tonoi*). This, fairly plainly, is an echo of Aristoxenus' proposition (xxi), that two notes which differ in their positions in the *pyknon* cannot melodically be placed on the same pitch. The

[17] For a careful study of the treatise see Solomon 1980a.

most important general point to be noted in connection with this fact is that proposition (xxi), with its direct bearing on modulating sequences, is demonstrated by an appeal to one of his theorems about simple, non-modulating sequences, in fact to the first. The rules of sequence that Aristoxenus has worked out can be applied directly, it appears, to cases well beyond those to which they initially referred.

To gain an impression of the consequences of the present rule, take two instances of the Greater Perfect System, in its enharmonic form, for example, and place them a perfect fourth apart. Over the span of an octave plus a tone they inhabit the same range of pitch; and within that compass there are eight points at which the pitch of a note in the higher *tonos* coincides with that of a note in the lower. Wherever they do, the two notes in each pair have the same relation to the *pyknon*. (See Figure 6. This will be completely true only if we adopt Aristoxenus' thesis that the lower boundary of the tone is the lowest note of a *pyknon*.)

This, then, as Cleonides says, is an eminently melodic modulation. Consider next the case where the two *tonoi* are placed very close together, only a semitone apart. Here they will occupy the same range over only a tone less than two octaves; and here again there will be notes in one *tonos* which share their pitches with notes in the other. But these coincidences are fewer (there are only four); and wherever they occur, the note in one system is the highest note of a *pyknon*, while in the other it is the lowest (see Figure 7). Modulation through a semitone, by these criteria, is thoroughly unmelodic.

One of Cleonides' remarks in this passage seems to mean that the most straightforwardly melodic modulations are those through concordant intervals and through the tone. Modulation through a fourth fits the Aristoxenian criteria admirably, as we have seen. Modulations through a fifth or through a tone offer less common ground between the two systems, but there is no conflict between the relevant notes' positions in the *pyknon*. Modulation through an octave is the extreme case, in which, over the range which they share, every note in one system coincides with a note in the other, and all the coinciding notes are identically related to the *pyknon*. This is hardly a modulation at all, as Ptolemy points out.[18] The privileged status assigned to these modulations explains, I think, why Aristoxenus claims that the *tonoi* must be placed a semitone apart. The point is that it is possible to arrive at a series of pitches separated by semitones, across the span

[18] Ptol. *Harm.* 58.21–59.29, especially 59.12 ff.

(i) GPS in higher *tonos*		**(ii) GPS in lower *tonos***
nētē hyperbolaiōn		
paranētē hyperbolaiōn		
tritē hyperbolaiōn		
nētē diezeugmenōn	----- ***barypyknos*** -----	*nētē hyperbolaiōn*
paranētē diezeugmenōn	----- ***oxypyknos*** -----	*paranētē hyperbolaiōn*
tritē diezeugmenōn	----- ***mesopyknos*** -----	*tritē hyperbolaiōn*
paramesē	----- ***barypyknos*** -----	*nētē diezeugmenōn*
mesē		
		paranētē diezeugmenōn
		tritē diezeugmenōn
		paramesē
lichanos mesōn		
parhypatē mesōn		
hypatē mesōn	----- ***barypyknos*** -----	*mesē*
lichanos hypatōn	----- ***oxypyknos*** -----	*lichanos mesōn*
parhypatē hypatōn	----- ***mesopyknos*** -----	*parhypatē mesōn*
hypatē hypatōn	----- ***barypyknos*** -----	*hypatē mesōn*
proslambanomenos		
		lichanos hypatōn
		parhypatē hypatōn
		hypatē hypatōn
		proslambanomenos

Figure 6 Modulation through a perfect fourth

(i) GPS in higher *tonos*		**(ii) GPS in lower *tonos***
nētē hyperbolaiōn		
		nētē hyperbolaiōn
paranētē hyperbolaiōn		
tritē hyperbolaiōn		
nētē diezeugmenōn	***barypyknos* ----- *oxypyknos***	*paranētē hyperbolaiōn*
		tritē hyperbolaiōn
		nētē diezeugmenōn
paranētē diezeugmenōn		
tritē diezeugmenōn		
paramesē	***barypyknos* ----- *oxypyknos***	*paranētē diezeugmenōn*
		tritē diezeugmenōn
		paramesē
mesē		
		mesē
lichanos mesōn		
parhypatē mesōn		
hypatē mesōn	***barypyknos* ----- *oxypyknos***	*lichanos mesōn*
		parhypatē mesōn
		hypatē mesōn
lichanos hypatōn		
parhypatē hypatōn		
hypatē hypatōn	***barypyknos* ----- *oxypyknos***	*lichanos hypatōn*
		parhypatē hypatōn
		hypatē hypatōn
proslambanomenos		
		proslambanomenos

Figure 7 Modulation through a semitone

of the octave, by moving from a given point through steps of a fourth and a fifth (the step of a tone need not be used); but no pitches closer together than that can be reached in this way. Nor can pitches separated by intervals which are not multiples of a semitone. Thus an arrangement of thirteen *tonoi* spaced by semitones gives the maximum number of *tonoi* that can be located in the octave in such a way that any of them can be reached from any other through some sequence of modulations of the privileged sort. In this way they are all coordinated into a single scheme.

All this presents an attractively coherent picture, but a potentially damaging objection must be addressed. I have argued that Aristoxenus intends the 'rules of sequence' articulated in Book III to be applicable across the board, whether the melody modulates or not. But I have not tried to substantiate this by examining those rules one by one in connection with the modulating sequences, and if I did, some of them would certainly break down. No doubt the theses about successions of *pykna* and ditones would survive unharmed, but if modulation through a tone is acceptable, as Cleonides says, the rules about the number of tones that can occur in succession will certainly not be maintained.

Two comments on this issue are in order. First, so far as I can see, definite anomalies will arise only when the rules are specified in terms of the sizes of intervals and nothing else, not when they concern the routes that can be taken from notes specified by their relations to the *pyknon*. If that is true, it reinforces the impression we gained earlier that the formulations dealing only with the sizes of intervals are inadequate if taken literally, and are designed to guide us to the second and more rigorous mode of presentation. Secondly, suppose it to be the case that certain modulations, acceptable by Cleonides' criteria, involve sequences of intervals which by the standards of Book III are unmelodic. Cleonides himself shows how this situation could be accommodated. In his exposition, 'melodic' and 'unmelodic' are not absolute predicates of modulations, such that either a modulation is fully melodic or it is to be rejected altogether. There are more and less melodic modulations. In that case it is at any rate conceivable that a modulation which played fast and loose with some rules of sequence might accord satisfactorily with others, and while being less melodic than a modulation through a fourth, for instance, would not put a composition that used it altogether beyond the bounds of theoretical respectability. There is evidently a grey area here, where a composer may be justified in trying some risky manoeuvres. He must recognise the implications of what he is doing, but can find scope for the exercise of his own musical judgement.

One further feature of Book III confirms its conceptual continuity with the issues about the *tonoi* that were evidently addressed in the sequel. It ends, as we have seen, with the beginning of a demonstration of the thesis that there are just three melodically acceptable ways of arranging the constituent intervals of a perfect fourth. Aristoxenus tells us himself, when commenting on the work of Eratocles, that when this has been established, along with comparable demonstrations of the arrangements of the fifth and of the ways in which the two structures can be melodically combined, the theorist will be in a position to determine the number and form of the melodic arrangements of the octave; but that if these preliminary steps are neglected, there will apparently be 'many times seven' ways in which the octave's intervals can be arranged (6.19–31).[19] There can be little doubt that the next part of the text contained these steps and that the reasoning was based on the conclusions already set out in the surviving theorems; and the result that will have been reached, as the later compilers confirm, is that there are seven arrangements of the intervals in an octave. Eratocles, in fact was right, though his procedure was inadequate.

It seems fairly clear that if they are considered simply as a self-contained group, the propositions about arrangements of intervals belong to the study of *systēmata*, which Aristoxenus announces in Book II as the fourth topic with which harmonics should deal (36.15–37.7). It is in this context that he makes one of his dismissive comments about earlier theorists' treatment of 'the seven octachords which they called *harmoniai*' (36.30–2). But when introducing the fifth topic, that of the *tonoi*, he also hints at a connection of some sort between them and the *systēmata* examined under the previous heading. 'The fifth part concerns the *tonoi* in which the *systēmata* are placed when they occur in melody' (37.8–10). The remark suggests that the *tonoi* somehow map out pitch-relations between different forms of *systēma*, and not merely between thirteen instances of the same type.

What connection, then, might there be between differences of *tonos* and differences in the 'arrangements', 'forms' or 'species' of the octave and the other concords? It is an exceedingly difficult issue. Elsewhere I have given a

[19] While there are only seven melodically acceptable arrangements of the intervals of an octave made up of four quarter-tones, two ditones and one tone, as in an enharmonic scale, the different arrangements of these magnitudes that are mathematically possible are in fact 15 times as many; there are 105 of them. (I am indebted to Dr Jonathan Barker for the mathematical wizardry involved in calculating this number.) Severe constraints have plainly been imposed by purely musical considerations which Eratocles neglected; and all the premises needed to establish them are demonstrated in the theorems of Book III.

moderately detailed account of it which I still think is on the right lines;[20] here I shall use a broad brush, tackling only a few detailed points of evidence and interpretation. To begin at an extremely impressionistic level, there are two ways in which a modulation of *tonos* can be conceived. In the first place it is a transposition of the same series of note-functions and intervals to a different pitch-level, as (approximately) in our notion of 'key'. In the second, we focus instead on a single range of pitch, and register the fact that the pattern of notes and intervals by which it is structured after the modulation is different from the one that characterised it previously.[21] From the perspective of a composer of melodies, the latter conception is arguably much more musically enriching than the former. According to the first approach, a modulation only transposes a section of the composition into a different pitch-range, and nothing has been structurally altered; the melody is merely broken down into segments in which its underlying scale-pattern inhabits different levels of pitch. According to the second, a modulation is a transition between different ways of arranging intervals within the musical space that the melody uses, and so opens up the possibility of exploiting new and unexpected relations between its elements.

The species of the octave constitute seven different ways in which a given octave of pitch can be organised, without any change of genus. We might then not unreasonably expect a theory of *tonoi* to be connected with them in some way. There is fairly compelling evidence, too, that the *harmoniai* under which melodies were classified in the fifth century and the earlier fourth, and from which melodies were deemed to derive their characters, could be recast without too much distortion by a tidy minded theorist (that is, by Eratocles), so as to become the seven octave-species (see pp. 43–52 above). Given the pivotal role of the *harmoniai* in the musical thought of the period in which he grew up, a second very plausible expectation would be, then, that Aristoxenus would pay some attention to questions about the relations between them, and the ways in which a melody can modulate from one to another. The statement at 37.8–10 which I quoted above points strongly to the same conclusion.

Only one Greek authority, however, makes a clear and deliberate connection between the *tonoi* and the species of the octave. This is Ptolemy, writing in the second century AD, and from a thoroughly anti-Aristoxenian

20 See Barker 1989a: 17–27.

21 A shift from one modern key to another, if considered purely melodically, can also of course be conceived in the second of these ways as well as the first. Thus when we modulate from C major to G major, the range between c and c′ is initially structured by the interval-series t,t,s,t,t,t,s (where 't' stands for 'tone' and 's' for 'semitone'), and after the modulation it has the new form t,t,t,s,t,t,s.

stance. He rejects the system of thirteen *tonoi* spaced by semitones, insisting that since the role of a *tonos* is to project a distinctive octave-species onto a given range of pitch, and since there are only seven octave-species, so too there are only seven *tonoi*. If we translate his mathematically expressed conclusion into Aristoxenian terms, they are spaced, from the lowest upwards, at intervals of tone, tone, semitone, tone, tone and semitone.[22]

Despite Ptolemy's denunciation of Aristoxenus' views, there is common ground between them. In the first place, both theorists make the distances between the *tonoi* independent of the genus of the scales that are employed. *Tonoi* stand at the same distances apart no matter what sizes of intervals they contain. Ptolemy's scheme of spacings looks superficially like part of a diatonic series, but in fact that has nothing to do with the matter. The essential consideration, for him as for Aristoxenus, is that corresponding notes in all the *tonoi* should be capable of being reached by moves through concordant intervals (*Harm.* 63.12–64.15). Secondly, the series of *tonoi* at which he arrives is precisely that of Aristoxenus with 'superfluous' elements subtracted. The names he assigns to his seven *tonoi* are exactly the same as those which the Aristoxenian compilers give to the *tonoi* standing the same distances apart in their sequence. It looks very much as though what Ptolemy gives us is not 'Aristoxenus rejected' but 'Aristoxenus purged'.

It would be strange, after all, if Ptolemy's conception of *tonos* and Aristoxenus' were not in fairly fundamental ways alike. Unless Ptolemy and his opponents were at some level discussing the same subject, the Ptolemaic polemic would have no tendency to show that the Aristoxenians were wrong; it would show only that they were talking about something else. Ptolemy's central point against them is that six of their thirteen *tonoi* are redundant, since they do not do what it is assumed they should, that is, to map onto a given octave of pitch a structure different from that produced by any other *tonos*. If Aristoxenus and his followers did not share some version of that assumption, Ptolemy would at best be belabouring them for the wrong reason.

Suppose, then, that this assumption, or something close to it, was indeed shared by all parties to the debate.[23] How are we to account for the fact that Aristoxenus posited thirteen *tonoi*, not the Ptolemaic seven? One answer

[22] See especially Ptol. *Harm.* II.10. His discussion of *tonoi* runs from II.7 to II.11; II.5 is also relevant.

[23] It must be conceded that most of the Aristoxenian sources show little sign of recognising the assumption, and their accounts can, for the most part, be read merely as descriptions of a framework for transposition. The strongest indication that traces of the other way of conceiving changes of *tonos* survived into the later Aristoxenian tradition is in Aristides Quintilianus, the difficulties and confusions of whose chapters on *tonoi* and modulation (*De mus.* I.10–11) can hardly be unravelled unless both conceptions are implicitly at work. See the notes *ad loc.* in Barker 1989a: 421–30.

might be that his exposition implied, misleadingly, that what are in fact two different projects amounted to the same thing, and that the creation of a complete set of transposition keys would at the same time provide, without redundancy, a complete system of structure-transforming *tonoi*. Perhaps he failed to make explicit or even to notice the fact that the considerations bearing on these two projects are not identical. The confusions in Aristides' account (n. 23 above) might point in this direction.

The second possibility is more interesting. I noted above that Ptolemy's names for his seven *tonoi* are the same as those given by Aristoxenians to those that stand in the same relations in their own scheme. Let us leave aside for the moment Aristoxenus' thirteenth *tonos*, an octave above the first. Now the names assigned to the remaining five *tonoi*, those absent from Ptolemy's system, have some curious features.[24] One is that each of these *tonoi* is given two names, 'higher Mixolydian *or* Hyperiastian', 'lower Hypolydian *or* Hypoaeolian', and so on. The second name, in each case, places the relevant *tonos* in either the Iastian or the Aeolian family, and none of the *tonoi* shared with Ptolemy fall into those groups. Their families are Dorian, Phrygian, Lydian and Mixolydian, and only one of them is given a second name (Ptolemy's Mixolydian is named by the Aristoxenians as 'lower Mixolydian *or* Hyperdorian'). The first name designates the *tonos* as a higher or lower version of some other, which always lies a semitone away from it and is always one of those shared with Ptolemy.

The correct way of construing these dry facts is of course debatable. It is quite likely, in particular, that the second set of names for these *tonoi* was not devised by Aristoxenus himself, but at a later stage of the tradition.[25] But it seems to me that the evidence points to at least two significant conclusions. First, the Aristoxenians recognised that their 'redundant' *tonoi* fell into a different category from the others, even if the distinctive 'family names' which mark the fact were relatively late additions. Secondly, and from a different perspective, each of them is essentially the same as one of its neighbours, distinct from it only in that it lies a semitone above it or below. The respect in which it is 'the same' can only be the one that Ptolemy points out, that the octave-species which it and its neighbour project onto the relevant range of pitch are identical. In a comparable sense the two

[24] The names said to have been used by Aristoxenus are set out at Cleonides 203.4–204.8, Arist. Quint. 20.9–21.1.

[25] Thus Arist. Quint. (n. 23 above) speaks of a lower Lydian *tonos* 'which is now called Aeolian', a lower Mixolydian 'which is now called Hyperdorian' and so on. Cleonides, by contrast, does not say 'now', only 'also'. For discussion of this and other issues to do with the *tonoi*, see for instance Winnington-Ingram 1936: 48–80, Gombosi 1939 and 1951, Solomon 1984.

outermost *tonoi* in the system are also the same but different, different in pitch but identical in arrangement.

What this suggests is that the Aristoxenian conception of the role of the *tonoi* was essentially tied up, like Ptolemy's, with their relation to the octave-species. Where they differ is that Aristoxenus allows, as Ptolemy does not, that there is substantial musical value in a system that makes room for transpositions of 'key' across the complete range of semitones, and so creates two differently pitched positions for any one octave-species. Only one species has a unique pitch-level, and it is the Dorian, whose fundamental status (as a *harmonia* or an octave-species) had long been thoroughly entrenched in the Greek tradition, and which serves as the fixed point of reference to which all others are related.

The amount of space I have given to this issue is ridiculously disproportionate to the task I set it, that of indicating a way in which the closing, fragmentary passage of Book III may have been put to use in the sequel – an enormous sledge-hammer for a very small nut. But plainly Aristoxenus' treatment of the *tonoi* is of interest in its own right, and it was worth making this rudimentary attempt to recover its outlines. I think also that the considerations we have been reviewing, together with my suggestions about the relation of the theory of *tonoi* to the principal theorems of this book, allow us to guess with some probability at the manner in which the topics in the missing part of the *El. harm.* were expounded. That is, it seems likely that Aristoxenus continued to write in a theorematic mode. Certainly the fragmentary passage at the end of Book III, the first step in the study of *systēmata*, sets out in that form. The links we have found to connect Book III's theorems, and the conclusions about arrangements of *systēmata*, with Aristoxenian doctrines about the *tonoi* and modulation give some modest but not contemptible support to the view that his propositions on those subjects too were cast as 'demonstrations'.

I recognise, of course, that the evidence is far from conclusive. But if it is a good guide, a large part of Aristoxenus' 'advanced course' in harmonics will have been presented theorematically. Perhaps the whole of it was, after the preliminaries of Book II. We have indeed been prepared for this state of affairs by the comments he makes, early in the second book, about points of methodology which need to be borne in mind when one is about to tackle 'the elements' (43.25–44.2; see p. 193 above). They are precisely those that are pertinent to an attempt to establish the science's conclusions through logical reasoning based on well-founded axioms.

In that case the missing part of the *El. harm.* may have been admirably rigorous, but it will probably have been at least as hard to master and as stylistically bleak as the surviving portion of Book III. Possibly that does something to explain why, in its original form, it became neglected and ultimately lost, generations of summarisers having found in it nothing more useful than a collection of conclusions, which they more or less faithfully recorded. If our speculations are on the right lines, however, the ingredient which disappeared from the tradition was the cement that held the whole enterprise together as a coherent, integrated system. What was abandoned was the argumentation showing how propositions dealing with any one of the various topics are logically related to those concerned with others, and how all of their demonstrations depend, eventually, on the same small handful of fundamental, interconnected principles or axioms. That is Aristoxenus' way of justifying his thesis that all melodic phenomena are expressions of a single nature, and of articulating the 'astonishing orderliness' that it displays. To recycle a phrase I used earlier, the theorems of Book III do indeed amount to a part of the 'triumphant dénouement' of his scientific exposition. They are not, however, the closing *tutti*, only the opening, thematic phrases of a long and contrapuntally intricate finale.

CHAPTER 9

Contexts and purposes of Aristoxenus' harmonics

Aristoxenus set out to convince his hearers and readers that students of harmonics should adopt the methods of research, the systematic approach, the standards of accuracy and the coherent argumentative strategies that are proper to a science in the Aristotelian mould. If he is to be believed, no previous theorist in the empirical tradition had come close to fulfilling these requirements, and he may very well be right, since the picture of their work which I tried to paint in Chapters 2–3 suggests that they had no such aim in mind. Aristoxenus, by contrast, set himself to the task of transforming harmonics into something it had never been before, a science whose credentials were as recognisable and legitimate as those of any other and which could now take its proper place among the sciences of nature. His repeated insistence on this objective in the *Elementa harmonica* suggests that he saw it as the most important aspect of his project; and in that case the audience he was addressing must have been composed of people to whom such issues mattered. Aristoxenus, we may conclude, was trying, above all, to persuade his fellow philosophers and scientists in the Lyceum that in his hands the previously trivial discipline of empirical harmonics had developed into one which deserved their intellectual respect.

These conclusions are uncontroversial, unoriginal and unsurprising, and nothing that I say in this chapter is intended to cast doubt on them. The question I want to raise is whether they tell the whole story. Aristoxenus' work in harmonics may have had other purposes too, and need not have been designed only to enlighten philosophers of science, along with the select few who found the topic entrancing in its own right. We have seen that almost all the theorists whom he regarded as his predecessors were musicians, and I have argued that they used their quantitative analyses and their geometrical diagrams of scales in teaching their apprentices to understand the styles of the composers whose works they studied as models or inspirations for their own. Given that Aristoxenus regarded his harmonic investigations as continuous with theirs, though massively superior, we can begin by asking whether there is any evidence that he too intended them to be used, or even used them himself, in training the composers of the future. His strongly held views about the excellence of the music of the past and the decadence of contemporary practices would have given him compelling reasons for doing so.

HARMONICS AND MUSICAL COMPOSITION

Unlike his predecessors, Aristoxenus – so far as we know – never worked as a professional musician, but he had grown up in a musician's family and had received a musician's training, and would have been very adequately qualified to teach the art of composition if he had so wished. We have seen already that he included harmonics among the disciplines that a composer should learn, though they needed a secure grasp on many other matters too, and that at least some theorists in his own time were also composers and performers.[1] From this very broad perspective it seems possible that he designed his teaching in harmonics for an audience which included people intending to enter the profession, as well as the philosophers and scientists of the Lyceum. The *Elementa harmonica* certainly contains material which could plausibly be reckoned useful to them, as for instance its analyses of scale-forms in each of the genera, its specifications of melodic and unmelodic sequences of intervals, and (in parts now missing from our text) its study of admissible and inadmissible modulations between *tonoi*, not to mention its reflections on *melopoiia*, 'melodic composition', itself.

But plainly these vague considerations are not enough. They show at best that Aristoxenus *might* have intended his lectures and treatises for this purpose, and the evidence that he actually did so is thin; there are, I think, only two passages in the *Elementa harmonica* on which we might build a more positive case. *Melopoiia*, as we have seen, is the last of the 'parts' of harmonics listed in the introduction to Book II, at 38.17–26. No details are given, and the summary treatments which appear in Cleonides and Aristides Quintilianus inspire no confidence that apprentice composers would have got much practical help from Aristoxenus' remarks.

But that may not reflect his own estimate of their value; and his statement at the end of the passage in Book II should make us pause. It is in melodic composition, he says, that harmonics reaches its *telos*. The phrase is ambiguous. It might mean only that once the topic of *melopoiia* has been covered, there are no more; that is where the discipline ends. Very often, however,

[1] See pp. 102–3 and 91–6 above.

and especially in Aristotelian philosophy, *telos* has a stronger meaning. It is the 'end' of a process in the sense that it is its goal or fulfilment, that for the sake of which it exists. If that is what it means here, Aristoxenus might be telling us that students of harmonics will have achieved their goal when they have learned how to compose melodies; that is not only where the discipline ends but what it is for. But this would be too hasty a conclusion. Even if *telos* here means 'purpose', the purpose might lie in an intellectual understanding of *melopoiia* and not in a practical competence, an understanding which equips the student to appreciate and discuss existing compositions intelligently, rather than to produce his own. In that case it is a purpose perfectly in tune with the elite refinements of the Lyceum, and need have little or nothing to do with the concerns of the 'technicians' who worked for a living in the musical trade.

Our interpretation of the second passage will depend, once again, on the meaning of a single word, and the word appears in a sentence which poses other problems too; but in this case it is much harder to avoid the conclusion that Aristoxenus is envisaging the act of composition as his goal. In his preface to Book II, Aristoxenus complains that people have misunderstood some of the things he has said in the past about the extent of music's ethical significance, and have completely overlooked others. We shall return shortly to the substantial points he is trying to make; only one incidental feature of his comments need concern us immediately. These people, he says 'have misunderstood what we said in our public lectures (*deixeis*): that we are trying to make (*poiein*) each of the *melopoiiai*, and in general that one kind of music damages the character while another kind benefits it' (31.22–6). In the present context the crux is the word *poiein*. I have translated it as neutrally as possible, but in fact in musical contexts of this sort its sense is precise and unambiguous; it invariably means 'to compose'. Aristoxenus could hardly have said more clearly that the purpose of his project, or one of its purposes, was to produce compositions. Translators (including an earlier version of myself) have found the implications of his statement embarrassing and have done their best to conceal or avoid them. Some have fudged the issue by devising a phrase which might allow readers to infer that Aristoxenus is talking about producing analyses of types of composition, rather than actually composing.[2] Others have tacitly re-written Aristoxenus' sentence to make it mean what they think it should, rather than what it

[2] E.g. Macran 1902, 'we aim at the construction of every style of melody', where the choice of the word 'construction' rather than 'composition' seems calculated to suggest a theorist's representation of form and structure. Macran does not comment on the matter in his notes.

does.[3] We need to abandon these evasive strategies and face the problem squarely.

It might be argued that although *poiein* does indeed mean 'to make' or 'to create' and in contexts like this would normally mean 'to compose', what Aristoxenus says he is aiming to 'make' are not in fact compositions or melodies. He specifies the object to be made as 'each of the *melopoiiai*' (*tōn melopoiiōn hekastēn*), and it would be absurd as well as linguistically inaccurate to take this as meaning 'each and every tune'. Here, as elsewhere in the *Elementa harmonica*, *melopoiiai* are not individual compositions but styles of composition;[4] and since 'making' a style, whatever it may mean, is plainly not identical with 'making' a piece of music, Aristoxenus is not necessarily committing himself and his students to the task of composing. The relevant question, then, is what can be meant by 'trying to make (or 'create' or 'compose') each of the styles of melodic composition', if it is not to involve composing pieces in each style, and I confess that I cannot answer it. Certainly one can analyse and describe such styles without composing anything, but such activities cannot be captured by the simple verb *poiein*, versatile though it is. I think we should take Aristoxenus' statement in its most obvious sense; the goal he has set himself (along with his students or collaborators, if his 'we' is a genuine plural) involves composing melodies of every sort.

But we still should not jump to the conclusion that this was even one of his goals when he was writing and talking about harmonics. There remains one possible escape-route, quite apart from the fact that it would be unsafe to hang large claims about the purposes of his harmonic theorising on a single word set in a problematic context. The sentence we have been considering is an intruder in the *Elementa harmonica.* Aristoxenus says that he made his remark in the course of his *deixeis*, and I take these to have been analogous to the *epideixeis* presented by the sophists. That is, they were public lectures designed to advertise the speaker's expertise in some area of thought or skill, to attract people's interest in the subject, and

[3] Da Rios 1954, 'cerchiamo di mostrare, riguardo ad ogni composizione ed alla musica in generale, quel che può nuocere e quel che può giovare al carattere'; Barker 1989a, 'What we are trying to do is to show for each kind of melodic composition and for music in general that such and such a type damages the character while such and such another improves it', a version often repeated in later publications, e.g. Gibson 2005: 209 n. 127. Tempting though they may be, these translations are impossible. They pretend that Aristoxenus' verb *poiein* means 'to show', 'mostrare', rather than 'to make', and that is plainly wrong. (They also construe the syntax of the sentence in a different way from the one involved in Macran's translation and in the translation I offer in the text above; it would be perfectly acceptable, indeed compulsory, if Aristoxenus had written 'to show', but must be rejected if we give *poiein* its proper meaning.)

[4] This is clearly the sense of the term at 23.4, 23.13, 35.15.

perhaps to entice some of them to become fee-paying students. It is clear from the recorded titles of his works that Aristoxenus was knowledgeable about virtually all aspects of music, and it is most unlikely that he confined himself to the topic of harmonics on these public occasions. Hence his purposes in the *deixeis* to which he refers need not have been at all the same as in his specialised lectures and treatises on harmonics. Perhaps in some of his public appearances he set out to show that on the basis of the knowledge he possessed, he was able to compose melodies in any style he chose, and demonstrated his ability in live performance – performing more impressively, no doubt, than the charlatans lambasted in the Hibeh papyrus (see pp. 69–73 above). Whether that is a plausible conjecture or not, we cannot properly infer anything about his intentions in the *Elementa harmonica* from what he reports himself as saying about his objectives in an entirely different setting.

HARMONICS AND MUSICAL CRITICISM

The hypothesis that Aristoxenus followed his predecessors' example and used his teaching in harmonics to help aspiring composers to master their craft was worth considering, but there is nothing solid to support it and I think it should be abandoned. We nevertheless have persuasive evidence that the subject was one which in his view should be studied by others as well as the scientifically minded philosophers mentioned at the beginning of this chapter. Aristoxenus tells us at the beginning of the *Elementa harmonica* that harmonics deals with

> everything that bears upon the study of *systēmata* and *tonoi* . . . Matters studied at a higher level, when *poiētikē* is already making use of *systēmata* and *tonoi*, no longer belong to this science, but to the one that embraces both this science and the others through which everything to do with music is studied; and this is the accomplishment of the *mousikos*. (1.19–2.7)

I discussed these remarks in Chapter 6 (pp. 138–40 above), and drew the conclusion that the all-embracing 'science' or branch of knowledge to which they refer is not the specialised province of composers or other practitioners of the musical arts, but belongs to anyone who deserves the title *mousikos* in its wider sense. Such a person will, as we say, 'understand' music; he will have a proper appreciation of the qualities of the music he hears and will be equipped to make sound critical judgements about its merits and defects; in short, he will be an authoritative connoisseur. There is nothing here to suggest that he must also be a practical musician, and we shall find reasons for thinking that typically at least, he is not. The passage gives the impression that a *mousikos* with these impressive credentials must have mastered a great many disciplines and that harmonics is among them. Other passages will confirm this conclusion, and this will significantly expand the range of persons who could benefit, in Aristoxenus' view, from instruction in harmonics, and for whom his treatises and lectures are likely to have been designed. In the remainder of this chapter we shall explore the boundaries between the science itself and this broader 'accomplishment', in an attempt to discover more precisely what 'connoisseurship' or 'critical judgement' involves, and to identify the contributions that can and cannot be made to it by harmonics and the other technical disciplines.

We may note as a preliminary that if those who pass judgements on pieces of music need to be competent in harmonics, the discipline will have compelling claims, in the Greek context, to a place in the 'liberal *paideia*' of the cultivated elite. Our texts provide ample evidence of the liveliness of public debate about the merits of musical performances, especially in Athens, and of the vociferous vigour with which audiences expressed their opinions; and they also reveal a progressive parting of the ways between popular taste and that of the educated fourth-century intelligentsia.[5] For the latter, the capacity to rise above the masses' love of virtuosity and melodramatic effects, and to form musical judgements on an aesthetically and intellectually sophisticated basis became a mark of their superior status. In so far as he represents harmonic theory as one of the essential instruments in such a judge's intellectual tool-box, Aristoxenus is continuing the project of rescuing it from the hands of the mere 'craftsmen' with whom it had begun, and recommending it to the attention of the cultured upper classes for whom in earlier times it had seemed as irrelevant as learning how to cook.[6]

Aristoxenus says very little about these matters in the *Elementa harmonica* itself. In that work, when he mentions patterns of vocal movement which, as a harmonic scientist, he finds unacceptable, he calls them 'unmelodic' (*ekmelēs*); those that are acceptable are 'melodic' (*emmelēs*). The former, to put it crudely, break formal rules, and the latter do not. But a melody which is in this sense 'correct', *emmelēs*, is not necessarily what we would call a 'good' piece of music. It may be boring, or revoltingly sentimental, or gratuitously rhetorical in expression, or may suffer from any number of other aesthetic faults. There is just one passage in which he allows himself

[5] For a thorough study of the issues and the evidence see Wallace 1997.

[6] Cf. Aristotle, *Pol.* 1339a.

a couple of judgements of this latter sort. A form of melodic composition that separates the notes *mesē* and *lichanos* by a complete ditone is 'not the most despicable (*phaulotatē*) but just about the finest (*kallistē*)'. Nowadays, he goes on, it is largely neglected and unappreciated, and most composers place *lichanos* higher in its range, closer to *mesē*, because of their relentless pursuit of 'sweetness' (*glykainein*) (23.4–17; see p. 128 above). Their musical ideal, he insinuates, is insipidly saccharine.

Aristoxenus' comments plainly cannot be justified on the basis of the propositions of harmonics alone. He says nothing to suggest that these contemporary compositions are any less 'melodic' than those in the ancient style he admires. It is instructive that this is the only point in the *Elementa harmonica* where the adjective *kalos* occurs.[7] Its absence from the rest of the text points to deliberate linguistic self-discipline on Aristoxenus' part, since *kalos* is not only the term most frequently used in Greek to convey approval in aesthetic (and many other) contexts, but is a very common element in the vocabulary of almost any writer. I am as sure as one can be about such things that he excludes it from his repertoire because of its deeply entrenched role in evaluative musical judgements that go beyond the scope of harmonics. Our principal source for his writings on such judgements is probably not always quoting his words verbatim, but leaves little room for doubt that he represented them, *inter alia*, as judgements about what, in music, is *kalos*.

The source is the Plutarchan *De musica*. Precisely how much of its material on this topic comes from Aristoxenus' lost works is debatable, but I shall not debate it here. I shall confine myself to a single, extended passage in which his footprints are unmistakable, and only a hyperbolically cautious commentator – a veritable Descartes brooding over his lump of wax – would question its origins. It runs from Ch. 31, where Aristoxenus is named as the authority on whom the writer is drawing, to Ch. 36 (1142b–1144e). The passage does not form a single, continuous argument. It jumps without warning from one issue to another, and themes which in one place are broached and then dropped are reintroduced, several pages later, as if they had never been mentioned. The compiler has apparently collected bits and pieces from one (or possibly several) of Aristoxenus' writings which discussed matters to do with musical judgement and understanding, and has

7 I exclude a handful of instances of the adverb *kalōs*, in which it has no aesthetic resonance and merely means 'well', as in 'having enumerated well', *kalōs exarithmēkotes*, 32.31. It is intriguing that the adjective *kalos* is also remarkably rare in Aristotle's *Poetics*. I am grateful to James Porter for pointing this out to me; the reasons may perhaps be analogous to those which apply to Aristoxenus, but I cannot pursue the matter here.

simply placed them side by side. The fact that no detectable attempt has been made to forge them into a coherent, progressively developing exposition encourages the belief that the individual excerpts have been fairly faithfully transcribed; they have not been modified, at any rate, to fit an argumentative pattern imposed by the compiler, since there is none. But it means also that the order in which ideas are presented is a poor guide to such connections of thought as there may be between them. The route I shall follow therefore ignores the arrangement of the Plutarchan text, though for the most part it respects what seem to be the boundaries of the individual fragments. No doubt other equally enlightening pathways could be found. I have divided the material up, rather roughly and artificially, under four headings.

TECHNICAL KNOWLEDGE AND PRACTICAL ANALYSIS

The principal question with which this section is concerned is how far an understanding of the technical musical disciplines can take us towards a capacity for musical judgement, and what difficulties stand in their way. The passage preserved at the beginning of *De mus.* 34 (1143e–f) sets off abruptly on a theme familiar from the early pages of the *El. harm.* (11.7–25). It asserts that 'our predecessors' studied only one of the three genera, the enharmonic, and within that genus only *systēmata* spanning an octave; and 'a person who has advanced only as far as this knowledge would not fully understand the subject-matter of harmonics'. The next sentence seems to be about to tell us what a complete grasp on harmonic matters would involve, but it turns out to refer to the qualifications of a person whose knowledge transcends harmonics altogether. It is a person 'who can follow the individual sciences and the whole body of music, along with the way in which its elements are mixed and combined; for a person who is only an expert in harmonics [a *harmonikos*] is in some way circumscribed'.

It looks as if a step in the discussion has dropped out, and that the original line of thought was roughly this: 'a person who studies only the enharmonic does not grasp the whole subject-matter of harmonics; analogously, a person who understands only harmonics lacks a full understanding of music'.[8] The focus of the next part of the discussion, at all events, is on the business of forming a judgement (*krisis*) about a piece of music in all its dimensions, not just about those aspects of it with which harmonics is concerned. It

[8] The matter is complicated by the punning play made here, as in the *El. harm.*, with ambiguities in the words *harmonia* (the enharmonic genus *or* the whole subject-matter of harmonics) and *harmonikos* (an expert in harmonics *or* a person who studies the enharmonic genus). Possibly the apparent elisions in the argument arise from this *jeu d'ésprit*, and are due to Aristoxenus himself.

is here, in the remainder of Ch. 34 and the whole of Ch. 35 (1143f–1144c), that we meet the thesis I considered earlier (p. 173), that perception and thought (*dianoia*) must 'run alongside each other' in the formation of such judgements; neither should dash ahead or lag behind.

The task involves an intricate complex of perceptions and interpretative 'thought'.

> For three elementary components must always impinge on the hearing at the same time, a note, a [rhythmic] duration and a syllable or articulate sound [lit. 'letter']. From the course of travel (*poreia*) through notes we can discern the pattern of attunement (*to hērmosmenon*), from that through durations we can discern the rhythm, and from that through articulate sounds and syllables we can discern what is said. (1144a–b)

At any moment in time, then, the ear must attend simultaneously to elements of at least three different sorts,[9] each of which must be construed as a staging-post in a specific 'course of travel'. All three journeys take place at the same time and must be tracked simultaneously by the ear; and the listener must also, and again simultaneously, extract from their successions of movements, through interpretative *dianoia*, the harmonic and rhythmic patterns into which they are organised and the verbal meanings carried by their sequence of articulate sounds. The time in which all this is to be done is precisely the time in which the performance takes place, and there is no question of analysing the behaviours of the three elements in separate episodes, one after another, as one might if one were studying a written score. 'Since they all go forward together, we have to direct the attention of perception to all of them at once' (1144b).

'But this is clear too,' the writer continues, 'that unless perception can separate out each of the things mentioned, it will be unable to follow them individually and to grasp what is faulty in each of them and what is not' (1144b). This might mean either of two things. One is that perception must separate out each individual note, duration and articulate sound from the continuum to which it belongs; alternatively, it must separate out the three individual strands of progression from the complex in which they all appear together. No doubt Aristoxenus would subscribe to both theses, but in view of the way this statement is contrasted with its predecessor (we have to attend to all three strands at once; but we must nevertheless separate out the 'things mentioned'), it is probably the latter that is uppermost in his mind. Though every element in any one form of progression is

[9] When the object to be judged is a fully realised musical product, a composition as actually performed on some occasion, it will have several other aspects to which the critic must also simultaneously pay attention.

fused, when it strikes our hearing, with elements in the other two, we shall not be competent musical critics unless we can perceptually disentangle them, perceiving each in the context of its own mode of 'journeying', independently of the others.

Aristoxenus next comments that before we can 'follow' the individual progressions and assess their merits, 'we must know about continuity (*synecheia*)' (1144b). So far as the melodic strand of the music is concerned, this brings us back squarely into the territory of harmonics, since in the *Elementa harmonica* the study of *synecheia* is a fundamental part of the science. It deals with the ways in which it is melodically 'natural' for notes and intervals to follow one another, and in Aristoxenus' treatment it culminates in the formulation of the 'rules of progression' enshrined in the theorems of Book III.[10] Aristoxenus compares the fact that intervals in a melody cannot succeed one another in just any random order with the fact that not just any syllable can succeed any other in speech (27.15–33); and in the *Rhythmics* he compares both those cases with that of rhythm, where, once again, only certain arrangements and sequences of elements, in this case temporal durations (*chronoi*), are acceptable, 'conformable to perception and capable of being organised in accordance with the nature of rhythm' (*Rhythm.* 2.8). There are therefore natural constraints on the ways in which temporal durations or articulate syllables can succeed one another to form 'continuous' sequences in rhythm or speech, just as there are constraints on the sequences of notes and intervals that can enter into continuous *melos*.

No one can be a competent judge of music unless they understand what natural *synecheia* amounts to in each of these three dimensions, and such understanding is an essential ingredient in the capacity for critical judgement (*kritikē dynamis*). 'For good qualities and their opposites do not arise in this or that separate note or duration or articulate sound, but in a continuous sequence of them; and this is a mixture (*mixis*) of the incomposite elements, dependent on the way in which they are used' (*De mus.* 1144b–c).[11] Plainly an adequate grasp on these three modes of

[10] See especially *El. harm.* 27.15–29.1, with pp. 161–4 above.

[11] Here I follow Volkmann's edition (Leipzig 1856) in placing *tōn* after *kata tēn chrēsin*, rather than before it as in the MSS. The word *mixis*, 'mixture', seems to be used here inconsistently with the way it is treated at 1143b, where it is a mixture of elements occurring simultaneously (a melodic pattern and a rhythm), and is explicitly distinguished from a combination of elements into a sequence, which is called *synthesis*, as in the *El. harm.* Its appearance here may be merely a slip of the pen (either Aristoxenus' or the compiler's); or possibly some material has been omitted, in which the discussion returned from the sequential *synthesis* of items in each separate dimension to the simultaneous *mixis* of melodic, rhythmic and lexical elements.

continuity demands, from Aristoxenus' perspective, a thorough mastery of the relevant branches of musical science.

We were right to conclude, then, that expertise in these technical disciplines is not just an intellectual goal to be pursued for its own sake as an abstract exercise in 'academic' learning, but is also capable of being applied in a more practical arena, where individual pieces of music are subjected to consideration and judgement. It is indeed essential if such judgements are to be intelligently made. The fact that the objective of scientific investigation, simply as such, is understanding (*xynesis*), and that understanding is something private, 'hidden somewhere in the soul' (*El. harm.* 41.15–16), does not prevent it from being deployed in the service of wider ends. I have suggested that these wider ends were principally those of pure aesthetic connoisseurship, but Aristoxenus may have been hoping also to instruct people with more direct responsibilities for the well-being of the musical arts. In the institutionally competitive context of Greek music-making at its highest cultural level, the prestige and economic welfare of musicians and other matters impinging directly on their lives regularly depended on the adjudicators' exercise of 'critical judgement' in awarding the prizes. Aristoxenus' readers will inevitably have understood him as alluding, at least in part, to the qualifications which these official judges ought to possess.

If they lack the complex and finely honed powers of perception and interpretative discrimination that are called for, or do not understand the principles of musical *synecheia* in each of the relevant domains, such judges will be ill equipped to assess the 'good qualities and their opposites' that compositions and performances display. If the Greek of the phrase I have quoted were translated word for word, it would read 'the well and the oppositely', which is scarcely English at all, but perfectly acceptable Greek. Evidently it signals an evaluative judgement; something is done well or badly. But it remains to be asked what sort of evaluation is envisaged. When Plato writes of such matters, a few decades before Aristoxenus, he insists in a similar way that those who are to 'judge music intelligently' must have 'acute perception and understanding of rhythms and *harmoniai*', and must 'understand correctness in melodies, the correctness of the Dorian *harmonia*, for example'. He must therefore have a sound grasp on the elements from which melodies are composed, without which he cannot know whether or not they are 'correct' (*Laws* 670b–c, with 669a–b).

This sounds remarkably like Aristoxenus; and though when the details are examined it turns out that they have rather different modes of judgement in mind, both, at a general level, are talking about an evaluation that is possible only for technical experts, whose ears and minds have been rigorously trained in the musical disciplines. Such people, unlike those who have merely learned, by rote and without understanding, how to 'sing to an accompaniment and step in rhythm' (*Laws* 670b–c), are fully qualified to decide whether or not a piece of music is, in Plato's language, 'correct'. But he very deliberately distinguishes judgements about correctness from those of another sort which an adjudicator or critic must make, those that differentiate a piece of music that is *kalos*, 'fine', 'noble' or 'beautiful', from one that is not. Such judgements cannot be made, like the others, on the basis of technical expertise alone (670e). Despite certain divergences between Plato's approach to these issues and those of Aristoxenus,[12] they draw very similar distinctions between two forms of judgement, to which Aristoxenus gives no less weight than his predecessor.

TECHNICAL KNOWLEDGE AND CRITICAL JUDGEMENT

The point is made clearly at the beginning of Ch. 36 of the Plutarchan *De musica*.

> The next thing to notice is that those who have expert knowledge of music are not as such fully equipped for the business of critical judgement (the *kritikē pragmateia*). For it is impossible to become a complete *mousikos* and *kritikos* just on the basis of those things which are treated as parts of musical expertise as a whole, for instance from experience with instruments and in singing, from the training of perception – I mean that which leads to understanding of attunement (*to hērmosmenon*) and again of rhythm – and in addition from the sciences of rhythmics and harmonics and the disciplines dealing with instrumental accompaniment and diction, and from any other such disciplines as there may be. (1144c–d)

Aristoxenus is in no doubt that these disciplines and forms of experience can enhance our capacity to perceive the various kinds of element and progression at work in a composition, along with the techniques used in its realisation and execution, and that they give us the capacity to scrutinise authoritatively the extent to which the composer's use of his ingredients conforms to the natural ordering of *melos* and rhythm, and to assess the performer's skills. In this sense people who are thoroughly at home in all the 'parts of musical expertise' will be competent judges of what is done

[12] Most importantly, Plato's account of these judgements turns on his thesis that all music is *mimēsis*, 'imitation', a thesis that is forcefully articulated at *Laws* 668a and again at 668b–c, and to which the subsequent arguments repeatedly appeal. There is no trace of it in Aristoxenus. Secondly, Plato's conception of *to kalon*, 'fineness' in music, is ethically loaded; the best music is the most morally edifying. I shall argue that Aristoxenus' conception of the relation between musical and ethical properties is rather different, and that the judgements with which he is mainly concerned are independent of assumptions about the effects of music on human character.

well and what is not. In Plato's terms, they understand what is correct, and will unfailingly recognise both it and its opposite in all aspects of a musical work, as it unrolls in a performance. But for Aristoxenus as for Plato, as this passage shows, these qualifications by themselves will not equip people fully for the business of musical judgement.

Aristoxenus now tries to explain why this is so – why, as he puts it, 'it is impossible to become a *kritikos* on the basis of these accomplishments alone' (1144d). By a *kritikos* he evidently means someone qualified to pass judgement on the true worth of the music he hears, though in what this 'true worth' consists, over and above correctness, we have not yet been told. He does so by drawing a distinction between those objects of judgement which are *teleia* and those which are *atelē*. These terms have various uses, but the context makes it clear that he is treating something as *teleios* if it constitutes an end or purpose (*telos*) in its own right, and as *atelēs* if it is not. Things that are *teleia* are ends in themselves; those that are *atelē*, but are nevertheless appropriate objects of musical judgement, do not have that status, but are 'things which contribute to those ends and occur for their sake' (1144d).

Examples are offered by way of clarification. Each musical composition, the complete work which will be sung, or played on the pipes or the kithara, is an end in itself. So too is its actual performance, the episode of playing or singing through which it is expressed, considered in its entirety. One can find instances of items that are only means to such ends in the various 'parts' of a performance, those elements in it, we might say, which contribute to its overall quality (1144d).

> Thus when one hears an aulete one can judge whether the pipes sound in concord or not, and whether their articulation is clear or the opposite. Each of these factors is a part of auletic performance; neither, however, is its goal (*telos*), but they exist for the sake of the *telos*. For over and above these and other such things one will pass judgement on the *ēthos* of the performance, assessing whether it has been executed in a manner appropriate (*oikeion*) to the composition which was entrusted to the executant, and which he has chosen to realise and perform. Just the same applies to the emotions indicated in a composition through the composer's art. (1144e)

The distinction drawn here between means and ends is tolerably clear. Clarity of articulation on the pipes,[13] for instance, is not, in the context

[13] The word translated 'articulation' is *dialektos*, roughly 'conversation'. It may possibly allude to the way in which the aulete's two pipes 'talk' to one another, in which case it might be translated 'interplay'. Compare 1138b, where we are told that the *dialektoi* of musical accompaniments were more complex in the old days; at Aristotle, *De anima* 420b8, *dialektos* is one of the features which instruments such as lyres and auloi share with the human voice.

of a performance, an objective pursued and valued for its own sake, but is something used in the service of an effective rendering of the piece being played. Aristoxenus says little at this stage about the way the distinction applies in the sphere of composition. But we may suppose that the manner in which a modulation between *tonoi* is negotiated, for instance, will be serving its proper purpose if it contributes to the aesthetic and emotional impact of the composition as a whole. A person who treated it as admirable in itself would in effect be regarding it as an isolated exercise in compositional technique, rather than as an element in a complete artistic creation. Aristoxenus' thesis is that expertise in the special musical disciplines equips us only to judge elements of a composition or performance in this second way. It gives no insight into the extent to which these elements and their uses contribute to or detract from the 'character' (*ēthos*) of the work.

Two distinct classes of musical 'ends in themselves' are recognised here, the composition and the performance. One might argue that this is misleading, and that one of them should be regarded as subordinate to the other. The relation could be viewed from either of two perspectives. The performance might be regarded as the means by which a composition is translated into sound and presented to the hearing as music; alternatively, one might treat the composition as just one of the resources on which the performer draws in creating a fully realised musical artefact. Aristoxenus adopts neither position. Although one of the criteria by which a performance is to be judged is its 'appropriateness' to the composition, both the composition and the performance are capable of being appreciated as distinct, fully complete and autonomous works of art. They are the final products of two distinct and autonomous arts or skills, *poiētikē*, the art of the 'maker' or composer, and *hermēneia*, the art of the performer or interpreter.[14] For each of these artists (the same person may of course take both roles) there is a specific type of product which is the completion of his work, and can be assessed for its own intrinsic value, not merely as a contribution to the work of the other.[15]

The technical musical disciplines provide no access, then, to the attributes we should consider when assessing a piece of music as an end in

[14] *Hermēneia* is a term used regularly of performance, in suitable contexts, but carries with it the sense 'interpretation' or 'expression', which is rooted in its etymology and at the heart of most of its uses.

[15] In a well-known passage at the beginning of the *Nicomachean Ethics*, Aristotle points out ways in which one art or skill may be subordinate to another, where the latter makes use of the former's products (as bridle-making, for example, is subordinate to horse-riding). He asserts that in all cases the 'ends' of the skills higher in such a hierarchy are 'more choice-worthy' (*hairetōtera*) than those of the lower. But he does not deny that those of the lower are nevertheless genuine 'ends' for the crafts and craftsmen whose completed products they are. See *N.E.* 1094a9–18, and cf. 1112b11–24.

itself, for its own intrinsic worth. Suppose it is agreed that two compositions or two performances are equally 'correct' in their deployment of the resources on which they draw. It is tempting to sum up the kinds of judgement involved at the next, apparently non-technical level as those which conclude that one of them is nevertheless a 'better' piece of music than the other. That sort of evaluation is certainly among those with which Aristoxenus is concerned. But if such judgements have any substantial basis, they must depend on the judicious critic's sensitivity to attributes of some sort that the music possesses; and they are attributes which do not fall under the scrutiny of the technical or scientific expert as such. The critic must therefore be able to detect these attributes and identify them for what they are, as a preliminary to the formation of any summary evaluative judgement. What, then, are the attributes in question?

CRITICAL JUDGEMENT, *ĒTHOS* AND THE 'APPROPRIATE'

We have been told (1144e) that the critic of a performance will assess its *ēthos*, and will decide whether it is appropriate (*oikeion*) to the composition being executed. The way these points are couched leaves room for uncertainty as to whether the writer is referring to two different judgements, one about *ēthos*, the other about 'appropriateness', or whether he means that a judgement about a performance's *ēthos* is by that very fact a judgement about its appropriateness to the composition.[16] Perhaps both interpretations are relevant. First, the *ēthos* of the performance will be 'judged' (*krithēsetai*) in the sense that it will be discriminated and identified. Secondly it will be 'judged' in the sense of being evaluated, and such evaluation consists, partly or wholly, in an assessment of the appropriateness of this *ēthos* to the composition in question.

Little more can be gleaned about these matters from this passage by itself. If we now go back, however, to earlier chapters in this Aristoxenian stretch of the *De musica*, the two vital conceptions at work here, *ēthos* and appropriateness, reappear in key roles in a closely related context. These chapters are concerned with judgements directed at compositions (about which the passage we have considered drops only a few hints) rather than performances; but the themes and underlying ideas are similar, and the earlier chapters put a little more flesh on the bones we have so far disinterred.

[16] The Greek form of expression points to the latter reading, if a firm choice is to be made. To gain a glimpse of its structure, simply omit the words 'and will decide' from my paraphrase in the first sentence of this paragraph.

Chapter 33 begins by stating a view that is by now familiar. Knowledge of harmonic science provides an understanding of the genera of attunement, of intervals, *systēmata*, notes, *tonoi* and modulations between *systēmata*; 'but beyond that point it is unable to advance' (1142f). To illustrate what 'advancing beyond that point' would involve, Aristoxenus refers briefly, as he never does in the *Elementa harmonica*, to a specific musical example, a piece of work by a popular composer. He does not provide a notated score, and readers are evidently expected to be familiar with the piece; it seems to have been very well known.[17] 'One cannot even seek, through harmonics, the ability to decide whether the composer chose appropriately (*oikeiōs*) in the *Mysians*, for example, in choosing the Hypodorian *tonos* for the beginning, or the Mixolydian and Dorian for the finale, or the Hypophrygian and Phrygian for the middle section' (1142f).

Here again, as at 1144e, we meet the notion of appropriateness, *oikeiotēs*, in a context concerned with critical judgement. The fact that the term is not just a casual visitor to this context is confirmed by its pivotal role in the discussion that follows. Harmonic science cannot tackle questions of the kind under scrutiny because, Aristoxenus says, 'it is ignorant of the power (*dynamis*) of appropriateness (*oikeiotēs*)' (1143a). That is, it has no way of defining *oikeiotēs* or of recognising its presence where it exists and its absence where it does not, and can tell us nothing about the difference that its presence in a composition makes to the music's qualities. Close on the heels of *oikeiotēs* comes the concept we found associated with it at 1144e, that of *ēthos*. 'Neither the chromatic nor the enharmonic genus comes bringing with it the power (*dynamis*, perhaps here merely 'attribute') of complete *oikeiotēs*,[18] in accordance with which the *ēthos* of the melody that has been composed is made evident to perception. That, rather, is the artist's task' (1143a). A few lines later, after some similar remarks about rhythms, Aristoxenus adds: 'for we always speak of appropriateness with an eye to some *ēthos*' (1143b).

Let me summarise the new points that emerge from these passages. First, the quality of *oikeiotēs* is not brought to a composition by the presence in it of structures that a harmonic scientist can define, but by the way in which the 'artist' (*technitēs*), the composer, has worked with these materials. It is Philoxenus' choice of certain *tonoi* for specific sections of his dithyramb,

[17] If the standardly accepted emendations to the texts of ps.-Plutarch and Aristotle are reliable, the work was a dithyramb by Philoxenus; see Aristotle, *Pol.* 1342b7–12.

[18] My translation follows an emendation suggested in a note *ad loc.* in the Loeb edition of Einarson and De Lacy 1967, but not printed in their text. The MSS give the sense 'the complete *dynamis* of *oikeiotēs*'.

for instance, that is to be judged appropriate or inappropriate. Secondly, by making his piece, or a section of it, 'appropriate' in some sense that needs to be unravelled, the composer makes its *ēthos* 'evident to perception', aesthetically present to the sensitivity of the well-qualified listener. Finally, the statement quoted from 1143b seems to mean that when we call a composition or some part of it 'appropriate', the statement in this form is incomplete; it always means 'appropriate to such and such an *ēthos*', or more fully 'appropriate to the musical expression or evocation of such and such an *ēthos*'.

In Plato and Aristotle, each distinguishable kind of element in a composition is assigned an *ēthos* of its own, regardless of its association with other musical ingredients.[19] This is a thesis which Aristoxenus is apparently challenging. It is not, for example, the enharmonic or the chromatic genus as such that is appropriate to a specific *ēthos*; if it were, harmonics itself might be expected to cast more light on the matter than apparently it does. The *ēthos* of a piece arises from some aspect of the way in which enharmonic or chromatic *systēmata* and other such elements are treated by the composer. The next part of the text in the *De musica* expounds this position a little more fully, using for the second time a particular example.

I say that the cause of this [i.e. of a composition's *ēthos*] is either some combination (*synthesis*) or some mixture (*mixis*) or both.[20] In the work of Olympus,[21] for example, the enharmonic genus was placed in the Phrygian *tonos* and mixed with the *paiōn epibatos*.[22] It was this that created the *ēthos* of the beginning of the *nomos* of Athena; for when one adds to it the way in which the melody and rhythm are composed, and the skilful modulation of rhythm from paionic to trochaic, this constitutes what Olympus produced.[23] And yet while the enharmonic genus and the Phrygian *tonos* were maintained, and with them the entire *systēma*, the *ēthos*

[19] See e.g. Plato, *Rep.* 398c–401a, *Laws* 669b–670a, Aristotle, *Pol.* 1340a–b, 1341b.

[20] Here, as the sequel shows, *mixis* is the blending of elements occurring simultaneously, and *synthesis* is their combination in a temporal sequence. See also n. 11 above.

[21] Olympus seems to be semi-legendary, but is thought of by Aristoxenus as a historical figure, a great aulete and composer from the distant past (perhaps the seventh century) who brought the music of the pipes to Greece, and who was the 'founder of the noble music that is truly Greek' (1135b). Compositions reputed to be his still survived in Aristoxenus' time. See especially 1133d–f, 1134f–1135a, 1137a–e, and cf. pp. 98–101 above. For some earlier references see Euripides, *Iph. Aul.* 573–8, Plato, *Ion* 532b, Aristotle, *Pol.* 1340a.

[22] This is the name of a rhythm; for details see Arist. Quint. 37.5–12, and for a comment, 82.29–83.2.

[23] The MSS read 'this constitutes the enharmonic genus of Olympus'. The words *enharmonion genos*, 'enharmonic genus', which in the Greek are the last in the sentence, may have been added by an interpolator or by the compiler himself. But they can hardly have been written by Aristoxenus. The enharmonic genus, as has explicitly been said and is in any case obvious, is just one ingredient in the complex described. If those words are deleted, the expression that remains, *to Olympou*, is acceptable Greek; its literal sense is roughly 'the thing of Olympus'. Given the presence of the noun *ēthos* earlier in the sentence, the implicit meaning would be 'the *ēthos* of Olympus' composition'.

underwent a great change; for the section called '*Harmonia*' in the *nomos* of Athena is far distant in *ēthos* from the introduction. (1143b–c)

The *ēthos* of a composition and of each of its parts arises, then, in the case under discussion, from the ways in which melodic and rhythmic elements are associated with each other, either simultaneously or in sequence, and not from any one of them by itself. The *ēthos* can be substantially altered if just one ingredient is changed while the others are held constant. In another, more detailed analysis of a style associated with Olympus, Aristoxenus indicates a different way in which *ēthos* can be affected, this time by the roles in which certain notes or parts of a *systēma* are and are not used. Composers of this type of music used the note *nētē diezeugmenōn* in the accompaniment, but 'in the melody it did not seem to them appropriate (*oikeia*) to the *spondeiakos tropos* [the style of "libation-music"]' (1137c). *Nētē synēmmenōn*, similarly, was used in accompaniments, 'but anyone who used it in the melody would have been ashamed of the *ēthos* that its presence produced' (1137d). Again, in Dorian compositions, though not in others, they always refrained from using the tetrachord *hypatōn*, in order to 'preserve the *ēthos*', and 'out of respect for its nobility (*to kalon autou*)' (1137d–e).[24]

It is significant that Aristoxenus makes no attempt to specify which kinds of combination and mixture *in general* will produce an *ēthos* of a particular sort. Neither in these passages nor elsewhere does he offer an expert's recipes for the generation of musical 'characters'; in contexts of this kind he always relies on examples. It seems clear, in fact, that in his view such recipes or definitions cannot be given, at any rate in a way that would allow one composer's effects to be replicated, in a different composition, by another. 'No one,' he says, 'can imitate the style of Olympus' (1137b), even though technical analysis can bring out many of the ingredients essential to it, as is shown in the discussion I have just been sketching (1137d–e).

This, I think, is part of what he has in mind when he repeatedly insists that though expertise in musical technicalities is an essential prerequisite of critical judgement, it is not enough by itself. There are no rules which a scientist can discover and formulate, specifying those technically identifiable mixtures and combinations of elements which will unfailingly make a composition appropriate to the expression of a given *ēthos*. Just after his remarks on the *nomos* of Athena, he adds a comment about the limitations of a person who 'knows the Dorian' in the way that a student of harmonics

[24] In this passage Aristoxenus may be recording an account given by his predecessors, as in the closely related passage at 1134f–1135b, discussed on pp. 98–101 above.

does, but who lacks the capacity 'to judge (*krinein*) the appropriateness of its use'. Such a person 'will not know what he is doing. He will not even preserve the *ēthos.* In connection with Dorian compositions it is even a matter of some doubt (*aporeitai*) whether harmonics is capable, as some people suppose, of distinguishing one Dorian composition from another' (1143c). The point of these comments seems to be that Dorian pieces vary in *ēthos*, so that a composer who adheres strictly to the technical requirements of Dorian melody will not thereby 'know what he is doing', that is, what musical effects he is producing, and may shift from one *ēthos* to another without realising it; and that it is at best very doubtful whether people with expertise in harmonics are thereby equipped to pick out the features which make one Dorian composition significantly different from another.

Aristoxenus has said a good deal to emphasise, elucidate and justify his thesis that harmonics and similar disciplines can tell us nothing about the *ēthos* to which a composer's treatment of his materials is appropriate. But the surviving material leaves us almost completely in the dark about the nature of the 'accomplishment' which gives a competent critic his purchase on issues to do with *ēthos*, the criteria on which he will formulate his critical judgements, or the insights which enable a good composer to select and intertwine his harmonic and rhythmic resources so as to create the *ēthos* at which he is aiming. There is just one passage which may shed some light on the matter, though it hardly amounts to a blaze of illumination.

I mentioned earlier (p. 102 above) a little story told by Aristoxenus which the Plutarchan compiler has preserved, the story of Telesias of Thebes. Let us remind ourselves of its outlines. He was trained in his youth in the 'finest' (*kallistē*) kind of music, that of Pindar, Pratinas and other composers from the past; but in later life he became so enraptured by the allure of the 'complex, theatrical music' of more recent times that he came to despise the 'noble' (*kalon*) style in which he had been brought up, and set out on a thorough study of the most elaborate and innovative compositions of Philoxenus and Timotheus. But when he tried his hand at composing in both styles, the Pindaric and the Philoxenian, he could achieve no success in the latter. 'The reason was the excellent (*kallistē*) training he had as a boy' (1142b–c).

I shall come back later to the 'moral' drawn from this anecdote in the text. For the present let us reconsider the dynamics of Telesias' experience, as the narrative depicts them. It was at quite an advanced age, 'past the prime of life', that he became besotted with music in the new, theatrical style. In a blatant piece of authorial 'spin', the writer in fact says that he was 'completely deceived' by it (*exapatēthēnai*). He then 'learned thoroughly' (*ekmanthanein*) the most advanced works of Timotheus and Philoxenus. This means, I think, not merely that he learned them by heart, but that he subjected them to meticulous study. Earlier, we are told, he had 'laboured well' at all branches of musical education, and as a result he was presumably well qualified to examine these works in close detail. We may imagine him as having analysed their scale-patterns and modulations, for example, their ways of concatenating different rhythmic units, and so on.

Then he set out to compose on the basis of the same techniques, but somehow it always went wrong. The residue of the 'excellent training he had as a boy' stood in the way of success. But what sort of obstruction it created is less than clear. Plato's Socrates in the *Republic* or Aristotle in the *Politics* might have said that it made him recoil in distaste from the goal he had set himself, since it did not measure up to his ingrained standards of what is *kalon*.[25] That might conceivably be what Aristoxenus meant, but it is not the most natural reading of the narrative, which says nothing to suggest that Telesias lost his appetite for the project. He did not turn away from his objective, but try as he might, he could not attain it.

The story also makes it clear that his failure did not arise from a shortage of technical competence, in any straightforward sense; his repertoire of knowledge and skills was impeccable. What he seems to have lacked are the aesthetic instincts which guided Philoxenus and Timotheus towards an effective deployment of their ingredients in their characteristic manner, and it was his early training that made him unable to acquire them. Now the passages we considered earlier imply that music is 'effective' in so far as it is well adapted to the expression of some *ēthos*, and in one place this idea is linked to its capacity to convey specific emotions (1144e). This need not mean that the feelings expressed in Philoxenian and Pindaric compositions were invariably quite different; either composer, presumably, could convey pathos or cheerfulness or exaltation. The fundamental distinction lay in the fact that these emotions were expressed in different ways, corresponding to the composers' different styles (*tropoi*). Pindaric music no doubt represented erotic passion, for example, by different means and with more restraint than Philoxenian, and in that sense the two will have differed in *ēthos*.

These points suggest the following conclusions. Despite Telesias' new-found admiration for music in the modern style, his early training had made him incapable of perceiving which selections, combinations and mixtures of melodic and rhythmic elements would convey specific emotions in the

[25] On the influence of early musical training on evaluative judgements of this sort see especially Plato, *Rep*. 401b–402a, Aristotle, *Pol*. 1340a–1341b.

up-to-date manner, and would invest his compositions with an authentically Philoxenian *ēthos*. In his feeling for the relation between an emotion and the musical resources appropriate to its expression, he remained, for all his efforts, obstinately Pindaric. I use the word 'feeling' advisedly. In intellectual terms, if one is to move from the immediate appreciation of a work's musical effects to an understanding of the ways they have been produced and could be reproduced in a new composition, what is called for is some kind of induction or abstraction, a movement from the particular to the general. From perceiving certain kinds of effect created by specific means in a particular case, we move to the thought that relevantly similar 'combinations and mixtures' will produce relevantly similar effects when introduced into another piece of music. But this quasi-mechanical representation of the process is a travesty, and it will not work. What counts as a 'relevant' similarity? Aristoxenus has announced repeatedly that the gap between the resources used by a composer, which are open to technical analysis, and the *ēthos* which his deployment of these resources creates cannot be bridged by scientific or technical knowledge. A sensitive and experienced critic, or a composer well in command of his style, will have a 'feeling' for the kinds of musical construction which, within this style, will be 'appropriate' or effective conveyors of this or that *ēthos*. The story tells us that such feelings arise from consistent training and familiarisation in our malleable years, and that once established they will obstruct our best attempts to adapt ourselves to novel forms of musical expression.

EVALUATIVE JUDGEMENT, *ĒTHOS* AND 'ETHICS'

Let us imagine a critic who has a well-trained and discriminating ear, who is thoroughly competent in musical analysis, and who has in addition the qualities needed to grasp the *ēthos* of a composition and to appreciate the means by which and the style within which it has been created. There remains another task which we might expect such a paragon to undertake, that of formulating a judgement on the composition's overall merits. Is it a good composition or not?

In a passage we considered earlier (1144e), where it is the qualities of a performance rather than a composition that are to be assessed, there is a determinate *ēthos* to which the item under scrutiny *ought* to be appropriate, that of the composition 'entrusted' to the performer. If it is appropriate to it, the performance is to be commended; the critic apparently has a firm basis for evaluation. It is not obvious that there is a similarly fixed point of reference by which he can judge the virtues of a composition. We have been told that its *ēthos* is revealed to the listener by its *oikeiotēs*, which apparently means that we identify its character by grasping what it is appropriate for. Its *ēthos* becomes clear to us, perhaps, when we recognise that it is appropriate to a particular style's representation of certain emotions, or to the creation of a specific atmosphere, such as that of pastoral innocence or resolute martial endeavour. It still remains to be asked whether such perceptions can give grounds for evaluating a composition positively or negatively, and if so, by what criteria.

Socrates argues, in the *Republic*, that what we would call the strictly 'musical' elements of a piece, its *harmoniai* and rhythms, should 'follow' the words of the poem that is set (*Rep*. 398d, 400d). In Aristoxenus' language, it is to the *ēthos* expressed or conveyed by the words that all other aspects of the composition should be made 'appropriate'. (To judge by a passage of Plato's *Laws*, this is at least partly because, when melody or rhythm is taken in isolation, without words, it is 'extremely difficult to grasp what is intended'.[26]) At a deeper level, all the ingredients of a composition 'follow the character (*ēthos*) of the soul', and 'good diction and good *harmonia* and gracefulness and good rhythm follow good character' (*Rep*. 400d–e). A composition is ultimately to be adjudged good or bad on the basis of considerations to do with the quality of the human dispositions whose character its musical ingredients 'follow', and these considerations are of an essentially moral or ethical sort. The principal question in this part of the *Republic* is whether the relevant dispositions are those suited to the 'best' kind of people, those who will become rulers or 'guardians' in the ideal community. The *Laws* takes a wider and more pluralistic view, raising issues about whether a piece is suitable for men or women or slaves, for example (*Laws* 669c, 802d–e); but the fundamental grounds for judgement on its merits remain firmly centred on its affinities with modes of human character, and its effects on the moral dispositions of its hearers.[27] Even when the Athenian in the *Laws* is discussing what we might conceive as the artistic coherence of a work, the ways in which its ingredients are or are not accommodated to one another and integrated into a well-rounded whole, its coherence or the lack of it is represented in ethical terms (see especially 669b–670a).

Against this background, questions of two sorts need to be raised about Aristoxenus' views on the assessment of music. First, Greek critics, like

[26] *Laws* 669e. But the thesis raises tricky problems, since if the melodic element as such, for instance, has no readily discernible *ēthos*, it will apparently remain 'extremely difficult' to decide whether or not it is appropriate to the words in cases of the kinds envisaged in the *Republic*, where words are present. Plato does not satisfactorily resolve this conundrum.

[27] On the ways in which music can influence human character see especially *Rep*. 401b–402a.

their modern counterparts, often did make broad evaluative judgements about music; and when they declared that one composition was 'better' than another, they meant not merely (or not at all) that the former and not the latter was technically 'correct', but that it had greater value as a work of musical art. Does Aristoxenus suppose that such judgements can be justified, and that some pieces of music are objectively better than others? Secondly, does he suppose, like Plato, that the greater value of the 'better' work is always to be construed in ethical terms, or as Aristotle does in the last book of the *Politics*, that it is constituted by its usefulness for any of several social purposes? Or does he think that there are intrinsically musical kinds of excellence, which do not depend on a composition's salutary ethical effects or its social utility? If he took this last view he might still believe that a piece of music can be morally edifying or socially useful, but the judgement that it is musically admirable would not in itself be a judgement to that effect.

The answer to the first question is straightforwardly 'Yes'. Certainly Aristoxenus believed that some music is good and some is despicable. Evidence for this view is easily found, but before we can interpret it in detail we need an answer to the second question; and that is much harder to resolve.

Book II of the *El. harm.* begins with a preface on the importance of prefaces (30.9–32.4), part of which is well known to students of ancient philosophy. It recalls a story which Aristotle was fond of repeating about Plato's 'lecture on the good'. Plato's audience expected to be told something of practical value to their lives, which would direct them towards 'wealth, health, power, in short some amazing felicity'; and they were seriously put out when what they actually got was a disquisition on mathematics and metaphysics (30.16–31.3). It is much better, Aristoxenus concludes, to follow Aristotle's example and explain clearly in advance what one's topic is and where its limits lie.

Aristoxenus himself, it appears, had previously had experiences a little like Plato's. Members of his audiences, too, had sometimes misconstrued what he was offering them.

> Mistakes can be made in either of two directions. Some people suppose that the discipline is massively important, and that by listening to a discussion on harmonics they will become not only *mousikoi* but better in their characters. They have misunderstood things that we said in our *deixeis*: 'we are trying to create each kind of melodic composition', and of music in general, 'one sort damages people's characters, while another sort benefits them'. They not only misunderstood that, but completely failed to notice our qualification, 'in so far as music can provide such benefits'. (31.16–29)

Others, he goes on, fall into the opposite error of supposing that the subject is of no importance at all; but we can leave them aside.

Several taxing difficulties lurk among the details of these remarks, but they do not affect the aspects of their meaning that are most relevant here.[28] Aristoxenus' hearers were quite wrong to imagine that studying harmonics would make them better people; it can do nothing of the sort. What he had actually asserted was quite different, that some kinds of music – not some courses of study – are ethically damaging and some beneficial. The association of this statement with his reference to 'each kind of *melopoiia*' suggests that it is the manner of composition rather than the technical resources employed that differentiates the harmful from the edifying; and that seems consistent with the position adopted in the Plutarchan excerpts.

But Aristoxenus seems to have been no more than lukewarm in his convictions about the influence that music can exert upon character. The qualification which some members of his audience allegedly overlooked is patently designed to damp down expectations about the extent of its capacity to edify or corrupt. Some familiar views on the matter – we may think of the views sometimes attributed to Damon,[29] or those set out in Plato's *Republic* and *Laws*, or those of Aristoxenus' own misguided hearers, or perhaps even those of Aristotle in the *Politics* – apparently struck him as exaggerated. If that inference is correct, it calls into question the straightforwardness of the relation between his remarks here and those in the Plutarchan treatise, where considerations to do with *ēthos* are central to musical judgement. The identification of a composition's 'character' evidently tells us something important about its musical qualities,[30] but it may imply little if anything about its potential effects on the 'characters' of those who listen to it. In that case, though such moral efficacy as it has might still be a consequence of its *ēthos*, its *ēthos* itself can hardly consist in its power to affect human character for good or ill. We need – after a couple of brief preliminaries – to examine the *De musica*'s references to *ēthos* rather more closely.

28 There are various problems of syntax, some of which I have resolved, following Macran, by taking the particle *hoti* as the counterpart of a modern quotation-mark. We have already discussed the more substantive puzzle posed by Aristoxenus' use of the verb *poiein* at 31.23; see pp. 231–3 above.

29 Plato at *Rep.* 400b–c seems to guarantee that Damon had things to say about the ethical significance of rhythms. On the other hand the notion that he set out comparable ideas about the ethical characters or influences of the various *harmoniai*, a notion that was current in later antiquity and has been built up by modern scholars into an elaborate 'Damonian theory of *ēthos*', seems to me to rest on the flimsiest of foundations. This is not the place to set out my grounds for this heresy, however. I hope to do so elsewhere in due course.

30 At 1143d, in particular, it is that 'for the sake of which' a combination of elements is put together.

I noted earlier that there are passages, in Aristoxenus and elsewhere, in which the *ēthos* attributed to a musical phenomenon can have no direct connection with 'ethics'. We would be more inclined to label it as an 'aesthetic' property of the item in question.[31] We have seen also that the Plutarchan discussion links *ēthos* with the expression of emotion, though it does not explore the connection; and Greek writings are full of allusions to the power of music to create emotional atmosphere and to work upon the feelings of its audience. Some authors, but not all, construe this emotional power as the source of its influence on character, and we need to discover, if we can, whether Aristoxenus belongs in either of these camps.

But there is another equally important distinction to be drawn. Quite commonly, in both English and Greek, 'aesthetic' properties are indeed specified in terms which refer, directly or obliquely, to human emotions. We may describe a composition or some part of it as cheerful or gloomy or ecstatic, and so did audiences and writers in antiquity. But we should not automatically assume that these descriptions refer to the effects that the music has on the feelings of its hearers. Listening to 'cheerful' music is more like being in the company of a cheerful person than like being made to feel cheerful oneself; it may have that effect or it may not, and the music's 'character' does not depend on the outcome. We tend to regard it as somehow inherent in the music itself, regardless of the way in which any listener responds to it, though of course there is some connection between the two. We need therefore to ask first whether the Aristoxenus of the *De musica* conceives *ēthos* as belonging to a composition 'inherently' or only in its effects on the listener, and secondly whether the character it possesses or the effect it produces is of an ethically significant sort.

Neither question is answered explicitly in the text. On the first issue, however, such indications as there are point to the view that *ēthos* is inherent in the music. There are, to begin with, no references in the relevant passages to any listener's emotional response. When Aristoxenus is discussing the *ēthos* of Philoxenus' *Mysians* or Olympus' *nomos* of Athena, for example, he does not suggest that the critic discovers it by introspection, examining the emotions set up in him by the music, or by studying these compositions' effects on their audiences. (Many other writers take precisely that route; compare for instance Aristotle's remarks about the *harmoniai* at *Pol.* 1340a–b.) It is identified through an act of judgement (*krinein*) directed at the music itself. Secondly, he appears to treat such judgements as objectively true or false. The critic's task is to recognise the *ēthos* of each composition,

[31] The most clear-cut case is at *El. harm.* 48.32.

or of each of its parts, and to detect the means by which it is produced, and there is no hint that two equally competent critics might justifiably arrive at different conclusions about it because of their different emotional responses. On the contrary (and this is a third point), the *ēthos* to which a composition is appropriate can be discerned only by someone armed with the special though non-technical faculty of critical judgement, *to kritikon*, which seems to be treated as an instrument enabling him, and not just any emotionally responsive listener, to arrive at the correct diagnosis.

Finally, when Aristoxenus mentions the emotions (*pathē*) at 1144e, he does not attribute them to the listener, or describe them as 'effects' of the music; he says that they are 'indicated' or 'signified' (*sēmainomenōn*) through the composer's art. His skilful combinations and mixtures of musical sounds have made them 'indicative' of specific emotions. Aristoxenus' expression suggests an analogy with speech or writing, where the meaning of a series of sounds or letters is often designated by the same verb, *sēmainein*. It is what the utterance 'signifies', and this significance is there to be found in the words, which possess it whether their hearers understand them or not. The emotional significance of a piece of music, similarly, is inherent in it regardless of its audience's inattention or obtuseness.

We come, then, to the question whether the conception of *ēthos* embedded in these texts carries connotations of a recognisably ethical kind. Aristoxenus' apparent reluctance in the *Elementa harmonica* to credit music with much influence on human character does not settle the matter, and neither does the fact that the passages we have studied in the *De musica* nowhere allude to its moral effects. If the *ēthos* of a composition is inherent in the music and does not depend on its impact on an audience, it remains possible that the attribution of an *ēthos* to it conveys ethical approval or disapproval of the music, almost as if it were itself a moral agent.[32] The fact which I mentioned earlier, that Aristoxenus and others sometimes attribute *ēthos* to musical items in contexts where there can be no suspicion of an ethical meaning also turns out to be irrelevant. The items in question are always mere musical elements – the chromatic genus, for example – taken in isolation from any composition; and Aristoxenus in the *De musica* is

[32] I have known a philosopher (but I will preserve his anonymity) who maintained, seriously and passionately, that the music of Richard Strauss is morally corrupt. He certainly did not mean that listeners exposed to it were thereby corrupted; he believed, in particular, that his own intimate familiarity with the Straussian oeuvre had left his character unscathed. He found in the music itself qualities analogous to those constituting depravity in a human being; and he argued that a composition's wickedness, like that of a villainous person, may or may not rub off on its associates. Whether that happens or not, the music remains a proper object of moral disapproval.

adamant that the kind of *ēthos* he is discussing never belongs to them, only to the product produced from them through the composer's art.

It is hard to find an unambiguous answer to the question in the *De musica* itself, partly because Aristoxenus gives no descriptions here of the *ēthē* of the compositions he mentions. We cannot tell from these passages what kinds of language he would have thought suited to the task. Other writers in the Aristoxenian tradition, however, record a broad classification into three types. One kind of *ēthos* is *diastatikon* or *diastaltikon* ('expansive', perhaps 'elevating'), a second is *systaltikon* ('contracting', with the implication 'depressing'), and the third is *hēsychastikon* ('calming', 'peaceful'). The fullest account is in Cleonides.

> The *diastaltikon ēthos* of melodic composition is that through which magnificence is indicated (*sēmainetai*), and manly elevation of soul, and heroic deeds and emotions appropriate to such things. These are used especially in tragedy, and in other works that maintain this type of *ēthos*. The *systaltikon ēthos* is that through which the soul is drawn together into abasement (*tapeinotēs*) and unmanliness. This condition is in harmony with erotic emotions and dirges and laments and the like. The *hēsychastikon ēthos* of melodic composition is that which brings calmness of soul and a free and peaceable condition. In harmony with it are hymns, paians, encomia, songs of instruction (*symboulai*)[33] and similar pieces.[34]

On general grounds we can expect Cleonides to have stayed close to an authentically Aristoxenian source, and with some qualifications the same is true of Aristides Quintilianus in the part of his treatise where the same classification is sketched. This expectation is supported by various similarities of detail between Cleonides' presentation and that of the excerpts in ps.-Plutarch. In both places, and again in the relevant passages of Aristides, *ēthos* is assigned to melodic compositions, *melopoiiai*, not to individual aspects of a work such as the genera or *tonoi* of its *systēmata*. Both authors introduce the notion of appropriateness, *oikeiotēs*, and both use the verb *sēmainein* to designate what music does in 'indicating' or 'expressing' something. There is, however, at least one potentially disturbing difference. Cleonides' descriptions of the second and third *ēthē* consist wholly of statements about their effects on the soul (as do Aristides' descriptions of all three), and I have emphasised the fact that such statements are absent from the Plutarchan excerpts. But the problem can be resolved without saddling

[33] The musical sense of *symboulai* is uncertain, but cf. the related expression used at Isocrates 2.42 to mean 'didactic poems'.

[34] Cleonides 206.6–18. Compare Arist. Quint. 30.12–15, together with the parallel classification of styles of rhythmic composition at 40.14–15. What seems to be the same classification, rather differently expressed, appears at Ptol. *Harm.* 99.18–25, and again at Bacchius 304.18–20.

Aristoxenus with an 'audience response' theory of *ēthos*. I suggest that in describing the first *ēthos* on his list, the *diastaltikon*, Cleonides has preserved something of the wording of the original; here he says nothing about psychological effects, and speaks, like the Plutarchan passage, in terms of what a composition 'signifies'. In the second and third descriptions, by contrast, he has paraphrased more loosely, as does Aristides, lapsing into language that is common in other (and especially Platonist) writings, but which I believe Aristoxenus avoided.

The question we need to resolve is whether the terms in which the *ēthē* are characterised can or must carry ethical connotations. All three of the principal terms are unusual, quasi-technical coinages, though *hēsychastikon* is formed from a very common root in a familiar way, and there is no difficulty in understanding it. The other two are trickier. *Diastaltikon* is formed from the verb *diastellein*, which normally means 'to expand' or 'to divide'. It seems to be used very rarely in reference to human feelings or dispositions. Its counterpart *systellein*, however, from which *systaltikon* is derived and which literally means 'to draw together', is not uncommonly used to mean 'to humble' or 'to oppress', or 'to make (something or someone) low' in the sense 'despicable'.[35] Here, evidently, we have usages with immediate emotional and ethical force; and even if there was no precedent, before Aristoxenus, for recruiting a derivative of its converse, *diastellein*, to a parallel function at the opposite pole, the linguistic innovation would have been perfectly intelligible.

The language of Cleonides' explanatory expansions of these words carry unmistakable evaluative resonance. The nouns *megaloprepeia* ('magnificence' of character) and *tapeinotēs* ('abasement'), for instance, and the adjectives *andrōdēs* ('manly') and *anandros* ('unmanly') are familiar from the ethical vocabulary of Plato and Aristotle, and all of them are as much judgemental as descriptive; the first and third express approval, the second and fourth contempt. The crucial points, however, are that their descriptive meanings carry essential reference to attributes of human character and action, and that under their judgemental, evaluative aspect they place people and actions high or low on a scale of intrinsic moral worth. *Megaloprepeia* of character is admirable simply as such, and *tapeinotēs* is despicable, regardless of whether they enhance or diminish their bearers' value on some other scale, making them more or less companionable, for instance, or more or less socially useful.

[35] See e.g. Plato, *Lysis* 210e4, Eur. *H.F.* 1417.

In that case, when the elements of a composition are so combined and mixed that the complex becomes 'appropriate to some *ēthos*', it thereby becomes an indicator or signifier of some condition primarily ascribable to a human character in its moral dimension. In a human being such a condition is praiseworthy or the reverse. Here it is conceived as being so also in a piece of music, probably in a sense parallel to that in which a poem will be morally contemptible if it 'signifies' attitudes or emotions which, in a person, would be evidence of contemptible character. Critical judgement in the musical domain, as in the poetic, is the capacity to perceive this kind of significance in an artistically crafted complex of sounds.

The question whether an evaluative judgement is 'ethical' or 'aesthetic' is arguably one that I should not have raised in the first place. The Greek language marks no clear borderline between these kinds of value, and perhaps we should not try to foist the distinction on Aristoxenus. A more apposite distinction is the one I outlined earlier, between theorists who treat the 'excellence' of a composition as lying in its capacity to enhance the characters of its hearers, and those who locate it in the intrinsic significance of the music itself. For theorists of the latter sort, among whom I include Aristoxenus, the quality possessed by a 'fine' (*kalos*) composition, for example, rests on a meaning inherent in it, worked into it by the composer's exercise of his musical talents. It is what makes the composition 'good music', and in discovering its presence we are making a properly musical judgement, not one grounded in observations or theories of its psychological or sociological effects.

By these criteria, what we would call the 'aesthetic' or 'strictly musical' quality of a work consists at least partly in its signification of a character that evokes moral approval or disapproval. This helps us to appreciate aspects of the story of Telesias which we have so far ignored, and to understand the moral drawn from it in the text. The music in which Telesias had been trained in early life was *kalos* (noble, fine, beautiful) in the highest degree. The charms of the 'theatrical and complicated' music which later attracted him were 'deceitful'; and they tricked him into 'despising those fine (*kalōn*) things in which he had been brought up'. To despise what is fine is a sign of serious ethical-aesthetic misjudgement. Derivatives of the adjective *kalos* reappear in the dénouement of the narrative, where the obstacle to his success in Philoxenian composition is 'the most excellent (*kallistē*) training he had as a boy', and in what I have called its moral. 'Thus if a person wishes to treat music finely (*kalōs*), let him take the ancient style as his model' (1142c–d).

The story could have been told without any hint that one style was 'better' than the other, and could still have made the point that a reliable sense of how musical effects are produced in a particular idiom depends on one's early training. But Aristoxenus has made clear from start to finish his admiration for the old and distaste for the new, largely by insistently attaching *kalos* and its cognates to the former and pointedly avoiding them in references to the latter. In the context of ethical and aesthetic evaluation, *kalos* is the ace of trumps. If one kind of composition is *kalos* and the other is not, there is no room for hesitation about which ought to be preferred and which rejected, or for regarding the choice as a matter of personal taste. The weight of such judgements, which imply that a preference for what is not 'beautiful' (an important component in the meaning of *kalos*) is not an acceptable option, derives from the word's ethical loading. In making such a choice one would show oneself to be blind to the qualities that make a *person* admirable.

Our sense of what is *kalon* is instilled in us through early training and experience. If we are to understand its nature in the musical sphere more fully, we cannot do so through purely musical studies. Nor, presumably, could we do so in the realm of the graphic arts, for instance, just by studying painting and drawing. It is an attribute whose nature is not intelligible, even in one domain, through concepts proper to that domain alone, but incorporates values applicable everywhere in human experience.[36] Hence for Aristoxenus, as for Plato, the quest for musical excellence takes us beyond the bounds of music itself. The 'moral' of the story of Telesias reads, in full: 'Thus if a person wishes to treat music finely and with judgement, let him take the ancient style as his model, and let him supplement this music with the other branches of learning, and set philosophy over himself as his teacher; for it is philosophy that has the capacity to judge the proper measure for music, and how it should be used' (1142c–d).

The Plutarchan compiler uses sources of several kinds, and one might reasonably wonder whether the reference to philosophy in this remark genuinely comes, like the rest, from Aristoxenus; it may strike one, perhaps, as having too much of the smell of Platonism about it. But I think such doubts would be misplaced. The sequel to the statement, like the tale of Telesias which precedes it, is unquestionably Aristoxenian, and once this compiler has set off on a passage derived from a major source he generally sticks to it without intruding material from elsewhere. Secondly, Aristoxenus was

[36] The thesis that it could be grasped through any one of the special sciences would thus fall foul of what I have called the 'same domain' rule. See p. 168 above.

himself a philosopher as well as a musical theorist; we have seen that he plainly shared some of Plato's views; and though some of Plato's philosophical ideas, especially in metaphysics, were rejected in the Aristotelian milieu in which he was embedded, a great deal of his teaching was absorbed and recognised as authoritative. If (as is often suggested) the present passage owes a good deal to Plato, and in particular to his *Phaedrus*,[37] there is nothing surprising about that. If the appeal to philosophy does indeed come from Aristoxenus, it will bring us back to a point I made when introducing this phase of our investigation (p. 234 above), that he is assigning the whole business of critical judgement on music to an intellectual elite, and that since harmonics is part of the critic's essential equipment, its arcane technicalities can no longer be left in the province of professional musicians alone. Its role as an instrument of aesthetic appreciation as well as its new guise as a reputable science should force it on the attention of people for whom philosophy, too, is a serious and worthwhile pursuit. We cannot blame Aristoxenus for the fact that he was if anything too successful in giving his brand of harmonics this intellectually elevated profile, and that in the hands of his 'scholastic' followers in later centuries it lost virtually all its connections with the realities of musical practice. But that is another story.

[37] See for instance the notes to the passage in Einarson and De Lacy 1967.

PART III

Mathematical harmonics

CHAPTER 10

Pythagorean harmonics in the fifth century: Philolaus

With the work of Aristoxenus this phase of the empirical tradition of harmonic theory has run its course. We now turn the clock back a hundred years or so to consider our earliest evidence about the mathematical form of the discipline, which originated with the Pythagoreans of the fifth century or conceivably with Pythagoras himself in the sixth. Evidence from later writers gives us a fair general picture of the approach to harmonics which they adopted and whose outlines I sketched in Chapter 1; I shall not repeat the programmatic points I made there. But on matters of detail and on the work of individual Pythagoreans of the early period the late sources are often unreliable. For the most part I follow Burkert in treating Aristotle as the most authoritative of our sources on the subject, together with a few fragments from the work of the late fifth-century Pythagorean Philolaus which most modern commentators take to be genuine. If they are, Philolaus is the only Pythagorean harmonic theorist before the fourth century of whose work we have solid and significant details.[1]

The prize exhibit is a short passage from Philolaus' essay *On Nature* quoted by Nicomachus and Stobaeus. In modern editions it is sometimes printed as a continuation of Philolaus fragment 6, sometimes as a separate item; I shall refer to it as frag. 6a. It has been much discussed,[2] but I have a serious purpose in re-examining it at some length here. Although much of what is regularly said about it is true, one important aspect of it seems to have been overlooked, or at any rate underplayed, and it has a crucial bearing on the relation between Philolaus' approach to harmonics and those of other theorists in both the main traditions. In order to unearth this feature and

[1] For the most thorough modern investigations of the fragments' claims to authenticity and of Philolaus' work as a whole see Burkert 1972, especially 218–98 and 386–400, Huffman 1993, and more briefly Kahn 2001 ch. 3. There is a good deal to be said for Burkert's view (broadly shared by Huffman and Kahn) that much of the information retailed by Aristotle also came from Philolaus' treatise.

[2] It appears as the second paragraph of DK 44B6. Detailed discussions of it will be found in Burkert 1972: 389–94 and Huffman 1993: 145–65.

to explore its implications we need to look closely at the whole fragment, even though this will involve repetitions of points made often by others before. The translation below is as nearly literal as I can make it, and with its thicket of transliterations and other barbarous coinages it may be barely intelligible as it stands. But it will be best to have it in front of us as a point of reference, in a version which does not smooth out the peculiarities of its diction. The first part of my discussion will try to elucidate them.

FRAGMENT 6A: PRELIMINARY ANALYSIS

The size of *harmonia* is *syllaba* and *di' oxeian*, and *di' oxeian* is greater than *syllaba* by an epogdoic. For from *hypata* to *messa* is a *syllaba*, from *messa* to *neata* is a *di' oxeian*, from *neata* to *trita* is a *di' oxeian*; and what lies between *trita* and *messa* is an epogdoic. *Syllaba* is epitritic, *di' oxeian* is hemiolic, and *dia pasan* is duple. Thus *harmonia* is five epogdoics and two *diesies*, *di' oxeian* is three epogdoics and a *diesis*, and *syllaba* is two epogdoics and a *diesis*.

The expressions *harmonia*, *di' oxeian* and *syllaba* refer to the octave, the perfect fifth and the perfect fourth. Hence at the most straightforward level the first sentence means simply 'an octave is a fourth plus a fifth', as indeed it is. But Philolaus' language needs closer inspection. His terms do not carry any reference to numbers, as do 'octave', 'fifth' and 'fourth' (which indicate that these intervals span the ranges between any given note of a scale and the eighth, fifth and fourth notes in order from it respectively; rather similarly, the regular Greek term for the fourth is *dia tessarōn*, 'through four [notes or strings]', the fifth is *dia pente*, 'through five', and the octave is *dia pasōn*, 'through all'). Philolaus' expressions *syllaba* and *di' oxeian* seem to come from the language of musicians, rather than philosophers or scientists, and *harmonia* inhabits both spheres. Both these facts are important. *Syllaba* means 'grasp'. According to the likeliest explanation, it referred originally to the group of strings which lie under a lyre-player's fingers in what we might call their 'starting position', poised over the four lowest strings on the instrument. *Di' oxeian* means 'through the high-pitched [strings]', that is, those that complete the span from the fourth string to the top of the octave. Thus in the musical scheme that Philolaus' description reflects, the most basic arrangement is one in which the pivot between the octave's main structural components is a fourth from the bottom and a fifth from the top.[3] The point that needs emphasis here is that when the terms are interpreted in this way, their direct reference is to components of the attunement which

[3] For an explanation of these terms along the lines I have sketched see Porph. *In Ptol. Harm.* 97.2–8.

lie in specific positions, and not to the sizes of intervals regardless of where they are placed.

The noun *harmonia* is one we have already met in several contexts. It is cognate with the verb *harmozein*, 'to fit together', 'to join', as in the work of a carpenter. It is the 'fitting-together' of diverse elements into a unity, and as such it plays a significant role in Presocratic cosmology, notably in Heraclitus and Empedocles[4] and in a crucial fragment of Philolaus himself, frag. 6, which we shall consider later. But it also figures prominently in Greek musical vocabulary, as we have seen, and in Philolaus it seems to form a bridge between the musical and cosmological domains. In one of its musical applications, and the most relevant here, a *harmonia* is an attunement, an integrated pattern of relations into which a collection of notes is 'fitted together' by a musician when he tunes his strings. Philolaus' attunement spans an octave, the range whose bounding notes are most perfectly coordinated with one another; and he knows the term regularly used by later writers to refer to that interval, *dia pasōn* (*dia pasan* in his Doric dialect). But he uses that expression only once in the fragment, at a point where he is simply identifying this interval's ratio. What his more prominent deployment of the word *harmonia* emphasises is the coherence and unity of the relation, not its dimensions. (Even his alternative expression, *dia pasan*, does not point directly to the interval's size or ratio, any more than do *syllaba* or *di' oxeian*, but only to its comprising the whole range covered by the attunement's notes.)

Hypata, *messa*, *trita* and *neata* are the Doric forms of the note-names which appear as *hypatē*, *mesē*, *tritē* and *nētē* in the Attic Greek of most theorists; from here onwards I shall use the words' more familiar forms. In the system which Philolaus describes, *hypatē* is the lowest note of the octave and *nētē* the highest; *mesē* is a fourth above *hypatē* and a fifth below *nētē*; *tritē* is a fourth below *nētē* and a fifth above *hypatē*; and *tritē* is higher than *mesē* by a whole tone (though Philolaus does not put it in that way, and his language poses problems to which we shall shortly return). These relations are set out in Figure 8. We have already become familiar with the division of an octave into two subsystems, each spanning a fourth and separated by a tone; it is normal in Greek theory, and other writers give the same names as Philolaus to the notes which form their boundaries, with one exception. The note in the position of Philolaus' *tritē* is usually called *paramesē*,

[4] See especially Heraclitus frag. 51, Empedocles frags. 26.11–14, where the cosmic principle of Love is called *Harmoniē* (cf. frags. 22, 35, and references elsewhere to Love under various names), and frags. 71, 96, 107.

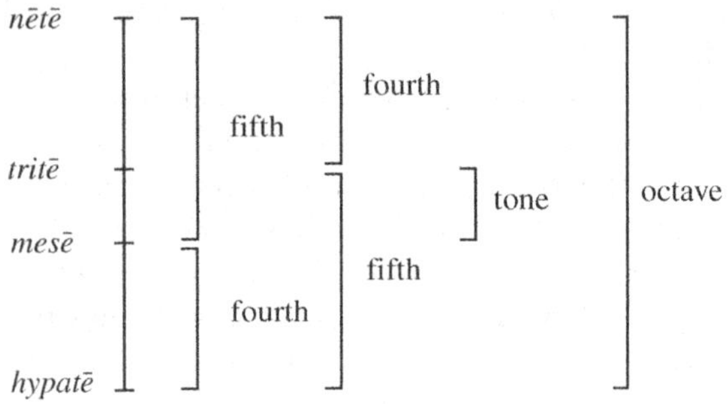

Figure 8 The *harmonia* of Philolaus

and *tritē* is differently placed. This oddity in Philolaus' usage has implications which will also be considered in due course.

I must explain one other group of expressions before we go on. The uncouth terms 'hemiolic', 'epitritic' and 'epogdoic' are crudely Englished versions of Greek adjectives referring to ratios. *Hēmiolios* means 'half-and-whole', and specifies the relation in which one item is one-and-a-half times the other, in the ratio 3:2. Expressions formed by adding the prefix *epi-* to an ordinal adjective (e.g. to *tritos*, 'third', or *tetartos*, 'fourth') mean 'a third in addition', 'a quarter in addition', and so on. Thus when the ratio between two terms is *epitritos*, 'epitritic', the larger amounts to the smaller 'and a third of the smaller in addition', and the ratio is 4:3. When it is *epogdoos*, 'epogdoic', the larger exceeds the smaller by one eighth of the smaller, in the ratio 9:8. The remaining ratio which Philolaus mentions, the 'duple' or 'double' (*diplasios*), is of course the ratio 2:1.

With these preliminaries behind us we can move on to consider the passage's overall structure.[5] It appears to be conceived as an argument, or perhaps as two separate arguments; the second sentence (given my punctuation) is linked to the first by the connective 'for' (*gar*), and the final sentence is introduced by the word 'thus' (*houtōs*). Any adequate interpretation must account for this impression of reasoning from premises to conclusions.

The position of the connective 'for' shows that if the first two statements amount to an argument, the first states the conclusion, the second the grounds on which it is based. Now every significant term in these sentences, with one exception, belongs to the vocabulary of musicians; only 'epogdoic' is a mathematician's expression, alien to the language of musicians and musically educated amateurs. If we set questions about this mathematical

[5] I shall postpone discussion of Philolaus' special way of using the word *diesis*.

intruder aside for a moment, Philolaus seems to be trying to convince anyone with an ordinary person's knowledge of music of the truth of his initial assertion. Because we all know that 'from *hypatē* to *mesē* is a *syllaba*, from *mesē* to *netē* is a *di' oxeian*', and so on, we can see that 'the size of *harmonia* is *syllaba* and *di' oxeian*, and *di' oxeian* is greater than *syllaba* by an epogdoic'. The argument is perfectly adequate. The pattern of relations set out in the second sentence, and apparently taken to be common knowledge, guarantees that a *syllaba* plus a *di' oxeian* is an octave; and a glance at Figure 8 will make it obvious that *di' oxeian* exceeds *syllaba* by an epogdoic.

All that is plain and simple, but there are two features of the argument that call for comment. The first is that despite its use of a mathematical expression referring to a ratio, nothing in it depends on the identity of this ratio or any other. No mathematical calculations are involved at all, and any arbitrary label for the interval between *mesē* and *tritē* would have served Philolaus' purpose equally well. His reasoning in these sentences would have been unaffected if he had called it the 'tone', *tonos* or *toniaion diastēma*, as other theorists do. I do not know why he chooses the more recondite expression, which is unlikely to have been used by musicians themselves. It is just possible that the more familiar *tonos* had not yet been adopted as this interval's name, though we would expect musicians to have had some word by which they could refer to it, and we know of no other.[6] It may be relevant that when the term *tonos* does acquire this meaning, it is normally treated as identifying an interval by reference to its size, and not (except in special contexts) by its position in the system; and the practice of assigning sizes to intervals, as I have already said, was itself a product of the theorists' work, not something inherited from traditional musical language. Whatever the explanation of Philolaus' choice may be, he must at some earlier stage in his essay have specified the ratios of *di' oxeian* and *syllaba* as 3:2 and 4:3 respectively, and shown that 3:2 exceeds 4:3 in the 'epogdoic' ratio, 9:8. From that moment on, he could use the word *epogdoos* either to refer to the relevant interval's ratio, or simply as a convenient stand-in for the (perhaps non-existent) musical name of the interval itself; and its function in the first two sentences is plainly the latter.

My second point about the first two sentences is this. If Philolaus' conclusion is obvious to anyone familiar with the pattern of concords formed by the four fundamental notes, it seems strange that he should need any sort of argument to prove it, especially one such as this, which would be incomprehensible to readers who lacked that elementary piece of musical

[6] For a brief and lucid discussion of the history of the word *tonos*, see Rocconi 2003: 23–4.

knowledge. I suggest that the reason is to be found in the fact which I mentioned above, that the familiar musical designations of the concords do not identify them by their sizes but by their locations in the system, and that the practice of assigning sizes to intervals was a theorist's innovation. Part of what Philolaus is doing is to persuade his audience that this practice makes sense, and that once we have accepted it we shall see that the interval between *tritē* and *nētē*, for instance, can intelligibly be described as the 'same' interval as the one between *hypatē* and *mesē*. In the language of ordinary musical discourse, *syllaba* is simply the name of the lower of these intervals. But as soon as we have introduced the notion of an interval's size, as Philolaus does in the first sentence, the name can be applied equally, as it is in the second sentence, to any interval with the same compass.

We shall return to this suggestion later, but first we must review the rest of the fragment. The next sentence is straightforward; it states, without argument, the ratios of the fourth, fifth and octave (it is here, where nothing but the ratios is in question, that we meet the standard term for the octave, *dia pasōn*). As I have pointed out, Philolaus' use of the term *epogdoos* presupposes that the ratios of the fourth and the fifth (and if theirs, then surely that of the octave) had been established in an earlier passage. This sentence, then, looks like a reminder, perhaps prompted by the immediately preceding reference to the epogdoic. If it has a function in the reasoning of the passage we have, it does not leap to the eye. This too is an issue we shall revisit later.

The final sentence, with its introductory 'thus', has the air of expressing a set of conclusions for which the necessary premises have just been provided. The conclusions are that *harmonia*, the whole structured complex of an attunement spanning an octave, amounts to five epogdoics and two *dieseis*,[7] the fifth or *di' oxeian* to three epogdoics and a *diesis*, and the fourth or *syllaba* to two epogdoics and a *diesis*. Plainly, however, these propositions cannot be inferred from the earlier part of the fragment without addition, if only because the *diesis* has not previously been mentioned. Three questions must be answered if we are to reconstruct Philolaus' train of thought. First, what, in his usage, is a *diesis*? Secondly, how can the surviving statements of the fragment most economically and plausibly be supplemented, so as to provide premises which will underpin his conclusions? Thirdly and crucially, in which of its roles does the term *epogdoos* appear here? Is its specification of a ratio relevant to the argument, or is it functioning (as

[7] In Philolaus' Doric, the plural of *diesis* is *diesies*, whereas in the Attic of most other theorists it is *dieseis*. I shall use the Attic form, as I do for the names of the notes.

in the first sentence) merely as a label for a musical interval, in a piece of reasoning to which ratios make no contribution?

The term *diesis* is common in musicological writings and we have met it before. It always refers to a very small interval, but its scope is not restricted to intervals of any one size. When a writer uses it to pick out one size of interval in particular, it may be anywhere between (roughly) a quarter-tone and a semitone, and only the context or an adjective qualifying the noun will reveal which it is. Sometimes the author makes it clear that the intervals he designates as *dieseis* come in several sizes.[8] In Philolaus' fragment, however, the *diesis* must plainly be an interval of just one size, since it is the interval by which a perfect fourth exceeds two whole tones (two 'epogdoics'). Given that the ratio of the fourth is 4:3 and that of the tone is 9:8, it turns out to be the so-called 'minor semitone', slightly less than exactly half a tone, whose ratio is 256:243.[9]

Did Philolaus know this cumbersome ratio? Archytas and Plato in the next century certainly did, and it has sometimes been argued that Archytas' very *recherché* handling of it points to an earlier and more straightforward context for its use, which is assumed to be that of Philolaus.[10] I am not altogether convinced by this argument, though for reasons I shall give later I think its conclusion is true. Let us suppose that it is, that Philolaus had established it earlier in the treatise, and that when he mentions the *diesis* here, its ratio is relevant to the argument. In that case it must also be relevant that the ratio of the tone is 9:8, as Philolaus' term for it indicates, and the argument must be of a mathematical sort.

Such an argument could be reconstructed, but it would be moderately complex, and in order to give Philolaus appropriate premises we would need to make quite substantial additions to the statements we have. There is a simpler solution. The word *diesis* is almost certainly drawn, like others in this passage, from the language of musicians. Outside musical contexts it means a 'putting through' or 'letting through', and its musical use may be based on the image of a pipe-player 'letting the melody through' almost imperceptibly, perhaps without any detectable movement of his fingers, from one note to another that is very close by.[11] We can interpret Philolaus' usage as referring most directly to the passage through the tiny 'space' remaining between the penultimate note of the system and the lowest,

[8] See e.g. Aristotle, *Metaph.* 1053a12–17, discussed on pp. 350–3 below.

[9] In order to distinguish Philolaus' rather unusual use of the term from that of other theorists, I shall continue to italicise *diesis* when it occurs in a Philolaan context, and not otherwise.

[10] See e.g. Burkert 1972: 388–9, Huffman 2005: 420–2.

[11] Cf e.g. West 1992a: 255 n. 42, Rocconi 2003: 11 n. 35.

hypatē, after we have moved down the scale from *mesē* through two steps of a whole tone each.[12] In that case, in order to justify the conclusion, we need make only one addition to the statements made in the first two sentences: '*syllaba* is greater than two epogdoics by a *diesis*'.[13] We can then construe the mathematical meaning of *epogdoos* as irrelevant to the present stretch of reasoning, as it is (so I have argued) in the opening sentences; and taken together with the supplement I have suggested, the contents of those sentences are sufficient to justify the conclusion. From this perspective the third sentence, where the ratios of the concords are identified, can be no more than a parenthesis. We can elide it, and the passage becomes a single, coherent argument which makes no essential reference to ratios at all. But we shall shortly have to consider it again.

There is another small but significant pointer to the conclusion I have drawn. When Philolaus mentions epogdoics in his final sentence, he puts the adjective in the neuter plural, *epogdoa*. If there is a noun to which the adjective is implicitly attached, it cannot be *logoi*, 'ratios'; the only plausible candidate is *diastēmata*, 'intervals'. *Diastēma* is a musical rather than a mathematical term, and we have a report (whose credentials are respectable though not unchallengeable) about Philolaus' use of it. According to Porphyry, he used it to refer to the 'excess' (*hyperochē*) of one item over another;[14] and in context this means that when one interval differs from another by a certain *diastēma*, we reach the one from the other by a process of addition. We get from a fourth to a fifth by 'adding' a certain amount, a whole tone, and the tone is the 'excess' of the fifth over the fourth. Ratios do not behave like this; we do not add 9:8 to 4:3 to reach 3:2 – the notion is nonsensical, and this is part of Porphyry's point. Hence once again, if we infer from Philolaus' grammar that he is implicitly referring to *diastēmata*, we must conclude that he is not relying at this point on the measurement of intervals in terms of ratios.

Given the Pythagoreans' well-documented championship of ratios in musical analysis and this fragment's seminal place in the mythology of orthodox scholarship in Greek harmonics, this is a remarkable result. It is clear enough that Philolaus was well acquainted with a ratio-based method and used it himself, but in the reasoning of this passage ratios take a back

[12] It cannot be demonstrated from this fragment alone that the *diesis* is the lowest of the three intervals between *mesē* and *hypatē*. But every other theorist who considers systems of this type in their basic form locates it in this position.

[13] I postpone consideration of the grounds on which Philolaus could have justified this assertion. Given that he had previously done so, such a statement would fit perfectly at the end of the second sentence. It is tempting to conjecture that Philolaus put it there, and that it somehow got lost in the course of the passage's transmission, but I shall not commit myself to so impertinent a guess.

[14] Porph. *In Ptol. Harm.* 91.11–13.

seat. His arguments require us to do nothing more complex than to recognise that talk of the 'sizes' of musical intervals makes sense, that some fundamental intervals in the system are equal to others, and that larger intervals can be constructed by adding smaller ones together. This looks much more like the 'empirical' theorists' procedure, and like theirs it is the direct offspring of musical experience, not of mathematics. In that case Philolaus' approach is a hybrid between two perspectives which were later treated as incompatible. But that fact, if it is one, is not after all so very startling. We have no indication that they were construed in that light until the last decades of the fourth century (see further pp. 362–3, 390–1 below).

So far as I am aware, this reading of the fragment is new, and no doubt it will be controversial. It also leads me to recant a position I have adopted in the past, that ideas attributed to Philolaus in two passages of Boethius should be rejected as spurious.[15] The principal reason for dismissing them is precisely that they mix a ratio-based analysis of intervals with one which represents intervals as having 'sizes', where these sizes are not expressed as ratios but as amounts to which simple magnitudes are assigned; and they imply that we can construct larger intervals from smaller ones simply by adding these magnitudes together. Although I would still maintain that Burkert's reasons for accepting Boethius' testimony are inadequate by themselves,[16] the reading of frag. 6a which I have offered suggests that the mixture of approaches involved in Boethius' accounts may have a genuinely Philolaan pedigree, despite its mathematical confusions. We must therefore take a look at this material; and this will lead back to some further thoughts on the passage we have been examining.[17]

[15] My only published comment on this issue (so far as I can recall), Barker 1989a: 38 n. 36, is noncommittal, but that was mere cowardice. Until very recently I was convinced that Boethius should not be believed.

[16] Burkert 1972: 394–400. His main argument depends on the assumption that Nicomachus, on one of whose lost works Boethius' account was based, found these aspects of Philolaus' theories in the same source as frag. 6a, which he quotes in a surviving treatise (Nicom. *Harm.* ch. 9). But that assumption cannot be guaranteed. Cf. Huffman 1993: 370, and n. 17 below.

[17] Huffman 1993: 364–74, gives a careful assessment of the two passages' claims to authenticity. He recognises the attractions of the hypothesis that one of them (Boeth. *Inst. mus.* 3.8) reflects early Pythagorean ideas, but contends, rightly in my view, that it is closely related to the other (*Inst. mus.* 3.5), which he takes to be spurious (366). But though his discussion of 3.5 succeeds in casting suspicion on it, I do not think it conclusive. The argument he regards as clinching the case (373–4, cf. 362–3) focuses on the passage's association of the unit with the point, the number 3 with the 'first odd line', and the number 9 with the 'first odd square'; and it depends on Burkert's attempt to show that the sequence 'point-line-plane-solid' belongs to a Platonist and not to a Pythagorean repertoire (Burkert 1972: 66–70, cf. 23–4, referring especially to the evidence of Aristotle, *De caelo* 299a2 ff.). The issues are too intricate to pursue here; I merely record the fact that I am not persuaded that they compel us to reject Boethius' report. Burkert himself did not deploy his arguments in an assault on its authenticity, as Huffman does; as I noted above, he regarded its evidence as reliable.

THE EVIDENCE OF BOETHIUS

The first of Boethius' two reports[18] alleges that Philolaus identified the whole tone with a specific number of 'units', 27, and that he divided this into two unequal parts, one called the *diesis*, comprising 13 units, the other called the *apotomē*, which is 14. He credits Philolaus with two arguments, of which the second seems to explain how he arrived at these numbers, while the first is designed to show that the numbers 13 and 27 have special significance. Only the second concerns us at this stage. In outline, it is this. The ratio of the *diesis*, conceived as the remainder of a perfect fourth after two whole tones (each in the ratio 9:8), is 256:243. The difference between these numbers is 13. We now construct a note a whole tone above the upper boundary of the *diesis*; this upper boundary is represented by the number 243, since smaller numbers are here assigned to higher pitches. The number to which 243 stands in the ratio 9:8 is 216, and the difference between 243 and 216 is 27. If we attach this number to the whole tone and 13 to the *diesis*, there will be 14 left for the larger part of the tone, the *apotomē*.

According to Boethius, Philolaus produced this remarkable piece of reasoning when considering how the whole tone can be divided in two, and it strongly suggests that he could find no way – in terms of ratios – of dividing it into equal parts. That is to be expected, since the ratio 9:8 cannot be divided into equal sub-ratios of integers, as Archytas later demonstrated (see pp. 303–5 below). But we need not suppose that Philolaus was equipped with a proof. Nor need we imagine that he computed the ratio of the *apotomē*, which in fact is 2187:2048. Instead, he used the ratio of the *diesis* as a basis for representing the 'size' of each of these intervals by a single number, as though it were a simple quantity or a 'distance' between two notes and could be measured as so many units. This is of course mathematically absurd.[19] But it fits well with Philolaus' combination of the two systems of measurement in frag. 6a. It attempts, indeed, however incoherently, to integrate them even more closely, by deriving the single numbers which represent the intervals' sizes from the terms of the ratios themselves.

[18] *Inst. mus.* 3.5, reproduced as DK 44A26.

[19] Its absurdity becomes obvious when we recall that the *diesis* 'is' 13 units because its ratio is 256:243 and 13 is the difference between this ratio's terms, and that 27 'is' the whole tone because its ratio is 243:216 and 27 is the difference between these numbers. If we divide this tone into 13 parts for the *diesis* and 14 for the *apotomē*, the *diesis* will run either from 243 to 230, or from 229 to 216; and patently neither 243:230 nor 229:216 is equal to 256:243. (This way of illustrating the point comes from Burkert 1972: 396.)

Boethius' second report claims that Philolaus offered four definitions, which can be paraphrased as follows.[20] (i) A *diesis* is the interval by which the perfect fourth (ratio 4:3) exceeds two whole tones. (ii) A *komma* is the interval by which a whole tone (ratio 9:8) exceeds two *dieseis*. (iii) A *schisma* is half a *komma*. (iv) A *diaschisma* is half a *diesis*. It is again obvious that Philolaus is not thinking in terms of ratios alone. The ratio of the *komma* can be computed; it is 531441:524288, but this – in Burkert's phrase – is 'pure frivolity'.[21] Boethius has already told us, in fact (*Inst. mus.* 3.5), that Philolaus identified the *komma* with the unit, 1, as being the difference between a *diesis* (13) and an *apotomē* (14). No pairs of integers will specify the ratios of the *schisma* or the *diaschisma*. If these vanishingly small intervals had any place in Philolaus' thought, they must have been conceived simply as quantities or linear distances.

All this is very odd, but if Boethius' account is reliable, there is no great mystery about the roles in which Philolaus deployed these peculiar calculations and definitions. They need have nothing to do with the analysis of different varieties of musical scale, the chromatic and enharmonic systems, as Burkert (following Tannery) suggested;[22] Boethius says nothing to point in that direction, and if the conclusions I shall sketch later about Philolaus' overall agenda are correct, he had no reason to be interested in scales of those sorts. What the passage at *Inst. mus.* 3.8 indicates is that they are wholly focused on the division into equal halves of the elementary intervals of the system outlined in frag. 6a, the tone and the *diesis*. Half a *diesis* is a *diaschisma*, and half a tone is two *diaschismata* plus one *schisma*; and in order to describe the *schisma* we have first to define the *komma*. None of this can be understood in terms of ratios. All the material in Boethius seems, then, to be continuous with the contents of the fragment quoted by Nicomachus and Stobaeus. Its primary concern, like that of the fragment, is with the business of measuring intervals in a musical system against one another as simple quantities.

BOETHIUS AND FRAGMENT 6A

The question why Philolaus was concerned with the business of 'halving' the tone and the *diesis* is one I shall postpone; there are other matters that we must consider first. The simple quantities involved in Boethius' reports are derived, paradoxically, from ratios – specifically the ratio of the *diesis* –

[20] *Inst. mus.* 3.8, printed as an appendix to DK 44B6.
[21] Burkert 1972: 395. [22] Burkert 1972: 398; cf. Tannery 1904b: 224–5, Huffman 1993: 365.

and ratios are also mentioned in frag. 6a. Let us now return to that fragment and ask whether the ratios figure there, despite my earlier comments, in more than a merely parenthetical role. I have suggested that part of its purpose is to drive home the possibility of shifting from the old practice of referring to intervals by their positions in the system to one which identifies them by their 'sizes', in some sense of that expression. We have seen also that Philolaus must have offered some kind of demonstration, earlier in his treatise, of the ratios of the intervals he calls (in the relevant sentence) *syllaba*, *di' oxeian* and *dia pasōn*, and of the one that 'lies between *tritē* and *mesē*', the 'epogdoic'. If Boethius is to be trusted, he had also worked out the ratio of the *diesis*. If these intervals have determinate ratios, then there can be intervals elsewhere in the system which are relevantly 'the same'; they have the same ratios and therefore the same 'sizes'. The third sentence of the fragment, which alludes to the ratios, will then serve as a reminder that this approach has a solid basis, and reinforces the conceptual shift which the passage is negotiating.

The fragment's arguments, however, are couched mainly in terms familiar to a musically educated audience that is innocent of mathematics, and presses these terms into service in the new environment of sizes and measurement in ways that make no direct appeal to ratio-theory. The vocabulary of the Boethian passages, by contrast, with its *apotomē*, *komma*, *schisma* and *diaschisma*, goes far beyond the regular terminology of musical discourse, and the exposition contains strange manipulations of numbers. But the essential conceptual apparatus used there is a direct extension of that of frag. 6a. In both cases, despite the references to ratios, each interval is treated primarily as an item of such-and-such a size, and intervals can be added together, divided in half and so on, just like the linear 'distances' involved in the empiricists' system of measurement.

If this reading of the fragment and Boethius' reports is on the right lines, it smooths the path between them and strengthens the case for regarding the latter as reliable. It also opens up a new, though inevitably hypothetical perspective on Philolaus' bizarre approach to the quantification of intervals, which is obvious in the Boethian passages but which is also detectable – or so I have argued – in the fragment. Scholars have generally treated the curiosities expounded by Boethius either as displaying only mathematical incompetence, or as evidence of Philolaus' obsession with 'number-symbolism', or both. My suggestion, starting from the reinterpretation of frag. 6a, is that it represents an attempt to fuse two quite different ways of measuring intervals, both of which were current among theorists but still in their infancy in Philolaus' time, one set in terms of ratios and characteristic of 'Pythagorean' harmonics, the other staying closer to musical experience and used principally by theorists engaged professionally in musical practice. It undermines neither the accusation of mathematical incompetence nor the thesis, to which I shall return, that Philolaus credited certain numbers with symbolic meaning. It merely adds that a further explanation for his manoeuvres can be extracted from the history of harmonics itself; and it posits that the relation between the two approaches was less clear-cut and less adequately understood in this period than is usually supposed.

I have still not explained why (in my opinion) Philolaus undertook the project whose problematic remnants we find in the passages we have been considering, and why he thought it necessary to pursue these strange calculations. Any suggestion about the details must be premised on a view about the overall purposes of his foray into harmonics, and I shall say little about it at this stage. There is broad agreement between most scholars on the fundamental points; I think the consensus is correct, and I shall return to it later. To put it in its simplest terms, it is that Philolaus' harmonic investigations should be understood, at least primarily, as a contribution to cosmology and not to the study of music for its own sake. His analyses of musical relations are only means to the greater end of understanding reality as a whole; and if, as I have suggested, he takes a leaf out of the empiricists' book, that is only because he saw it as a useful addition to his cosmological tool-kit.

THE MUSICAL STRUCTURE OF PHILOLAUS' ATTUNEMENT

There is one intrinsically musical issue, however, that needs to be clarified first. The language of frag. 6a shows that Philolaus is drawing on his own and his readers' knowledge of features of a form of attunement used regularly in real music-making. Commentators from Nicomachus onwards have inferred that although it certainly spanned an octave it contained only seven notes; by the standards of the eight-note octaves treated as the norm by later theorists, one element is missing. The inference is based on Philolaus' use of the name *tritē* for the note a tone above *mesē* and a perfect fourth below *nētē* (elsewhere the note in this position is called *paramesē*). *Tritē* means 'third [note or string]', and in other sources invariably refers to the third note counting down from *nētē*. If it has the same sense here, there can be only one note instead of the usual two between the boundaries of the fourth at the top of the octave;[23] see Figure 9.

[23] Nicom. *Harm.* 9.253.3–254.2 Jan, cf. e.g. Huffman 1993: 153–6, 165.

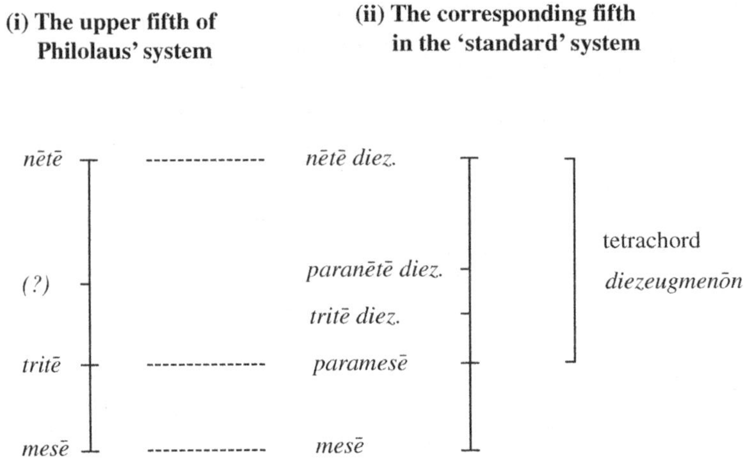

Figure 9 The 'third note' in Philolaus' *harmonia*

There is nothing very startling about this conclusion. Although instruments with eight strings (and sometimes more) were quite common among musicians of this period, we know from vase-paintings and allusions in poetry that there were only seven on the traditional tortoiseshell lyre, on which schoolboys still learned the rudiments of the art.[24] Anyone with enough education and leisure to take an interest in Philolaus would certainly have known the names of this instrument's notes and strings, even if they had only a concert-goer's more distant acquaintance with the elaborate instruments used by professionals.

Nicomachus also assumes – and most modern commentators agree – that the scale built into Philolaus' attunement proceeded, with one exception, through steps of a whole tone or a *diesis* each. The exception is the interval above *tritē*, where a note that would appear in the eight-note system is missing; and this, Nicomachus tells us, amounted to a *diesis* plus a tone, roughly a minor third.[25] This diagnosis, too, is plausible. Diatonic octave-systems in which the fourths at the top and the bottom are divided, reading downwards, into steps of tone, tone, *diesis* are well known elsewhere;[26] and

[24] Cf. e.g. the Homeric *Hymn to Hermes* 49–51, Pindar, *Nem.* 5.22–5, Ion frag. 32 West. Reputedly ancient seven-note systems are also mentioned in some theoretical and historical sources, e.g. [Arist.] *Probl.* 19.7, 32, 47, [Plut.] *De mus.* 1140f; Nicomachus himself describes another version of such a system in *Harm.* 3, cf. *Harm.* 5.

[25] Nicom. *Harm.* 9.253.6–7.

[26] Most relevantly in Plato, *Tim.* 35b–36b, [Eucl.] *Sect. can.* propositions 19–20.

Philolaus' scheme would differ from those only in so far as it omitted one note. Perhaps he gave a full description of this attunement's structure in a passage which has been lost. But we should note that there is nothing to prove it, either in frag. 6a or in Nicomachus. Nicomachus cites no authority for his statements apart from the fragment itself, and we may reasonably suspect that they represent only his attempt to interpret it. The fragment's last sentence quantifies the octave, fifth and fourth in terms of whole tones ('epogdoics') and *dieseis*, but it does not say that the system's scalar steps were of these sizes and no others. It cannot itself be construed as a list of these steps, if only because it quantifies the octave by reference to seven intervals and there can be only six steps in his seven-note system. The evidence leaves open the possibility – though we may think it far-fetched – that the scalar steps of the attunement which Philolaus had in mind were not all tones or *dieseis*; and in any case it does not guarantee that he analysed any system in all its details.

But let us suppose that he did, and that it followed the diatonic pattern I have specified. It remains true that the focus of the fragment is on the three concords, octave, fifth and fourth, which give the attunement its fundamental structure, on their sizes (whether conceived as ratios or as 'distances') and the quantitative relations between them, and on the way in which the fourths and fifths interlock to form the octave-attunement's integrated and symmetrical skeleton. We may pertinently add that if the attunement took the form that Nicomachus attributes to it, any substantial emphasis on its minutiae would have undermined the impression of complete symmetry conveyed by the fragment's second sentence, since the pattern of intervals into which the fourth between *hypatē* and *mesē* is divided is not perfectly replicated in the fourth between *tritē* and *nētē*. It seems to me that a good deal (not quite all) of the information retailed by Boethius fits neatly into the same agenda.

If the system's symmetry is an important consideration, we should be able to specify its mid-point, the fulcrum upon which equivalent structures at the top and the bottom are balanced. In its centre is the interval of a tone between *mesē* and *tritē*, in the epogdoic ratio 9:8; and 9:8 cannot be factorised into two equal ratios of integers (see p. 272 above). But there is an interval, the *diesis*, which is very close to half a tone, and which plays a part in the attunement's analysis. According to our present hypothesis it is a recognisable interval of a musical scale, and its ratio is known. On the basis of the terms of this ratio, 256:243, Philolaus converts the *diesis*, quite improperly, into a simple quantity represented by the number 13, and by a similar strategy he identifies the tone with the number 27. The larger remnant of the tone, amounting to 14 units, is then named as the *apotomē*, and the difference between the *diesis* and the *apotomē*, one unit, is called the *komma*. If we divide the whole tone in half, then, we are cutting it into segments amounting to a *diesis* plus half a *komma* each. Doggedly pursuing this line of thought, Philolaus assigns a name also to half a *komma*; it is the *schisma*. Hence we can identify the absolute centre of the system as lying at a point higher than *mesē* by a *diesis* and a *schisma*, and lower than *tritē* by the same amount.

This reconstruction of Philolaus' reasoning is obviously speculative, but it accounts for almost every ingredient of Boethius' reports. It falls short by failing to explain why he also provides a name, *diaschisma*, for half a *diesis*; and it says nothing about his reflections on the significance of the numbers 27 and 13. I shall comment briefly on the second of these points in due course, but I confess that I have no plausible explanation of the first. My interpretation has the substantial advantage, however, that it places these manoeuvres in precisely the context to which Boethius assigns them, that of the division of the tone. Rival interpretations, as I have said, have located them in another setting, the analysis of enharmonic and chromatic scales. In that project there will be a role for the *apotomē* (in the chromatic) and for the *diaschisma* (in the enharmonic).[27] But there will be none for the *komma* and the *schisma*; and there is no independent evidence in Boethius or anywhere else that Philolaus attempted an analysis of more than one variety of scale. Again, if it is agreed that his primary interests were cosmological rather than musical, it will not have been the special features of the various different types of scale that captured his attention, but their similarities, the shared features which gave all of them their structural coherence and distinguished them from arbitrary jumbles of notes and intervals. If we can extrapolate from the consensus of later writers, that coherence is created, in every case, by exactly the symmetrical interweaving of fourths and fifths, tying together the opposite ends of the octave, which Philolaus places at the heart of his enterprise.

HARMONICS AND COSMOLOGY

Let us now try to place all this in the context of Philolaus' cosmology. The only direct textual evidence that frag. 6a belongs with the more familiar and more explicitly cosmological frag. 6 (the first paragraph of DK 44B6) is very weak, consisting in nothing but the fact that they are quoted together

[27] For details see Burkert 1972: 398.

as a continuous passage by Stobaeus. But we are now in a position to see that they do indeed hang intelligibly together, whether or not frag. 6a followed frag. 6 immediately in Philolaus' treatise (which I am inclined to doubt), along with certain other Philolaan fragments (notably frags. 1–4), the material in Boethius and much of Aristotle's testimony. Students of Presocratic philosophy will be no strangers to the first half-dozen fragments of Philolaus; they have been discussed many times, and here I shall pick out only a few essentials.[28]

According to Diogenes Laertius, frag. 1 was the opening sentence of Philolaus' book. 'Nature in the *kosmos* was fitted together (*harmochthē*, from a verb cognate with *harmonia*) from things that are unlimited and things that impose limit, both the whole *kosmos* and all the things in it.' Frag. 2 offers an argument to show that things cannot all belong either to the class of 'unlimiteds' or to that of 'limiters'; and frag. 3 sketches another reason why they cannot all be 'unlimiteds'. Frag. 6 returns to the same ideas. It tells us that none of the things that there are, and of which we can have knowledge, could come into existence if the 'being' of the things from which the *kosmos* is constituted did not include both limiters and unlimiteds. Philolaus continues:

> But since these principles (*archai*) are not alike or of the same kind, it would have been impossible for them to be coherently integrated (*kosmēthēnai*) with one another if *harmonia* had not come upon them, in whatever way it came into being. Things that were alike and of the same kind had no need of *harmonia*, but it was necessary for things that were unlike, and not of the same kind or rank, to be held together by a *harmonia* of the sort that would hold them together in a *kosmos*.

Earlier Presocratics too had made use of the idea that material objects and the universe as a whole are made up of different or opposed ingredients fitted together or 'harmonised' in some special way; and medical writers more or less contemporary with Philolaus sometimes expressed this relation in explicitly musical terms, or assimilated musical *harmonia* to a universal principle through which mutually hostile elements are amicably integrated into a unity.[29] The language in which Philolaus expresses this thought is unusually abstract, and he does not clarify his remarks – at least, not in this passage – by mentioning examples of these limiters and unlimiteds. We can make some reasonable guesses about unlimiteds, however, and a strong

[28] For judicious brief discussions of these and other fragments see Kirk, Raven and Schofield 1983: 324–8, Kahn 2001: 23–38; see Burkert 1972 and Huffman 1993 for fuller treatments and bibliographies.

[29] See the passages from Presocratic philosophers cited in n. 4 above. For allusions in medical sources see the Hippocratic *De victu* (*On Regimen*) 1.8.2, a passage we shall look at briefly below, and the speech attributed to the doctor Eryximachus in Plato, *Symposium* 185e–188e, especially 187a–e.

candidate for the role of limiter is number. Length, for instance, conceived simply as such, is unlimited and inadequate to define any particular thing; it becomes something determinate only when pinned down by a number.[30]

I do not think we should hesitate to connect frag. 6a and the reports by Boethius closely with the ideas of frags. 1–6. The notion of an interval in the dimension of pitch picks out nothing determinate. It is unlimited in the sense that it has no definite boundaries; it can be as large or as small as you like. For any actual interval to come into being, this unlimited 'amount' must be limited by number, whether the numbering involves ratios or quantified linear distances. But the association of limit with the unlimited can give rise to a coherent whole (*kosmēthēnai*) only if '*harmonia* comes upon it'; and in the context of frag. 6a *harmonia* is musical attunement.[31] The fragment shows us how intervals determined by number are 'locked together', as frag. 6 has it, into a well-ordered and symmetrical whole.

Before glancing again at Boethius' reports in this connection, I want to add a word or two about a passage in a work of a very different sort, the Hippocratic treatise *On Regimen* mentioned in n. 29. It is generally assigned a date around 400 BC.[32] The writer is discussing the development of the human foetus.

> When it has moved to a different place, if it attains a correct *harmonia* containing three concords, *syllabē, di' oxeiōn* and *dia pasōn*, it lives and grows using the same nourishments as before. But if it does not attain *harmonia*, and the low-pitched (*barea*, lit. 'heavy') elements do not become concordant with the high-pitched (*oxea*, 'sharp') in the first concord or the second or that which runs through all (*dia pantos*), if just one of them is faulty the whole tuning (*tonos*) is useless. (*On Regimen* 1.8.2)[33]

The passage's resemblances to Philolaus leap to the eye. It echoes his musical terminology (transposed into the Attic dialect), which is uncommon elsewhere, and its use of the word *harmonia* (which recurs in 1.9) exactly parallels Philolaus' deployments of it in frags. 6 and 6a. *Harmonia* is simultaneously the principle which must govern relations between diverse elements in the developing foetus if it is to become a living whole, and a structure

[30] Compare frags. 3–4, where we are told first that nothing could be known if all things were unlimiteds, and secondly that all knowable things possess number and could not be known if they did not.

[31] This is a slightly different perspective from that of Eryximachus at Plato, *Symp.* 187a1–c5, where the hostile elements to be harmonised are high and low pitch, but the underlying thought is of the same order.

[32] See e.g. Joly 1960: 203–9.

[33] I follow here the text printed in the Budé edition of 1967. It involves several emendations, of which the most important is in the list of concords in the first sentence. The words '*syllabē, di' oxeiōn*' were suggested by Bernays in 1848 and independently by Delatte in 1930 to correct the MSS text *syllēbdēn diexiōn* (or a slight variant), which makes no good sense. Like most commentators I regard the emendation as certain.

spanning the compass of an octave, properly organised into substructures spanning a fourth and a fifth. It seems rather likely that the author had read Philolaus *On Nature*, and was drawing on it directly. What is particularly striking is that unlike many later non-musical authors who exploit notions in 'Pythagorean' harmonics, he makes no allusion to ratios; that aspect of frag. 6a is completely elided. For the writer of *On Regimen* it was apparently the musically described structures of an octave-attunement, not its interpretation in mathematical terms, that could provide a useful model for the organisation of a healthy embryo.[34] If Philolaus' priorities could be read in this way in his own time, this passage gives further support to my contention that the principal focus of his argument, in that part of his work, was on relations between intervals conceived in the manner of musicians rather than mathematicians, and that his approach had much in common with that of the empiricists.

We return now to Boethius' reports. I have explained how I interpret the bulk of them in relation to Philolaus' picture of a *harmonia* as a fully integrated structure, but there is one more aspect of the account at *Inst. mus.* 3.5 that calls for comment. Besides revealing how the numbers 27 (for the tone) and 13 (for the *diesis*) are derived from the intervals' ratios, Boethius tells us that Philolaus invested these numbers with significance in their own right. According to Philolaus,

> the tone had its origins in the number that constitutes the first cube of the first odd number, for that number was greatly revered among the Pythagoreans. Since 3 is the first odd number, if you multiply 3 by 3, and then this by 3, 27 necessarily arises; and this stands at the distance of a tone from the number 24, the same 3 being the difference. For 3 is an eighth part of the quantity 24, and when added to the same it gives the first cube of 3, 27.[35]

As for the *diesis*, its number is 13 partly because this is the difference between the terms of its ratio, 256 and 243, but also 'because the same number – that is, 13 – consists of 9, 3 and unity, of which unity holds the place of the point, 3 the first odd line, and 9 the first odd square'.[36]

Philolaus' main strategy here is clear enough; he aims to trace these musical numbers back to an origin in the number 3.[37] The Pythagoreans' interest in the symbolic meanings of numbers is well known, and I shall

[34] Hence it is misleading to use this passage, as Burkert does (1972: 262), to confirm that 'the numerical ratios' play a part in Hippocratic embryology.

[35] This translation is that of Bower 1989 with slight modifications.

[36] Translation from Bower 1989.

[37] Given that preoccupation and the use he makes of squares and cubes, he can hardly have failed to notice that one of his key numbers, 243, is the product of the square and the cube of 3. The other, 256, is based similarly on the numbers 2 and 4. It can be factorised as the square of 4 multiplied by itself, or as the cube of 4 multiplied by the square of 2, or in several other such ways.

not attempt any general exploration of that bewildering territory; but a few points deserve emphasis. First, the Greeks were finding non-mathematical significance in specific numbers long before the time of the Pythagoreans, whose speculations in this area probably involved more reflection on pre-existing ideas than autonomous invention.[38] Secondly, the number 3 seems to have been among the most important in this wider cultural context, and so too, in at least one rather striking setting, were its square and cube.[39] Finally, notions connected with these and other numbers in Pythagorean and other sources can be placed on a continuum from the less to the more abstract and mathematical. Thus we find a widespread association of 3 with the male and 4 with the female;[40] at a slightly more abstract level comes the link between the number 3 and completeness, since it has a beginning, a middle and an end;[41] while Philolaus' excavations of this number from 27 and 13, and his identification of it with the first odd number and the first odd line, are based in genuine arithmetic (and a species of geometry), even if his motivation has more impressionistic cultural roots.

The number 3 and the unit, in terms of which Philolaus analyses 27 and 13, also join 2 and 4 as elements in the celebrated *tetraktys* of the decad, the numbers 1, 2, 3, 4, adding up to the perfect number 10.[42] I need not harp on this theme, which is discussed by every modern writer on the Pythagoreans; and readers will hardly need to be reminded of Aristotle's sarcastic comments about the importance of this perfect number in Pythagorean astronomy.[43] But it is perhaps worth underlining Aristotle's remark in the same passage that the Pythagoreans linked to their account of the organisation of the universe 'everything they could find in numbers and in *harmonia* that agreed with the attributes and parts of the heaven'. This allusion to *harmonia* in the context of comments on the role of the number 10 points to a connection between this number and those involved in musical analysis; and it surely lies in the fact, repeatedly mentioned in later sources, that it is the numbers involved in the ratios of the primary concords, 2:1, 3:2 and 4:3, that come together to form the *tetraktys* of the decad.

Given that Aristotle drew much of his information about Pythagoreans from Philolaus, this would seem to be another instance of Philolaus' fusion

38 See Burkert 1972: 474–9, with his references to the fascinating study of Homeric numerology in Germain 1954.

39 Burkert 1972: 474–5. The numbers 3 and $3 \times 9 = 27$ appear in the context of the cult of the Erinyes at Soph. *OC* 479 ff.

40 Burkert 1972: 475–6.

41 See Aristotle, *De caelo* 268a10 ff.

42 On ways of connecting 2 and 4 with Philolaus' numbers, see n. 37 above.

43 Aristotle, *Metaph.* 986a3–12.

of the mathematics of ratio with thoughts about individual numbers. Very probably these ideas were widespread in the earlier Pythagorean tradition. The connections between the *tetraktys* and the harmonic ratios may be associated with the 'pebble-diagrams' attributed to Eurytus and others, in which the number 10 is represented by an array of ten dots or pebbles set out in rows to form an equilateral triangle, with one dot at the top, two in the row below it, three in the next and four in the last; and the ratios of the concords appear in the relations between the numbers of dots in adjacent rows.[44] This is a long way from anything we would consider soberly as 'mathematics'. The link between the ratios of *harmonia* and the hypothesis that there are ten principal bodies in the cosmos (which Aristotle derides) does not lie in the supposition that there are ten elements in a *harmonia*, integrated in determinate ratios; there are not. It has nothing to do with a ratio-based 'harmony of the spheres'.[45] Pride of place is given, once again, to numbers extracted individually from the ratios, not to the relations expressed by the ratios themselves.

But there is another twist in this curious maze. A passage in Nicomachus' treatise on arithmetic asserts that some people, 'following Philolaus', adopted the name 'the harmonic mean' (*mesotēs harmonikē*) for the mathematical mean otherwise known as 'subcontrary' (*hypenantia*), and offers a rather odd explanation for Philolaus' usage which we need not investigate.[46] The mean in question can be defined as follows. When we have three terms, A, B, C in descending order of magnitude, B is the harmonic or subcontrary mean between A and C if the fraction of C by which B exceeds it is the same as the fraction of A by which A exceeds B. In the example most commonly used by Greek writers, A is 12, B is 8 and C is 6. A second report, in a work of Iamblichus, asserts that Philolaus among others 'is found to have made use of musical proportion (*mousikē analogia*)'.[47] Musical proportion involves four terms, including between the two extremes both the harmonic mean and the arithmetic mean, which is such

[44] See e.g. Kahn 2001: 31–2.

[45] It should follow that when Aristotle wrote his famous account of just such a ratio-based 'harmony' at *De caelo* 290b12 ff., he was not thinking of a theory held by Philolaus or perhaps by any Pythagorean (he does not actually say who its champions were), though most commentators conclude that he was. But inferences of that sort are risky in this treacherous territory, and although we might find it hard to square the notion that there are ten celestial bodies with the thesis that they form a coordinated pattern of attunement, or to accommodate smoothly to one another the ways in which numbers and ratios are treated in the two contexts, it is perfectly possible that all these ingredients and numerical manipulations could co-exist comfortably in Philolaus' methodologically malleable mind.

[46] Nicom. *Arith.* 2.26.2.

[47] Iambl. *In Nicom.* 118.23; this and the passage of Nicomachus are printed together as DK 44A24.

that if we have three terms as above, A − B = B − C. The proportion is regularly illustrated by the series 12, 9, 8, 6, in which 9 is the arithmetic mean between 12 and 6, and 8 is their harmonic mean. The reliability of these reports has been carefully assessed by Carl Huffman, who concludes that there is nothing suspect or improbable about what Iamblichus says, but that Nicomachus' statement (which he takes to imply that Philolaus originated the use of the term 'harmonic mean') is open to serious doubt.[48] This scepticism about Nicomachus rests on foundations which I think can be undermined, but that issue need not concern us; if Philolaus understood the notion of musical proportion he must certainly have known of the subcontrary mean, whether or not he was the first to call it 'harmonic', since the latter is involved in the definition of the former.

In the case of Iamblichus too, Huffman's assessment is not conclusive, as I think he would agree. But its grounds are quite strong; let us assume that he is right. There is no difficulty in giving musical proportion a place in Philolaus' harmonic constructions, as we know them from frag. 6a. When the relations between *hypatē*, *mesē*, *tritē* and *nētē* in its octave attunement are specified as ratios, the ratio of *nētē* to *tritē* is 4:3, that of *tritē* to *mesē* is 9:8, and that of *mesē* to *hypatē* is 4:3. These relations are exemplified in the sequence 12, 9, 8, 6, and the whole construction is a specimen of musical proportion.

So far we have met no difficulties. But when we ask what role the concept of musical proportion might have played in Philolaus' scheme, we may hesitate about the answer. Means and proportions have serious work to do in the harmonic constructions of Archytas and Plato in the next century; their systems are grounded in the principle that notes (conceived as the terms of ratios) earn their places in a coherent pattern of attunement by intervening between other notes as means of appropriate sorts (see pp. 302–3 below). But the coherence of Philolaus' system, so far as our direct evidence takes us, rests on no such mathematically sophisticated foundations. It depends, as we have seen, on a conception of symmetry which makes no appeal to these means and proportions (of which we hear nothing in the other Philolaan fragments and testimonia), and in which the ratios themselves play only a secondary role. It appears to have been assembled partly from the culturally shared repertoire of musical knowledge (frag. 6a), and partly from piecemeal, unsystematic and mathematically naïve manipulations of numbers.

[48] Huffman 1993: 168–71.

Perhaps, then, Huffman's doubts about Nicomachus' report are justified; but we can still accept what Iamblichus says, though not for Huffman's reasons. On what I take to be the likeliest interpretation of his statement, it is not only plausible but uncontroversially true. He does not say that Philolaus, or anyone else whom he names, understood the notion of musical proportion or even mentioned it, but only that it is something 'they are found to have used'. This need mean no more than that if you look at the system of ratios in Philolaus' *harmonia*, you will find that the terms are related in musical proportion, as indeed they are.[49] We even have a test case for this interpretation, since another writer in Iamblichus' list of those 'found to have used' musical proportion is 'Plato in the *Timaeus*'. Terms in the relevant proportion certainly figure in Plato's harmonic construction at *Tim.* 35b–36b, as they do in Philolaus, but the word *analogia*, 'proportion' does not occur in the passage; and though a 'proportion' involving four terms is mentioned earlier (32a–c), its description at 32b makes it clear that it is not 'musical proportion', and it is not called by that name. Iamblichus is merely identifying systems whose elements fit into this proportional pattern. He does not say that the conception was available to these systems' authors, still less that they understood and explicitly deployed it; and if that is what he meant to imply, the evidence of the *Timaeus* shows that he is not to be trusted.

My overall conclusion, then, is that the usual interpretations of Philolaus' work in harmonics are mistaken at least in their emphasis. I agree that his contributions to the science are to be placed in the context of his cosmology; and he was certainly familiar with the use of ratios in musical analysis and knew the ratios of certain intervals. But so far as our evidence goes, he did not treat ratios as the only acceptable representations of musical intervals, as later Pythagoreans and Platonists did, or even as the most illuminating ones. Nor did he require that a coherent system of attunement should be generated by any mathematical principle of proportion. He used ratios to underpin the (relatively new) idea that intervals can be identified by their sizes, and as a starting-point for identifying certain intervals with particular numbers. His approach combines calculations involving ratios with the 'linear' conceptions of practical musicians and empirical theorists,

[49] Cf. Huffman 1993: 169. 'That Philolaus knew of and used the "musical proportion" (12, 9, 8, 6) is very probable in the light of the musical theory found in F6a. In fact when Iamblichus says Philolaus is "found using" the musical proportion he may well be referring exactly to F6a.' So he may. But if frag. 6a is the basis of Iamblichus' remark, we have no reason to accept Huffman's inference that Philolaus 'knew of' this type of proportion and deliberately used it. The presence of musical proportion in the *harmonia* is determined independently, as it turns out, by the facts of musical practice.

and it is the latter that play the most important part in his account of the system's integration and symmetry.

None of this conflicts with the testimony of Aristotle. He explicitly mentions the Pythagoreans in connection with ratio-based harmonics only once, in the course of his long exposition of Pythagorean ideas at *Metaph.* 985b23 ff.[50] Here his reference to the 'ratios of the *harmoniai*' (985b31–2) appears in a context preoccupied with their emphasis on numbers, not on relations between numbers, as the basic elements in reality; and the discussion runs on seamlessly to their contention that there must be ten principal bodies in the universe, since 10 is the perfect number. A remark a little later in the *Metaphysics* (987a22–5) seems to reflect a slippage between ratios and individual numbers very similar to those we have noted in Philolaus. The Pythagoreans' definitions are superficial, he says, because 'they supposed that the first [item, perhaps number] to which a given definition belonged was the essence (*ousia*) of the thing, as if one supposed that double and the dyad are the same, on the grounds that the double always belongs first to the number 2'.[51] In general Aristotle is sharply critical of the Pythagoreans, especially of their misunderstandings about number; whereas he takes a positive view of the credentials of mathematical harmonics as such, as we shall see in Chapter 13, and exploits some of its propositions for his own purposes. It does not necessarily follow, of course, that what he knew of specifically Pythagorean (or Philolaan) harmonics was infected with their confusions about number in wider metaphysical or cosmological contexts, but I can see no reason why it should have been an exceptional case, somehow quarantined from the contagion.

50 The ratios of the concords play an essential role in the theory of the harmony of the spheres discussed by Aristotle at *De caelo* 290b12 ff., and when he mentions 'certain people' who hold this theory he may well have Pythagoreans in mind. But as I noted above, he does not say so.

51 This strategy is not identical with the one used by Philolaus in defining the *diesis* as the number 13, but their affinities are obvious.

CHAPTER II

Developments in Pythagorean harmonics: Archytas

Archytas of Tarentum, according to Aristoxenus, was the 'last of the Pythagoreans' in the continuous tradition stretching back to the founder. The precise dates of his birth and death are not known, but his life seems to have spanned almost exactly the same period as Plato's (427–347 BC), and if the seventh Platonic letter is genuine they were personal friends.[1] Certainly there are close connections between their writings on harmonics, though the evidence about the relation is not always easy to interpret.

Archytas was by all accounts a remarkable man, distinguished simultaneously as a philosopher, a mathematician, an inventor of ingenious gadgets, a statesman and a military commander, and admired also for his personal qualities, his kindness, resourcefulness, self-control and affection for children. He counts as a heroic figure in the early history of mathematical harmonics, which he raised to new levels of conceptual and technical sophistication and channelled in unprecedented directions. Only a few fragments of his writings survive, along with reports in various (and variously reliable) later sources, but they are enough to allow us to reconstruct a coherent general outline of his approaches and ideas, and to piece certain parts of his work together in some detail.

Half a millennium later, in the most accomplished of all Greek essays in the mathematical style of harmonics, Ptolemy speaks of Archytas with evident admiration. Though Ptolemy's general attitude to his predecessors is less contemptuous than Aristoxenus' and his comments on them less consistently vituperative, he rarely mentions them except to criticise; and though his approval of Archytas is also undercut by criticisms and reservations, he writes of him more warmly than of any other theorist. I have argued elsewhere that he borrows from Archytas rather more freely than he

[1] See Plato, *Ep.* 7, 338c–339d. For a thorough examination of everything to do with Archytas' life and work see Huffman 2005. I refer to him repeatedly in this chapter, and our differences over certain points should not disguise the fact that his is by far the best study of Archytas that we have. There is a useful shorter survey in Kahn 2001: 39–48.

admits, and may have regarded Archytas' work, in important respects, as the one true ancestor of his own.[2] It is from Ptolemy that we learn most of what we know about the details of Archytas' harmonic analyses, and about some of the principles which helped to determine the form they took. The evidence is therefore far removed in time from the original, and Ptolemy himself was probably dependent on an intermediate source.[3] There is nevertheless no good reason to doubt its authenticity, though we must of course be careful to distinguish, where we can, the data which Ptolemy had at his disposal from the inferences he drew from them and the opinions he expressed about them.

ARCHYTAS AND PTOLEMY'S 'PRINCIPLES OF REASON'

> Archytas of Tarentum, of all the Pythagoreans the most dedicated to music, attempted to preserve that which is in accordance with reason, not only in connection with the concords but also in the divisions of the tetrachords, on the grounds that commensurability between the differences is intrinsic to the nature of melodic intervals. (Ptol. *Harm.* 30.9–13)[4]

Earlier passages of the *Harmonics* allow us to give this initially obscure remark a clear interpretation. In saying that Archytas 'attempted to preserve that which is in accordance with reason', Ptolemy means, in the first place, that the ratios he assigned to the intervals of an attunement were not chosen merely as those which seemed to fit the musical data most closely, but were required to conform to the dictates of a 'rational' principle. In the language of this work, a 'rational' principle is one grounded in mathematics, and we can infer that in Ptolemy's opinion Archytas constructed his analyses, in part at least, on the basis of mathematical considerations.

That is a little vague; mathematical considerations are very various. But we can be more precise. Ptolemy means also, as the sequel shows, that the principle Archytas tried to apply approximated to his own notion of harmonic rationality, in short, that it was identical with or closely related to the central principle governing his own analyses of musical systems. A 'difference' (literally 'excess'), in this context, is the difference between the terms of a ratio expressing the size of a musical interval. The items with which such a difference must be commensurable are the terms of the

[2] Barker 1994b, cf. Barker 2000a: 65–7, 120–8.

[3] The most likely source is a musical theorist named Didymus, who can tentatively be dated to the middle of the first century AD, and who appears to have had direct access to a work by Archytas himself; see p. 438 below.

[4] Where references to Ptolemy's *Harmonics* are given in this form they are cited by page and line of Düring's edition. References in the form 'II.13' are by the work's book and chapter numbers.

ratio themselves; and it is commensurable with them if it constitutes a unit by which each term can be exactly measured. It must therefore be an integral factor of each. This is precisely the condition which must be met, on Ptolemy's own view, by the ratio of every incomposite melodic interval, that is, every individual step of a well-formed scale.[5] Ptolemy explains earlier that the Pythagoreans agreed that a condition of this sort must apply to the ratios of concords.[6] Thus the difference between the terms of the ratio of the perfect fourth or the perfect fifth, for example (3:2 and 4:3), which is in each case 1, is an integral factor of both the ratio's terms. In ratios such as 5:3, 9:5 and so on, this condition is not satisfied. It entails, in fact, that every appropriate ratio, when expressed in its lowest terms, has the form n + 1:n, so that the two numbers are successive integers. Mathematicians call such ratios 'superparticular'; the corresponding Greek adjective is *epimorios*, and I shall call them 'epimoric' (for a more technical definition of 'epimoric' see n. 7 below).

Ptolemy's statement implies that in the opinion of theorists other than Archytas, this condition must be fulfilled by the ratio of any interval which is genuinely a concord, and that Archytas' originality lay in his extension of the principle to 'melodic' intervals as well. This implication needs to be qualified in at least three ways. First and straightforwardly, not all the concords recognised by the Pythagorean theorists Ptolemy mentions have epimoric ratios, as Ptolemy knew and had previously explained; the ratios of some of them are multiple.[7] But this is a quibble. Ptolemy is abbreviating the point in the interests of his immediate focus of attention, to which only epimoric ratios are relevant. Secondly, if Ptolemy means that Archytas' position was an advance on that of his predecessors, there is no independent evidence that the principle (in the form 'all concords must have ratios that are either multiple or epimoric') was current before the time of Archytas himself. There are passages in Plato and in Aristotle that may hint at it;[8] but

[5] Ptol. *Harm.* 1.7, 1.15. [6] *Harm.* 1.5.

[7] *Harm.* 11.18–20, 12.1–5. The ratio of an octave plus a fifth is 3:1, and that of a double octave is 4:1. These ratios are called 'multiple' (in Greek *pollaplasios*), since the larger term is an exact multiple of the smaller; and here it is not the difference between the terms that constitutes the unit by which they are measured, but the lower term itself. The ratio of the octave itself, 2:1, is also multiple rather than epimoric, even though its terms are successive integers. An epimoric ratio is defined as one in which the larger term is equal to the smaller plus a unit-part (one half, or one third, and so on) of the smaller; whereas in the ratio 2:1 what has to be added to the smaller to produce the larger is not a part of the smaller but the whole of it. Ratios such as 9:5, which are neither multiple nor epimoric, are usually called *epimereis* in Greek, which I shall anglicise as 'epimeric' (in the Latin-based vocabulary of mathematicians they are 'superpartient'), or are described by the phrase 'number to number'.

[8] Plato, *Tim.* 35b–36b (where the expressions 'multiple' and 'epimoric' do not occur, but the only ratio mentioned which does not have one of those forms is explicitly distinguished as a ratio of 'number to number'), Ar. *De sensu* 439b–440a, where the concords, like pleasant colours, are said to 'depend on the best-ratioed (*eulogistoi*) numbers'.

it first appears explicitly in surviving sources in the Euclidean *Sectio canonis*, which dates from the end of the fourth century at the earliest.[9] It is from this treatise, I believe, that Ptolemy drew much of his information about what he thought of as non-Archytan Pythagorean theory; and if he supposed that a direct enunciation of the principle predated Archytas (he does not explicitly say this) he may well have been wrong. Such general theoretical axioms are remote from what little we know of fifth-century Pythagorean thought, even from that of Philolaus, and it is at least as likely that it originated with Archytas himself. Thirdly, as Ptolemy goes on to emphasise, if Archytas extended the principle to melodic as well as concordant intervals, he did not do so consistently. This fact, which stands out obtrusively from the 'divisions of the tetrachord' to which we shall shortly turn, may reasonably make us hesitate to accept Ptolemy's contention that Archytas 'attempted' to assign epimoric ratios to all melodic intervals, and yet for some reason failed.

Whether it is applied only to concordant intervals or also to melodics, the principle can hardly stand without some argumentative backing. Purely arbitrary principles are obviously unacceptable, and this one needs justification urgently, since in either form it has awkward consequences. If it is restricted to concords, it requires that one interval which Greek ears recognised as concordant (as did the logic of some harmonic theorists) is in fact no such thing; the ratio of the octave plus a fourth is 8:3, which is neither multiple nor epimoric. If it is extended to melodic intervals a number of other difficulties will arise; most straightforwardly, one of the intervals involved in a simple structure repeatedly treated as fundamental by mathematical theorists, including Plato and the author of the *Sectio canonis* and perhaps Philolaus before them, must be rejected as improperly formed. The interval is the one which, together with two whole tones in the ratio 9:8, fills out the span of a fourth (ratio 4:3), and whose ratio is 256:243.[10] By the criterion which Ptolemy attributes to Archytas and adopts himself, it cannot be a 'rationally' formed melodic interval, and the structure in which it plays a part is flawed.[11]

9 [Eucl.] *Sect. can.* 149.12–24. In most respects the approach taken to harmonics in the *Sect. can.* differs substantially from that of Archytas, as it is presented by Ptolemy and in other sources to be considered below; but it certainly draws on his mathematical work in proposition 3, and may do so elsewhere. The treatise is discussed in Ch. 14 below.

10 We have already met this interval in connection with Philolaus, and its ratio is specified in Boethius' reports about him, which (so I have argued) deserve to be taken seriously. There is no doubt that Archytas knew the ratio, as we shall see below. But the evidence for Philolaus and Archytas comes from much later sources, and the first surviving text to quantify the interval as a ratio is Plato *Tim.* 36b2–5.

11 For Ptolemy's own very different treatment of a system of this sort see *Harm.* 39.14–40.20.

The requirement that musical relations should conform to mathematical principles of any sort reflects the idea that the quality we perceive in some pitch-relations and not in others, on the basis of which we call the former and not the latter 'musical', is the audible expression of a privileged variety of mathematical form. The notes of a melody or of the scale on which it is based, or those bounding a melodically acceptable interval, stand in relations to one another which, as we put it, make musical sense. They are not merely different from one another but are also in some way akin, coming together as elements in a coherent unity. It is the translation of this intuition into mathematical language that generates the requirement of 'commensurability'. The terms of a ratio, corresponding to the notes of an interval, can come together coherently only if the relation between them can be grasped as intelligible. This means that they must be capable, in Plato's phrase, of being 'measured against one another'; and this is possible only if the unit by which each term is measured is the same. This unit, furthermore, must exist as one of those elements in the ratio whose auditory counterparts are detected by the musical ear, since the perception of an interval's musicality need not depend on our comparing it, or its notes, with anything outside itself. Out of all this emerges the requirement that the ratio of a concordant or a melodic interval must be either multiple or epimoric. If it is multiple, the smaller term is the measure of the greater. If it is epimoric, the difference between the terms is the measure of both. If it is neither, the ratio contains no component which can serve as the unit of measurement for both terms, and the relation between them, from that perspective, is uncoordinated and unintelligible.

I have argued elsewhere that this account corresponds to Ptolemy's own justification for his mathematical principles.[12] I cannot prove that Archytas reasoned along the same lines; but if he did indeed adopt, in any form, the principle of 'commensurability' which Ptolemy attributes to him, it is hard to see how considerations of any other sort could have led him to it.[13] If Ptolemy's evidence is anywhere near the mark, however, it brings to light one very important point, regardless of the nature of Archytas' reasoning. An enquiry within which the forms available to the ratios of musical intervals are determined by mathematical axioms involves a much more sophisticated conception of science than does an unembroidered

[12] Barker 2000a: 74–87.

[13] No fourth-century writer preserves an argument of this sort, though there may be traces of it in Aristotle's treatment of ratios at *De sensu* 439b–440a, and cf. [Ar.] *Problems* 19.41. The argument offered at *Sect. can.* 149.17–24 for the principle as it applies to concords seems, on the face of it, trivial, and can in any case have no bearing on questions about melodic intervals; see pp. 375–8 below.

attempt to correlate recognised intervals with the ratios that are judged, by empirical means, to correspond most closely to them. This is a science which goes beyond a formal interpretation of the facts, to account for them on the basis of high-level abstract principles. In Aristotle's terms, it is a science which is not content simply to record 'the fact that . . .', but articulates also 'the reason why . . .'; and it is significant that Aristotle marks the distinction between 'empirical' and 'mathematical' harmonics in precisely this way.[14]

ARCHYTAS' DIVISIONS OF THE TETRACHORD: MATHEMATICAL PRINCIPLES AND MUSICAL OBSERVATIONS

In the passage from which our short excerpt was taken, Ptolemy goes on to set out, and to criticise, three 'divisions of the tetrachord' which, so he says, Archytas articulated. He names them as enharmonic, chromatic and diatonic, and they do indeed correspond, in their general outlines, to the schemata which are given these names in sources from Aristoxenus onwards. Whether Archytas himself designated them in this way we cannot tell, but that is unimportant. In either case, assuming that they are authentic, they are the earliest analyses we have which bring together three different types of system matching the three genera of Aristoxenian theory. If Philolaus' *harmonia* has affinities with any of the genera, it is (in Aristoxenus' terms) diatonic,[15] and so is that of Plato's *Timaeus* (pp. 319–21 below). Elsewhere in Plato, and in Aristotle, the various patterns of attunement are designated in a completely different way, as Dorian, Phrygian and so on, corresponding to distinctions drawn by the *harmonikoi* within the framework of structures most closely related to the enharmonic; and the *harmonikoi* discussed by Aristoxenus, as we have seen, dealt only with enharmonic systems. The remarks of the author of the Hibeh papyrus fragment, though he mentions all three names, betray the absence of any clear distinction between chromatic and diatonic. So far as we can tell from the surviving evidence, Archytas' detailed and fully quantified analysis of the three harmonic systems was unprecedented.

In Aristoxenian theory, an enharmonic tetrachord fills up the span of a perfect fourth with two very small intervals (quarter-tones) at the bottom, followed by a single step of a ditone. The two lowest intervals in a chromatic

[14] Ar. *An. post.* 78b34–79a6; see pp. 353–61 below.

[15] Or, just possibly, enharmonic; see Winnington-Ingram 1928. But I have argued (above pp. 276–8) that Philolaus' analysis may not presuppose any particular way of filling in the concordant intervals between the attunement's fundamental notes.

tetrachord are larger than those of the enharmonic, but when taken together they too occupy less than half the compass of the fourth. The upper interval is correspondingly smaller than it is in the enharmonic, but must occupy more than a tone and a quarter. In diatonic tetrachords the space occupied by the two lowest intervals together amounts to at least half the span of the fourth and commonly to more; hence the highest interval must always be a tone and a quarter or less. With certain qualifications to be noted below, Archytas' three schemata fit these criteria. Their intervals are of course expressed as ratios, rather than as Aristoxenian tonal distances. The ratios of intervals in his tetrachords, reading from the top downwards, are as follows:

Enharmonic: 5:4, 36:35, 28:27;
Chromatic: 32:27, 243:224, 28:27;
Diatonic: 9:8, 8:7, 28:27.[16]

Despite their general affinities with the Aristoxenian patterns, these divisions differ from his in several ways, and present other features too that seem puzzling. I do not claim that the list which follows is complete. (i) Most of Aristoxenus' divisions, though not all, include two intervals that are equal. None of Archytas' does. (ii) In the Archytan divisions, and in no others known to us, the lowest interval in each genus is the same. (iii) In those of the Aristoxenian divisions which gained widest currency, the two highest intervals in diatonic, taken together, are equal to the sum of the two highest intervals in chromatic, and also to the highest interval in enharmonic; all three magnitudes are ditones. In Archytas, the same equality holds between the overall spans of the two upper intervals in diatonic and chromatic (though the span in question is not exactly two whole tones), but the highest enharmonic interval is smaller. (iv) Aristoxenus enunciates the rule that the central interval of the three in a division can never be smaller than the lowest, a principle in which he is followed by Ptolemy, and to which the great majority of divisions known to us conform.[17] Archytas' enharmonic breaks this rule.[18] (v) Finally and most strangely, the ratios of the two highest chromatic intervals are strikingly anomalous, since unlike all the others in these systems, neither is epimoric. It is this peculiarity that Ptolemy has in mind when he comments that Archytas failed in his

16 For a meticulous study of these systems and of the passage of Ptolemy where they are set out and discussed (*Harm.* 30.3–32.23) see Huffman 2005: 402–28.

17 See Aristox. *El. harm.* 52.8–12, Ptol. *Harm.* 32.7–10.

18 So does the chromatic attributed by Ptolemy to Didymus in the tables of *Harm.* II.14 (see also Ptolemy's comment on the division in the preceding chapter, at *Harm.* 68.27–9). But such divisions are exceedingly rare.

attempt to achieve 'commensurability' in the ratios of melodic intervals (*Harm*. 30.13–14, 32.1–3).

One might set about explaining these features of the divisions in either of two ways, or through a combination of both. One approach would treat them as consequences of some mathematical operation through which the divisions were generated; the other would seek to account for them empirically and historically, as characteristics of genuine musical systems current in Archytas' time, ones that were later modified or abandoned. The latter strategy would presuppose that Archytas had set himself, like the *harmonikoi* but in a different way, to specify the structures of attunements actually used by contemporary musicians, rather than to derive a collection of purely theoretical systems from abstract mathematical principles. At least part of his intention, on this reading, would have been to represent the data of real musical practice in mathematically intelligible terms. Even if he conceived his results as significant also in a metaphysical or cosmological context, he would nevertheless have treated the perceptual data as essential evidence to which his conclusions must conform. This hypothesis is supported by the fact that despite the differences I have noted, there are very close correspondences between Archytas' divisions and some of those described, in different language, by Aristoxenus.[19] It is encouraged also by the various apparent anomalies in the divisions, which a completely theoretical scheme might be expected to eliminate, and again by the broadly approving attitude of Ptolemy, for whom the perceptual data constituted a crucial control on the work of mathematical construction. The same conclusion might be drawn from the mere fact that Archytas' divisions seem designed to accommodate all three of the main categories into which Aristoxenus and his successors divided the melodic systems of musical practice, including the hitherto vaguely conceived chromatic, and that his diatonic abandons the pattern ($9:8 \times 9:8 \times 256:243$) derivable from a simple manipulation of concords and their ratios, which was the central, if not the only point of reference for most metaphysically minded theorists.[20]

[19] They are examined in detail in Winnington-Ingram 1932, cf. Huffman 2005: 412–14.

[20] The interval of a tone (ratio 9:8) can be constructed in practice by moving from a given note through a perfect fifth upwards followed by a perfect fourth downwards, or the reverse. When two such tones have been constructed in succession, the *leimma* (ratio 256:243) will be left as the residue of a perfect fourth taken from the original note. The procedure is useful because, as Aristoxenus says, it is much easier to construct concords such as the fifth and the fourth accurately by ear than it is to construct discords such as the tone or the *leimma* (which Aristoxenus treats as a semitone); but it cannot be used to construct, for instance, the quarter-tones of Aristoxenus' enharmonic or the one-third tones of one of his forms of the chromatic, or any of Archytas' divisions. It is mentioned and explained in several theoretical sources (see especially Aristox. *El. harm*. 55.3–56.12, [Eucl.] *Sect. can*. prop. 17), and was very probably used by musicians themselves. The Greeks called the procedure *lēpsis dia symphōnias*; I shall refer to it as the 'method of concordance'.

I am tolerably confident that this diagnosis is part, at any rate, of the truth. If that is so, it is important, not just because these analyses, so construed, would allow us a glimpse of what early fourth-century melodies were like, but because it would mark a turning-point in the story of Pythagorean harmonics, a shift from a focus on exercises in mathematical cosmology to a direct engagement with the details of musical practice. At the same time one must not lose sight of the fact that mathematical considerations also had a part to play in these divisions' construction,[21] and hence that their peculiarities may not all arise straightforwardly from the attempt to provide a faithful representation of the 'facts'. The principle of 'commensurability' to which Ptolemy refers probably figured among these considerations (though in a slightly different form, as I shall try to explain); so almost certainly did another general mathematical thesis which will be outlined later (pp. 302–3).

Even if these divisions are designed to reflect patterns of tuning in contemporary musical currency, they cannot be supposed to replicate them exactly. For one thing, musicians' tuning-practices are variable, and the theorist's descriptions of them are not. For another, neither the instruments available to Archytas nor the discriminations of the human ear could be so finely calibrated as to guarantee that such-and-such an interval's ratio is exactly the one designated in his divisions, and not one marginally larger or smaller. It would be impossible, by purely observational techniques, to establish that the ratio of the middle interval in enharmonic, for instance, even in just one paradigmatic performance, was precisely 36:35 (and not e.g. 72:71 or 107:105). Archytas' figures must have been chosen, from among those within the range to which the perceptual evidence guided him, to fit a pattern determined by assumptions of a mathematical sort.

Let us turn now to the five features of the divisions which I listed above. (i) The fact that no two ratios in any one division are equal is almost inevitable if all the ratios are epimoric. There is only one way in which the ratio of the fourth, 4:3, can be divided into three sub-ratios all of which are epimoric and two of which are equal, and it is too remote from musical usage to be a candidate for serious consideration.[22] If we assume, then, that the non-epimoric ratios of Archytas' chromatic are exceptional cases calling for special explanation, this characteristic of his divisions emerges

[21] At least, I take it to be a fact, as I shall explain below, and I shall try to soften the impact of Huffman's criticisms of my position (2005: 416–17).

[22] It factorises 4:3 as 8:7 × 8:7 × 49:48, which could only be construed as an impossible version of a diatonic division, containing two intervals rather larger than a tone, and a residue a good deal smaller than a quarter-tone. (The general formula for factorising n+1:n into three epimorics two of which are equal is $(n + 1{:}n) = (b + 1{:}b) \times (b + 1{:}b) \times (b^2{:}b^2 - 1)$, where $b = (2n + 1)$. Thanks, here again, to Dr Jonathan Barker for lightening my mathematical darkness on these matters.)

directly from his otherwise consistent attribution of a privileged status to epimorics, supplemented by only the most rudimentary piece of musical knowledge. (ii) The equality of the lowest intervals in all three divisions cannot be explained on purely mathematical lines. Other theorists assign equal intervals or ratios to the lowest positions in certain forms of the chromatic and the diatonic, but the corresponding interval in enharmonic is always smaller; and those who represent intervals as ratios have no difficulty in constructing an enharmonic division in which this condition is met and all the ratios are epimoric, even if they agree with Archytas that its highest ratio is 5:4.[23] If there was a theoretical basis for Archytas' contention that the lowest interval is the same in all three genera, we have no idea what it was. It seems more likely that it reflected what he believed himself to have discovered, empirically, in the performing practices of contemporary musicians.[24]

(iii) Three quite different considerations can be brought to bear on the third issue, and all of them, I think, are sound. When Aristoxenus assigned the same span, a ditone, to the highest interval in enharmonic and to the sum of the two highest intervals in paradigmatic forms of each of the others, these equalities presumably represented his interpretation of aspects of familiar contemporary tuning-procedures (though this point will shortly be qualified). The fact that Archytas' highest enharmonic interval is smaller than the relevant composite intervals in diatonic and chromatic has one boringly obvious explanation. Given that the lowest interval in each of the divisions is the same, it is straightforwardly impossible to equate just one of the remaining intervals in enharmonic with the sum of the remaining two intervals in the others. It also seems easy enough to account for the fact that the highest enharmonic interval is not a ditone ($9:8 \times 9:8 = 81:64$) but slightly less ($5:4 = 80:64$). So small a difference looks marginal, and 5:4 is not only the epimoric ratio that approximates most closely to that of the ditone, but is satisfyingly simple. It is only to be expected that the next epimoric in order after those of the fifth and the fourth, 3:2 and 4:3, would find a significant role in a system grounded, at least partly, in mathematical conceptions.

[23] Ptolemy's enharmonic is $5:4 \times 24:23 \times 46:45$; that of Didymus is $5:4 \times 31:30 \times 32:31$. See the tables of Ptol. *Harm.* II.14.

[24] One might wonder whether conventions differed in his native Tarentum from those established elsewhere, in Athens, for example. But there is no evidence to support such a guess, and the fact that musicians were constantly on the move from one centre to another, and competed in the major festivals against others from quite different parts of the Greek world, points at least to a broad similarity between practices in different areas, even if there were minor regional variations. For discussion of the lives led by musicians see Bélis 1999, and more briefly Barker 2002a, ch. 8.

The third consideration is altogether more interesting. When Aristoxenus writes about the enharmonic, he recognises that the form he describes and treats as authentic is not acceptable to most contemporary listeners, and does not exactly match the tuning-pattern used for enharmonic melodies by most contemporary musicians. It is nevertheless, he asserts, the finest of all melodic systems, as is clear to those who have immersed themselves in the 'first and second of the ancient styles' (*El. harm.* 23.1–12). Aristoxenus' 'genuine enharmonic', then, is a historical reconstruction rather than a contemporary reality. The 'first and second ancient styles' in which he locates it are probably to be identified with those he attributes elsewhere to the aulete Olympus and his successors, who date (in so far as they are not merely figures of legend) from the seventh century and the early sixth.[25]

By these standards what passed for enharmonic music in Aristoxenus' time hardly deserved the name,[26] and the fact that Archytas' enharmonic does not tally completely with his may be a sign that it reflected fourth-century practice more faithfully. Certain additional details that Aristoxenus provides give firmer outlines to this initially vague suggestion. What musicians in his time found unacceptable, it turns out, was not the minuteness of its quarter-tones, as one might have expected,[27] but the size of its highest interval, the ditone.

> It is not surprising that those who are familiar only with the currently prevailing type of melodic composition exclude the ditonal *lichanos*;[28] for the majority of people nowadays use higher *lichanoi*. The reason for this is their constant passion for sweetening (*glykainein*). An indication that this is their goal is that they spend most of their time working in the chromatic, and when they do, occasionally, approach the enharmonic they force it towards the chromatic, and so distort the melody. (*El. harm.* 23.12–22)

These people's 'distortions' of the enharmonic thus involve a reduction in the size of its upper interval, so that it sounds more like a chromatic system, and this practice arises from their pursuit of 'sweetness'.

These observations can be applied directly to the enharmonic of Archytas. Its upper interval is smaller than the ditone of the 'noble and ancient' enharmonic that Aristoxenus prefers. So far as quantitative descriptions reveal, whether they are Aristoxenian or Pythagorean, the difference is very slight. Aesthetically, however, it is not negligible. What makes one interval

[25] [Plut.] *De mus.* 1134f–1135b. [26] But cf. *El. harm.* 48.15–20.

[27] This is, however, the objection raised by unnamed musicians and theorists in a passage almost certainly derived from Aristoxenus at [Plut.] *De mus.* 1145a–c, and by a number of later writers.

[28] That is, the location of *lichanos*, the second-highest note from the top of the tetrachord under discussion, at a ditone below its highest note, *mesē*.

strike the ear as harsher, and another as smoother or 'sweeter', is the larger number of 'beats' or 'interferences' set up between the notes bounding the former. The number of beats generated in a given time from the interaction of pitches whose frequencies are related in epimoric ratios of small numbers is much less than that from those in more complex ratios involving larger terms. It is this phenomenon that allowed ancient musicians, and allows modern ones too, to distinguish such intervals as the perfect fifth and the perfect fourth, by ear, with great precision from their near neighbours. The relative infrequency of the beats they set up explains why Greek theorists described the notes involved in these concordant intervals as blending smoothly together, in a way that discordant pairings did not. By this criterion, an interval in the ratio 5:4, that of the upper interval of Archytas' enharmonic (and of a modern major third), will be noticeably 'sweeter' than a true ditone in the ratio 81:64. Tiny though the mathematical difference is, it takes only a moderately sensitive ear to appreciate its musical effect.[29]

(iv) Most Greek constructions are consistent with Aristoxenus' rule that the central interval of the three in the tetrachord is never smaller than the lowest. Archytas' enharmonic is not. This anomaly stems directly from the constraints imposed on the other two intervals in the division. Archytas was apparently convinced that the lowest interval in each genus was the same (see (ii) above). It must therefore appear in diatonic as well as in enharmonic and chromatic; and it is impossible to construct a plausible diatonic division in which the lowest interval is smaller than one in the ratio Archytas attributes to it, 28:27. His estimate of this interval's size in diatonic is in fact smaller than that of any other Greek theorist. Given that the lowest interval in his enharmonic must be no smaller than that, the second constraint, that its highest interval must come close to the compass of a ditone, ensures automatically that the middle interval will be smaller than the lowest. Hence even if there were already in Archytas' time a tendency towards the practice formalised in Aristoxenus' enunciation of the rule, it was apparently (in Archytas' opinion) over-ridden, in the case of the enharmonic, by these other considerations.

(v) We come, finally, to the peculiarities of Archytas' chromatic ratios, 32:27 × 243:224 × 28:27. Their oddity is accentuated by the fact that they approximate closely to those of an easily constructible chromatic division

[29] For some additional points on this matter see Huffman 2005: 412–13, with his references to Winnington-Ingram 1932.

which uses only epimoric ratios, 6:5 × 15:14 × 28:27.[30] An explanation of the strange-looking Archytan ratios is therefore urgently needed, and Ptolemy supplies one. If he and his source are to be trusted, the explanation is Archytas' own. 'He locates the second note from the highest in the chromatic genus from the one that has the same position in diatonic. For he says that the second note from the highest in chromatic stands to its counterpart in diatonic in the ratio of 256:243' (*Harm.* 32.2–6).

The explanation makes good sense. Let us assume that the tetrachord under consideration is the one focused on by most theorists, the one bounded by *mesē* at the top and by *hypatē mesōn* at the bottom. Immediately above *mesē*, in a typically constructed octave, lay a whole tone, followed by another identical tetrachord. The ratio of the highest interval in Archytas' diatonic tetrachord is that of a whole tone, 9:8; above it, then, lay another interval of the same size. The ratio 256:243, by which the second note of Archytas' chromatic is lower than its counterpart in diatonic, is that of the so-called *leimma*, the residue of a perfect fourth when two whole tones have been subtracted. Hence Archytas' second chromatic note could be reached by a very simple method, by tuning downwards through a perfect fourth from the note immediately above *mesē* (that is, *paramesē*). This is just the sort of method that a practical musician might be expected to adopt. Granted, then, that Ptolemy's explanation echoes that of Archytas, as his 'for he says' implies, we can infer that Archytas based his assessment of the highest chromatic ratio on his observation of musicians' tuning-procedures. The ratio is that of a perfect fourth less the whole tone lying above the tetrachord, i.e. 9:8 × 256:243 = 32:27. Once that is established, the ratio of the middle interval is determined by straightforward arithmetic, since that of the lowest interval is treated as a constant, 28:27. It is the ratio by which 4:3 exceeds 32:27 × 28:27, which is 243:224.

Both mathematical ideas and empirical observations seem therefore to be at work behind these features of Archytas' divisions, and in the most interesting cases the two coincide or combine. The mathematician's preference for epimoric ratios,[31] especially those whose terms are small numbers, comes

[30] This is the 'soft chromatic' described by Ptolemy (*Harm.* 35.4–6). Its highest interval exceeds that of Archytas' chromatic by an interval in the ratio 81:80, which is quantitatively marginal; and by the criteria outlined above the Ptolemaic version is perceptibly sweeter.

[31] I must insist that this *is* a mathematician's preference, and that purely musical considerations cannot explain all the major features of Archytas' analyses, though they are certainly involved as equal partners along with mathematical ones. What Huffman (2005: 416) describes as Burkert's '*reductio ad absurdum* of the thesis that Archytas was striving for superparticular ratios' is nothing of the sort (and Burkert himself did not regard it in that light), since it actually presupposes that Archytas deliberately adopted epimoric ('superparticular') ratios in most places; and that cannot be

together with contemporary musicians' predilection for 'smooth' or 'sweet' intervals such as the concords and the 5:4 major third, and with musically straightforward methods of tuning an instrument. The explanation of the highest chromatic interval that we have unearthed from Ptolemy's remarks suggests the possibility of accounting in a similar way for the ratio of the lowest interval, 28:27, which is common to all three attunements. Imagine, once again, that the specimen tetrachord is placed next to a whole tone in the ratio 9:8, but that this time the tone lies below the tetrachord instead of above it (as it does if the tetrachord is the one between *paramesē* and *nētē diezeugmenōn*). The interval formed by combining this tone with the interval at the bottom of an Archytan tetrachord is another epimoric whose terms are relatively small numbers; it is 7:6 ($9:8 \times 28:27 = 7:6$). This interval (the 'septimal' minor third, by contrast with the slightly larger and commoner version of this interval, in the ratio 6:5) is no doubt less easy to tune precisely than a perfect fourth or a major third. But in this case the musician has a second point of reference to help guide his ear. An interval in the ratio 7:6 falls short of a perfect fourth by an interval whose ratio is another simple epimoric, 8:7. Hence if he begins from the note a tone below the tetrachord and tunes upward through a perfect fourth, the upper boundary of the Archytan tetrachord's lowest interval will lie at a pitch which divides this fourth in the ratios 7:6 and 8:7. These are the two largest epimorics into which the fourth can be divided; and the musician can home in on the correct pitch for the note he is tuning by listening for the moment at which its relations with *both* the bounding notes of the fourth reach their maximum of 'smoothness'. There is scope for the possibility, then, that Archytas identified the ratio of the lowest interval of his tetrachords at least partly on the basis of his observation of performers' tuning-procedures, interpreted in the light of his mathematical assumptions.

At first sight, Archytas' divisions not only abandon the principle that all relevant intervals should have epimoric ratios (since two of the chromatic intervals do not), but also show no special preference for ratios whose terms are small numbers (since even if we leave the chromatic ratios on one side, those of the ratios 28:27 and 36:35 are hardly 'small' in this context). It

a mere accident. All it shows for Huffman's purposes is that it would have been theoretically possible for Archytas to find epimorics very close to the ratios he actually uses in chromatic, and that of course is true, as I have explained. If the reconstruction of his procedure I gave above is correct, his reasons for not making these adjustments were musical, not mathematical, and to that extent Huffman is right. But that does nothing to account for Archytas' exclusive choice of epimorics at every other point in his systems.

turns out, however, that if the scheme is expanded from a single tetrachord to a complete octave, every note in all three genera can be located from some other note in the same genus through an interval in a small-number epimoric ratio, one whose terms lie within the range of the numbers 1 to 9; and when the divisions are set out on that basis, every ratio of the form n + 1:n within that range, from 2:1 to 9:8, figures at least once in the pattern. By moving through a series of intervals in epimoric ratios, with every move beginning and ending on a note included in the system, one can reach every note of each system from any starting-point whatever; they are all linked by an unbroken chain of epimorics. By this method, in fact, one can reach all the notes of all three systems from any starting-point in any of them, since some of the moves will bring us to the 'fixed' notes which all of them share. I shall not go into the details here, which can readily be grasped from diagrams printed in two of my previous publications.[32]

I stand open to correction, but I do not see how the fact that all three systems fit so smoothly into this matrix can be regarded as a mere fluke. It suggests that Ptolemy's attribution to Archytas of a principle grounding acceptable divisions in epimoric ratios is well founded, but that his criticism is misplaced. Archytas need not have been trying, and inexplicably failing, to give epimoric ratios to every interval between adjacent notes in his divisions. His system is consistent with a different though related principle, that every note in each genus must be constructible, mathematically, as a term in an epimoric ratio with some other note in the same division. While there are connections, as we have seen, between small-number epimorics and efficient methods of tuning, it seems at least very likely that the mathematical requirement was an influential partner in Archytas' process of construction; what musicians actually did was interpreted in the light of

[32] Barker 1989a: 46–7, presented again in a slightly different way in Barker 2000a: 124. Huffman 2005: 417, argues that even so Archytas cannot have used the principle that all melodic intervals must be constructible by moves through superparticular (epimoric) or multiple ratios, since Ptolemy's evidence shows that he identified the position of one chromatic note by another method, which involves the non-epimoric ratio 256:243. So he does (p. 299 above); but that misses the point. What I meant (or since memory can mislead, at any rate what I mean now) is not that he used the 'epimorics only' principle as the basis of a method of discovery, but as a principle against which conclusions reached by any means whatever needed to be tested. In this particular instance, according to the account I gave earlier and whose outlines Huffman accepts (2005: 417–18), he 'discovered' the position of the chromatic note by inference from his observation of musicians' tuning-practices; but according to my present hypothesis he would have modified his conclusion if it had proved impossible, on reflection, to reach the position it assigned to that note through small-number epimorics. Of course this cannot be proved, and the 'epimoric map' given in my diagrams is my own construction, not that of Archytas. But I would argue that the pattern it shows is far too neat and comprehensive to be coincidental.

what they 'ought' to be doing, where this 'ought' reflects the perspective of philosophical and mathematical rationality.

THE THREE MATHEMATICAL MEANS

Another brief extract from Archytas' work points in a similar direction.[33] 'There are three means in music,' it begins. 'One is arithmetic, the second geometric and the third subcontrary, which people call "harmonic".' It goes on to define each of these mathematical means, and to pick out some of the properties of the relations generated when a mean of each sort is inserted between two other terms. Let me briefly recapitulate the basic facts about these means, which we have already glanced at in connection with Philolaus (pp. 283–4 above). The arithmetic mean, B, between two numbers A and C (where C is the smaller) is such that A–B = B–C. The geometric mean is such that A:B = B:C. The harmonic or subcontrary mean has a more complex definition. It is such that A–B is the same fraction of A as B–C is of C (e.g. if A–B is one quarter of A, then B–C is one quarter of C, as it is if A is 20, C is 12, and the harmonic mean between them, B, is 15).

As we noted earlier, the simplest musical application of this system of means is in the construction of concords. When an octave is divided in the familiar way into two perfect fourths separated by a whole tone, the structure is demarcated by four notes, of which the second note stands to the first in the ratio of the fourth, 4:3, the third note stands to the second in the ratio of a tone, 9:8, and the last note stands to the third in the ratio of a fourth, 4:3. (Hence the ratio of the third note to the first and of the last note to the second is that of the fifth, 3:2, and the ratio of the last to the first is that of the octave, 2:1.) The smallest whole numbers which capture this arrangement are 6, 8, 9, 12.[34] Of these four numbers, 9 is the arithmetic mean between 6 and 12 (since 9–6 = 12–9), and 8 is their harmonic mean (since 12–8 is one third of 12, and 8–6 is one third of 6). The geometric mean does not appear directly in this construction, but the terms of the ratios of a series of octaves (2:1 for the octave, 4:1 for the double octave, 8:1 for three octaves, and so on) form a geometric progression, 1, 2, 4, 8 . . ., so that 2, for example, is the geometric mean between 1 and 4 (since 2:1 = 4:2).

Intervals in the ratios 3:2, 4:3 and 9:8 can therefore be constructed mathematically by the insertion of the arithmetic and harmonic means between

[33] DK 47B2 (Archytas frag. 2 in other modern references), quoted at Porph. *In Ptol. Harm.* 93.6–17. For a full discussion see Huffman 2005: 162–81.

[34] This sequence of numbers appears repeatedly in later sources in connection with these calculations.

terms in the ratio defining the octave, these terms themselves being part of a geometric sequence. All four of the other epimoric ratios which map out Archytas' divisions, those in the series from 5:4 to 8:7, can be constructed in a similar way, in two easy steps, by introducing means between the terms of the ratios of the lesser concords, the fifth and the fourth. One can use means of either the arithmetic or the harmonic type; in each case the ratios between the middle term and the extremes will be the same, in reverse order (just as the insertion of an arithmetic mean between terms in the octave-ratio 2:1 gives 4:3 at the top and 3:2 at the bottom, while the insertion of the harmonic mean places them the other way round). Thus the ratio of the fifth, 3:2, can be represented equally well as 6:4. The arithmetic mean between its terms is 5, whose insertion generates the ratios 6:5 and 5:4. Inserting the harmonic mean between the terms of the same ratio (in this case most helpfully represented as 15:10) will give the same results in the opposite order.[35] An arithmetic mean can be placed between the terms of the ratio of the fourth, 4:3, if it is expressed as 8:6, where the arithmetic mean is 7, and this gives us the remaining ratios in the Archytan scheme, 8:7 and 7:6. To find an easy way of inserting the harmonic mean instead, first multiply the terms of the ratio 4:3 by 7, and the same results in the opposite order will again appear.[36] Hence every one of the ratios needed to define these divisions can be constructed by introducing arithmetic and harmonic means between the terms of the ratio of the octave, and then locating either arithmetic or harmonic means between the terms of the ratios of the smaller concords, into which the octave was divided in the first stage of the operation.[37]

ARCHYTAS' THEOREM ON THE DIVISION OF EPIMORIC RATIOS

These observations increase the probability that Archytas' analysis was driven in part by an impulse towards mathematical systematisation, and

35 In the sequence 15, 12, 10, the number 12 is the harmonic mean between 15 and 10. 15:10 = 3:2, 15:12 = 5:4, and 12:10 = 6:5.

36 In the sequence 28, 24, 21, the number 24 is the harmonic mean between 28 and 21, since 28–24 is one seventh of 28, and 24–21 is one seventh of 21. 28:21 = 4:3, 28:24 = 7:6, and 24:21 = 8:7.

37 Huffman 2005: 169–70, rightly emphasises the fact that in frag. 2 Archytas does not only give the definitions of the 'three means used in music', but also explains that when the mean is arithmetical, the interval (*diastēma*) between the larger terms is smaller than that between the smaller terms; when it is geometrical the two intervals are the same size; and when it is harmonic the interval between the larger terms is the greater of the two. As Huffman points out, Archytas does not explain 'what we are to make of these comparisons'. But we might guess that one of his reasons for mentioning them was that he noticed and was intrigued by the fact that the insertion of arithmetic and harmonic means between terms in the same ratio always generates the same pair of ratios in the opposite order, as in the cases anatomised above.

there would be nothing surprising in that. He was renowned for his work in mathematics, which extended far beyond issues relevant to musical theory.[38] He also worked out a mathematical proof of a theorem that bears directly, and very importantly, on musical issues, the proposition that there is no mean proportional, 'neither one nor more than one', between terms in epimoric ratio.[39] Here a 'mean proportional' is a geometric mean. Between two terms in epimoric ratio, A and B, there is no intermediate term, X, such that A:X = X:B; nor is there a series of terms, X, Y, . . . , Z, such that A:X = X:Y = ··· = Z:B. The proof applies, of course, to epimoric ratios wherever they occur, not just in musical contexts. But it is most obviously relevant, and has the most striking consequences, in the field of harmonics,[40] and the fact that Archytas thought it significant enough to deserve formal proof is another sign of the privileged position he attributed to epimoric ratios in this domain. Its upshot, as Greek theorists understood it, is that no interval whose ratio is epimoric can be exactly halved, or divided into any number of sub-intervals all of which are equal;[41] and it was of major importance (though it was often misconstrued) both in the development of mathematical approaches to the division of intervals, and in later controversies between Aristoxenians and Pythagoreans. It was held to show, among other things, that it is impossible to divide the span of a tetrachord (a fourth, ratio 4:3) into equal sub-intervals, or into intervals all of which are multiples of the same unit; a tone (9:8), similarly, cannot be divided into any number of equal parts. In that case there must be something seriously wrong with the analyses of Aristoxenus, and indeed of the *harmonikoi* too, who talk blithely of quarter-tones, half-tones and the like, and suppose that a tetrachord in enharmonic, for example, contains two intervals of a quarter-tone each and another eight times that size, so that the whole fourth is divisible into ten equal segments.

38 See especially DK 47A14, quoted from Eutocius, which records a complex and sophisticated solution to the problem of constructing two mean proportionals between given terms, a question to which Archytas had reduced the famous 'Delian problem' of doubling the size of a cube. Other ancient references are given at DK 47A15. For a masterly discussion of this bewildering material see Huffman 2005: 342–401.

39 He is credited with the proof at Boethius, *Inst. mus.* III.11 (DK 47A19), and there is a closely related version at [Eucl.] *Sect. can.* proposition 3 (for some musical applications see propositions 10, 16, 18). For further discussion see pp. 351, 356, 380 below; and cf. Burkert 1972: 442–7, Knorr 1975 ch. 7, Huffman 2005: 451–70.

40 Both of the full statements of the proof preserved in ancient sources occur in musicological texts, not works of pure mathematics; so do most of the briefer references to it.

41 See e.g. *Sect. can.* props. 16 and 18, cf. Theo Smyrn. 53.1–16, 70.14–19 (in both of the passages in Theo Smyrn. the proof is misunderstood), Ptol. *Harm.* 24.10–11.

Others before Archytas had stumbled on some of the more obvious and awkward consequences of the proposition that Archytas proved; we have already seen Philolaus struggling manfully if incompetently to quantify a division of the tone (and, if my reconstruction is correct, the octave) into equal parts.[42] But there is no evidence and little likelihood that any of Archytas' predecessors had constructed a theorem to prove the proposition, or had even conceived or stated it as a truth about epimorics in general. Our reports about his theorem show beyond doubt, as Huffman says, 'that Archytas understood the demands of a rigorous deductive proof'.[43] It seems equally clear that he thought of harmonic theory as a discipline in which rigorous proofs of a mathematical sort should play a significant part.

HARMONICS, PHYSICAL ACOUSTICS AND MUSICAL PRACTICE

Archytas' harmonic constructions, in their mathematical guise, fit smoothly with his attempt to represent the pitch of sound as a quantitative variable;[44] what we perceive as a higher pitch is, in physical fact, a more rapid and vigorous movement (see pp. 27–9 above). The ratios assigned to musical intervals can therefore be straightforwardly understood as ratios between the speeds at which the notes forming the interval are transmitted through the air. The theory has obvious flaws and was adjusted in various ways by later writers. It nevertheless enables musical thought to cross a scientifically important boundary, that between the contents of our sensory and aesthetic awareness and an objectively real state of affairs outside us and independent of us. It provides the impressions we receive in our musical experience with an intelligible basis in the world accessible to the quantifications and measurements of a physicist; and the patterns we perceive as networks of musical relations find an objective counterpart in the dynamic interplay of aerial travellers, some faster, some slower, intertwining in precise schemes of mathematical order. It is their well-choreographed dances that are presented to our ears as attunements and melodies.

Pythagorean musical theory had been associated from the outset with cosmological and semi-scientific ideas. What mattered to Philolaus and his

[42] Huffman's comment (2005: 418) that Philolaus' procedure 'showed that the octave cannot be divided in half' is slightly misleading. The construction given in frag. 6a evidently *does* not so divide it, but if my reading of Boethius' evidence is correct he did not think the task impossible, and did his best to provide resources by which the outcome of such a division could be quantitatively expressed.

[43] Huffman 2005: 470.

[44] Archytas frag. 1, on which see Huffman 2005: 103–61, with the copious references to other discussions which he provides.

predecessors, so far as we can tell, was the mathematically specifiable system of order which certain fundamental musical relations exemplified and to which harmonic analysis gave access; and there are no good grounds for attributing to them an interest in the musical phenomena as such. Whatever may have been true of Hippasus and his other predecessors, Philolaus himself seems to have been uninterested even in the nature of the niche occupied by sounds in the world of matter and movement. Archytas' work opened the way for richer and more detailed explorations of such abstract patterns of order, most notably by turning the spot-light on the special status of epimoric ratios, by demonstrating techniques for manipulating them in the construction of harmonic divisions, by proving these ratios' resistance to equal division, by his classification and definition of the three 'musical means' and by his deployment of these means in his analyses of attunements.

But his studies in physical acoustics point also to a scientific interest in sound and pitch themselves, and reinforce the impression given by his tetrachordal divisions that he was concerned, much more directly than earlier Pythagoreans, with the domain of the audible for its own sake.[45] I have argued that the divisions were designed to reveal the mathematical organisation inherent in real musical practice; where I differ from other modern commentators, it is usually because they give even more weight than I do to the role of strictly musical considerations in Archytas' work. No matter whether their view or mine is nearer the mark, we are bound to conclude that Archytas pointed mathematical harmonics in an entirely new direction, and one, we must add, in which rather few of his successors seem to have followed him; most of them reverted to a more abstract approach, detached from the phenomena of musical experience, allied to the theories of Philolaus and heavily influenced by Plato.[46] It may be Archytas that Aristotle has in mind when he remarks that the role of mathematical harmonics is to explain the facts which empirical harmonics records; it could hardly do that if the system whose structures it explains, by grounding them in mathematical principles, were only the 'rational' constructions of abstract theory.

[45] A passage to be considered in Ch. 13 suggests that Archytas, like Philolaus and Plato, may also have thought of harmonics as relevant to issues in cosmology and metaphysics; see pp. 329–38. But the grounds of this hypothesis are not secure enough to carry much weight.

[46] The principal exception in later antiquity is Ptolemy (second century AD); there are traces of an interest in bringing mathematics into connection with musical realities in the work of Didymus (first century AD) and perhaps Eratosthenes (third century BC). I have discussed these people's approaches elsewhere: see Barker 2000a on Ptolemy, 1994a on Didymus, 2003 on Eratosthenes, on whom see also Barker and Creese 2001, Creese 2002: 126–56.

Archytas would also appear, in this respect, to be an appropriate target for Socrates' comment in the *Republic*, that the Pythagoreans concern themselves too much with things that are audible (531a1–3, cf. c1–2). Socrates' second and related criticism, however, seems wide of the mark where Archytas is concerned. 'They do not ascend to problems, to investigate which numbers are concordant and which are not, and in each case why' (531c2–4).[47] Whatever Socrates means by 'concordant' here,[48] the quest for principles which govern the ratios of well-formed intervals and complexes of intervals was as much a part of Archytas' enterprise as the analysis of contemporary tuning-systems. Plato was certainly aware of the fact, at least when he wrote the *Timaeus*, since in that dialogue he adopts one of Archytas' own central theoretical conceptions and redeploys it for his own purposes, and hints at his recognition of another (pp. 320–1 below). The *Republic*'s remarks about 'Pythagoreans' tell us, in the end, rather more about Plato's agenda than they do about the ideas of the people criticised; and about Archytas in particular, I think, they tell us no more than we knew already. They can indeed be misleading. We shall return to them in their own context in Chapter 12.

[47] Huffman 2005: 414 seems to conflate the two criticisms, and finds Archytas to be an appropriate target for both. But the distinction should be preserved; see further pp. 315–18 below.

[48] This poses a substantial problem; see p. 316 below.

CHAPTER 12

Plato

We have already called on Plato's help from time to time, and that is as it should be, even if we set aside his gigantic stature in the Western philosophical tradition. Once due allowance has been made for the fact that the conversations in his dialogues are fictional (though most of the characters are not), and for his own attitudes, prejudices and philosophical aims, the dialogues are an unparalleled source of information about the cultural milieu inhabited by elite Athenians and intellectually eminent visitors to the city in the late fifth and early fourth centuries, about the beliefs they held and the issues they discussed, and about the ways in which their ideas were expressed and debated. It is only to be expected, too, that music should figure prominently as a topic in these conversations, in view of the central place it held in every Greek's cultural experience. In fact, however, Plato shows rather little interest in it in his earlier work. There are various passing allusions, and a small handful of passages which from other perspectives have real theoretical interest; but in the context of a study of harmonic science, none of them has much to offer.[1]

MUSICAL ETHICS IN THE *REPUBLIC* AND THE *LAWS*

With the *Republic* the scene changes. Musical imagery becomes abundant, and sometimes does serious philosophical work; and there are two major set-pieces on musical topics. They treat the subject in quite different ways, and are segregated from one another in different parts of the dialogue. There is the passage on harmonics in Book VII, parts of which we have

[1] Leaving aside occasional, quite casual references to music (e.g. *Lysis* 209b4–7), there are only four passages in dialogues written (probably) before the *Republic* that have any musicological substance; they are *Lach.* 188c–d, *Prot.* 326a–b, *Symp.* 187a–e, *Phaedo* 85e–86d. It is worth noting that none of the statements in these passages is put into the mouth of Socrates himself, and that he discusses only one of them (very critically), at *Phaedo* 92a–95a. None of them provides much grist for our present mill, but for brief comments on the *Symposium* passage see pp. 72 n.4, 280 n.31 above. I examine them in a different context in Barker 2005a, chs. 3–4.

already reviewed and to which we shall shortly return. Much earlier, in Book III, in the context of a discussion of children's education, there is an elaborate examination of the ways in which different melodic and rhythmic styles reflect different dispositions of the human soul, and of the powerful influence they can exert, for good or ill, on the development of people's characters. The ethical significance of a melody is made to depend, in this passage, on the characteristics of the *harmonia*, the pattern of attunement (named here as Mixolydian, Lydian, Dorian, Phrygian and so on), which provides its framework of notes and intervals. It is these *harmoniai*, not the individual melodies as such, which are the bearers of ethical attributes, 'imitating' desirable or undesirable psychic dispositions and drawing their hearers' souls into their own likeness. Socrates therefore proposes a drastic purge of the *harmoniai* (and corresponding purges of musical instruments and of rhythmic structures too), banning from his ideal city all but the most edifying of them, Dorian and Phrygian. Plato's last work, the *Laws*, does not point an accusing finger at any specific *harmoniai* or varieties of scale; but it is if anything even more vehement in its condemnation of musical forms and practices that have a tendency to corrupt the souls of their executants and hearers. All the subtle complexities and aesthetically pleasing sophistications introduced into music by 'modern' composers are to be rejected, in favour of a noble simplicity that is supposed to have characterised the music of a lost golden age.[2]

Plato's Athenian spokesman in the *Laws* insists that musical practices in the city must be closely monitored and controlled, and that those who are to sit in judgement on them must be very thoroughly qualified for their role. Among other things, they must have a first-rate technical understanding of the elements of musical compositions and their inter-relations, at a much higher level than that of the citizen-amateurs who regularly performed in choruses, and even than that of most composers (*Laws* 670a–671a). But though the passage makes it clear that these people must master the kinds of analysis provided by harmonic scientists, it tells us nothing of their details. In Book III of the *Republic* there is no allusion to harmonic theory at all. In Socrates' ethically motivated witch-hunt among the *harmoniai*, the culprits are detected by their ethical or emotional resemblance to conditions of the soul. His arguments have sometimes been thought to presuppose an

[2] Plato projects his musical ideal onto the period before the Persian Wars, around 500 BC (*Laws* 700a–701b, with 698a–699e). But his golden age of music never really existed, of course, and certainly not at the time he imagines it. Far from being ruled by strict, stable and orderly norms, the musical world of the early fifth century was a ferment of innovation and controversy. For a brief and helpful survey see West 1992a: 341–55.

analysis of the structures of these *harmoniai*; but he gives none, and there is no reason to believe that Plato had such an analysis in the back of his mind and was implicitly relying on it. Socrates poses here as a musical ignoramus ('I do not know the *harmoniai*', *Rep.* 399a), deferring to the greater expertise of his companion Glaucon; and his arguments are based on Glaucon's identification of the *harmoniai* which match certain very impressionistic descriptions ('mournful', 'sympotic' and so on), without the least reference to structural considerations. No harmonic technicalities are involved, and the arguments should be understood in their own terms. The one point at which technical issues are raised and even Glaucon's knowledge proves inadequate has nothing to do with harmonics; and here Socrates shelves the matter as one on which they must consult the real expert, Damon.[3]

Plato's reticence about the details of harmonic theory in the *Laws* may be due simply to the understandable judgement that the context did not call for them; here he is not even attempting to identify precisely which musical attunement-systems should be approved of and which condemned. He needs only to point out that an adequate understanding of the ethical and socio-political significance of music in its various forms depends partly on a knowledge of musical 'theory', and specifically of harmonic analysis. The *Republic* is another matter, since the situation calls for the best arguments that can be mustered to support Socrates' purge of the *harmoniai*, and yet the arguments he puts forward against some and in favour of others do not depend at all on representations of the different structures which distinguish them. But if there is a reason for this over and above the contextual ones which might reasonably be adduced, I think it is straightforward. Of the two existing approaches to harmonic theory distinguished in *Republic* Book VII, Socrates and Glaucon regard that of the empiricists with undisguised contempt, and though the Pythagoreans are also criticised, Socrates seems to imply that their methods could be adapted to better effect. If either approach is to be preferred, then, it is theirs; and it has very close affinities with Plato's own major foray into harmonics in the *Timaeus*, which we shall examine below. But at the time the *Republic* was written, it was only the work of the *harmonikoi*, the despised empiricists, that could have had any bearing on Socrates' concerns in Book III. No pre-Archytan Pythagorean or other mathematical theorist, so far as we can tell, had even begun to dissect

[3] *Rep.* 400b–c. The topic here is rhythm, not *harmonia*; and though *Rep.* 424c and remarks by Nicias at *Lach.* 180c–d suggest that Socrates had a genuine (if slightly ironic) respect for Damon's intellectual, musical and educational attainments, I do not think he can be credited with any work in harmonic analysis at all, still less that the passage in *Rep.* III depended on it (see p. 47 above).

different forms of the scale, let alone to examine the differences between the *harmoniai* used by contemporary musicians. Philolaus' cosmologically oriented study of the basic outlines of an octave-attunement had nothing to offer in this context, and even Archytas' mathematical description of systems in the three genera, if Plato knew of it at this time, gave no purchase on the distinctions between the *harmoniai* which Socrates discusses. In that case there simply was no form of harmonic analysis which he could appropriately have used.

The purpose of harmonics in *Republic* Book VII and the *Timaeus* is to guide us towards an understanding of the principles governing the structure of reality as a whole, and to provide a form of understanding which will help us to restore our own distorted souls to their original perfection. The notion that the analysis of musical systems can and should lead to enlightenment in such mentally vertiginous and apparently non-musical areas as these was already current in the Pythagorean tradition, as we have seen, and that fact partly – but only partly – explains Plato's adoption of Pythagorean concepts and procedures in this second phase of his musical thought. Soberly considered, however, it is an extraordinary hypothesis. In its Platonic version it makes sense only in the context of his complex arguments and bold conjectures in metaphysics and epistemology, and it cannot be detached without serious loss from its settings in the dialogues in which it plays a part. I cannot do justice to these matters in a book of this sort. The brief sketch of the theory's philosophical environment that I shall offer is intended only as a rough guide for non-specialists, from which experienced students of Plato may prefer to avert their eyes. For anyone who wishes to pursue the issues accurately and in depth, enough has been written about them by scholars and philosophers down the centuries to occupy a reader for several lifetimes.[4]

THE PHILOSOPHICAL CONTEXT OF THE *REPUBLIC*'S DISCUSSION OF HARMONICS

The passage on harmonics in *Republic* Book VII, like the musical reflections in Book III, is part of a discussion of education; but the two educational programmes are very different. Book III was concerned with character-training, not with 'academic' learning, and those to be trained were children. Book VII turns to the intellectual education, in adult life, of a small, hand-picked elite

[4] For those who do not have a lifetime to spare, Annas 1981 is still in my view the best introduction available.

who – if their talents and application match up to the task – are to become philosophers, and rulers of the perfect city which Socrates and his friends are mentally constructing. In this city the authority of the philosopher-rulers or 'guardians' is absolute. But Plato was under no illusion that the total concentration of authority in a few hands is a panacea for political ills. He knew from personal experience as well as from reflective thought that a malign or misguided dictatorship is a catastrophe, worse even, in his opinion, than a democracy, for all the latter's light-headedness and lack of principle. It is essential, then, that the rule of the philosophers should be based on the firmest of foundations. The combination of Machiavellian skills, worthy intentions and plausible opinions which in the normal run of things can earn someone a reputation as a good statesman will by no means be enough to ensure that what the city gains from their government will genuinely benefit it. If they are to construct and preserve what is best, their grasp on which is best must have the unshakable authority of absolute knowledge.[5]

According to the *Republic*'s reasoning, anything whose nature can be known must, in the first place, be real; secondly, it must possess its attributes without qualification, independently of circumstances and of the enquirer's perspective; thirdly it must be eternal and eternally unchanging. Suppose I claim to know, absolutely and unshakably, that something has such and such a property, for instance that the cement-mixer in my barn is orange. If there is no such thing (the machine is a figment of my imagination or has been stolen), my claim is empty. If it does exist, its orangeness depends on the lighting conditions and the colour-sensitivity of the viewer; it is not an objective, knowable fact. Its paint will in any case flake off and it will cease to be orange. It will rust; one day, perhaps, it will have been recycled into dog-food cans, and no proposition about my cement-mixer will have any purchase on reality. But no proposition whose truth-value can change or evaporate can be known absolutely and without qualification. In this very strict sense there can be no knowledge of individual cement-mixers or apples or bottles of wine, or indeed of anything to which we gain access through our senses.[6]

Can there, then, be any genuine objects of knowledge? Two considerations in particular seem to have encouraged Plato to give this question an affirmative answer. One is the example provided by mathematics. Propositions such as '2 + 2 = 4', or 'the triangle's internal angles are equal to two

5 For an account of Plato's relevant experiences, perhaps from his own hand, see Plato, *Ep.* 7, especially 324b–326b; cf. *Rep.* 555b–576b on democracy and tyranny and on the forms of human character which correspond to each of them, *Gorg.* 515b–519d on the failings of various famous Athenian statesmen.

6 See particularly *Rep.* 476e–480a.

right angles' seem immune to the diseases infecting statements about the world presented to our eyes and ears. The number 2 and its attributes are independent of time, change and perspective, and the triangle discussed by mathematicians is not the imperfect and ephemeral diagram scribbled on a blackboard but something quite different, 'The Triangle', or possibly 'triangularity itself'. These 'objects' exist nowhere in the perceptible sphere; they are accessible only to the mind. Nevertheless they are real. The number 2 is not a figment of our imagination, and when we make true statements about it they are true objectively and eternally. Big Bangs and collapses of universes may come and go; the number 2 and its unchanging nature do not.

Secondly, Plato's dialogues are concerned above all with questions about values and virtues. 'What is courage?' asks Socrates in the *Laches*; 'What is piety?' in the *Euthyphro*; 'What is self-control (or moderation)?' in the *Charmides*; 'What is virtue?' in the *Meno*. In the *Republic* the core question (there are many others) is 'What is justice?' The entitlement of any particular person, disposition, action or mode of behaviour to be called 'just', 'virtuous', 'courageous' and so on is of course endlessly debatable. But Plato apparently thought that such debates make sense only if there is something which justice or courage or virtue really is, something to whose definition a person's actions or character may more or less imperfectly and temporarily correspond, or in whose nature they 'participate', as the *Republic* and other dialogues of that period express it. We must know what justice, for example, really is, before we can coherently make judgements about the extent to which this or that action exemplifies it. Such judgements can qualify as true or false only if there is something determinate which justice is, and though different people may have different views about its nature, as indeed they do, their views can be understood as competitors for the truth only if there is some one thing which each of them is attempting, felicitously or otherwise, to describe. Such entities as these, which Plato calls 'forms', are as objectively real, as knowable and as inaccessible to anything but the mind as are those discussed by mathematicians. Whatever their natures may be, they possess them absolutely, independently of us and our opinions, changelessly and eternally. They are fully qualified, then, as potential objects of knowledge. To find a way in which human minds can attain this knowledge is the task of a philosopher.

If a philosopher is to know what justice or virtue really is, it is not enough that he or she[7] should be able to grasp the truth about its nature; they must also understand why it is so, and why, demonstrably, it cannot be otherwise.

[7] For Socrates' defence of his thesis, outrageous by contemporary standards, that some women may have the abilities appropriate to the rulers or 'guardians' of his ideal city, and should therefore be educated and trained in the same way as the corresponding men, see *Rep.* 451b–457b.

It is by coming to grasp the reason why something is so that a person can 'tie down' a true opinion about its nature, as the *Meno* puts it, and convert it into knowledge (*Meno* 97d–98a). The philosophers must therefore pass beyond realities of the sorts I have mentioned to some higher principle or truth which determines that they are what they are, and explains why they must be so. In their dealings with mathematics they must also come to understand that when considered in isolation from higher levels of philosophical reasoning it is an insecure field of study, since it relies on axioms whose truth cannot be demonstrated or explained within mathematics itself. So long as its theses depend on undemonstrated postulates, which Plato calls '*hypotheseis*', it cannot carry the stamp of knowledge. What disqualifies it is not that the truth of its propositions is in serious doubt, but that they are not 'tied down'; and neither they nor propositions about values and virtues can be so tied until they are shown to be consequences of a higher principle which is grasped 'unhypothetically'. That is, when we understand it fully we shall see, without reference to anything else, that it cannot be otherwise; the act of understanding it provides its own guarantee or 'tether'.[8]

Since Plato represents reality as a single, rationally unified system, the natures of all its components are ultimately determined by that of a single nature which stands above them all. It is, so Socrates says, superior even to reality itself (*Rep.* 509b); it is that on which the nature of reality depends. There is thus only one unhypothetical principle, to which all genuine knowledge must be tied. Plato calls this highest of all beings 'the good', or 'the form of the good'. He nowhere attempts to provide a full account of its nature,[9] but its designation plainly implies that reality and goodness are indissolubly linked. The real is a manifestation of the good.[10] We know why reality is as it is if and only if we know why it is best that it should be so; and we can know that only if we understand fully what the good itself is. The *Republic*'s philosophers can therefore have no knowledge of values, or indeed of anything else, unless they have an unshakable grasp on this highest of all truths. But such understanding cannot be acquired easily or quickly. It can come only after a long and arduous programme of intellectual training, and even then only to a few gifted individuals,

[8] See *Rep.* 509d–511e, particularly 510c ff.

[9] Cf. *Rep.* 506b–e, and on the complexities and difficulties involved in an approach to an understanding of it, 531d–541b.

[10] This is not to be construed as an expression of innocent optimism; Plato is under no illusion that human beings live in 'the best of all possible worlds'. The world we inhabit is unstable and imperfect; it is only the eternal reality of the forms that is wholly and immutably good.

exceptional in both mind and character. Before it advances to anything we would recognise as 'philosophy', this training demands the mastery of five mathematical disciplines, studied in a set order: arithmetic, plane geometry, solid geometry ('stereometry'), astronomy and finally harmonics.[11]

The main purpose of these studies is to accustom the trainee philosophers to thinking about realities accessible to the mind alone, wholly detached from any application to the unstable, mutable world presented to the senses. Arithmetic, as here conceived, is not about numerable pluralities of material objects but about numbers, simply as such. Geometry is not to do with the measurement of areas of land, but with entities such as the square and the circle in their own right. Platonic astronomy, more surprisingly, is not the study of the visible stars, but is – as we might put it – a purely theoretical enquiry concerned with the relations between abstract points in motion.[12] The task of the science of harmonics, correspondingly, is not to study sounds or to anatomise the music we experience through our ears, 'seeking the numbers in audible concords', but to examine relations between numbers; its central purpose, we are told in words I have quoted before, is to 'ascend to [mathematical] problems, to investigate which numbers are concordant [with one another] and which are not, and in each case why' (531c2–4).

HARMONICS IN THE *REPUBLIC*

Glaucon comments that this is an extraordinary (or perhaps 'superhuman') project, *daimonion pragma* (531c5). Like the preceding review of existing approaches to harmonics, both Pythagorean and empirical, his remark suggests that nothing comparable had hitherto been attempted. Socrates agrees, but insists that harmonics, so conceived, is 'useful in the quest for the beautiful and good, but pursued in any other way it is useless' (531c6–7). I have already outlined one reason for its usefulness which is repeatedly emphasised in the text; when the mathematical disciplines are treated as they should be, without reference to the objects of sense-perception (which are semi-real at best), they habituate the mind to the study of genuine realities, those that are changeless and eternal, and whose existence and nature can be grasped only in thought. But this is only part of the story. The other part is less explicit in Socrates' reflections, and it is in his comments on harmonics that it comes closest to the surface.

[11] The programme is introduced and described at *Rep.* 521c–531c.
[12] See e.g. Mourelatos 1980, 1981.

Harmonics, as described here, is not just the study of relations between numbers. Its enquiries turn on the distinction between numbers that are 'concordant' (*symphōnoi*) with one another and those that are not; and it is only at this point that it seems to have any plausible connection with the musically oriented harmonics of other theorists, or to acquire any right to the name. There is a difficulty here about the meaning of the adjective *symphōnos*. Socrates applies it to numbers, but this extended usage can be understood only in so far as we are clear about its meaning in the context from which it has been taken, that is, in its application to musical sounds. We need to know which attribute of sonorous relations it is that Socrates is transferring to relations between numbers. I do not think that in this passage it is the attribute which the word regularly designates in technical writings, that which marks off 'concords' from other classes of melodic relations. The sense on which Socrates' extended usage relies is more general, as it is in some other Platonic passages and often in non-technical literature of other sorts; it means something like 'sounding well together', and applies to notes in any relation that can occur as an element in a genuine musical melody.[13] Only on this interpretation, as it seems to me, can Platonic harmonics be reckoned a substantial science, let alone a *daimonion pragma*. If Socrates' 'concordant numbers' were merely the ratios corresponding to musically 'concordant' intervals, *symphōniai* in the technical sense, these intervals are very few and had long ago been assigned their mathematical counterparts. There was no mystery in Plato's time about 'which numbers are concordant' in that sense, though no doubt the question why they are so seemed a good deal thornier. But if the distinction, conceived musically, is more like that between 'melodic' and 'unmelodic' relations, the domain to be considered is vastly enlarged, and no one in the fifth century had even attempted to establish the mathematical identities of all the relations that fall into each category, or the criteria by which the matter is to be decided.[14]

But Socrates is not talking about music. He is talking about numbers, and about relations which hold between them simply as numbers. His distinction between relations that are *symphōnoi* and those that are not is

[13] See e.g. [Hom.] *Hymn to Hermes* 51, Aristoph. *Birds* 221, 659; the noun *symphōnia* has this broader sense at Plato, *Symp.* 187b, *Crat.* 405d.

[14] The studies of Archytas (Ch. 11 above) are evidently relevant here. The *Republic* cannot refer to them directly, since they postdate its dramatic scenario; and we do not even know whether the *Republic* was composed before or after them, or whether Plato knew about them when he wrote it. If he did, we might read the passage as a covert allusion to Archytas' work, and as expressing carefully qualified approval. Huffman 2005: 423–5 gives a very clear account of the import of the necessary 'qualifications', though this is too feeble a word for his picture of the distance between Archytas' studies and those that Plato's Socrates requires.

therefore to be understood as one that applies within mathematics itself, and as intrinsic to relations between numbers as such. This means that although the word used to convey the distinction is imported into the mathematical sphere from outside, from the language of musicians, the distinction itself is not (unlike, say, the distinction between numbers which are and those which are not the numbers of London buses). It belongs to mathematics as solidly and securely as the distinction between odd and even numbers.

This has a crucial implication. The judgement that certain relations are 'melodic', musically admissible, while others are not, is at least in significant part an evaluative one. Universally, in Greek discussions of these matters, musical relations are conceived as in some sense 'better' than unmusical ones, and it is because they are better that they can play a part in the construction of things of beauty. Relations of other sorts are excluded because they are in a corresponding sense defective. Plato's distinction between what is *symphōnos* and what is not carries with it the same evaluative loading; the point of distinguishing 'concordant' pairs of numbers from others (rather than classifying them along some other lines), and of grasping what it is that makes them so, is to establish a clear borderline between 'better' relations and 'worse'. It follows that mathematics itself, or at least the branch of it which Plato calls 'the study of *harmonia*', is not as one might have supposed a value-free zone. It distinguishes superior relations between numbers from inferior ones, and if pursued as it should be it will reveal also what it is about them that makes some of them 'musical' while others are not. Our perception that some audible relations belong properly in the domain of music, and that others are unsuited to it, is no more than a distant and derivative echo of this fundamental dichotomy in the realm of number itself.[15]

Plato's mathematical harmonics (perhaps better described as 'harmonic mathematics') therefore has something to teach us about value, that is, about what is and what is not objectively good. No doubt the direct contribution it can make to the philosophers' quest for the nature of the good itself is quite modest. At this stage they are still in the 'hypothetical' domain of mathematics, and they still have to tackle the business of high-level philosophical reasoning which Plato calls 'dialectic', from which alone a full understanding of the unhypothetical first principle can, in the end, emerge. But it is a step along the way, enabling the philosophers at least to grasp that goodness is not a construct of social or aesthetic convention, but

[15] For other indications of the evaluative strand in the mathematics of Plato's time see Barker 1994b, and on Plato in particular see Burnyeat 1987 (especially 238–40), and 2000.

inheres in formal, intelligible relations embedded in the structure of reality independently of human preferences, traditions and needs; and it enables them also to establish, in a preliminary, 'hypothetical' way, what some of these privileged relations are.

In the *Republic* Plato states none of the principles or propositions of his harmonics; we get no glimpse of the science in action. A justly famous and poetically delicious passage in Book x conjures up the picture of a musically ordered universe, in terms which are closely linked to the Pythagorean doctrine of the 'harmony of the spheres'. It is a literary *tour-de-force*, but its language is that of allegorical myth, not of science, and for all the ingenuity of devoted commentators it gives no purchase to detailed mathematical analysis, at least from the perspective of harmonics.[16] What the *Republic* lacks in this respect, however, is amply supplied in the *Timaeus*, in a passage which attracted more discussion in later antiquity than almost any other in the Platonic corpus, and which spawned an entire genre of writing in the field of harmonics.

HARMONICS IN THE *TIMAEUS*: THE SOUL OF THE UNIVERSE

The *Timaeus* begins with a recapitulation of the discussions of the *Republic* (*Tim.* 17a–19a), and so announces itself as some sort of continuation of the project initiated there. After a few pages it abandons the familiar Platonic dialogue form, and the bulk of it is a monologue spoken by the (real or fictional) Pythagorean, Timaeus of Locri, after whom it is named. What it offers is an elaborate account of the physical universe and its contents, premised on Plato's theory of forms and therefore presented as no more than a 'likely story' (29b–d), since according to that theory nothing in the perceptible and material realm can be certainly known. Its analyses and explanations of the origins and the structure of the cosmos and everything within it are grounded in mathematics, drawing on all five of the disciplines incorporated into the philosophers' education in the *Republic*. As a study of the workings of the physical universe it is continuous with the work of Presocratic cosmologists; and in its treatment of mathematics as the key to an understanding of the world it is a development of the Pythagorean tradition. But it is vastly more detailed and sophisticated than anything we know from those sources.

[16] *Rep.* 616b–617d. Something can indeed be made of it from an astronomical point of view, but that is another matter. It allows no secure inferences about the pattern of musical relations at which it hints.

Timaeus represents the universe as the product of a divine agent, not a creator in the Old Testament sense but a 'craftsman' (*dēmiourgos*), who fashions a pre-existing chaos into a system organised in the most perfect possible way. The universe he constructs is a living being, whose soul animates and sets in motion its bodily parts. This must be so, since the visible cosmos is perpetually in motion, most notably on the grand scale in the unfailing cycles of the stars and planets, and no mere body, in Plato's view, can move itself, or for ever sustain, through its own agency alone, a movement imparted to it from elsewhere. Only soul is a self-mover. The soul of the universe, furthermore, transmits to its bodily members not only movement but rationally intelligible order, detectable in its most spectacular form, once again, in the complex but regular and beautiful patterns woven by the movements of the celestial bodies. Hence the soul which the craftsman builds for the universe is itself a paradigm of rational order, whose self-movements are integrated within a perfectly, 'harmoniously' regulated structure.

The word 'harmoniously' is not chosen at random. It is most importantly in Timaeus' account of the construction of the soul of the universe (or the 'World Soul', as it is commonly known) that the conceptions of harmonic science come into play. The process by which the World Soul is built is described as if the craftsman were a skilled metal-worker.[17] He takes certain metaphysical stuffs, whose nature, fortunately, we need not examine, and fuses them into a compound as if blending metals into an alloy (35a1–b1). He then forms the compound into a strip, upon which he marks off lengths through a mathematically structured process of division.

This is how he began to divide. First he took away one part from the whole, then another, double the size of the first, then a third, one and a half times the second and three times the first, then a fourth, double the second, then a fifth, three times the third, then a sixth, eight times the first, then a seventh, twenty- seven times the first. Next he filled out the double and triple intervals, once again cutting off parts from the material and placing them in the intervening gaps, so that in each interval there were two means, the one exceeding [one extreme] and exceeded [by the other] by the same part of the extremes themselves, the other exceeding and exceeded by an equal number. From these links within the previous intervals there arose hemiolic, epitritic and epogdoic intervals [i.e. intervals in the ratios 3:2, 4:3 and 9:8 respectively]; and he filled up all the epitritics with the epogdoic kind of interval, leaving a part of each of them, where the interval of the remaining part had as its boundaries, number to number, 256:243. And in this way he had now used up all the mixture from which he cut these portions. (*Tim.* 35b4–36b6)

[17] See particularly Zedda 2000.

All this is less complicated than it may at first appear. The process of division has three stages. In the first, the craftsman marks off lengths based on two geometrical progressions, a series of doubles, 1, 2, 4, 8, and a series of triples, 1, 3, 9, 27. Their combination gives the sequence 1, 2, 3, 4, 8, 9, 27. Secondly, he 'fills out the double and triple intervals (*diastēmata*)' by inserting two means between the terms bounding each such interval; these means are clearly defined, and are the harmonic and arithmetic means of Archytas' classification (pp. 302–3 above). If this phase of the division is to be represented in whole numbers its terms must be multiplied by 6. Thus the series of doubles becomes 6, 12, 24, 48, and the series of triples becomes 6, 18, 54, 162. The insertion of means into the series of doubles gives **6**, 8, 9, **12**, 16, 18, **24**, 32, 36, **48**, and their insertion into the series of triples gives **6**, 9, 12, **18**, 27, 36, **54**, 81, 108, **162**. Combining these two sequences into one we get **6**, 8, 9, **12**, 16, **18**, **24**, 27, 32, 36, **48**, **54**, 81, 108, **162**.

It is already clear that the division has something to do with music. This is shown not just by Plato's use of the term *diastēma*, 'interval' (which in any case has a non-musical sense that is perfectly appropriate in this context, meaning a 'gap' or 'distance' separating two things), and by his incorporation of Archytas' three kinds of mean into the structure.[18] It emerges from the shape of the division itself. The numbers from 6 to 24 represent the fundamental notes of an attunement spanning two octaves, each divided in the regular way to give perfect fourths at the top and the bottom, separated by the interval of a tone. Thus 12:6, for instance, is the ratio of an octave (2:1); 12:9 and 8:6 are each equivalent to 4:3, the ratio of a fourth; the intervening ratio, 9:8, is that of a tone. The sequence from 24 to 48 yields another octave with the same structure, except that as well as the two terms marking the inner boundaries of the fourths, 32 and 36, it includes also the term 27, which lies a tone from 24, in the ratio 9:8. Beyond the number 48 there lies another tone (54:48 = 9:8), which is best construed as a disjunction between two octaves, the next of which is bounded by 108 and 54. Only one term, 81, divides this fourth octave; 81:54 is the ratio of a perfect fifth, 3:2, and 108:81 is that of a fourth, 4:3 (the 'missing' term would be 72, since 72:54 = 4:3, 108:72 = 3:2 and 81:72 = 9:8). Finally, beyond this octave, we find 162:108, equivalent to 3:2, another perfect fifth. The entire structure spans four octaves, plus the tone separating the third octave from its successor and the concluding perfect fifth, in all four octaves and a major sixth.

[18] The geometric mean is not explicitly mentioned, but intermediate terms in the original series of doubles and of triples are the geometrical means between their neighbours.

The division's links with music might still, at a pinch, be reckoned coincidental, fortuitous consequences of a purely mathematical operation. Its third step, however, leaves no room for doubt. When an 'epitritic' interval (a perfect fourth in the ratio 4:3) is 'filled up with the epogdoic kind of interval' (these intervals are tones in the ratio 9:8) together with a residue in the ratio 256:243, what is generated is a tetrachord, divided into two tones and the small interval approximating to a semitone which Philolaus called the *diesis* and which later became known as the *leimma*.[19] Tetrachords of this form reappear times without number in post-Platonic sources as a representation of a diatonic division. We have seen also that such tetrachords can be accurately tuned in practice in the simplest possible way, through the 'method of concordance', and that this fact was apparently known to Archytas, though the ratios of his own diatonic are different (p. 299 above). There can be no motive for this division of each 'epitritic interval' into two epogdoics and a *leimma* except to capture the shape of a recognisably musical structure. From a purely mathematical perspective it looks both awkward and arbitrary, and the cumbersome 'number to number' ratio of 256:243 seems embarrassing in the company of this extensive array of ratios that are otherwise multiple or epimoric (cf. pp. 288–90 above). Another passage in the *Timaeus*, which we shall examine shortly, confirms not merely that the system has musical connotations, but also that for Plato's purposes in the dialogue it was essential that it should.

It is not, however, a description of any musical scale or attunement that was ever used in practice, if only because it has far too large a range. Most Greek analysis takes place within the compass of two octaves; Philolaus and the *harmonikoi* (and Archytas too, so far as our evidence takes us) were apparently content with one; and the range of any one instrument on which an attunement had to be formed can rarely have exceeded an octave by much. Even according to Aristoxenus' more generous estimate, the maximum compass of a single instrument or of the human voice is around two octaves and a fifth, and is certainly less than three octaves (*El. harm.* 23.22–32). Only one of the surviving musical scores has a range of more than an octave and a fifth, and few of them even approach that span.[20] The compass of the *Timaeus* 'scale' is not determined by musical

[19] The noun *leimma* is derived from the verb *leipein*, and means 'remainder', 'that which is left'. Plato himself does not use the noun, but the theorists who coined it probably did so on the basis of this passage, where it is described as the interval or distance (*diastasis*) that is 'left', *leiphtheisa* (this is the passive past participle of the same verb).

[20] For the (quite extraordinary) exception, a fragmentary score on a papyrus of the second century AD, with a compass of more than two octaves, see Pöhlmann and West 2001: 134–37. I am grateful to an anonymous reader for drawing it to my attention.

considerations at all, but by the metaphysically motivated requirement that its two primary sequences of terms should include both square and cubic numbers. It is to be considered musical in an abstract, mathematical sense, in virtue of its perfect proportionality, completeness and integration, to which human musical constructions can only distantly approximate; and the reasoning that generates it is the product of something that at least closely resembles the non-empirical, rationalistic harmonics adumbrated in the *Republic*.

These points do something to explain another feature of the division which would make it unhelpful in the context of an enterprise directed to the analysis of literally musical systems, that is, that from a musical point of view it is in certain respects underdetermined. There is at least one crucial general issue about which Plato leaves us entirely in the dark, since he does not tell us which way up we should read his scale, whether smaller numbers represent higher or lower notes. It is natural to assume that they are lower, so that they correspond to the slower speeds of movement with which lower pitches are linked in Archytas' acoustic theory, a variant of which reappears in the *Timaeus* itself (67b). But that interpretation generates an anomaly in the structure, since either the tone falling between the numbers 24 and 27 or that between the numbers 48 and 54 must lie within a tetrachord (the other can be construed as a disjunction); and whichever it is it will lie in the wrong place, at the bottom of a tetrachord, in the position occupied in a regular diatonic system by a *leimma*. We can resolve this problem by reading the scale in the other direction. But if we do so, the resulting correlation of larger numbers with lower notes is hard to justify except by reference to the relative sounding-lengths of strings or pipes; and this is precisely the sort of 'empirical' reference which Plato could be expected to avoid.

There are other uncertainties too. Timaeus has explained, in outline, how the open fourths in his system are to be 'filled up'. But he does not specify the order in which tones and *leimmata* are to be placed, and he says nothing explicitly about the subdivisions of the two open fifths (between 54 and 81 and between 108 and 162), or about that of the minor third between 27 and 32. These gaps in his account can be repaired, but the fact that he does not complete the task himself is significant. He is not interested in the task of making his construction correspond at every point to the shape of a system that could be used in practice. Nor does it matter, from the perspective of his project, whether high notes are associated with large numbers or with small, since the issue simply does not arise. There are no notes or pitches in his *harmonia*; there are only numbers.

When the craftsman has finished marking out his strip, he splits it lengthwise in two, and then, after fitting the two pieces together in the form of the letter X, he bends each of them round into a circle, and sets the two circles revolving. But the inner of the two circles, it now turns out, has itself been divided into seven concentric circles of different sizes, sizes which are once again determined by the two sequences of numbers involved in the first stage of the division, 1, 2, 4, 8 and 1, 3, 9, 27. These seven circles revolve in the same direction at various speeds, and their motion is also subject to that of the single outer circle (*Tim.* 36b–d). There are many problematic details here, but I shall ignore them. The general significance of the arrangement becomes clear in the immediate sequel, when the craftsman fastens bodily substance to these incorporeal circles, to be carried around by their revolutions. 'And it [the bodily substance] becomes the visible body of the heavens, while the other [the being whose construction has been described] is a soul, invisible but possessed of rationality and *harmonia*, the best of all the intelligible and eternal entities that have been brought into being by the best of agents' (36e–37a). More details are provided at 38c–e. The seven inner circles are those that carry the visible bodies we call moon, sun, Venus, Mercury, Mars, Jupiter and Saturn, while the undivided outer circle carries the fixed stars. As in Aristotle's account of a theory of the 'harmony of the spheres' in the *De caelo* and implicitly in *Republic* Book x, harmonics becomes fused with astronomy; and underlying both is the mathematics of 'concordant numbers'.

THE *TIMAEUS* ON HARMONICS AND HUMAN PSYCHOLOGY

A little later in the dialogue, Timaeus describes the way in which mortal creatures, including humans, were originally constructed. The craftsman himself builds their souls, on the same pattern as that of the universe though from an inferior alloy (41d4–7), but he entrusts to lesser gods the task of contriving bodies for them and fitting bodies to souls (41a7–d3). But when the mortal bodies are formed and the revolving circles are bound into them, the inrush of nourishment and sense-impressions throws the soul into confusion, so that the orderly structure on which its rationality and its grasp upon truth depend is distorted almost (but not quite) to breaking point.

> The double and triple distances, three of each, and the means and linkages made up of hemiolic [3:2], epitritic [4:3] and epogdoic [9:8] ratios, while they cannot be totally destroyed except by the agent who bound them together, bend into twisted shapes of all sorts, and inflict every kind of breakage and ruin on the circles, so far as that is possible, so that they scarcely hold together, and move, but move irrationally, sometimes forwards, sometimes sideways, sometimes upside down. (43d4–e4)

It is like the condition of someone standing on his head and seeing things on his right as if they were on the left (43e4–8).

Musical mathematics thus enters human psychology. It is the structure of double and triple distances, and of the means which 'link' their terms and generate the ratios of the lesser concords and of the tones on which the divisions of tetrachords are based, that give the soul its rationality; and when they are disturbed the consequence is irrational confusion. It can be no coincidence that Timaeus' description of these distortions and their consequences evokes echoes of the ways in which fifth-century writers, especially the comic dramatists, depicted the corruptions and disfigurements inflicted on music by irresponsible modern composers. There are the same 'twists and turns' (*strophai*, elsewhere *kampai*) and 'ruinations' (*diaphthorai*), and the same image of seeing things the wrong way round.[21] Plato's implicit association of these musical malpractices with the psychological chaos into which the soul is initially plunged by its association with the body carries an incidental but clear message about music itself: to mistreat the structures proper to genuine melody and attunement is to abandon rationality.

The soul's confusion is at its most acute in the early years of mortal life, stirred up by the violent influxes and effluxes of nutriment and the onrush of bodily growth, and by the onslaughts of intense and unfamiliar sensations. Later, these disturbances lessen, and the soul's cycles can be restored to their proper condition. But complete recovery is possible only if the soul, at this stage, absorbs the 'true nourishment of education' (on all this see 43a–44c). The *Timaeus* does not revisit the detailed educational programmes of the *Republic*, but it does make it clear that music is among the most valuable resources on which we can draw when attempting to restore our souls to rationality and health.

In the relevant passage, Timaeus outlines the principal benefits we gain from our possession of sight and hearing. We have been equipped with sight not, primarily, for the humdrum purpose of finding our way about the world, but rather

> in order that we may see the revolutions of mind in the heavens, and apply them to the cycles of our own thought, which are akin to those others but disturbed while they are undisturbed, and by learning them thoroughly and engaging in reasonings true to what is naturally correct, and by imitating the altogether unwavering revolutions of the divine, we may establish soundly their wavering counterparts in ourselves. (47b6–c4)

[21] See particularly the famous passage of Pherecrates' *Chiron* quoted at [Plut.] *De mus.* 1141d–1142a (Pherecrates frag. 155); compare e.g. Aristoph. *Clouds* 968–72, *Thesm.* 49–69.

Voice and hearing have been given to us for a similar purpose. For one thing, they are the prerequisites of *logos* (which means both 'speech' and 'reason', as well as 'ratio' in appropriate contexts). But secondly,

> that part of music which can be deployed by the voice and directed to the hearing is given for the sake of *harmonia*. And *harmonia*, which has movements akin to the revolutions of the soul within us, is not reckoned useful, by anyone who treats the Muses intelligently, for the sake of irrational pleasure – as is nowadays generally supposed – but as having been given by the Muses as an ally in the attempt to bring the revolution of our soul, which has become ill-attuned (*anharmoston*), into proper order and concord with itself. (47c7–d7)

From a combination of two other passages of the *Timaeus* (70d7–72b5 and 79e10–80b9), it is possible to excavate quite a detailed account of the physiological and psychological processes which mediate the transition from the arrival of musical sounds at our ears to a rational appreciation and absorption of their divinely ordered patterns of movement. The extraction of this account from the text is a moderately intricate business; I have attempted it elsewhere and shall not now retrace all my steps.[22] Summarily, the upshot is this. The part of the soul upon which sounds, musical or otherwise, are initially registered is irrational, incapable of understanding. But it is not merely a recipient of physical stimuli. It is the locus of emotional reaction, and it could be no such thing unless the stimuli somehow presented themselves to it as meaningful. What Timaeus tells us is that 'messages' sent to it either through the channels of the senses or from the soul's rational part are transformed *en route* into 'images', quasi-pictorial likenesses of things that may be horrible and terrifying or sweet and delightful. (The agent of transformation is the liver, from whose surface the stimuli are reflected, as if from a mirror, but one which converts them from impulses into images. A rough modern analogue is the screen of a computer's monitor, which receives coded patterns of electronic impulses and presents them to our eyes as something entirely different, text and pictures.) Music, then, is received by the 'irrational soul' in the form of images or likenesses. At this level, as we are told in the *Republic* and the *Laws*, it is a fabric of *mimēseis*, 'imitations' of things other than itself.[23]

But the process does not end with the emotional responses of this part of the soul. The movement of sound through the body is circular, beginning in the head, travelling down to the liver and to the irrational soul which is housed nearby, and then back, transformed, to the head, the locus of the soul's rational part. The task of the rational soul is to interpret the

[22] Barker 2000b. [23] See e.g. *Rep*. 401a, *Laws* 668a–c.

imagistic 'phantasms' transmitted to it from below; it is like a *prophētēs* (one who 'speaks out clearly') who places comprehensible interpretations on the divinely inspired but inchoate and apparently crazy utterances of a *mantis*, the utterances, for example, of the priestess of the Delphic oracle. To ignorant people, says Timaeus at 80b5–8, musical combinations of sounds give pleasure (*hēdonē*); but to intelligent people they give delight, *euphrosynē*, 'because of their imitation, in mortal movements, of the divine *harmonia*'. They can give no such delight to people who fail to appreciate the connection of these sound-patterns with the *harmonia* of the World Soul. It follows that these 'intelligent' people are those who have mastered the science of harmonics in its Platonic form and have recognised its cosmological meaning.[24] The rational part of their souls, if of no one else's, is equipped to interpret the images conjured up by the impact of music on the irrational soul, and it is only their souls whose revolutions can be fully restored to their rational and harmonious order.

The *Timaeus* therefore implicitly assigns to mathematical harmonics, coupled with astronomy, a crucial role in the business of human life. It is the instrument through which we can regain the perfection which our souls lost when they were harnessed to the bewildering paraphernalia of bodily existence. In this light it no longer seems strange that later Platonists and Platonising 'Pythagoreans' (Nicomachus, Theon of Smyrna, Plutarch, Proclus and others) devoted so much study to the passage describing the musical structure of the World-Soul, or that harmonics played so large a part in their philosophical and mathematical reflections.

HARMONICS IN THE IVORY TOWER

The intellectual milieu into which Plato's harmonics inserted itself in its own time was quite different from that inhabited by the work of the *harmonikoi*. The ideas of the *harmonikoi* were propounded in the first instance for the benefit of practical musicians, and may also have been sketched, in outline, in public places for anyone to hear. Plato had views about the proper training of composers and performers, as we see in the *Laws*, but that is not the business in which he was engaged in the *Republic* or the *Timaeus*; and he did not teach in public. The Academy was a small, private institution for the intellectually ambitious. Some of the dialogues,

[24] At 47c Timaeus speaks of 'that part of music which can be deployed by the voice and directed to the hearing', implying that there is another 'part' of music inaccessible to these human faculties. This must be the music of the World-Soul and of its bodily organs, the stars and planets. We cannot hear or sing it, but a mind trained in mathematical harmonics can grasp its nature and significance.

particularly those written before the Academy's foundation, were certainly intended for a wider audience; but Plato's later writings, including the *Timaeus*, make no concessions to the uninitiated, and would have been unintelligible to all but a very few. Though Archytas' harmonic studies engage more closely and deliberately with the realities of human music-making, they too belong in the context of mathematical and philosophical enquiries relevant and accessible only to specialists. (There is no evidence, and no probability, that Archytas was a teacher in the mould of the sophists. His public persona was that of a statesman and a military commander.)

We are entering a new world here, one in which dedicated philosophers, scientists and mathematicians discourse with one another in language, and for purposes, beyond the imagining of outsiders. Such discourse is familiar enough in the modern world, where almost every advanced discipline is barricaded behind its own preconceptions, obsessions and jargon, where attempts in recent decades to open the frontiers between specialised university departments and research institutes have regularly failed, and where the guardians of academia typically discount efforts to communicate the experts' ideas to a non-specialist audience as negligible popularisations. But before Plato's time it had no precedent. Back in the fifth century even the most elevated intellectual enterprises were public property. Anyone could hear the sophists touting their wares, and buy more elaborate versions of them if they had some disposable funds; anyone could listen to Socrates' conversations and appreciate a caricature 'Socrates' on the comic stage; the reflections of sophists and Presocratic philosophers resurfaced repeatedly in publicly performed poetry and drama; and one could buy a copy of Anaxagoras' cosmological treatise for a few pence in the market-place. During the lifetimes of Archytas and Plato, and at least partly as a consequence of their work, mathematical harmonics joined the higher échelons of philosophy, mathematics and the natural sciences in withdrawing itself from public presentation and debate. Specialists talked to specialists, but to virtually nobody else. The work of Aristoxenus in the next generation, as we have already seen, took empirical harmonics a long way down the same path.

CHAPTER 13

Aristotle on the harmonic sciences

The only sustained discussion of musical issues in Aristotle's surviving works is in the last book of the *Politics*. Like the conversations in Book III of Plato's *Republic*, to which it is in part a response,[1] it focuses on the value of music in the life of a city and its citizens, and it says nothing about the musical sciences. It alludes several times, however, to the work of unnamed experts in musicology, at least some of whom were Aristotle's contemporaries.[2] This suggests that he had some acquaintance with up-to-date studies by musical specialists, perhaps including their work in harmonics; and references to harmonic science, and to concepts used by its exponents, are scattered here and there in his other writings. Almost all of them are brief. It seems fairly clear that Aristotle made no substantial contributions of his own to the subject, and that it was marginal to his main areas of interest. I shall argue, too, that despite the confidence of his various pronouncements, his grasp on some of its concepts and procedures was a little uncertain.

A study of his remarks pays dividends none the less, for three main reasons. First, slight though they are, they contribute rather more than has generally been recognised to our knowledge of mathematical harmonics in the fourth century, especially in its Pythagorean form. Secondly, they are relevant to an understanding of certain aspects of Aristotle's own thought, since he occasionally makes use of ideas drawn from mathematical harmonics in his studies of non-musical topics. Thirdly, we have seen that the ideas which he developed about scientific method in general, and about the structure of scientific knowledge, subsequently formed the backbone

[1] The views of Socrates in the *Republic* are explicitly mentioned and discussed at 1342a–b. It can plausibly be argued that other parts of the passage also address Plato's views.

[2] Aristotle attributes relevant ideas and investigations to 'those who have philosophised about this [musical] education' (1340b5–6), to 'some contemporary musical specialists (*mousikoi*) and those philosophers who are well versed in musical education' (1341b27–9), to 'some philosophers' (1341b33), to 'those who are involved in philosophical activity and in musical education' (1342a31–2), to 'people who work in this field', which is here evidently musical scholarship rather than philosophy (1342b8–9), and to 'some musical specialists' (1342b23–4).

of the influential approach to harmonics devised by Aristoxenus; and they can also be used to cast light on the procedures of another major treatise, the Euclidean *Sectio canonis*. I outlined the gist of Aristotle's ideas on these matters in Chapter 4 above, and shall not dwell further on their general characteristics; but his own view of their bearing on harmonic science is very different from that of Aristoxenus. His statements on this issue and on a handful of related concepts call for some comment. I shall therefore divide this chapter into three parts, corresponding to the three areas of interest I have identified.

AN ARISTOTELIAN FRAGMENT ON PYTHAGOREAN HARMONICS

I shall not reconsider here the passages on Pythagorean harmonics and philosophy which I touched on in Chapter 10 above, and which are in any case well known. Instead, I shall try to supplement what can be gleaned from them by mulling over the rather less familiar contents of a fragment from one of Aristotle's lost works, and attempting to reconstruct the outlines of the context in which it was originally set. It shows clearly that he was familiar with at least some of the work of Archytas, and with methods of analysis characteristic of the Pythagorean tradition, and I shall suggest that it gives us some reason to believe that he had studied them quite extensively. It is less obvious that he had absorbed their ideas and procedures with perfect accuracy. The passage, as we have it, begins with a direct quotation but continues as paraphrase, and its latter part contains vexing confusions. Our assessment of Aristotle's mastery of its topic will depend at least partly on whether we attribute these confusions to him or to the author of the paraphrase. In either case, however, we can extract valuable information from it about the preoccupations of mathematical theorists in his time or a little earlier.

The passage appears in the Plutarchan *De musica* at 1139b–f; the fragment itself is at 1139b, and runs as follows.

> *Harmonia* is celestial, and its nature is divine, beautiful and wonderful. In potential (*dynamis*) it is four-fold, and it has two means, the arithmetic and the harmonic; and its parts, magnitudes and excesses are revealed in accordance with number and equal measure; for melodies acquire their structure in two tetrachords.[3]

The compiler does not tell us which of Aristotle's works contained it, and commentators' views on the issue have differed. Rose assigned it to his *Eudemus* and Ross to his *De philosophia*; Gigon simply includes it among

[3] It is printed in the collections as frag. 47 Rose, frag. 25 Ross and frag. 908 Gigon.

the fragments whose work of origin is not named in our source, and does not commit himself to an opinion about the work's identity. Before considering its contents in detail I should like to make some comments on the question of its origins; and in due course I shall offer a suggestion of my own. If I am right, the nature of its Aristotelian context may have a substantial bearing on our interpretation of the fragment itself.

The fragment and the discussions associated with it occupy Chapter 23 of the *De musica*. It is the second in a run of four chapters (22–5) which stand out sharply from their surroundings. The bulk of the work is taken up with discussions of the history of musical styles and practices and of cultural issues connected with them; these draw occasionally on musical theory of an empirical or Aristoxenian sort, but never on mathematics. Chapters 22–5, unlike any other passages in the *De musica*, are concerned entirely with mathematical harmonics and its philosophical or cosmological applications. Chapter 22 discusses Plato's construction of the World-Soul in the *Timaeus*; Chapter 23 is the Aristotelian passage; and Chapters 24–5 give brief expositions of ideas apparently drawn from Pythagorean musical-cum-mathematical cosmology. It is true that the compiler makes an attempt to connect these passages with the preceding material. In Chapter 17 he argues that Plato's exclusion of all *harmoniai* except Dorian and Phrygian from his educational scheme in the *Republic* was not due to his ignorance of the others, or of the fact that Dorian and Phrygian were sometimes used in music unsuited to his purposes. Chapters 18–21 are a loose continuation of this theme, purporting to show that reputable 'ancient' composers who used only a limited number of rhythmic and melodic forms did so by deliberate choice, and not because they were unaware of the existence of other possibilities. The compiler then begins Chapter 22 by saying that he has shown that Plato did not reject other kinds of music out of ignorance, and will now demonstrate that he was also well versed in harmonics. But after the discussion of material from the *Timaeus*, Plato disappears from sight, never to re-emerge, and the exploration of themes in mathematical harmonics continues under its own momentum.

The compiler of the *De musica* used a number of sources, some of which he identifies by name, and all of which with a few very minor exceptions come from the fourth century or earlier. It seems clear that he adopted the practice of using the same source as the basis for quite long stretches of his own text, with only occasional brief intrusions from elsewhere; thus the bulk of Chapters 3–10, for instance, is derived (in my view) from Heraclides,[4]

[4] For a contrary opinion see Gottschalk 1980: 134 n. 22, and for an attempt to justify my own position see Barker (forthcoming).

and Chapters 31–9 (by scholarly consensus) from Aristoxenus. I would reckon it an odds-on bet that the self-contained discussion of mathematical harmonics in Chapters 22–5, so markedly different from anything else in the *De musica*, is also paraphrased from a single fourth-century treatise. Before presenting my suggestion about that treatise's author and identity I want to point to certain features of its four chapters which give further support to the hypothesis of a common origin.

In Chapter 22 the compiler sets off, as we have seen, by promising to show that Plato was well versed in harmonics. He continues:

> Thus in the passage on the creation of the soul in the *Timaeus*, he demonstrated his dedication to mathematics and music in the following words [from *Tim.* 35c–36a]: 'And after that he filled out the double and triple intervals by cutting off portions of his material and inserting them inside those intervals, in such a way that within each interval there were two means.'

He then lists and briefly defines the arithmetic, harmonic and geometric means, but he does not refer to them again in the chapter. Nor does he say anything about the in-filling of the triple intervals, or about any of the complexities in the remainder of the *Timaeus*' account. All he does is to explain in both musical and mathematical terms how the octave between *hypatē mesōn* and *nētē diezeugmenōn* can be divided in the familiar way by the insertion of two intermediate notes, *mesē* and *paramesē*. The arithmetical work is done by representing the octave-ratio as 12:6, and by assigning the numbers 8 and 9 to the intermediate notes. The rest of the discussion is an exploration of the symmetrical patterns formed within the octave by this elementary division, rather in the manner of Philolaus frag. 6a (p. 264 above). All the intricacies of the *Timaeus* passage have been elided, and we are apparently to be convinced of Plato's deep understanding of harmonics solely on the grounds that this rudimentary construction can be extracted from his account.

Chapter 23 turns to Aristotle, and it too is concerned only with the octave, which is named both in the opening quotation and at the end of the passage by the word *harmonia*; this again is reminiscent of Philolaus. Like its predecessor, this chapter identifies no notes or intervals within the octave between *hypatē mesōn* and *nētē diezeugmenōn* except those marked out by the 'fixed' notes *mesē* and *paramesē*, giving the boundaries of its upper and lower tetrachords; and it too makes use of the sequence of numbers 12, 9, 8, 6. But it supplements Chapter 22 by explaining (with some confusions which we shall consider below) how these results are reached by the introduction of the arithmetic and the harmonic mean between terms in the ratio of the octave, and by describing (again with some uncertainty of touch) the relations between the intervals constructed through the insertion of means of each type. Here we seem to move beyond the Philolaan model to observations set out by Archytas in frag. 2 (pp. 302–3 above) and exploited in the *Timaeus*.

The first sentence of Chapter 24 is evidently designed to run on directly from the last sentence of Chapter 23, since the latter provides its grammatical subject, which is again *harmonia*. It describes the 'parts' of *harmonia* in the unmistakably Pythagorean language of 'limit' and the 'unlimited' which is central to Philolaus frag. 6, and as even, odd and even-odd (that is, as the product of an even and an odd number). Here too *harmonia* is at least primarily exemplified in the structure of the musical octave, expressed in terms of number and ratio, since its even, odd and even-odd components are once again the numbers 12, 9, 8 and 6. As in the preceding chapters, the author refers only to the octave's division into its concordant substructures, and does not go on to analyse the divisions of its tetrachords in specific varieties of scale. Finally, the brief Chapter 25 touches on the idea that our senses are manifestations of *harmonia*, and that this is true above all of sight and hearing, which are 'celestial and divine', and which reveal *harmonia* to us with the help of light and sound. These notions may seem strange, but there are several parallel passages. The general thesis about the senses finds a partial echo in Aristotle's *De anima*; and very similar propositions about the 'celestial' status of sound and hearing and their value in revealing the secrets of *harmonia* appear in Plato's *Timaeus*, in another fragment of Aristotle, and later in a fascinating passage of Ptolemy's *Harmonics*; there are affinities, too, with remarks in the work of Archytas and in Plato's *Republic*.[5]

The focus on the basic mathematical structure of an octave-*harmonia* gives these chapters of the *De musica* a continuous theme, and they progress in an orderly way from one aspect of it to another. Thus an excerpt from the *Timaeus*, shorn of the complexities of its context, a quotation from Aristotle and recognisable ingredients of the work of Philolaus and Archytas have been brought coherently together. The result looks like an elaboration of the harmonic analysis of Philolaus frag. 6a, combined with the metaphysics and cosmology of frag. 6, given a deeper conceptual foundation through

[5] Aristotle, *De anima* 426a–b; Plato, *Tim.* 47b–d; Aristotle frag. 48 Rose, frag. 24 Ross (both editors assign the fragment, rightly in my view, to the same work as that to which they assign the fragment quoted in *De mus.* ch. 23, the *Eudemus* according to Rose, the *De philosophia* according to Ross); Ptolemy, *Harm.* 93.11–94.20. The ideas expressed in these passages seem also to be connected, though rather more distantly, with the representation of astronomy and harmonics as 'sister-sciences' in Archytas frag. 1, which is echoed by Plato at *Rep.* 530d and reappears again, ingeniously transformed, at the end of the passage in Ptolemy's *Harmonics*.

Archytas' account of the means, and extended into other areas touched on by Archytas and Plato. In short, it is an exploration of the way in which the octave's structure was represented and put to metaphysical use in a developed form of Pythagorean theory whose components were all in place by the second or third decade of the fourth century. There is nothing elsewhere in the *De musica* to encourage the thought that the compiler himself was capable of sifting and integrating passages and ideas from four or more original texts to produce this selective and well-constructed account; we can be confident, I suggest, that in line with his procedure elsewhere he was simply paraphrasing and summarising part of a single existing treatise. The chapters' contents gives us no reason to doubt that his source came from the same period as those he relied on in the bulk of the rest of the work, that is, from the fourth century, and it is a reasonable guess that it came from the same stable as his other major authorities, that is, from a writer in the Lyceum.

In that case the writer is likely to have been Aristotle himself, the latest of those to whom the passage explicitly or implicitly refers. The compiler regularly mentions his major sources (though he does not always tell us where their contributions begin and end), and none of the others named in the *De musica* is a plausible candidate. There is nothing intrinsically improbable about the hypothesis. We may even be able to identify the work on which the *De musica* draws, since there is one lost Aristotelian treatise into which a discussion of this sort would seem to fit perfectly. It was an essay which examined parts of the *Timaeus* in tandem with aspects of the work of Archytas. There is some doubt about the exact form and meaning of its title (recorded in slightly different versions by Diogenes Laertius and Hesychius) and we have very few clues about its contents, but it evidently made some attempt to relate aspects of Archytas' thought to ideas set out in the Platonic dialogue.[6] Since we know so little about it I cannot weigh up the merits of my suggestion by confronting it with detailed evidence from elsewhere, and to that extent it can be no more than a guess. I can only say that it strikes me as plausible; and if it were correct it would give us a glimpse of the strategy adopted in the treatise. It would suggest, for example, that Aristotle was not trying to bring out contrasts

[6] The evidence about this work and another Aristotelian essay on Archytas is assembled as testimonium A13 in Huffman 2005, and discussed on pp. 581–94. Huffman gives its title as *A Summary of the Timaeus and the Works of Archytas*; but we should be wary about the word 'summary', which is not guaranteed by the Greek, and also about the implications of the phrase 'the works', which might suggest a treatment of the whole body of Archytas' oeuvre. According to the ancient sources Aristotle's treatise was in only one book, and it seems unlikely to have had so compendious a scope.

between the two bodies of work he was considering but rather to establish common ground, and to extract from them a compendious representation of 'Pythagorean' philosophy in its most up-to-date form.

The hypothesis that these chapters are paraphrased from a work by Aristotle remains a strong one, in my view, whether or not it was the essay on the *Timaeus* and Archytas. In that case, since they are apparently designed to expound a central theme of Pythagorean mathematical harmonics and to sketch some of the metaphysical uses to which its exponents put it, it would be unsafe to assume that the direct quotation from Aristotle at the beginning of Chapter 23 is an expression of his own views. Most of it is soberly analytic and (given a little further clarification) could have been asserted by anyone who had mastered the relevant mathematics. But the rhapsodic opening sentence, '*Harmonia* is celestial, and its nature is divine, beautiful and wonderful', seems to hint at connections between harmonics and the study of the heavens which elsewhere Aristotle vehemently rejects. More probably, I think, the fragment is of a piece with all the other material in the passage surrounding it, and was part of his exposition of theses he found in his Pythagorean sources. He would hardly have taken the trouble to construct an account of their ideas if they had not struck him as interesting, but it does not follow that he shared them.

Of the four chapters in the *De musica* it is the one devoted to unravelling the meaning of the Aristotelian fragment that can contribute most for my purposes in this book. I shall now spend a little time on its details. After its opening sentence the fragment is densely compressed and bristles with technicalities. 'In potential it (*harmonia*) is four-fold, and it has two means, the arithmetic and the harmonic; and its parts, magnitudes and excesses are revealed in accordance with number and equal measure; for melodies acquire their structure in two tetrachords.' These statements must have served, in their enigmatic way, as an introductory summary of doctrine that would be explained in the sequel, and it seems probable that the discussion that follows in the *De musica* is a version of the original explanation.

The discussion establishes that what Aristotle calls the 'parts' of the octave-*harmonia* are its boundaries; the 'magnitudes' are the ratios 12:9, 9:6, 12:8 and 8:6; and the 'excesses' are the arithmetical differences between the ratios' terms. The 'magnitudes' 12:9 and 8:6 are 'equal in measure' (since each is equivalent to 4:3), as are 12:8 and 9:6 (equivalent to 3:2). 'Equality of measure' holds also, in a different but related sense, of the 'excesses' in the ratios 12:8 and 8:6, since 8 is the harmonic mean between 12 and 6; the difference between 12 and 8 is one third of 12, and that between 8 and 6 is one third of 6. The excesses in the ratios between the extremes and their arithmetic mean, 12:9 and 9:6, are 'equal in number'. The final remark seems designed to link the octave's mathematical structure to musical practice; the terms used in the mathematical analysis correspond to the boundaries of the tetrachords within which melodies are formed.[7]

Up to this point we are on firm and thoroughly Pythagorean ground, and there is nothing to suggest any misunderstanding on Aristotle's part; if the Plutarchan writer had not named the fragment's author, we would have had little hesitation in assigning it to a Pythagorean source. I shall not quote the whole of the subsequent paraphrase. It begins by attributing to Aristotle the statement that the 'body' of *harmonia* (which I take to be made up of the audible sounds in which its formal structure is instantiated) has parts which are unlike but concordant, and that its means are concordant with its parts (i.e. its boundaries) 'in accordance with numerical ratio'. It then identifies the various ratios and the intervals that correspond to them. There follows a statement whose cramped mode of expression owes more to a fondness for rhetorical counterpoint than to the pursuit of clarity, but whose content is elementary; it merely sets out basic features of the relations between the system's bounding terms and their arithmetic and harmonic means.

Difficulties start with the next pair of sentences. I translate them as they appear in the manuscripts.

> Aristotle reveals that their properties [those of the means and extremes] are such that *neatē* [the variant form of *nētē* used also by Philolaus] exceeds *mesē* by a third part of itself, and *hypatē* is exceeded by *paramesē* in the same way. Thus the excesses belong to the class of the relational; for they exceed and are exceeded by the same parts, since the extremes exceed and are exceeded by *mesē* and *paramesē* in the same ratio, the epitritic [4:3] and the hemiolic [3:2]. (1139d–e)

The initial problem is that one would naturally take the first sentence to refer to the harmonic mean between *nētē* (12) and *hypatē* (6); but their harmonic mean is *mesē* (8), and *paramesē* (9) has nothing to do with it. *Nētē* exceeds *mesē* by one third of itself, and *hypatē* is exceeded by *mesē*, not by *paramesē*, by the same fraction of itself. The obvious way of repairing the damage is to emend the MSS *paramesē* to *mesē*, as is done by Weil and Reinach. But we can adopt this manoeuvre only at the cost of making the latter part of the second sentence irrelevant; its reference to *paramesē* cannot be eliminated in the same way. If we follow Weil and Reinach, this remark must be struck out as an interpolation.

[7] What I have said so far about the fragment adds little to the notes *ad loc.* by Einarson and De Lacy 1967, with which I am in almost complete agreement.

But another and more interesting diagnosis is possible. If the closing statement is retained, its sense is that the ratio in which *nētē* exceeds *mesē* is the same as that in which *hypatē* is exceeded by *paramesē* (i.e. 3:2), and the ratio in which *nētē* exceeds *paramesē* is the same as that in which *hypatē* is exceeded by *mesē* (i.e. 4:3). This is of course true, and the point of stating it is presumably to bring out the system's structural symmetry, very much in the way that Philolaus had done (frag. 6a). What has happened in the first sentence, I suspect, is that a statement originally designed to specify the properties of the harmonic mean has been adapted to the purposes of the statement expressing the 'symmetry' thesis. The proposition lurking somewhere in the passage's history identifies *mesē* as the harmonic mean between *nētē* and *hypatē*, stating that *nētē* (12) exceeds *mesē* (8) by one third of itself, that is, of 12, and that *hypatē* (6) is exceeded by *mesē* by one third of itself, that is, of 6. As we have it in the MSS, by contrast, it must mean that *nētē* (12) exceeds *mesē* (8) by one third of 12, and that the excess of *paramesē* (9) over *hypatē* (6) is one third of 9. This amounts to the observation that the ratios of *nētē* to *mesē* and of *paramesē* to *hypatē* are equal. But the closeness of its language to standard ways of discussing the harmonic mean betrays its misunderstanding of the original statement underlying it.

The problems bedevilling the next part of the text (1139e–f) are even more acute and cannot be fully examined here; without emendation it is neither intelligible nor even grammatical. It is possible, however, that the distortions that have crept into the text arise once again from the superimposition of a new meaning on statements intended to convey another; and the passage has probably been further confused by later copyists' attempts to make sense of it. The underlying and superimposed senses are parallel to those in the earlier passage. This time the underlying statement identifies *paramesē* as the arithmetic mean between *nētē* and *hypatē*; *nētē* exceeds it, and *hypatē* is exceeded by it, by the same amount. The superimposed sense completes the evidence for the 'symmetry' thesis set out above; *nētē* exceeds *paramesē* in the same ratio as that in which *hypatē* is exceeded by *mesē*, a fact which the symmetry thesis presupposes, but which had not been expressed in the earlier passage's first sentence.

If this interpretation is on the right lines, the propositions expressed in both layers of the passage are true. From a mathematical perspective the relations identified in the superimposed layer are relatively trivial, but they are none the less striking and Philolaus had thought them important; they show that *harmonia* displays perfect symmetry of form. The underlying layer corresponds to Archytas' account of the harmonic and arithmetic means, and identifies their locations in the system; and its presence at some level of the text, as I have said, is unmistakably betrayed by its language. We are therefore faced with a choice. Commentators have regularly assumed that Aristotle's own version faithfully reproduced Archytan doctrine, and that the confusions (on my reading, the misinterpretations and superimpositions) are due to the author of the paraphrase. That may indeed be so. But since there can be no doubt that Aristotle, in his turn, was paraphrasing Archytas or a similar source (so that there are in fact three layers in the make-up of this passage rather than two), it is at least conceivable that the misunderstandings began with him.

The impulse to treat Aristotle as blameless and to fasten any confusions on the Plutarchan compiler is understandable. But there is one small piece of evidence that brings the case against Aristotle within the realm of possibility, though it does not amount to proof of guilt. In a passage of the *Politics* (1301b29–35) he distinguishes two kinds of equality. One is arithmetical, exemplified in the excesses of 3 over 2 and of 2 over 1. The other is equality 'in ratio', exemplified in the excesses of 4 over 2 and of 2 over 1, since the part (i.e. the fraction) of 4 by which 4 exceeds 2 is equal to the part of 2 by which 2 exceeds 1. In the first sentence quoted from the Plutarchan paraphrase, the 'superimposed' sense substitutes equality of ratio (12:8 = 9:6) for the relation created by the insertion of the harmonic mean (where 12–8 is the same fraction of 12 as 8–6 is of 6). In the *Politics*, apart from arithmetical equality, only equality of ratio is mentioned. The point is not that it neglects the 'harmonic' relation, but that the terms in which it describes equality of ratio are very similar to those in which Archytas described the harmonic mean, and which reappear in the Plutarchan discussion. The harmonic mean, according to Archytas, is such that 'the part of itself by which the first term exceeds the second is the same as the part of the third by which the middle term exceeds the third' (Archytas frag. 2). In the Plutarchan paraphrase, '*neatē* exceeds *mesē* by a third part of itself, and *hypatē* is exceeded by *paramesē* in the same way'. In the *Politics*, equality in ratio is exemplified by the excesses of 4 over 2 and 2 over 1, 'for 2 [the excess of 4 over 2] is the same part of 4 as 1 [the excess of 2 over 1] is of 2, since both are halves'.

The 'equality of ratio' which the *Politics* describes corresponds mathematically to the relation set up, in Archytas' classification, when a geometrical mean is introduced between two other terms. He describes this relation, however, in language completely different from Aristotle's: 'as is the first term to the second, so is the second to the third'. There is nothing wrong with Aristotle's formulation. But what we see now is that in discussing geometric equality and proportion, both he and the Plutarchan paraphrase use language reminiscent of Archytas' account of the harmonic, not the geometric mean. On this evidence, if Aristotle had wanted to convey the sense of the paraphrase's 'superimposed' thesis, he would have done so in much the same terms as those actually found in the text. The hypothesis that its origins there are with Aristotle himself must remain tentative, but it cannot be reckoned unthinkable.

ARISTOTLE'S OWN USES OF MATHEMATICAL HARMONICS

Let us turn now to territory where the ground is firmer, at least to the extent that we are dealing with the unadulterated Aristotle of the works that survive complete. Though he seldom engages closely with details of harmonic theory, he occasionally uses some of its concepts and conclusions to shed light on other topics. In the passage of the *Politics* where he discusses two types of equality, for instance, his topic is neither music nor mathematics. He is examining the different ways in which one might construe what people mean when they set up 'equality' as the goal of an attempt at political change,[8] and he uses the distinction between arithmetical equality and equality of ratio in order to rid their slogan of its ambiguities. But this passage does not allude explicitly to harmonics, and I shall put it aside. Several others might usefully be explored, but instead I shall select just one by way of example, and study it in a little detail.

In the third chapter of the *De sensu* Aristotle's topic is colour.[9] In the part we shall consider, he starts from the position that the fundamental colours are white and black; and claims that all other colours, red, green and so on, arise in one way or another from combinations of those two. One of Aristotle's main concerns is with the various kinds of 'combining' that can be involved. He looks at the case where the impression of a colour is produced by arrays of tiny black dots and white dots set side by side, the case where it is created by laying a thin film of white over a lower layer of black or vice versa, and the case in which white and black are fully blended together in something like a chemical fusion, where all parts of the mixture, however

[8] Early in the passage (1301b6) he refers to changes that transform political institutions as *metabolai*, which is a commonplace word, but is also the one regularly used for musical 'modulations'. It is possible that Aristotle had that sense in mind, and that in this context the noun should be understood as a metaphor. Its musical associations will in any case have been obvious to his original readers.

[9] A version of the discussion that follows was presented as a paper to the B Club in Cambridge in 2004, and I am grateful for comments made by members of my audience on that occasion.

small, are alike. The differences between these modes of combination need not concern us. Our starting point is Aristotle's supposition that in any of these scenarios, different colours are produced by combining white and black in different proportions; we get one colour when three units of white are put together with two units of black, for instance, and another when the ratio between the quantities is changed.

Aristotle's discussion is presented in three parts, each concerned with one of the sets of physical conditions I have sketched. In each of the three phases his comments on the relevant issues fall into two parts, one of which describes the kind of quantitative relation between white and black which produces colours of a superior sort, while the other contrasts that relation with one that is different and somehow inferior, and whose outcome, correspondingly, is a somehow inferior type of colour. These distinctions between better and worse relations do not depend on features of colours as such; they have nothing to do with the chemistry of pigments or the physics of light-refraction or the psychology of visual perception. They are of a mathematical sort, and can apparently be transferred, essentially unchanged, to any other domain in which relations between quantities play a significant role. Aristotle refers to mathematical relations between musical pitches as one parallel case; ratios of the same sorts are in play, and there is a similar distinction between better relations and worse.

Aristotle first introduces the ratios at 439b27 ff. Here he is discussing only the situation in which minuscule black dots and white dots are set side by side, in what we might call the pointillist scenario; but later he explicitly asserts that his remarks apply equally to the other kinds of case, those of superimposed colours and of genuine mixtures (440a12 ff., 440b18 ff.). The mathematical relations which generate each of the two categories of colour are the same in all three situations; and in that case we must suppose that the slightly varied descriptions he gives of each relation in different parts of the passage are intended to be equivalent. But the descriptions pose quite troublesome problems.

In connection with the first situation he says that the juxtaposition of particles of black and white colour will produce different colours when there are different ratios between the numbers of white and black particles, 'for they may lie beside one another in the ratio 3:2 or 4:3 or in accordance with other numbers'. This describes the 'superior' type of relation; of the other kind he says that they 'are in no ratio at all, but in an incommensurable (*asymmetron*) relation of excess and shortfall' (439b27–30). He then adds the following comment.

These colours [those of the better sort] are constituted in the same way as the concords. For the colours that are in the best-ratioed numbers, like the concords in their domain, are those of the colours that appear most pleasing, purple and red, for instance, and a few others of that kind; and they are few for the same reason that the concords too are few; while those that are not in numbers are the other colours. (439b31–440a3)[10]

In the second situation, where a black surface is seen through a film of white or vice versa, Aristotle says that many different colours will be produced 'in the same way as was previously stated; for there will be [in one kind of case] some ratio between the superimposed colours and the ones below, while others will be in no ratio whatever' (440a12–15). Finally, in the case where quantities of black and white are fused into a completely homogeneous mixture,[11] 'there will be many colours because the things that are mixed are capable of being mingled in many ratios; and some colours are in numbers, while in other cases there is only an excess [of black over white or the converse]' (440b18–20).

Thus where combinations of black and white produce the better kinds of colour, the colours are first described as being in some ratio of numbers, for instance 3:2 or 4:3, 'or in accordance with other numbers'. Secondly (in the same passage) we are told that they are 'in the best-ratioed (*eulogistoi*) numbers', like the concords, and that there are only a few such colours for the same reason that there are only a few concords. They are said, thirdly, to involve 'some ratio' between the amounts of white and black; and in their last appearance they are said to be 'in numbers'. We are given to understand that the sense of these various descriptions is in every case the same. Meanwhile the less attractive colours are said in the first passage to be 'in no ratio at all, but in an incommensurable relation of excess and shortfall', and are characterised shortly afterwards as 'not in numbers'. The formulations in the other passages simply repeat parts of these descriptions and introduce nothing new.

There is no great difficulty in construing 'in some ratio' and 'in numbers' (and their negative counterparts, 'in no ratio' and 'not in numbers') as equivalent expressions. At first sight the descriptions of the 'better' relations appear to apply to all ratios of integers without exception (all integers, and only they, can be described as *arithmoi*, 'numbers'), and if they did they

10 He follows this account of the difference between the better and worse relations with another, apparently offered as an alternative possibility; but since it does not reappear in the sequel I shall pass it by for present purposes. Aristotle seems not to have thought the possibility worth investigating further.

11 On 'complete mixture' of this sort see Aristotle, *De gen. et corr.* 1.10.

would give us an intelligible way of understanding the reference to cases where there is no ratio, but 'an incommensurable relation of excess and shortfall'. These would be cases where one of the quantities involved is greater than the other (it 'exceeds' the other and the other 'falls short' by some amount), but it is not possible to specify the relation between them as a ratio of integers; that would be true, for instance, of the relation between the lengths of the side and the diagonal of a square.

There are three interlocking complications, all arising from the first passage's comparison of the more attractive colours with the concords. First, we are told that there are only a few of these colours, just as there are only a few concordant intervals. But it is obviously not true that there are only a few ratios of integers; on the contrary, there are indefinitely many. If there are only a few relations of the relevant sort, they must correspond to some determinate small number of ratios of integers, and there must be some criterion by which they are marked off from the rest. Aristotle does not seem to provide a criterion which could play this role. Secondly, at this point in the text the musical concords and the attractive colours are not described simply as 'in ratios' or 'in numbers', but as 'in the best-ratioed numbers', which seems to mean something quite different. It also suggests that there is another class of relations which will still count as 'in numbers', but whose numbers come together in ratios of an inferior sort. They would be demarcated by their failure to meet the hypothetical criterion which I mentioned above; and the implication is puzzling for another reason too, since these relations would seem to be different from both the 'better' and the 'worse' ones described in the other formulations, and we hear nothing of a third, intermediate group of relations elsewhere in the discussion. Finally and rather similarly, if the cases where there is no ratio but an 'incommensurable excess' of one amount over the other are those where the relation cannot be expressed as a ratio of integers, it is not only the concords that will be excluded from this category, but also every one of the non-concordant intervals quantified by the mathematical theorists. The only intervals to which the description could apply are ones such as the exact half-tone; and in fact the theorists do not describe them in Aristotle's way, as intervals that can exist but are not in ratios of whole numbers. They simply deny that there can be such things. According to the regular interpretation of Archytas' theorem (pp. 303–5 above), a tone in the ratio 9:8 *cannot* be halved, since no interval whose ratio is epimoric can possibly be divided into equal parts.

We can reduce the number of major questions posed by these issues to two. What sort of relation is it that is 'in no ratio at all' but is constituted 'according to some incommensurate excess and shortfall'? And what is the criterion by which some determinate small number of ratios of intervals, those of the 'better' kind, can be distinguished from the rest? I think that the answers to these questions can be extracted from the resources of mathematical harmonics, and that there is indeed one answer that covers them both. From a historical perspective this answer is probably correct, and the fact that it will seem seriously inadequate when considered from a more abstract point of view should not lead us to reject it. Its shortcomings are symptomatic of anomalies in Greek mathematical harmonics itself.

Let us begin with my second question. By what licence are some ratios proclaimed as the best, and which are they? The clearest statement of an appropriate answer in writings on harmonics is in Ptolemy's account of ideas he attributes to 'the Pythagoreans', by which he certainly means Pythagoreans of the period we are considering here. Although one must be cautious in basing conclusions about fourth-century theories on so late (and in some respects so idiosyncratic) a source, there are good grounds for tracing these ideas back at least as far as Archytas, and for present purposes I shall take Ptolemy's evidence at face value.[12] What he says is that the Pythagoreans divided relations between musical pitches into two main classes, the concords and the discords, of which, so they said, the class of concords is the 'finer', *kalliōn*; and they divided ratios, correspondingly, into two primary categories. One of them 'is that of the so-called "epimeric" or "number to number" ratios, and the other is that of the epimorics and multiples; and of these the latter is better (*ameinōn*) than the former' (Ptol. *Harm.* 11.10–15). The sentence continues beyond this point, and I shall come back to it.

Epimoric and multiple ratios have already been mentioned frequently in this book, but a little recapitulation may be helpful. A ratio of the sort which Ptolemy identifies first is usually called 'epimeric' (*epimerēs*); his alternative expression for such ratios, 'number to number', is less common but was evidently current in the fourth century, since Plato uses it in the *Timaeus* (36b3). As conceived by Ptolemy and most other writers on harmonics, this category contains all the ratios that are neither multiple nor epimoric, ratios such as 7:4 or 15:11; Plato's example is the ratio of the *leimma*, 256:243.[13] We

[12] The issues are discussed in Barker 1994b; cf also Barker 2000a: 65–7.

[13] The Peripatetic philosopher Adrastus, around the end of the first century AD, is credited with a further classification of this motley crew of epimerics into subcategories with names that might almost have been coined by Aristophanes; we have the *pollaplasiepimereis* and the *pollaplasiepimorioi*, along with their ghostly Doppelgängers, the *hypopollaplasiepimereis* and the *hypopollaplasiepimorioi*, and Adrastus reserves the expression 'number to number' for ratios which fail to fit even under any of these descriptions (Theo Smyrn. 78–80). This is rollicking stuff, no doubt excellent sport for mathematicians, but we can ignore it and leave them to their fun.

are presented, then, with a pair of aristocrats in the kingdom of ratios, the multiples and the epimorics, and a plebeian gaggle including all the rest.

Ptolemy tells us that on the basis of the argument I have quoted, the Pythagoreans linked the 'better' class of ratios with the 'finer' class of musical intervals, the concords. At one level this association is obvious and by now very familiar; the ratios of the three primary concords are 2:1 for the octave, 3:2 for the fifth and 4:3 for the fourth, and of these 2:1 is multiple and the others are epimoric. But it soon emerges that Ptolemy means more than that; he means that these theorists adopted the principle that all concords *must* have multiple or epimoric ratios, and we have already seen that this principle was current in fourth-century mathematical harmonics. If the interpretation I offered in Chapter 11 is correct, Archytas extended its scope further, and insisted that not just the concords but all the intervals which form steps in any properly formed musical scale must be constructible through a procedure involving epimorics alone. Harmonic theory, then, has its elite ratios and its proletarians, and this obviously encourages the guess that Aristotle's 'best' ratios are precisely these elite ones, the multiples and epimorics. There is a very obvious objection to this hypothesis, but I shall leave it on one side until we have considered whether the same classification of ratios can give any help with our other question.

Aristotle speaks of relations that are constituted in no ratio whatever, 'but according to some incommensurate excess and shortfall (*kath' hyperochēn de tina kai elleipsin asymmetron*)'. What does he mean? Now in the course of Ptolemy's ruminations over notions he found in earlier Pythagorean forms of mathematical harmonics, his description of the feature which gives the aristocratic epimorics a higher status than their inferiors is almost an exact mirror-image of Aristotle's. The superior ratios are *en symmetrois hyperochais*, 'in commensurate excesses' (*Harm.* 16.13). A little later he credits Archytas with the thesis that 'melodic' intervals, that is, the individual steps of any scale, must be characterised by 'the commensurateness of the excesses', *to symmetron tōn hyperochōn* (*Harm.* 30.9–13); and it emerges that these expressions pick out precisely the class of the epimorics. If we take Ptolemy's expressions as a guide, we shall clearly be led to the conclusion that the relations which Aristotle describes as being in no ratio at all are in fact the epimerics.

In order to make further progress, we need to discover exactly what Ptolemy's expressions mean, and in what sense epimoric ratios have 'commensurate excesses'. Secondly, if these expressions specify the feature of epimorics which qualifies them for the ranks of the elite, we need to ask whether this feature has a counterpart in the case of the multiples, of such a sort that we can identify just one overarching consideration from which both classes of ratio derive their aristocratic credentials. The answers to both questions are most clearly and succinctly put, once again, by Ptolemy, in the later part of a sentence I quoted above. The Pythagoreans treated multiples and epimorics as the 'better' class of ratios, he says, 'because of the simplicity of the comparison, since in comparisons involving epimorics the excess is a simple part, and in the case of multiples the smaller term is a simple part of the greater' (*Harm.* 11.15–17). What he says here about epimorics reflects the fact that in such a ratio the 'excess', the difference between the two terms, is a 'simple part' (one half or one third and so on) of both the terms (see pp. 288–91 above). It can therefore serve as the unit by reference to which both terms are measured; and it is in this sense that it is *symmetros* with them. In ratios of the third, inferior sort, this is not the case; in the ratio 7:4, for instance, the difference, 3, is not a 'measure' of either 7 or 4. When we look at the multiples, it is obvious that the excess or difference cannot serve as a 'measure' except in the special case of 2:1; if we take the ratio 6:1, for example, the excess of the larger term over the smaller is 5, and this cannot be used as the measure of either 6 or 1. But in these cases the measure is the smaller term itself, of which, by definition, the larger is always a multiple.

These points lead to a more difficult question. In what sense does this feature of epimorics and multiples, the inclusion within them of an element by which the others can be measured, make them 'better' than the others, and why is it thought to underlie the superior kind of 'fineness' manifested audibly in the musical concords? Ptolemy's answer, which is hinted at in the phrase 'the simplicity of the comparison', is complex and hard to disentangle,[14] and is probably irrelevant to the Aristotelian context; I am reasonably confident that it is not one that he disinterred from fourth-century sources but a hypothesis of his own. The fourth-century answer is much more straightforward.

What gives the concords their special excellence, according to a whole series of writers from Plato onwards (*Tim.* 80a–b), including Aristotle (*De sensu* 7, 448a9 ff.), is the fact – as it is repeatedly said to be – that when the two notes of a concord are heard simultaneously, they are not perceived as two disparate items, isolated from one another. They blend together to form a single, unified sound which is identical with neither of them. When the two notes of a discord are heard at the same time, by contrast, no such well-integrated result is produced; they simply appear side by side as two separate sounds. To quote one clear definition from a later source,

[14] For discussion see Barker 2000a: 82–7.

intervals 'are concordant when the notes that bound them are different in magnitude [which in this author means "different in pitch"], but when struck or sounded simultaneously, mingle with one another in such a way that the sound they produce has a single form, and becomes as it were one sound. They are discordant when the sound from the two of them is heard as divided and unblended.'[15]

The emphasis of such accounts is always on the fusion and unification of a concord's elements to form a seamless whole. This is a special case of the same idea that gives us the 'harmonisation' of opposites through the influence of love in Eryximachus' speech in Plato's *Symposium* (185e–188e), and the synthesis of limiters with unlimiteds through *harmonia* in Philolaus frag. 6; more generally, it reflects the theme of diverse or even mutually hostile elements being integrated in a harmonious and admirable unity which runs through the Presocratic tradition from Heraclitus onwards. We can see how it links up with the ideas about multiple and epimoric ratios; the point is that they too are beautifully integrated complexes. The glue which holds together the greater and smaller terms is their 'common measure', the ingredient in the ratio that mediates between them, manifested in epimoric ratios by the 'commensurateness of the excesses' and in multiple ratios by the commensurateness of the terms themselves. In ratios of the third class, as in the case of the discords, there is no such ingredient to bring the terms into mutual agreement; as Aristotle says, the terms lie side by side 'according to some incommensurate excess and shortfall'.

Let us turn now to the obstacles in the way of this interpretation. The first is that Aristotle has described the substandard relations not only by reference to their incommensurate excesses, but as 'in no ratio at all'; and elsewhere in the passage the expression 'in no ratio' seems to be interchangeable with 'not in numbers'. The problem, obviously, is that neither description seems to fit the relations in the third class of ratios; whatever their deficiencies may be, they evidently are 'in ratios', and the ratios are defined by reference to whole numbers. The fact that Plato describes them precisely as relations 'of number to number' may well give us further cause for unease. As I said earlier, we would expect that 'in no ratio' and 'not in numbers' would designate genuinely irrational relations, like the relation between the diagonal of a square and its side; but we have seen that this cannot be what Aristotle meant.

This difficulty, however, may be illusory. There is at least one other relevant text which uses the 'no ratio' formulation, and spells out exactly

[15] Nicomachus, *Harm.* 12. 262.1–6.

the sense in which it is being used. A passage in the Aristotelian *Problems* discusses the intervals and the arithmetical relations created by doubling an octave, a fifth or a fourth.[16] The ratio of a double octave is 4:1; that of a double fifth is 9:4, and that of a double fourth is 16:9. All these are perfectly well-formed ratios of integers. But what the writer says is that the terms bounding the double fifth and the double fourth 'will have no ratio to one another (*pros allēlous oudena logon hexousin*)'. And he immediately explains what he means; *oute gar epimorioi oute pollaplasioi esontai*, 'for they will be neither epimoric nor multiple'. Here, then, we have the clearest possible evidence that the expression 'no ratio' could be used to refer to relations like 16:9 and 9:4, which by other standards would evidently count as ratios, ones belonging to the inferior 'number to number' class. (It may be worth noting that when Plato uses the 'number to number' formula at *Tim.* 36b3, he does not call the relation in question a *logos*, a ratio; but this by itself is not decisive, since in that passage not even ratios of the superior sorts are called by that name.)

In that case it is a reasonable hypothesis that Aristotle is using the expression 'in no ratio' in the same way as the writer of the Problem, to refer to these 'number to number' relations, all and any relations which are neither multiple nor epimoric. I do not know how this odd-looking usage arose; plainly, however, it existed. But there is still one glaring difficulty. When Aristotle describes the superior relations as 'in some ratio' or 'in numbers' he seems to be talking about relations in any genuine ratio whatever, that is, on our hypothesis, in any epimoric or multiple ratio. But when he describes them as those in the 'best-ratioed numbers', he seems to imply that there is a hierarchy even among the genuine ratios, and that the members of some sub-group of them are the best. That fits awkwardly with our impression that the contrasting group, the ones that are 'not in numbers', includes all and only the number-to-number relations; those of the genuine ratios which are not among the 'best' seem to be overlooked. Yet there is no doubt that those in the best-ratioed numbers amount only to a sub-set of the genuine ratios, not to all of them, since Aristotle tells us that there are only a few of them, just as there are only a few concords. We still have to locate the criterion by which some are marked off as the best, and we still have to account for the apparent absence of any reference to all the others.

At this point we come across a vulnerable spot in Greek mathematical harmonics. It was agreed on all sides that the concords are indeed few; within the span of the octave only three intervals are allowed to qualify for that

16 [Aristotle] *Probl.* 19.41.

title, the fourth, the fifth and the octave itself. It is also common ground, with no dissenting voices whatever, that the distinction between concords and discords is determinate and absolute. Notes related in the fourth, fifth or octave blend together to form a unity in which neither note is heard individually, and notes combined in any other relation do not; writers of this period recognise no borderline cases. Aristotle holds, similarly, that the number of the most attractive colours is small; and I think he is making the same claim here as when he says later that the 'species' (*eidē*) of colour are limited, that is, as I read it, that there is only a determinate number of them. He states this at the end of Chapter 3 of the *De sensu*, and tries to demonstrate it in a very obscure argument in Chapter 6. In Chapter 4, at 442a20 ff., it turns out that if we exclude white and black (since it is colours produced by combining them that are in question here) there are exactly five, which seem to be red, purple, green, blue and yellow. All other hues are mixtures or variants of these colour-species.

We might expect the harmonic theorists to have worked out some mathematical counterpart of their clear-cut aesthetic distinction between concords and discords, one that would objectively distinguish the 'best' ratios, those of the concords, from all other epimorics and multiples. But they did not, and they can hardly be blamed for their failure to do so, for the very good reason that no such mathematical distinction exists. In the case of the epimorics, for instance, there is no mathematical justification whatever for locating a sharp borderline anywhere in the series 3:2, 4:3, 5:4, 6:5 and so on. The Pythagoreans, of course, were impressed by the fact that the terms involved in the ratios of the basic concords are 1, 2, 3 and 4, summing to the perfect number 10 and forming the *tetraktys* of the decad. But this is not the sort of consideration that would have appealed to Aristotle; and even though it is true that when we hear intervals whose bounding pitches have frequencies related to one another as ratios of small numbers, they strike our ears as 'smoother' than those where the terms of the ratios are larger, there are no arithmetical grounds for separating out 2:1, 3:2 and 4:3 or any other small group as uniquely privileged cases, definitively fenced off from the rest. The problem is apparently so distressing that the theorists never bring themselves to mention it; it is the Medusa's head of mathematical harmonics and will turn you to stone if you look at it directly. The only writer in the tradition who says things which when squarely faced would unveil this ghastly truth is Ptolemy, and even he quickly shuffles them under the carpet.[17] Nothing can be done with it except to pretend that it does not

[17] Ptol. *Harm.* 15.6–17, 16.12–21, cf. Barker 2000a: 74–5, 80–2.

exist; and in that case it would be foolish to imagine that Aristotle, in this passage, was in a position to do anything else. The awkward gap which my interpretation leaves in his account falls precisely where we ought to expect one.

My overall conclusion, then, is that the relevant passages of *De sensu* Chapter 3 should indeed be understood against the background of the harmonic theorists' classification of ratios, together with their attribution of privileged status to epimorics and multiples; and that once some apparent difficulties have been resolved the most obtrusive problem that remains is not of Aristotle's own making. The theorists were unanimous in contending that the ratios of the concords have special status, and Aristotle took over this idea without inspecting its credentials more closely than the theorists did themselves. One might argue that even so he has been a little careless, since even if there really were a sharp cut-off point between the 'best' ratios and the others, it would come in the wrong place. He apparently needs five ratios of the superior sort, one for each of his species of colour, and if the alleged boundary is located where the harmonic theorists put it there can be only three. But this problem can be evaded. It is at least very likely that by Aristotle's time the harmonic theorists had extended their studies from relations within the single octave to the two-octave span which was subsequently treated as containing all possible harmonic relations. That extension is clearly visible in Book XIX of the *Problems*, most of which, I believe, belongs to the fourth century, and again in the *Sectio canonis*; and the latter identifies just the right number of concords for Aristotle's purposes, adding the octave plus a fifth (ratio 3:1) and the double octave (4:1) to the original three.

One might have expected the octave plus a fourth to be included too, but it is not. Though no one disputed that the octave plus fourth sounds like a concord, its ratio is 8:3, which is the wrong sort, neither epimoric nor multiple. The octave plus fourth, one might say, is Medusa's little sister. If its ratio and the fact of its concordance are put together and confronted directly, they will explode the principle that all concords must be epimoric or multiple, to which the *Sectio* explicitly subscribes and which is a foundation-stone of its reasoning. The author deals with this embarrassing point by the simplest of all possible strategies; he does not mention the relevant interval at all, and neither do the *Problems*. I shall argue in the next chapter that the *Sectio* was probably written around 300 BC, and it seems to be original only in its systematic organisation of pre-existing ideas. There is then no difficulty in assuming that Aristotle's harmonic sources also claimed that the number of concords is five.

ARISTOTLE ON THE METHODOLOGY OF HARMONIC SCIENCE

The issue of measurement, which occupied us at the outset of this study, makes a convenient point of entry to Aristotle's reflections on the concepts and methods of harmonic science itself. In the course of a combative discussion of the notion of a 'one' or a 'unity', he writes as follows.

It is obvious that the unit indicates a measure. In every domain what constitutes it is different, as for instance in *harmonia* it is a diesis, in length an inch [lit. 'a finger'] or a foot or something of that sort, and in rhythms a step or a syllable; similarly in weighing it is a weight of some determinate size.

He goes on to assert, *inter alia*, that if something is a 'one' or a 'unit' this implies that it is the measure of some plurality, and that the unit which is the measure for things of a given kind must itself be indivisible (*Metaph.* 1087b33–1088a4). The diesis is also mentioned in a very similar discussion earlier in the *Metaphysics*, as the indivisible unit and 'primary measure' in its domain (1016b17–24).

These references to a minimal and indivisible interval, the diesis, which functions as the measure of all others ('pluralities' of dieses), are bound to remind us of the 'smallest interval, by which measurement is to be made', mentioned in connection with empirical theorists at *Republic* 531a, and of the diesis or quarter-tone used for the same purpose in the work of the *harmonikoi* discussed by Aristoxenus. In the *Republic*, this interval's credentials as the unit of measurement depend on its being the smallest that the ear can detect, and in calling it 'indivisible' Aristotle seems to have the same thought in mind. At 1087b37–1088a3 he says that the unit of quantity in any domain is indivisible either 'in form' or 'in relation to perception', that is, so far as our senses can tell us. Of the examples he has given, only the syllable and perhaps the rhythmic step can be thought of as indivisible 'in form';[18] and it is hard to see how the units of measurement he has mentioned in connection with length and weight are indivisible even 'in relation to perception' (they are at best only indivisible within the system of measurement being employed). The notion that the musical diesis is indivisible in this latter sense reappears, however, in a passage of the *De sensu*, where Aristotle is trying to decide whether, and in what sense, a continuum can be conceived as having distinct parts. When we look at a millet seed, he says, our vision encounters the whole of it (and so, in a sense, all its parts); yet we cannot see a ten-thousandth part of it. Similarly, 'the

[18] An item is indivisible in form if it cannot be divided into other items of the same sort. A length is divisible into lengths, but a syllable is not divisible into syllables.

note inside the diesis escapes our perception, even though, when the *melos* is continuous, one hears the whole of it; and the interval between what is in the middle and the boundaries escapes perception' (445b32–446a4).[19] The diesis is indivisible in the sense that if a note were placed between its boundaries we could not distinguish its pitch from theirs, or perceive the interval by which it is separated from either of them.

There is no immediate difficulty, then, in relating these passages to the style of harmonic analysis practised by the *Republic*'s empiricists and by Aristoxenus' *harmonikoi*.[20] More complicated issues are introduced by another, closely related passage in the *Metaphysics* (1053a5–21). Here Aristotle says that in every context, people treat the smallest perceptible unit as the measure, and they suppose that they know something's size or quantity when they know it in relation to this unit. Once again he mentions the musical diesis as an example of such a unit, 'since it is smallest'. So far, everything is consistent with the other passages I have cited. But now he goes on: 'But the measure is not always numerically one; sometimes it is plural. The dieses, for instance, are two – not those assessed by hearing but those in ratios – and there are several vocal sounds by which we measure [in the science of phonetics], and the diagonal is measured by two units, as is the side, and so are all magnitudes' (1053a14–18).

Here, then, it is dieses 'in ratios', as conceived in mathematical harmonics, that are represented as units. Aristotle seems to mean that when larger intervals are expressed as ratios, their sizes need to be expressed in terms that refer to two different units of measurement. However this is understood,

[19] This is the best translation I can offer for a difficult sentence. On the interpretation I shall follow here, one makes the *melos* between the boundaries of the diesis 'continuous' by 'sliding' from one of them to the other; on the probable origins of the word *diesis* in the performing techniques of aulos-players see n. 11 to Ch. 10 above. (For discussion of some of the problems see Barker 2004: 108–12, Sorabji 2004: 135–6; we shall revisit the passage briefly in Ch. 15.) In that case we hear the whole continuum of sound, but we cannot identify any particular pitch between the boundaries of the diesis, or decide what interval lies between any such pitch and either of the bounding notes. If this is the picture which Aristotle is presenting, it again recalls the procedures mentioned at *Rep.* 531a–b; Aristotle's allusion to 'the note in the middle' may well be an echo of the 'sound in the middle' at *Rep.* 531a6–7. A sceptical critic might argue that Aristotle knew little of these matters except what he could derive from this passage of Plato.

[20] It may be objected that it is implausible to conceive the dieses of the *harmonikoi*, which are quarter-tones, as indivisible in precisely this sense, since a tolerably acute ear could indeed perceive a difference between a pitch lying within a quarter-tone's boundaries and the pitches of the boundaries themselves. The quarter-tone is not the smallest interval that the ear can detect, but the smallest that can be recognised as musically significant; the listener's ear will treat any smaller interval as a fudged attempt at the performance of a quarter-tone. The objection has some force. One might argue, however, that no smaller interval than the quarter-tone can be *identified* by perception; that is, that such intervals not only lack musical 'sense', but cannot be assigned any definite size, by the ear, in relation to other intervals. In that case they could neither be measured empirically nor be used as a unit of measurement for other intervals. To that extent at least, they 'escape perception'.

it need not undermine our interpretation of the previous passages, since his explicit distinction between the dieses 'assessed by hearing' and those 'in ratios' apparently implies that in the former case too the diesis is the measure, and that it does not fall into the special category where more than one unit of measurement is involved. But difficulties arise when we focus on mathematical harmonics itself, and ask what the two dieses are and how they serve as units of measurement.

The items to be measured must be intervals such as the octave, fifth, fourth and tone, and perhaps other melodic relations, all of them expressed as numerical ratios. Aristotle's statement that two units are required for their measurement suggests that he was aware of the proposition proved by Archytas (pp. 303–5 above), showing that no interval whose ratio has the form $n + 1:n$ can be divided into any number of equal parts; and he apparently assumes that all the intervals with which harmonics is concerned either have ratios of that sort, or else fall into some other class of which the same proposition is true. Hence larger intervals, expressed as ratios, cannot be measured as multiples of any one smaller interval. If they are to be measured by reference to smaller ones, intervals in at least two different ratios will be needed to function as 'units'. Thus the ratio of the octave, 2:1, for instance, cannot be divided (factorised) into any number of equal ratios of integers; but it can be represented as the product of the ratios of two different smaller intervals, the fifth and the fourth (3:2 and 4:3), since $2:1 = 3:2 \times 4:3$.

But this only illustrates the general idea. Plainly the fifth and the fourth are not dieses, and are not the 'smallest perceptible units' to which Aristotle refers. He is apparently talking about a pair of small intervals whose ratios are such that those of *all* other relevant intervals can be expressed through combinations of some number of each. Thus – again only by way of imperfect illustration – the ratios of the fifth and the fourth can be combined to express not only that of the octave, but for instance those of the double octave or the octave plus a fifth, the former as 3:2 taken twice together with 4:3 taken twice ($3:2 \times 3:2 \times 4:3 \times 4:3 = 4:1$), the latter as 3:2 taken twice and 4:3 taken once ($3:2 \times 3:2 \times 4:3 = 3:1$). Which small intervals are they, then, that can plausibly be regarded as minimal in the required sense, and can also serve jointly to measure all other musical intervals?

If the intervals fundamental to harmonic organisation are the concords and the tone, clearly they must be among the intervals that are measured by the dieses. These dieses must therefore measure the tone; and they must also be capable of measuring the amount by which the fourth exceeds two tones and the fifth exceeds three, that is, the so-called *leimma* (or Philolaan *diesis*) in the ratio 256:243. These conditions can only be met if one of the dieses is the *leimma* itself, and if the other is the interval by which the tone exceeds the *leimma*, that is, the *apotomē* as defined in the Boethian report on Philolaus (p. 272 above), whose ratio is 2187:2048. The tone will then be measured as one *apotomē* and one *leimma* (that is, its ratio of 9:8 will be factorised as 2187:2048×256:243), the fourth as two *apotomai* and three *leimmata* (since it exceeds two tones by one *leimma*), and so on.

To the extent that the calculations work, the interpretation makes sense, and no other candidates for the roles of these dieses can put up nearly so good a case. But it faces three obvious objections. First, no other source mentions a harmonic methodology which treats the *leimma* and the *apotomē* in this manner, as the basis of a system of measurement. Arguments from silence are risky, of course, but not altogether negligible. Secondly, neither this pair of dieses nor any other could provide units capable of measuring intervals in all the ratios involved in fourth-century mathematical harmonics. These dieses, for instance, are helpless in the face of many of the ratios in the divisions constructed by Archytas. We would have to assume that they were used only in connection with a division of the sort found in the *Timaeus*, based on a tetrachord containing two whole tones and a *leimma* (9:8×9:8×256:243), and with a limited number of its variants. The third objection carries the greatest weight. The central purpose of measurement is to express quantities in terms of some fundamental unit (or units). The diesis of empirical harmonics is fundamental in the sense that it is the smallest interval that our ears can identify. The *leimma* and the *apotomē* cannot be accredited on precisely those grounds, since they cannot both be the smallest; and both are substantially larger than either the quarter-tones of the *harmonikoi* or several of the intervals used in the Archytan divisions. The limits of musical hearing are in any case largely irrelevant in the context of mathematical harmonics. But neither do they have any fundamental status when considered mathematically. Plainly they are not mathematical minima, and their rebarbative ratios disqualify them from consideration as intuitively acceptable mathematical starting-points. Nor could their ratios be established, independently of any others, through experiments with relative lengths of an instrument's string. They can be assigned their identities, that is, their ratios, only through a process of derivation from the ratios of larger intervals, the octave, fifth, fourth and tone; it is knowledge of the ratios of these larger intervals that allows us to measure the *leimma* and the *apotomē*, not the other way round.

That is the central point. Right from the beginning, mathematical harmonics took the ratios of the octave and the lesser concords as its point of departure, and based its assessment of the ratios of other intervals on them. It invariably proceeded by dividing larger intervals into smaller ones, not by building up the former from combinations of the latter. It had, in fact, no use for the notion of a 'unit of measurement' in the sense that Aristotle envisages. The passage of the *Metaphysics* which prompted all this puzzlement is premised on the assumption that all measurement works from a least unit upwards. When that assumption is coupled with the fact that in mathematical harmonics most of the important intervals cannot be divided into equal parts, it will inevitably lead to a conclusion of the sort that Aristotle draws. But the assumption is false, and mathematical harmonics is one area in which its writ does not run. There is no point in beating about the bush; so far as this aspect of the subject is concerned, Aristotle did not understand what he was talking about.

Let us turn now to a second issue to do with the methodology of harmonics. In Chapter 4 I outlined the account of a science's structure which Aristotle offers in the *Posterior Analytics.* Here I want to draw attention to what I called the 'same domain rule' which played a pivotal role in Aristoxenus' conception of his science; 'one cannot demonstrate what belongs to one science by means of another'. In Aristotle's formulation, however, the rule is qualified. One cannot do this 'except when they [the two sciences] are so related that one is subordinate to the other, as things in optics are to geometry and things in harmonics are to arithmetic' (75b14–17).

This tells us that there are exceptions to the 'same domain' rule, and that harmonics is one of the sciences to which it does not straightforwardly apply.[21] Aristotle returns to the point in several other passages, again using harmonics as one of his examples. Since the point is not always made or elaborated in quite the same way, it will be helpful to have all the significant variants in front of us before we examine them. Ignoring merely tangential allusions and repetitions, there are four instances to be considered, of which the one just quoted is the first; I shall repeat it for the sake of putting all the evidence in one basket.

(A) One cannot demonstrate what belongs to one science by means of another, except when they are so related that one is subordinate to the other, as optics is to geometry and harmonics is to arithmetic. (75b14–17)

(B) [In order to know something scientifically, Aristotle says at 76a4 ff., we must know it on the basis of principles which hold of it as such; and in that case the middle term of a scientifically demonstrative argument must belong to the same kind as the other terms. He continues:] Alternatively, it is like the way things in harmonics are demonstrated through arithmetic. For such things

[21] For two illuminating discussions of these exceptional sciences see Lennox 1986, Hankinson 2005.

as these, similarly, are demonstrated; but there is a difference. For the fact that something is so belongs to one science, since the underlying kind is different,[22] but the reason why belongs to the higher science, to which the attributes belong in themselves. Thus it is clear even from these cases that one cannot demonstrate anything absolutely except from its own principles; but the principles of these sciences have something in common. (76a9–15)[23]

(C) The reason why differs from the fact in another way, when each is studied by a different science. Such cases are those which are so related to one another that one falls under the other, as optics is related to geometry, mechanics to stereometry [i.e. solid geometry], harmonics to arithmetic and observation of the stars to astronomy. Some of these sciences have almost the same names, as for instance mathematical astronomy and nautical astronomy, or mathematical harmonics and harmonics based on hearing. In these cases it is the task of the empirical scientists to know the fact, and that of the mathematical scientists to know the reason why. For the latter possess the demonstrations which provide the explanations, and often do not know the fact, just as people who study universals often do not know some of the particular facts, since they have not examined them. (78b34–79a6)

(D) One science is more exact than another, and prior to it, if it is the science both of the fact and of the reason why, and not of the fact alone, separately from the science which deals with the reason why. Again, if a science deals with something without treating it as inhering in an underlying subject, it is more exact than one that deals with it while treating it as inhering in an underlying subject, as arithmetic is more exact than harmonics. (87a31–34)[24]

The general drift of these passages is that facts falling within the domain of harmonics (or of one variety of harmonics) can be demonstrated, and thus explained, on the basis of principles belonging to a higher science of a mathematical sort. But the terms in which Aristotle refers to the two sciences are not always the same. In three of the four passages, (A), (B) and (D), he speaks of harmonics as subordinate, in this sense, to arithmetic. Passage (C) contains the same formulation; but it goes on, almost in the same breath, to identify the relation as that between two species of harmonics, 'mathematical harmonics' and 'harmonics based on hearing'. These expressions can only refer, respectively, to harmonics in the Pythagorean style and harmonics of the sort practised by the people Aristoxenus calls *harmonikoi*;

[22] English does not behave like Greek here, and no direct translation will quite capture the sense. Aristotle means that it is different from the kind that falls into the domain of the second science, which he has not yet mentioned in this sentence, and does belong in the province of the first.

[23] A literal translation of the end of the last clause, as it appears in the MSS, would be '. . . have the common', or '. . . have that which is common'. A tiny emendation, reading *ti koinon* for *to koinon*, would straightforwardly produce the sense that seems to be called for, '. . . have something in common'; but in the face of the unanimous MSS tradition I doubt that the change is justified.

[24] This second, very cumbersome sentence paraphrases the text rather than translating it. In Aristotle's Greek it contains just eleven words. I apologise, but can do no better.

and in that case, like some of Aristotle's statements about the diesis, they mark at least approximately the same distinction as that drawn in Book VII of the *Republic*.

Despite the close proximity of the references in (C), it is not clear whether Aristotle has the same distinction in mind when he speaks of 'harmonics' as subordinate to 'arithmetic'. It would be a strange way of expressing it. If he does mean that, 'harmonics' must be empirical harmonics, and mathematical harmonics is represented as coextensive with arithmetic, or as a part of it. Alternatively, 'harmonics' in these passages might refer to 'mathematical' or 'Pythagorean-style' harmonics, and his point will be that its propositions are demonstrated on the basis of principles proper to arithmetic, or rather *arithmētikē*, where this word has its regular sense, 'the science of number'. The evidence of (C) suggests that he may have confused or conflated these two lines of thought. Does he mean that empirical harmonics is subordinate to mathematical harmonics, or that mathematical harmonics is subordinate to arithmetic, or both? (I leave aside for a moment the more remote possibility that the relation holds between arithmetic and empirical harmonics. It will resurface shortly.)

If we are to do justice to Aristotle's various statements, the answer must be 'both'. Passage (C) guarantees that he is thinking at least of the relation between mathematical and empirical harmonics; the latter deals with the facts, the former with the explanations (78b39–79a3). Passage (D), by contrast, makes sense only if 'arithmetic' is given its ordinary meaning. It is more exact than harmonics because it deals with the items in its domain, numbers, 'without treating them as inhering in an underlying subject'. That is, it considers them simply as numbers, not as numbers 'of' or numbers 'inhering in' anything else. Harmonics, on the other hand, considers numbers only in so far as they characterise or inhere in pitched sounds (87a33–4). One cannot substitute 'mathematical harmonics' and 'empirical harmonics' for 'arithmetic' and 'harmonics' in this statement without making it unintelligible. Passage (D) does not show, by itself, that the relation between arithmetic and mathematical harmonics is precisely the same as that between mathematical and empirical harmonics in (C), where the higher science provides the explanations and the lower one the facts; but this is stated unambiguously in (B), and again in the first part of (C).

Aristotle's remarks can be combined into at least a superficially coherent picture if all three sciences are assimilated to a single hierarchy. The facts set out by propositions in empirical harmonics are demonstrated and explained by propositions in mathematical harmonics; and they in turn are demonstrated and explained by propositions in arithmetic. But since, on this interpretation, mathematical harmonics is not a fully autonomous science, and must call on arithmetic in the demonstration of its propositions, empirical harmonics will also depend, at one remove, on arithmetic for the explanation of its facts. In that case, when Aristotle speaks of harmonics as subordinate to arithmetic, 'harmonics' might after all be empirical harmonics, or simply harmonics in general, embracing both its variants. Passage (C) makes it clear, nevertheless, that Aristotle recognises the two harmonic sciences as distinct, representing one as concerned with the facts, the other as responsible (with the aid of arithmetic) for the explanations.

The relation that Aristotle apparently postulates between mathematical harmonics and arithmetic seems unproblematic, at least in outline. Harmonics of this sort depends fundamentally on operations with numerical ratios. It may call on principles which presuppose ways of classifying ratios by arithmetical criteria (as multiple, epimoric and so on), and upon classifications of means and proportions (geometric, arithmetic, harmonic). It sometimes appeals to quite sophisticated arithmetical theorems, such as the Archytan proof that no mean proportional lies between the terms of an epimoric ratio. The *Sectio canonis* in fact sets out a series of nine purely arithmetical theorems before proceeding to harmonic propositions, and draws on the former in all its proofs of the latter. Mathematical harmonics exploits principles proper to arithmetic at every turn, and there is an intelligible sense in which it is arithmetic that provides its explanations. Once it has been established, for instance, that the interval of a tone is in the ratio 9:8, it is argued that such an interval cannot be divided into equal parts *because* its ratio is epimoric, and *because* of the arithmetical theorem mentioned above.[25] At first sight all this seems clear, if a little uninteresting. There are in fact rather troublesome difficulties not far below the surface, as we shall see when we examine the *Sectio canonis* in its own right in Chapter 14; but we shall by-pass them here and turn to the supposedly parallel case of the relation between mathematical and empirical harmonics. Aristotle's account of this relation runs into choppy waters almost at once.

According to the account offered in passage (C), facts enunciated within empirical harmonics are explained by principles belonging to mathematical harmonics. Empirical harmonics, then, does not explain its own facts. Since these facts can nevertheless be explained, through demonstrations calling on principles of the higher science, and since all propositions that can be demonstrated are universal rather than particular, taking the form 'Every A is C', not 'This A is C', these facts must be generalisations grounded in

[25] See Proposition 16 in the *Sectio canonis*.

observation. Empirical harmonics, lacking the capacity to explain its own data, can amount to no more than a collection of such generalisations.[26] But it is hard to find any factual generalisation of that sort which could possibly be explained and confirmed on the basis of principles proper to the mathematical science.

Consider a candidate that may at first sight seem promising: 'The perfect fourth is concordant' (or 'Every interval that is a perfect fourth is concordant'). An appropriate demonstration of this fact, drawing on principles of mathematical harmonics, would take the following form:

(i) The perfect fourth is an interval whose ratio is of type f;
(ii) Every interval whose ratio is of type f is concordant; Therefore
(iii) The perfect fourth is concordant.

The problem here is straightforward. Mathematical harmonics, in any of its Greek versions, contains no principle corresponding to premise (ii). Theorists often adopted an axiom to the effect that every concordant interval has a ratio that is either epimoric or multiple, but that will not serve the purpose; many discordant intervals also have such ratios. There is no mathematically specifiable class of ratios ('type f') all of whose members are concordant.

Rather similar difficulties affect another plausible kind of example. It is a fact, adopted by mathematical and empirical theorists alike, that a perfect fourth added to (or compounded with) a perfect fifth makes an octave. One might suppose that this fact can be explained mathematically, roughly as follows (I ignore various intermediate steps that would be needed to plug all the logical gaps).

(i) The octave's ratio is 2:1, that of the fifth is 3:2, and that of the fourth is 4:3;
(ii) $3:2 \times 4:3 = 2:1$;
Therefore
(iii) A fourth compounded with a fifth makes up an octave.

The difficulty lies again in the first premise. Although it is fundamental to mathematical harmonics, it cannot be established by the principles of that science. The *Sectio canonis* does indeed contain arguments that purport to prove it, in Proposition 12. But Proposition 12 depends on Proposition 11, and Proposition 11 involves a logical mistake that cannot be repaired (see pp. 386–7 below). Even if it could, the first premise of our argument could not be established by Proposition 12 and still serve to demonstrate

[26] They may be stated in a different way, e.g. in the form 'The lowest interval in an enharmonic tetrachord is a quarter-tone.' But this is implicitly universal; it can be reformulated as 'Every interval which is the lowest in an enharmonic tetrachord is a quarter-tone.'

the conclusion we require, since Proposition 12 takes that conclusion as one of the assumptions from which it generates its proof of premise (i), and the putative demonstration would be circular.

The only way in which such a demonstration could persuasively be defended is by arguing that the first premise does not require demonstration or mathematical proof. Aristotle insists, after all, that the principles from which demonstrations are derived must be 'immediate', and do not call for demonstration in their turn. What will underpin the premise and allow it the status of a principle are repeated observations of the intervals produced from appropriately related lengths of a stretched string or a pipe.[27] As far as we can tell, it was in fact on observations of that sort, and not on mathematical reasoning, that theorists typically based their confidence in the relevant ratios; the *Sectio canonis*, probably at the end of the fourth century, offers the first (misguided) attempt at mathematical proof. This approach seems to avoid major difficulties; let us assume that any residual problems can be solved. But still it will not give Aristotle what he needs. No other propositions in mathematical harmonics, so far as I can see, could be established as principles in this manner, on the basis of observation alone. In that case the only demonstrations that could be offered would be ones that included premise (i), or a part of it, among their premises, and as a consequence the demonstrable part of empirical harmonics would be limited to a mere handful of elementary propositions. This example, then, exposes no logical flaws in Aristotle's position, but the position seems to be sustainable only at the cost of reducing the empirical science, in so far as the mathematical science can get any explanatory purchase on it, to triviality.

These examples have been hypothetical; perhaps we shall reach a better result by turning to propositions taken from an actual treatise in mathematical harmonics, with a view to discovering whether they are capable of explaining empirical facts. The *Sectio canonis* was composed, in my view, within a generation of Aristotle's death, and much of it seems to have been put together out of earlier sources.[28] It may therefore exemplify the kind of mathematical science that Aristotle had in mind, and it is the only continuous, full-length treatment of the subject that survives from the period. The bulk of it is set out as a connected series of propositions each of which is provided with a proof; and even if the proofs do not always match up to the strict requirements of Aristotelian demonstration, the work has plainly been

[27] On the route by which we can pass from empirical observations to principles, see pp. 110–12 above.
[28] The treatise and the problems surrounding its date are discussed in detail in Chapter 14 below.

conceived in the spirit of the *Posterior Analytics*, as an attempt to present a completed science (or a substantial part of one) in axiomatic form.

The first nine of the *Sectio*'s twenty propositions are purely arithmetical, and three others (which involve constructions rather than proofs) do not concern us here. These are the propositions proved in the remaining eight theorems.

Prop. 10: the interval of an octave is multiple [i.e. of multiple ratio].
Prop. 11: the intervals of a fourth and a fifth are both epimoric.
Prop. 12: the octave is 2:1; the fifth is 3:2; the fourth is 4:3; the octave plus a fifth is 3:1; the double octave is 4:1.
Prop. 13: the tone is 9:8.
Prop. 14: the octave is less than six tones.
Prop. 15: the fourth is less than two and a half tones, and the fifth is less than three and a half tones.
Prop. 16: the tone cannot be divided into two or more equal intervals.
Prop. 18: the *parhypatai* and *tritai* do not divide the *pyknon* into equal intervals.[29]

I have listed these propositions for one reason only; that is, to make it clear that not one of them can be understood as expressing any proposition in empirical harmonics (or 'harmonics based on hearing', as Aristotle describes it). None of the theorems proving them, therefore, can serve as a demonstration of a fact recorded by the empirical science. Propositions 10–13 specify the ratios of intervals, or the classes to which these ratios belong, and the ratio of an interval is not accessible to the ear. The remainder state conclusions which no empirical theorist of the period would accept. The claims made in Propositions 14 and 15 could, in principle, be tested aurally, but the only relevant fourth-century tests of which we know were taken, perhaps with some hesitation, to show that the first part of Proposition 15 is false.[30] In that case the second part of the proposition will be false too, since all parties accept that the fifth exceeds the fourth by a tone; and so is Proposition 14, since (again by general consensus) a fourth and a fifth together make up an octave. As to Propositions 16 and 18, there is plainly no way in which they can be construed as generalisations based on the evidence of the ear.

Nothing in the *Sectio canonis*, then, can support Aristotle's conception of the relation between the two forms of harmonics. Its mathematical

[29] The sense of this is that the two small intervals (according to empirical theorists, quarter-tones) at the bottom of an enharmonic tetrachord cannot be equal.

[30] Aristox. *El. harm.* 56.13–58.5.

reasoning neither proves nor explains anything in the domain of the empirical science. Statements made by Aristotle himself in passage (B) further undermine his position. It tells us that the 'underlying kind', that is, the subject of the conclusion demonstrated, belongs to the province of the empirical science, which deals with the fact; the attributes belong to the higher science which provides the explanation. The attribute assigned to the subject in the conclusion is therefore proper to the higher science, the subject itself to the lower. This seems to describe quite adequately the affiliations of subjects and attributes in at least Propositions 10–13 of the *Sectio*; the subjects are musical intervals, specified in terms familiar in empirical harmonics, and the attributes assigned to them are mathematical.[31] But despite the legitimacy of the subject-terms in the vocabulary of empirical harmonics, no such propositions will express facts that fall within its domain.

This result might be avoided if the term through which the demonstration links the subject and the predicate (the 'attribute') of the conclusion somehow belongs to both sciences. In the paradigmatic argument-form:

(i) Every A is B;
(ii) Every B is C;
Therefore
(iii) Every A is C,

the 'linking' or 'middle' term is B. This is perhaps what Aristotle has in mind at the end of Passage (B): 'the principles of these sciences have something in common'. But in that case, the premises which serve to prove and explain the (supposedly empirical) conclusion can belong to the higher, mathematical science only if A and C also belong in its domain, since both these terms figure in the premises. In that case the conclusion can also be read as a proposition in mathematical harmonics. Correspondingly, if the conclusion is an empirical proposition demonstrated from mathematical principles, not only its subject but its predicate too must fall within the empirical science's province. All three of the terms involved in the demonstration must in fact be 'common' to both sciences. But if every element in the demonstration has a foot in both camps, the whole argument can be understood either in mathematical or in empirical terms, and there seems to be no reason for introducing the slippery notion of 'kind-crossing' into any interpretation of it.

I think we must conclude that Aristotle's representation of the relation between mathematical and empirical harmonics not only fails to reflect

[31] In Propositions 12 and 13, where intervals are assigned their ratios, the ratio is expressed as an 'attribute' by means of an adjective; 'the octave is *diplasios*', 'the fifth is *hemiolios*' and so on.

the relation actually holding between them in the fourth century, but fails to make good sense. There is, however, at least one other way in which we might elucidate the thesis that mathematical harmonics explains facts catalogued in the science's empirical branch. Suppose that there is a set of mathematical principles from which the structure of a well-formed musical system such as a scale can be derived. The results of this derivation will be expressed mathematically, in the form of a sequence of ratios, as for instance those of the harmonic divisions of Archytas. So far we have nothing that could figure among the propositions of an empirical science. But suppose further that these sequences of ratios can be used to specify the relative lengths of a string or some comparable device from which notes can be produced, in such a way that (according to the postulates of mathematical harmonics) the intervals between them correspond to the ratios which the theoretical derivations produced. The resulting pattern of notes and intervals can then be assessed by ear, and perhaps some way might be found of describing it in empirical terms. If it strikes the musical ear as well formed, the thesis that it is well formed can become a proposition of empirical harmonics. But what explains its musical status will be the principles of mathematical harmonics from which its structure was derived.

This procedure is moderately close to one adopted, over four hundred years later, by Ptolemy. It is just possible that something like it was known in the fourth century, where its most probable locus is in the work of Archytas, though we have no clear evidence for it there or anywhere else. If it existed, the pattern of explanation it sets up might conceivably have prompted Aristotle's statements about the relation between the two forms of harmonics.

But this is hardly a promising hypothesis. Aristotle seems to regard the empirical science as going about its business of describing and cataloguing facts independently of its mathematical cousin, which intervenes only to provide explanations. Here, however, mathematical harmonics is the starting-point; the empirical scientist considers only constructions handed down to him from above, and the only constructions whose credentials can be mathematically explained are ones that have been drawn from that source. Further, as I have said, the notion that any such procedure existed in the contemporary repertoire is entirely speculative; and in any case the mode of explanation it offers seems resistant to representation in the form of an Aristotelian demonstration. I am no logician, and others may succeed where I have failed; but I see no way in which the trick can be done.

CONCLUSIONS

None of Aristotle's dealings with harmonics suggest that he thought of it as part of a musician's essential equipment, or as having any bearing on the wider concerns of musical critics and the educated public in general. It is a discipline pursued by specialists, partly for its own sake, and partly (at least in the mathematical form which captures most of Aristotle's attention) for any light that its concepts and conclusions may shed on problems in other scientific domains. His own scattered comments and discussions are unlikely to have interested anyone apart from dedicated scientists and students of metaphysics and logic. The passage containing the fragment which we examined at the beginning of this chapter shows that the Pythagoreans' version of the enterprise struck him as worth exploring in some detail. In other works he rejects their ways of exploiting its resources in the contexts of metaphysics and cosmology; but we see from his discussion of colours in the *De sensu* that he could treat relations between sounds as significantly parallel to relations between other objects of perception, and found mathematical harmonics itself, when shorn of peculiarly Pythagorean commitments, to be useful elsewhere as a source of illuminating analogies.

In that context his grasp on basic features of the harmonic theorists' work seems tolerably secure, and he is probably not responsible for the main problems generated by his treatment of it. But so far as his direct dealings with the intricacies of harmonics are concerned, my conclusions about Aristotle's understanding of the subject have been largely negative. In harping on his deficiencies I am not just indulging in the academic amusement of dancing on his grave. They are worth emphasising for a better reason, since they point to issues, connected for the most part with the relation between the two versions of harmonic science, that are both theoretically and historically problematic. In a broad sense, both sciences are attempting to establish the truth about the same subject. But they represent items fundamental to harmonic enquiry (notes, intervals and the like) in radically different ways; they differ in their methods of measurement, construction and analysis, and in the criteria by which they assess the credentials of putatively musical relations; and they reach incompatible conclusions. From the fourth century onwards they were regularly portrayed as rivals, and many theorists adopted positions entrenched in one camp or the other, explicitly advertising its merits and the defects of its competitor. But these belligerent gestures do not altogether conceal the fact that a great deal of diplomatic activity was going on behind the scenes, aimed at some sort of reconciliation between the warring factions; and both sides were routinely engaged also in piecemeal theoretical larceny, appropriating for their own purposes concepts, techniques and morsels of doctrine from the opposition's armoury. Aristotle's fumbling attempts in the *Metaphysics* to construe their methods of measurement in parallel ways, and in the *Posterior Analytics* to bring the two approaches into a coherent relationship with one another in the context of his theory of demonstration, are the first in a long and confusing line of bridge-building projects. Some are more impressive than others; none, in the end, is wholly successful.

CHAPTER 14

Systematising mathematical harmonics: the Sectio canonis

Up to the end of the fourth century, all our direct evidence about mathematical harmonics comes in the form of scraps. There are the fragments of Philolaus and Archytas, together with a few brief later reports on which we can reasonably rely; there are Plato's comments in the *Republic* and his psycho-musical construction in the *Timaeus*; there is a scattering of allusions and discussions in Aristotle and a couple of acid comments in Aristoxenus. Apart from the critique mounted by Theophrastus, which we shall consider in Chapter 15, there is very little else.

In earlier chapters I have tried to extract as much enlightenment from these bits and pieces as they can yield, and the amount is not negligible. But in some respects the absence of any complete treatise in the field leaves serious gaps in our knowledge. Quite apart from the loss of theories and arguments, we simply do not know what a 'complete treatise' of that sort would have looked like. We have nothing that unambiguously reveals the aims of such a work, the list of items that we might formulate as its table of contents, the way in which its propositions were combined with one another and integrated into a systematic whole (if indeed they were), or the style of presentation it adopted. It seems likely, in fact, that the various essays that once existed – those of Philolaus, Archytas and the mathematical theorists mentioned in the *Posterior Analytics*, for example – differed significantly in their aspirations and their modes of exposition; but we can say little about the overall structure and agenda of any of them.[1]

If its traditional dating is correct, the first example of a treatise of this sort that survives complete was composed around 300 BC. But this is a very substantial 'if'. Most scholars are prepared to accept that a continuous passage amounting to about two thirds of the whole can indeed be assigned to that period; but it has repeatedly been argued that the remainder cannot.

[1] For a well judged and intellectually satisfying (but inevitably hypothetical) attempt to reconstruct the programme of Archytas' work on harmonics, see Huffman 2005: 60–3.

These parts, it is said, must be later accretions, and the work was assembled in its existing form (that is, as the fullest MSS present it, and as it is printed in most modern editions) many centuries later. The difficulty is not one that we can responsibly ignore. We would be badly misled, both about the treatise itself and about the history of the science, if we took it to have been written, more or less as it stands, in Aristoxenus' lifetime or shortly afterwards, if in fact it emerged hundreds of years later from an entirely different intellectual environment.

Commentators' grounds for doubting that the whole text is an integrated product of the late fourth or early third century can conveniently be divided into two groups. One group fastens on details of the content of certain passages, which are held to be obtrusively anomalous. The other reaches similar conclusions by a different route, focusing primarily on aspects of the history of the text's transmission; the ways in which it is quoted and discussed by other authors are said to reveal that the version they knew was shorter, and lacked the parts whose origins are in dispute. This second group of contentions can be addressed without any close study of the treatise's contents, and after a few preliminary remarks I shall tackle it immediately. The first cannot be handled in the same way, and I shall deal piecemeal with the issues it raises, as they become relevant in the course of our section-by-section examination of the work. Although the grounds for my conclusions about these matters will not have been fully assembled until this chapter is complete, I shall resist the temptation to set the discussion out in detective-story mode, unveiling the solution (if such it be) only in the final scene. Such mystification might be mildly entertaining but would hardly conduce to clarity. I shall therefore come clean from the start; my view is that the reasons for denying the text's integrity as a document dating from about 300 BC are inadequate, and that though the positive reasons for taking the contrary view cannot be completely watertight they are strong and persuasive.

The Latin title by which the treatise is generally known is *Sectio canonis*; in Greek it is *Katatomē kanonos*, in English *Division of the Canon*. The Greek word *kanōn* refers, among other things, to a familiar device for constructing straight lines or measuring lengths, the ruler. In the present context it designates the ruler attached to or mounted on the sound-board of the instrument called the monochord (in later writings it is often used as the name of the instrument itself). As the name 'monochord' – which appears nowhere in this work – suggests, the instrument had a single string, stretched between two fixed bridges. A moveable bridge, placed at various positions on the sound-board underneath the string, determined the length of string that would be plucked and the pitch of the note it sounded. Marks on the ruler or *kanōn* indicated the positions at which the bridge should be placed successively to produce notes differing by predetermined intervals. Since the size of an interval depends on the ratio between the lengths of string that generate its bounding notes, the principles governing the procedure for marking out the *kanōn* are those of mathematical harmonics, in which each interval is correlated or identified with a ratio. To 'divide the *kanōn*' is to mark it out in this way, and a complete division of the *kanōn* is one that allows the instrument's user to play all the notes of a scale (or several scales), with all the intervals adjusted to the ratios deemed theoretically correct.

Only the last two paragraphs of the *Sectio* are directly concerned with such a division. They are preceded by a set of eighteen short arguments presented as theorems, and before the theorems comes a brief introduction. I shall call the theorems 'proposition 1', 'proposition 2', and so on; and though the two final paragraphs do not enunciate theorems, as their predecessors do,[2] I shall refer to them, for convenience, as propositions 19 and 20. The sequence of propositions falls clearly into two main parts (propositions 1–9, which are theorems in pure mathematics, and propositions 10–20, which introduce musical concepts and prove conclusions proper to harmonics); the second, explicitly musical group of propositions will need to be subdivided further, not always in quite the same way, as our examination of the work proceeds.[3]

THE *SECTIO CANONIS* IN PTOLEMY AND PORPHYRY

Questions about the history of the text's transmission, in both Greek and Latin versions, are far too complex to be treated fully here, and on these intricacies, along with many other matters, readers should consult the meticulous study by André Barbera.[4] I shall comment only on issues raised by the passages in later Greek writings which constitute the earliest evidence we have of the *Sectio*'s existence.[5] It will become clear that the writers in

[2] From a formal perspective proposition 17 is also a little different from the others; like the passage dealing with the division of the *kanōn*, it offers a construction, not a proof.

[3] The standard modern edition is Jan 1895: 148–66, reprinted with Italian introduction and notes in Zanoncelli 1990: 31–70. The most recent edition is Barbera 1991; see n. 4 below.

[4] Barbera 1991. The book includes a careful examination of the MSS and of the citations of the *Sectio* in other ancient writings; texts and translations of the treatise in its various versions; and surveys of previous scholars' views and arguments about the origins of its parts. I am grateful to Professor Barbera for his comments, in correspondence, on an earlier draft of some of this chapter's ingredients. I am pleased to discover that our positions are not as far apart as I had supposed, but he should not be held responsible for any errors or confusions still lurking in these pages.

[5] There are grounds for thinking that the *Sectio* was known to some writers who were at work rather earlier than those discussed here, specifically by Theon of Smyrna (early second century AD) and by his most important sources, Adrastus (late first century) and Thrasyllus (early first century). But the indications are too insecure to be relied on.

question assign it a date no later than the early third century BC, and I shall argue that despite some scholars' claims to the contrary, they provide no good reasons for believing that the work as they knew it lacked some parts of the text printed in modern editions. In fact they offer clear though not absolutely unchallengeable pointers to the opposite conclusion.

Porphyry, writing in the third century AD, quotes the first sixteen of the *Sectio*'s propositions as a single continuous passage which comes, he says, from a work by Euclid called *Division of the Canon*.[6] Some of the MSS of the treatise also ascribe it to Euclid, though others record it anonymously and others (quite implausibly) attribute it to a certain Cleonides.[7] If it could be shown that Euclid was indeed the author, the work would be firmly dated to around 300 BC; but doubts cannot be laid to rest so briskly. There is an independent tradition that Euclid wrote on music, but it is not a very solid one,[8] and the *Sectio*'s obvious similarities of language, presentation and method to Euclid's known writings might be products of later imitation rather than signs of genuinely Euclidean authorship. They might have been enough by themselves to persuade an editor or copyist to attach Euclid's name to manuscripts of the treatise, and to mislead Porphyry.

On the other side of the coin one might argue, first, that Porphyry's attributions of the many passages he quotes from earlier writers in this work are generally reliable.[9] Secondly, there is one passage in a pre-Porphyrian source, Ptolemy's *Harmonics* (12.8–27), which plainly draws on material from the *Sectio*. Ptolemy does not identify the work from which it is taken or name its author; he says only that the ideas and arguments it conveys are those of 'the Pythagoreans'. But this attribution is itself indicative, since most if not all of Ptolemy's material on the theorists to whom he gives this name seems to have come from sources preserving authentically fourth-century work.[10] Thirdly, the hypothesis that the *Sectio* was composed by an author other than Euclid who chose to imitate his manner is perfectly possible if he is held to have written at a date close to Euclid's own, but

[6] Porph. *In Ptol. Harm.* 99.1–103.25. The title and the ascription to Euclid are at 98.19.

[7] This is the same (but otherwise unknown) Cleonides to whom some of the MSS, and most modern commentators, attribute the Aristoxenian *Introduction to Harmonics* (*Eisagōgē harmonikē*) printed in Jan 1895: 179–207. On that work see Solomon 1980a, who discusses related issues about the *Sectio*'s authorship on pp. 368–73.

[8] The most significant source is Proclus, *In Eucl. Elem. I*, 69.3 Friedlein.

[9] The one case in which his ascription has repeatedly been disputed is that of the *De audibilibus*, which he attributes at *In Ptol. Harm.* 67.15 ff. to Aristotle. Here he may well be wrong (see especially Gottschalk 1968); but at least he has not assigned the work a date or an intellectual context very far removed from the true one.

[10] See Barker 1994b: 127–32. The fact that Ptolemy calls those responsible for these arguments 'Pythagoreans' should not be allowed to confuse the issue. It is merely his way of designating early exponents of a mathematical approach to harmonics, and need imply nothing about their other philosophical commitments.

much less persuasive if he is located in a later period, since Euclid's style is very different from that characteristically adopted by mathematical writers of later Hellenistic and early imperial times.[11]

No scholar, so far as I am aware, has found substantial grounds for doubting the integrity, as part or whole of a single work, of the segment of the *Sectio* attributed to Euclid by Porphyry, that is, the first sixteen propositions; nor have they found anything in this passage which conflicts with the hypothesis that it was written in the late fourth century or the early third.[12] The principal issue is whether the remaining parts of the text – the introduction, propositions 17–18, and the passage setting out the 'division of the *kanōn*' itself, propositions 19–20 – belong to the work in its original form. The fact that Porphyry presents us with a version from which these components are missing has been taken as evidence that they were not included in the text at his disposal, and the passages used by Ptolemy all fall within Porphyry's segment. It is arguable (though I shall dispute it) that such a text would make sense as it stands, without the rest, and when we come to consider the work's content and argumentation we shall see that it is precisely those remaining parts that contain the supposedly suspect features.

This reading of Porphyry does not convince me. When he quotes other people's writings, he does so to enhance his discussion of specific issues raised by Ptolemy's *Harmonics*, not to preserve whole works from the past for posterity. Take the case of the *Sectio*'s introduction. At the point where he quotes sixteen successive propositions from the treatise and attributes them to Euclid, Porphyry is commenting on the later part of Ptolemy *Harm.* 1.5 (12.8–27), whose own descent from the *Sectio* is beyond serious dispute. The introduction's exposition of theories in physical acoustics is quite irrelevant to this phase of Ptolemy's discussion, which is strictly mathematical in precisely the manner of those of the *Sectio*'s propositions which Porphyry

[11] These matters are discussed by F. E. Robbins in D'Ooge 1926: 28–34.

[12] Doubts of two kinds might be raised. First, if we are inclined to assume that the more ancient a text is, the more worthy it must be of our respect, the fact that the eleventh proposition contains a serious logical flaw (see pp. 386–7 below) might lead us to insist that it is the work of an inferior and therefore later writer; and since that proposition is pivotal to the whole, and the remainder cannot have been conceived without it, the entire treatise must be consigned to the intellectual swamps of later antiquity. Such snorts of outmoded prejudice should presumably be dismissed. Secondly, it is a fact that no echo of the treatise can confidently be identified in any writer before Ptolemy, in the second century AD, and this might constitute grounds for supposing it to have been written not much earlier than his time. But this point carries no weight; it merely puts the *Sectio* in the same condition as, for instance, the musicological works of Philolaus and Archytas, or the Peripatetic essay *De audibilibus*.

does quote. If he had been planning to present the introduction too, this was not the place to do it.

The fact is, in any case, that he does quote the introduction, virtually in full (a few phrases are missing), at the beginning of the same chapter of his commentary (90.7–22). Its purpose there is to introduce the notion of ratio, as it applies in musical contexts, and so to initiate a long discussion (90.24–95.23) about the relation between the concepts of *logos* (ratio) and *diastēma* (distance or interval). Evidently, then, the passage was one that Porphyry knew. A sceptical critic may object that he attributes it to no author, so that it remains possible that he regarded it as part of a different work. That is true. But the same consideration would count, to the same extent, against our attaching it to any author or text whatever; and the extent to which it should be allowed to count, I suggest, is zero. The anonymity of Porphyry's citation gives no positive reason for assigning the introduction to the *Sectio*, as he knew it, but equally gives none for denying the attribution. It is relevant, too, that the quotation is immediately followed in Porphyry's text by another which is also unattributed (90.24–91.1), but which we recognise as coming from the works of Euclid (*Elements* v, definition 3); and the fact that Euclid and the *Sectio* are much in Porphyry's mind at this moment is underlined by his quotation of two brief sentences from the *Sectio* itself shortly afterwards (92.29–93.2).

The absence of propositions 17–18 from Porphyry's long quotation provides, once again, no positive grounds for denying that they were part of the text on which he drew. The chapter of Ptolemy on which he is commenting, *Harmonics* 1.5, reviews certain basic theses, attributed by Ptolemy to the Pythagoreans, about the ratios of the concords and, incidentally, the interval of a tone. Porphyry's sixteen propositions are directly relevant to those topics, and to the treatment of them which Ptolemy is considering. Propositions 17–18 are not. They are concerned with relations specific to the enharmonic genus, with its incomposite ditones and so-called quarter-tones. These propositions have nothing to do with the issues of *Harmonics* 1.5, and it would have been quite inappropriate for Porphyry to include them; nor is there anywhere else in his commentary where they would naturally have found a place. Their absence tells us nothing about whether they were in his text of the *Sectio* or not.

The same point can be made about the 'division of the *kanōn*' in propositions 19–20, which is also missing from Porphyry's quotation. It is irrelevant to the immediate context in his commentary, and the method of construction it adopts is completely alien to Ptolemy's concerns, even when he sets about the task, later in the *Harmonics*, of devising his own divisions and criticising those of others. Porphyry had no good reason to include it. More positively, he himself provides a rather strong reason for believing that this material did indeed form part of the text he had in front of him. He cites the sixteen propositions as coming from 'Euclid's *Division of the Canon*', and it would be extraordinary if a work known by this title contained no division of the sort it announces. This way of referring to the treatise, which he uses also at 92.30, makes it virtually certain that whenever propositions 19–20 were composed, by Porphyry's time they were understood as an integral part of the work, and indeed as constituting its main agenda. The title indicates that the construction of the division is the treatise's primary purpose, what it is 'all about'.

I conclude, then, that it is a mistake to suppose that Porphyry and Ptolemy provide evidence for the existence of a 'short version' of the *Sectio*, which lacked the introduction and the last four propositions. On the contrary, these earliest allusions to the treatise give us good reasons for believing that it already contained the introduction and the division, and no solid grounds for doubting that propositions 17–18 were also included. Ptolemy's use of it suggests that he thought of it as originating in the fourth century, or very shortly thereafter, and Porphyry attributes it unambiguously to Euclid. Both of them, of course, might be wrong. We shall have to examine the work itself to find out whether or not the disputed parts of it fit comfortably within the intellectual context of that period, and whether, regardless of Porphyry's and Ptolemy's treatment, they can plausibly be regarded as elements of the text in its original form.

THE INTRODUCTION TO THE TREATISE

The introduction occupies only a page and a half of text, but by comparison with the spare, formal theorems it seems positively expansive. Its first part expounds a theory about the causes of sound and of differences in pitch. The writer then draws from that theory the conclusion that notes of different pitches are related to one another in numerical ratios, sketches a classification of ratios into three types, and finally offers grounds for the thesis that the ratios of all the concords fall under just two of the three headings. Some commentators have regarded this material as little more than a collection of miscellaneous jottings of a Pythagorean flavour, and as largely irrelevant to the subsequent theorems, which make explicit use of none of it except the closing contention about concords; and as a consequence they have argued that it cannot have been written by the same hand as the well-argued and systematically linked set of theorems that follows, or formed part of the same treatise.[13] This view strikes me as entirely mistaken. Despite its relative informality the introduction forms a tightly-knit, continuous argument. It is directed from start to finish to the task of establishing two quite specific conclusions; and these conclusions are indispensable to the work of the theorems, one as a necessary (though tacit) presupposition, the other as an explicit axiom. There is nothing haphazard or irrelevant about it, nor is there anything specifically Pythagorean.

The passage's argumentative structure will be seen at once if we set out its statements as a numbered sequence, together with indications of the logical relations between them. Numbering apart, what follows is a close paraphrase, almost a direct translation, with nothing substantial added or subtracted.

(i) If there were stillness and motionlessness, there would be silence; and

(ii) if there were silence and nothing moved, nothing would be heard. Then

(iii) if anything is going to be heard, impact and movement must first occur. Hence, since

(iv) (a) all notes occur when an impact has occurred; and

(b) it is impossible for an impact to occur unless movement has occurred previously; and

(c) of movements, some are more closely packed, some more widely spaced; and

(d) those that are more closely packed make the notes higher, and those that are more widely spaced make them lower; it is necessary that

(v) (a) some notes are higher, since they are composed of more numerous and more closely packed movements, and

(b) some are lower, since they are composed of fewer and more widely spaced movements. Hence

(vi) (a) those that are higher than what is required, when they are slackened [i.e. lowered in pitch], attain what is required by the subtraction of movement; and

(b) those that are lower than what is required, when they are tightened [i.e. raised in pitch], attain what is required by the addition of movement.[14] Hence it must be agreed that

[13] This was the opinion of Jan 1895: 115–20, and of Tannery 1904a.

[14] There is an ambiguity in the expression of this proposition and its predecessor. An alternative reading would be: '. . . those that are higher (lower) than what is required attain what is required through being slackened (tightened) by the subtraction (addition) of movement'. I cannot prove that the version adopted in my text is correct; but I think it is logically preferable.

(vii) notes are composed of parts, since they attain what is required by addition and subtraction. Now
(viii) all things composed of parts are related to one another in a ratio of number. Hence it is necessary that
(ix) notes, too, are related to one another in a ratio of number. Now
(x) of numbers, some are related in a multiple ratio, some in an epimoric, some in an epimeric. Hence it is necessary that
(xi) notes too are related to one another in these sorts of ratios. Now
(xii) of these [numbers], those in multiple and epimoric ratios are related to one another under a single name. And we know that
(xiii) (a) of notes, some are concordant, some discordant; and
(b) those that are concordant make a single blend out of the two notes, while
(c) those that are discordant do not. In that case it is reasonable (*eikos*) that
(xiv) since concordant notes make a single blend of sound out of the two, they are among those numbers which are related to one another under a single name, and so are either multiple or epimoric.

There are several points here at which one might question the logic of this argument or the credentials of its premises. We shall come to those in a moment. But the first and crucial point to notice is that it is nothing like a random miscellany. It is, unmistakably, an argument, a single, continuous and closely reasoned argument, directed throughout to the establishment of the conclusion given in (xiv). The author has evidently taken great pains to omit no necessary steps, though his expression of them is severely economical; and still more significantly, he has included nothing that is redundant – every statement and every inference has its part to play in the whole. Since the argument's conclusion, as we shall see, plays a major role in the reasoning of the main part of the treatise, the view that the introduction is irrelevant, like the view that it is haphazardly flung together, is not merely unproven but absurd.

One of the intermediate staging-posts in the argument, step (ix), serves two purposes. It contributes to the reasoning by which we reach step (xiv), but it also needs to be established in its own right before the theorems can proceed. There is nothing in our ordinary experience of sound or music to link intervals with mathematical ratios. If they are to be represented in that manner, as they are in the theorems, it must first be shown that regardless of the way we perceive them, the pitches of a higher and a lower note are in fact, as step (ix) puts it, 'related to one another in a ratio of number'. The first task of the *Sectio*'s introduction, then, is to demonstrate that this is so.

The route to this conclusion lies through a series of propositions in physical acoustics, treating sound as a movement or sequence of movements caused by impacts (presumably impacts made by some body on the air, though the writer does not say so). It is when pitched sound is conceived as physical movement, not as an auditory phenomenon, that point (ix) is taken to hold of it. In this respect the *Sectio*'s approach is similar to those of Archytas, Plato and Aristotle, and is as far as it could be from that of Aristoxenus; it would fall squarely into the class of those he considers irrelevant to harmonic science.[15] It is also worth reminding ourselves that although acoustic theories of this type probably originated in the context of Pythagorean thought, by the later fourth century they had been widely adopted by theorists and philosophers of several doctrinal persuasions. There is no mileage in the thesis that the *Sectio*'s introduction betrays Pythagorean affiliations of which the rest of the text is innocent.

The theory itself, however, differs from most of its known fourth-century predecessors. Plato seems to have modified Archytas' account, and Aristotle revised it further, but they share the view that sound is a movement transmitted through the air (or, if conceived as an audible phenomenon, is caused by such a movement), and that its pitch is dependent on the speed at which the movement is transmitted through the medium.[16] Many later writers adopt a similar position. Though the *Sectio* begins, like Archytas in frag. 1, from the thesis that a sound is a movement caused by an impact, it shifts at steps (iv)–(v) into talking of movements in the plural, and of each note as constituted by a sequence of movements, more or less closely packed together. When the movements follow one another in more rapid succession, the sound's pitch is higher.

The thesis that a sound which we hear as continuous is really constituted by a succession of impulses was probably inspired by observations of the behaviour of an instrument's string. When the string is plucked it oscillates to and fro; the writer conceives it as beating repeatedly on the surrounding air to make a sequence of separate impacts, and yet as long as its oscillations persist it emits an apparently continuous sound. Theories of this sort are found elsewhere too, though rather rarely, and if the *Sectio* dates from around 300 BC it is the earliest text to expound it unambiguously.[17]

15 See e.g. *El. harm.* 12.4–9, 32.19–28.

16 Archytas frag. 1, Plato, *Tim.* 80a–b, Aristotle, *De an.* 420a–b.

17 The clearest instances of such a theory, outside the *Sectio*, are at [Ar.] *De audibilibus* 803b–804a, and in a quotation from Heraclides by Porphyry at *In Ptol. Harm.* 30.1–31.21. The *De audib.* was probably written no later than the mid-third century (see Gottschalk 1968). It is uncertain whether the Heraclides cited by Porphyry is the well-known fourth-century philosopher or a later figure

Even fewer writers develop the idea, as the *Sectio* does, into a theory of pitch, where pitch depends on the rapidity with which impulses follow one another, not on the speed of the movements' transmission.[18] This theory too is likely to have been based on the observation of strings; the oscillations of a longer or slacker string are visibly less 'closely packed' than those of a shorter or tauter string, and the pitch it emits is lower.

As an explanation of differences in pitch, the *Sectio*'s hypothesis has advantages over ones based on speeds of transmission. It does not lead to the awkward consequence that differently pitched sounds originating at the same moment cannot arrive simultaneously at the ear; and it need not wrestle with the observation that objects in motion slow down as they move further from their source. It also seems well adapted to the purpose it will serve here, that of interpreting the thesis that the intervals between pitched sounds can be represented as ratios of whole numbers. A stretched string at rest is straight. We may suppose that it 'strikes' the air either at the instant when it departs from its position of rest, or when it reaches its position of maximum displacement. I suspect that the theory's proponents were impressed by the fact that in either case, the string will always have 'struck' the air a determinate number of times in a given period, a number that is in principle countable (though not of course in Greek practice). Hence there is always some ratio of whole numbers which corresponds to the difference between any two pitches. Each sound will have been constituted, during a given period of time, by a determinate number of 'parts' (cf. steps (vi)–(ix)).

There may seem to be a difficulty here. The pitch of a sound made up of 9 impulses per second, for instance, may be supposed to stand in the ratio 9:8, that of a tone, to one constituted by 8 impulses per second. Yet if they start simultaneously, at the moment when the latter's fourth impact occurs the former will not yet have made its fifth, so that both will have been constituted by four impacts and should apparently sound in unison. But the problem is soluble. Ratios become relevant when it is the relative pitches of sounds, not their absolute pitches, that are under consideration. If the first impulses of two different notes are simultaneous and their rates

who lived in the first century AD. Most commentators have taken the latter view (see Gottschalk 1980: 157), and I did so myself in Barker 1989a: 230. For indications that might point to the earlier Heraclides see Barker (forthcoming).

[18] The position of the *De audib.* on this matter is not entirely clear. In the passage cited in n. 17 it recognises that 'the impacts of air belonging to the higher notes occur more frequently', but this need not imply that higher pitch is *caused* by this greater frequency. At 803a the cause is the movement's greater speed. A passage of Theophrastus (frag. 716 Fortenbaugh) seems to direct some of its criticisms at theorists who take a view like that of the *Sectio* (the fragment will be discussed in Chapter 15); for other evidence of this position see the passages cited below in n. 19.

of impact are rationally related, there will follow a period of time during which their impacts do not coincide, and then a moment when they come back into phase. After that the pattern of non-coincidence will be repeated, and it is in that pattern that the difference between them consists. We may therefore stipulate that the numbers of impulses to be compared are those occurring between any two instants at which impulses of the two sounds coincide. In the case envisaged, there will always have been 9 of one and 8 of the other. There are indications that this approach was indeed sometimes adopted.[19] The supposition that there might be rates of impact such that after a first pair of simultaneous impacts the two would never again coincide goes well beyond the mathematical sophistication of these authors.

The last phase of the introduction's argument, steps (x)–(xiv), seems the weakest. It begins from a classification of ratios into three types, multiple (*pollaplasios*), epimoric or 'superparticular' (*epimorios*) and epimeric or 'superpartient' (*epimerēs*). Let us remind ourselves of what these terms mean. A multiple ratio has the form mn:n. In an epimoric ratio the greater term is equal to the smaller plus one integral part (one half, one third, and so on) of the smaller; when the ratios are expressed in their lowest terms this class includes all and only the ratios of the form n + 1:n, except 2:1, which is multiple. For present purposes an epimeric ratio is any ratio which is neither multiple nor epimoric. In the context of harmonics, the thesis that multiples and epimorics have a status that epimerics lack seems implicit in Archytas and Plato, and may go back to even earlier Pythagorean thought.[20] The conclusion with which the introduction ends, that all concords have ratios that are either epimoric or multiple, is taken as axiomatic by most later writers in the mathematical tradition of harmonics, and probably pre-dates the *Sectio*.[21]

It is obvious that the writer of the *Sectio* has set himself the task of persuading us that this conclusion is true, but the force of his argument seems questionable. It rests on two points. First, 'numbers in multiple and epimoric ratios are spoken of in relation to one another under a single name' (step (xii)); the implication is that those in epimeric ratios are not. Secondly, 'concordant notes make a single blend out of the two, while discordant

[19] Porph. *In Ptol. Harm.* 107.15–108.21, [Aristotle] *Probl.* 19.39.

[20] On Archytas see pp. 289–90 above, and on earlier Pythagoreans p. 272. Plato rather pointedly distinguishes the principal ratios involved in the construction of the World-Soul from that of the residual *leimma* of 256:243, describing the latter as a 'number to number' relation (*Tim.* 36b). In the surviving literature the term 'epimoric' itself occurs first at Ar. *Metaph.* 1021a2.

[21] See e.g. [Ar.] *Probl.* 19.41, Thrasyllus at Theo Smyrn. 50.19–21, Ptol. *Harm.* 11.1–20. As we saw in the previous chapter, it is probably the epimorics and multiples that Aristotle has in mind when he connects concords and pleasant colours with 'well-ratioed' numbers at *De sensu* 439b–440a.

notes do not' (step (xiii b–c)). This second thesis need not detain us; the idea that the two notes of a concordant pair mingle together to form an integrated whole, and that it is this that distinguishes them acoustically from discords, was familiar in fourth-century and later thought.[22] The first, too, has a straightforward interpretation, though more recondite ones have sometimes been suggested.[23] Its sense, I think, is simply that each individual multiple and epimoric ratio is represented in the Greek language by a one-word expression.[24] There are no such words for epimeric ratios. Later mathematicians developed a terminology which in principle allowed each of these ratios, too, to be represented by a single word, but there is no reason to think that this cumbersome terminology was current in the fourth century.[25] At that period such a ratio was normally named by a composite expression which named each number involved in the ratio separately, in the form 'n to m', where the counterpart of 'to' is *pros*.[26]

The two premises, then, are tolerably unproblematic; it is the argumentative use to which they are put that may strike a reader as odd. It appears to rest on the assumption that if something is an integrated whole, 'blended' into one as concords are but discords are not, then language must be capable of representing it by a single word. This suggestion of a perfect fit between language (specifically, the Greek language) and reality seems remarkably optimistic.[27] Perhaps, however, it is a little less naïve than that. The thought might be that since language, and in particular the deliberately contrived, intellectually based language of mathematicians, marks a clear distinction between two classes of ratio, representing each item in

[22] See e.g. Plato, *Tim.* 80b, Ar. *De sensu* 447a–b, 448a, *De an.* 426b, and in later sources Aelianus at Porph. *In Ptol. Harm.* 35.26–36.3 (including an elaborate illustrative analogy), Nicomachus, *Harm.* 262.1–6, Cleonides 187.19–188.2.

[23] A reading of this sort was first proposed by Laloy 1900: 236–41. For interpretations other than the one I give here, see Barbera 1991: 56 with n. 149. None of them, I believe, will adequately fulfil the function for which the *Sectio* employs the statement in question.

[24] Terms for multiple ratios are formed with the suffix *-plasios*, giving *diplasios* for double, *triplasios* for triple, and so on. Among the epimorics there is a special word for 3:2, *hēmiolios* ('half-and-whole'). The remainder are named by terms in which the prefix *epi-* is attached to an expression meaning 'third', 'fifth' or the like, as in *epitritos* ('a third in addition') for 4:3, *epogdoos* ('an eighth in addition') for 9:8, etc.

[25] See Nicomachus, *Arithm.* 1.22–3 for a flurry of these terms, including coinages such as *tetraplasiepitetramerēs* for the ratio 24:5.

[26] Thus in Plato, as I have said, epimerics are called 'number to number' ratios, *Tim.* 36b3. Sometimes a good many words are needed to express such a ratio's size; in the same passage it takes Plato eleven words to say '256:243' (*hex kai pentēkonta kai diakosiōn pros tria kai tetterakonta kai diakosia*).

[27] It is nevertheless a view that might have been derived from an innocent-minded reading of Plato and Aristotle, both of whom (the latter very explicitly) regularly take 'what is said' as the starting-point for philosophical exploration. Their position seems to be that although linguistic distinctions may sometimes turn out to be misleading, if carefully interpreted and attentively related to others they typically reflect the contours of reality accurately enough to guide us towards the truth.

one class as a unity and those in the other merely as loose associations, it will have placed the ratios under consideration here, those of the concords, in the class appropriate to them. Even so, it is hard to see the argument as conclusive, and it is interesting that the writer does not state it as if it were. In inferences drawn earlier in the introduction he makes liberal use of expressions conveying logical necessity; given what has so far been established, the author asserts, the next proposition 'must' be true.[28] At the final step, however, these claims to iron-clad necessity evaporate, and are replaced by the altogether milder statement that it is 'reasonable' or 'to be expected' (*eikos*) that the conclusion holds.

Granted that the entire argument of the introduction has been designed to generate this conclusion, and that it is the only thesis stated here that is explicitly relied on in the theorems (crucially in propositions 10–11), the author's slippage at this critical moment from logical necessity to mere 'reasonableness' seems disappointingly feeble. But it should not really surprise us. Very few Greek writers attempt the task of proving by rigorous argument that all concords *must* have multiple or epimoric ratios; and those who do not are well advised, since the task is impossible. It cannot be demonstrated logically or mathematically that the proposition is true, and in fact, from a musician's perspective, it is not. The interval of an octave plus a fourth was widely recognised as a concord, and yet its ratio, 8:3, is neither epimoric nor multiple (this problem was mentioned on p. 348 above and will be discussed further below).[29]

From another point of view, too, a non-demonstrative argument is appropriate here. Its location in the introduction rather than in the theorems indicates that its conclusion is logically prior to the latter's reasoning. It is to be regarded as a principle on which the theorematic demonstrations of this science will draw, not as a proposition which they themselves can prove. We saw in earlier chapters that Aristotle, and following him Aristoxenus, insist that the principles of a science cannot be demonstrated; anything that can be demonstrated *must* be demonstrated if the science is to complete its task, and 'anything that requires demonstration does not have the nature of a principle'.[30] The strategy adopted in the *Sectio* is thus consistent with

[28] Cf. 'must' (*dei*) in step (iii), 'impossible' (*amēchanon*) in (iv), 'necessary' (*anangkaion*) in the transition to (v), 'it must be agreed' (*phateon*) in the transition to (vii), 'necessary' (*anangkaion*) again in the transitions to (ix) and (xi).

[29] The closest approximation to a general demonstration is probably one that can be extracted, though with some difficulty, from the text of Ptolemy's *Harmonics*. But it depends on some debatable assumptions, and Ptolemy's efforts to rid himself of the problem of the octave plus fourth are less than convincing. See Barker 2000a: 74–87.

[30] Aristox. *El. harm.* 44.14–15, cf. Aristotle, *An. post.* 71b26–9.

Peripatetic theories of scientific demonstration, and its author would have been misguided if he had looked for demonstrative proof.

Nor could the writer have based his principle securely on an inductive survey of cases, where observations of a range of instances in which a concord's ratio is epimoric or multiple lead us to the (probable) conclusion that all of them are so. For one thing, such a survey would inevitably stumble over the case of the octave plus fourth, which from an empirical or aesthetic perspective provides a compelling counter-example. Secondly, our confidence in a principle does not rest merely on the discovery that it holds good in all instances so far observed, but on an insight into the nature of the item under discussion; and this may indeed lead us to revise our view of what counts as an 'observed instance'. Here the insight focuses on the status of a concord as a blended unity. It will be this, if anything, that entitles a theorist to deny that the octave plus fourth can be a concord, despite the auditory impression it creates; and in the light of the other considerations expounded in the *Sectio*'s introduction, it is this intellectually enlightening conception of what a concord is that commends the principle to our understanding. The argument shows that when a concord is conceived in this way, the principle need no longer be regarded as an arbitrary postulate, and can be comfortably associated with other facets of what is known about sound and pitch, and with established mathematical classifications of ratios. It has done, in fact, about as much as any such argument could.

PROPOSITIONS 1–9: THE MATHEMATICAL GROUNDWORK

The propositions in this group say nothing explicit about music or sound. It is true that the word *diastēma* appears in them repeatedly, and that in musical contexts it means 'interval'. In later theorems the first nine propositions will be used to generate conclusions about musical relations, and there the word unambiguously acquires its musical sense. But this is done by treating musical intervals as cases of just one of the kinds that fall under the designation *diastēma*, a term whose range of application is much broader. A *diastēma* is literally that by which two items of any sort are separated, a gap or a distance or a quantitative difference. The reasoning of the first nine theorems is purely mathematical and will hold of *diastēmata* in general. We shall postpone consideration of the question whether it is to be understood in geometrical terms (as concerned with relations between magnitudes such as distances) or in arithmetical terms (as concerned with the relations between numbered pluralities).

But the situation is not quite as simple as that. Although a *diastēma*, in the first nine theorems, might be a geometrical or an arithmetical relation, it is not the distance between two points or the difference between two numbers, but a ratio. Interpreted geometrically, it is the ratio between two lengths; taken arithmetically, it is the ratio between two numbers. Proposition 1, for instance, reads: 'If a multiple *diastēma* taken twice makes some *diastēma*, the latter is also multiple.' As the phrasing implies and the working of the theorem confirms, this means (where A, B and C are numbers or lengths) that if the ratio A:B is multiple, and if A:B = B:C, then the ratio A:C is multiple too. This whole group of theorems, in short, proves propositions in the mathematics of ratio; and since musical *diastēmata* are conceived in the sequel as ratios, these propositions can be used there, without modification, as premises in theorems about them. It is in fact possible to read *diastēma* as if it meant 'ratio' throughout.[31] In another respect too these theorems are clearly tailored to their task in the musically oriented second half of the treatise. They do not amount to a complete exposition of ratio-theory, and if regarded from the perspective of that discipline alone would amount to a rather haphazard compilation. They have plainly been selected for their bearing on issues in harmonics; they include all and only those that will be needed to prove theorems about musical relations in propositions 10–20.[32] If the work is in any sense a 'complete' account of some subject, the subject is mathematical harmonics, not the mathematics of ratio in general.

I shall not examine the workings of each theorem in this group in detail; most of them are quite simple. But they nevertheless provide a good deal of food for thought. The first five deal with *diastēmata* conceived under general descriptions, as multiple or epimoric. The main focus is on the former. Four of them consider situations in which a *diastēma* is 'taken twice', to produce a sequence of three terms, A, B, C, such that A:B = B:C. We have noted proposition 1, that if a multiple *diastēma* is taken twice the *diastēma* formed from the two *diastēmata* together is also multiple. Proposition 2 states the complementary thesis: 'If a *diastēma* taken twice makes a whole that is multiple, that *diastēma* itself will also be multiple.' Proposition 4 is the negative counterpart of proposition 1. 'If an interval that is not multiple is taken twice, the whole will be neither multiple nor epimoric.' (The second part of the conclusion, introducing a reference to epimorics that is absent from propositions 1, 2 and 5, depends on proposition 3; see pp. 381–2 below.)

[31] This usage would not be unprecedented. Archytas frag. 2 may be a doubtful case, but Aristotle, *Phys.* 202a18 is not.

[32] This overstates the case very slightly. There is one exception, proposition 4; I discuss it on pp. 381–2 below.

Proposition 5 completes the set. 'If a *diastēma* taken twice makes a whole that is not multiple, that *diastēma* itself will not be multiple either.'

These four propositions form a tightly linked group. Inserted among them is one which at first sight is quite different, proposition 3. 'In an epimoric ratio there is no mean number, neither one nor more than one, which divides it proportionally.' In more familiar language, there are no mean proportionals between two terms in an epimoric ratio. That is, if A:C is epimoric, there is no term B such that A:B = B:C, nor can A:C be broken down into any larger number of sub-ratios all of which are equal (so that e.g. A:X = X:Y = Y:Z = Z:C).

The proposition differs from the four surrounding it in two principal ways. Its subject is epimoric ratios rather than multiples, and it is expressed in strikingly different language, speaking of mean numbers and proportionality rather than of *diastēmata* taken twice. This seems a little strange, since a broadly equivalent proposition could have been formulated in the latter way, for instance on a pattern derived from proposition 1: 'If a *diastēma* taken twice makes some *diastēma*, the latter *diastēma* will not be epimoric.' To put it more generally, the two modes of presentation seem to differ in that proposition 3 is clearly conceived in arithmetical terms, alluding to the insertion of numbers between other numbers, whereas the theses to be proved in the other four theorems might, on the face of it, be interpreted against the background of geometry, where the *diastēmata* are ratios between lengths.

But this apparent difference is illusory. When one looks at the reasoning of those four theorems, it turns out that though the argumentation of propositions 1, 4 and 5 might, at a stretch, be construed geometrically, that of proposition 2 cannot. It is explicit, like proposition 3, in referring to numbers and relations between numbers. So too, we may recall, is the latter part of the introduction, whose theory that pitch depends on the numerousness of a sound's constituent impulses points directly towards an arithmetical interpretation of the relations between notes. Notes, it says, are made up of parts, and all such things are related to one another 'in a ratio of number'. It is numbers, not merely quantities, whose ratios it classifies into three groups; and its final step states that concordant notes are 'among the numbers that are related to one another under a single name'.

There are strong indications, then, that the author of the *Sectio* thought of his science as based in arithmetic, not geometry. André Barbera has provided persuasive arguments of other sorts for the view that where expressions in the mathematical theorems, and the line-drawings included in some MSS, seem to hint at a geometrical treatment, these have been overlaid by later hands on an originally arithmetical text, and I think he is right.[33] The point is of some importance, and not just for any contribution it may make to the history of mathematics. A geometrical approach to harmonic relations is found in Plato and is common in later sources. It comes naturally to theorists who take as their starting point the relations between the lengths of string or pipe from which notes are produced, or the dimensions of other sounding bodies. It is also well adapted to acoustic theories which make pitch dependent on the speed of a sound's transmission, since speed can be specified in terms of the distance travelled in a given time.[34] The fact that the *Sectio* addresses these matters from an arithmetical angle shows that its author is not thinking, in the first instance, of the ratios as they are exhibited on the monochord's *kanōn*; that topic, when he reaches it, will involve a transition from an arithmetical to a geometrical view-point. He is dealing with ratios between numbers inherent in the physical constitution of pitched sounds themselves; and his non-Platonic, arithmetical approach in the theorems is well suited to his equally non-Platonic acoustic theory.

From this perspective proposition 3 fits comfortably with the others. We may still be puzzled, however, both by the fact that it enunciates that proposition in terms quite different from theirs, and by the intrusion of a theorem about epimoric ratios in the middle of a sequence dealing with multiples. The latter difficulty may seem easy to resolve; proposition 3 stands where it does because it is called upon in the proof of proposition 4. That is true, but proposition 4 itself has curious features. First, it is the only one of the propositions about multiple ratios that draws a conclusion referring also to epimorics: a non-multiple *diastēma* taken twice makes a whole that is 'neither multiple nor epimoric'. Secondly, it amounts in reality to two separate theorems, since the proof about multiples is separate from the one about epimorics and independent of it. Thirdly, it underplays its hand so far as epimorics are concerned; as proposition 3 shows, no *diastēma* whatever, when taken twice, makes a whole that is epimoric. Finally, and most strangely, it is unique and anomalous among the nine mathematical

33 See Barbera 1991: 40–4. I accept the arithmetical interpretation with due humility, properly chided for previous errors by Barbera 1984a, and 1991: 41 n. 113.

34 A later writer, Aelianus, makes this point lucidly in a passage quoted at length by Porphyry at *In Ptol. Harm.* 36.9–37.5; earlier parts of the passage illustrate an approach based on the dimensions of instruments (33.19–35.12). Among fourth-century authors, only Plato makes his position clear. As one might expect from his 'speed' theory of pitch (*Tim.* 80a), he treats harmonics, in the *Republic*, as the mathematical study of audible movements; it is the acoustic counterpart of astronomy, the mathematical study of visible movements, and both are conceived as investigations in what we might call 'kinetic geometry'. See *Rep.* 527c–530d, and cf. the explicitly geometrical construction of the musically organised World-Soul at *Tim.* 34b–36d.

propositions in that it does no work; nothing in the later theorems depends on it.

Any explanation of these oddities will be speculative, but I suggest the following. In one of his mathematical sources the author found propositions 1–2, the part of proposition 4 which deals only with multiples, and proposition 5; the source presented them as a coordinated group concerned with multiple ratios. Their context was purely arithmetical. The theorem proving proposition 4's statement about multiple *diastēmata* was included for the sake of completeness, and the fact that it has no bearing on the *Sectio*'s special agenda was in that context irrelevant. The *Sectio*'s author took them over as a package. He realised that proposition 3, which is an indispensable basis for the musical theorems (it is the logical pivot of proposition 10, on which all the others depend), shows that the 'whole' referred to in proposition 4 cannot be epimoric any more than it can be multiple. He therefore extended the thesis of proposition 4 to make it include this point, adding a brief argument alluding to proposition 3; and this made it essential to insert proposition 3 before proposition 4. He seems to have overlooked the fact that neither part of proposition 4 has anything to contribute to his project.

On this hypothesis, though proposition 3 is crucial to later phases of the *Sectio*'s programme, it has been levered awkwardly and for rather poor reasons into a sequence of theorems among which it did not originally belong. As it happens, we have independent evidence about its origins, since Boethius attributes an almost identical theorem to Archytas.[35] My hypothesis would not entail that the four propositions on multiples came from a different author, though that is quite possible; but if they too were borrowed from Archytas, they probably appeared in another part of his writings. If my earlier arguments hold water, their context was purely arithmetical and was not shaped by the requirements of harmonics. That is unlikely to be true of proposition 3 itself.

We can therefore glean from the first five theorems a few clues about the way in which the *Sectio*'s author went to work. They suggest that he may not have been an innovative mathematician, but that he was an enterprising and in most respects (though not quite all) a skilful synthesiser. We shall find other evidence shortly of his familiarity with aspects of fourth-century mathematics that lay at some distance from harmonic theory. The task he seems to have set himself is that of selecting appropriate arithmetical

[35] *Inst. mus.* III.11; see pp. 303–5 above. Boethius' authority throughout Book III is almost certainly Nicomachus; for a summary of the evidence for this view see Bower 1989: xxiv–xxvii. For discussion of the versions in Boethius and the *Sectio* see Knorr 1975 ch. 7, Barbera 1991: 58–60.

theorems from an existing repertoire, and of linking them together, in alliance with an unusual and well-judged hypothesis about the physics of sound and pitch, as a basis for the systematic demonstration of propositions in harmonics. Most – possibly all – of these propositions were already familiar, as we shall see, and previous theorists had commended them through various kinds of observation and reasoning. But at least as far as we know, there had been no previous attempt to organise and demonstrate them as a coordinated group.

Propositions 6–9 can be disposed of quickly. All of them consider ratios with specific values. Summarily, theorems 6–8 prove that the ratio 2:1 is the product of the two greatest epimorics, 3:2 and 4:3 (proposition 6); that the product of 2:1 and 3:2 is 3:1 (proposition 7); and that when the ratio 4:3 is 'taken away' from the ratio 3:2, the result is 9:8 (proposition 8). The MSS offer two distinct proofs for proposition 6, one for each of the others. They are cumbersome but essentially straightforward, and all of them (with the possible exception of the first proof of proposition 6[36]) are conceived in arithmetical rather than geometrical terms.

Proposition 9 states that the product of six *diastēmata* in the ratio 9:8 is greater than 2:1. It proves this by direct calculation, showing that when seven numbers are found such that each (apart from the first) stands to its predecessor in the ratio 9:8, the seventh number is more than double the first. To make all the terms whole numbers they have to be large (the first is 262144 and the last 531441); otherwise the argument is unproblematic. It is particularly obvious here that the theorem is chosen, and must indeed have originally been articulated, for its bearing on an issue in harmonics, by the author of the *Sectio* or (more probably) by some predecessor. It is of little or no intrinsic mathematical interest; its function is to serve as the basis of a proof that the octave, whose ratio is 2:1, is less than six whole tones in the ratio 9:8 (see proposition 14).

One other feature of propositions 1–9 calls for comment here. Theorems later in the series sometimes draw upon earlier conclusions in the course of their reasoning, as one would expect, and they have been carefully arranged to make this possible.[37] It is rather more surprising that they draw also on results which are not proved and procedures which are not expounded in the text itself at all, and that they sometimes do so explicitly. 'We have learned (*emathomen*)', says the author in proposition 2, 'that if there are numbers in proportion, however many of them, and the first measures [i.e.

[36] On this issue see Barbera 1991: 42–4, 134–9, 267.

[37] Thus proposition 4 calls on the conclusions of propositions 2 and 3, and proposition 5 on that of proposition 1.

is a factor of] the last, it also measures those in between.' In proposition 9, similarly, 'we have learned how to find seven numbers in the ratio 9:8 to one another'. The proof of the thesis in proposition 2 and the construction called on in proposition 9 both appear in Euclid's *Elements*, as does the proof of a thesis relied on but not proved in proposition 3.[38]

When the writer of the *Sectio* says 'we have learned', however, he need not mean 'we have learned from Euclid's *Elements*', though of course he may. The *Elements* bring together a great deal of material that was already known to fourth-century mathematicians, and the *Sectio* may be referring to pre-Euclidean sources. That is not a decidable issue. What is intriguing is that its readers are assumed to be already conversant with some moderately sophisticated mathematics. The author writes, in fact, as if he and they shared a common background in the discipline, and had mastered the same range of propositions. There is no way of telling, unfortunately, whether the 'we' is that of teacher and pupils ('we studied these points last term'), or the collegial 'we' of an intimate coterie of mathematicians, or the elitist 'we' which assumes that all readers are educated fellow-intellectuals, or whether the expression indicates that the *Sectio*, as a text, was conceived as a sequel to other written texts that covered the matters in question. It shows, at all events, that despite the air of self-contained completeness which its theorematic approach conveys, the work's cogency depends partly on its insertion into a specific disciplinary context, and that its arguments feed on the environment which its readers are taken to inhabit. Mathematical harmonics, as the *Sectio* presents it, is not an isolated, free-standing science, but belongs with others as part of a wider intellectual enterprise.

PROPOSITIONS 10–13: THE TRANSITION TO HARMONICS

The first two theorems in this group seek to prove that the octave *diastēma* is multiple (proposition 10), and that the *diastēmata* of the fourth and the fifth are both epimoric (proposition 11). Proposition 12 provides arguments to demonstrate the values of the *diastēmata* or ratios of the various concords, 2:1 for the octave, 3:2 for the fifth, 4:3 for the fourth, 3:1 for the octave plus fifth, 4:1 for the double octave. Proposition 13 shows that the ratio of the tone is 9:8.

The reasoning of all these theorems relies on conclusions established in the preceding arithmetical propositions. But it is obvious that they cannot

[38] The proofs relevant to propositions 2 and 3 are at Eucl. *El.* 8.7 and 8.8 respectively; for the construction in proposition 9 see *El.* 7.2.

prove anything about octaves, fifths and other musical relations from the resources of arithmetic alone. Pure *arithmētikē*, the 'science of number', says nothing about such things. Information about them must be fed in from other domains; and the procedure must also have access to principles which allow propositions about musical phenomena, such as the octave, to be brought into logical connection with propositions dealing with multiple ratios and other such mathematical abstractions.

To form an impression of the kind of musical information that is introduced, consider proposition 10.

The octave *diastēma* is multiple. Let A be *nētē hyperbolaiōn*, let B be *mesē* and let C be *proslambanomenos*. Then the *diastēma* AC, being a double octave, is concordant. Hence it is either epimoric or multiple. It is not epimoric, since no mean falls proportionally in an epimoric *diastēma*. Therefore, since the two equal *diastēmata* AB and BC when put together make a whole that is multiple, AB is also multiple.

In order to make sense of this argument, we need to be familiar with the names of the notes, and to know the musical relations in which they stand; *nētē hyperbolaiōn* is an octave above *mesē*, and *mesē* is an octave above *proslambanomenos*. We must know also that the double octave is a concord. These are not very recondite pieces of knowledge; they belong to the most elementary level of harmonic theory in its empirical or descriptive guise. But they are plainly indispensable. The theorem calls, in addition, on two arithmetical results already proved, first that there is no mean proportional between terms in an epimoric ratio (proposition 3), and secondly that if the whole formed from two equal *diastēmata* is multiple, they are multiple too (proposition 2). The 'bridge' between the musical and arithmetical propositions is formed by the introduction's thesis that pitches are related to one another in numerical ratios, and by the principle it commends as 'reasonable', that the ratio of any concord must be either epimoric or multiple. They make it possible to cross the boundary between the two domains by tying a pair of musical conceptions, interval and concordance, to arithmetical categories. They themselves thus fall within the scope of neither of the two sciences at work elsewhere in the theorem, arithmetic and empirical harmonics. Neither of those sciences can demonstrate their truth, and nor can mathematical harmonics of the kind exemplified in the *Sectio*. Mathematical harmonics is constructed precisely through the formation of this bridge between the other two disciplines. Its theorems must therefore presuppose the bridging propositions and cannot be used to prove them.

Given the data taken from empirical harmonics, the arithmetical propositions and the bridging principles, the argument of proposition 10 is sound. Proposition 11 calls on resources of just the same sorts, and on the face of it its reasoning takes a course roughly parallel to that of proposition 10. But as commentators have repeatedly noted, it runs into trouble.

The *diastēma* of the fourth and that of the fifth are both epimoric. Let A be *nētē synēmmenōn*, let B be *mesē* and let C be *hypatē mesōn*. Then the *diastēma* AC, being a double fourth, is discordant. Hence it is not multiple. Therefore, since the two equal *diastēmata* AB and BC when put together make a whole that is not multiple, AB is not multiple either. But it is concordant; hence it is epimoric. The same demonstration applies also to the fifth.

This argument, like its predecessor, assumes familiarity with the named notes and the relations between them, and it assumes the knowledge that the double fourth (in modern terms, a seventh) is a discord. It relies on just one of the arithmetical theorems (proposition 5), and it seems to appeal twice to the bridging principle governing the ratios of concords. The second appeal, in the penultimate sentence, is legitimate; if the principle holds, and if AB is concordant but not multiple, it must be epimoric. The notoriously precarious move comes in the third and fourth sentences, where it is argued that since the double fourth is discordant it is not multiple. But the principle states only that all concords are either multiple or epimoric, not that all multiple *diastēmata* are concordant, and the inference from 'AC is discordant' to 'AC is not multiple' is unwarranted.

Scholars have sometimes tried to justify the move on the grounds that the writer, like most other harmonic theorists, concerns himself only with relations existing within the span of a double octave.[39] All the multiple *diastēmata* within that range are indeed concords (the octave, 2:1; the octave plus fifth, 3:1; the double octave, 4:1). It is only when we take the next step that we encounter the first interval in the series which by Greek standards is a discord; 5:1 is the ratio of an octave plus a major third. But the *Sectio* says nothing to indicate that the principle on which it relies is limited in this way, and its logic would be no less objectionable if it did. It could argue from the premise that all multiple *diastēmata* within the double octave are concords only if that fact were already established, and it can be established only by identifying the ratios belonging to the intervals involved. That is done in proposition 12. But the first step in proposition 12, showing that the ratio of the octave is 2:1, itself relies on the conclusion of proposition 11, and all its other steps depend on the first. The questionable move in

[39] See e.g. Barbera 1984a: 160–1.

proposition 11 would thus be 'legitimised' by reference to results which are derived from proposition 11's own conclusion, and the reasoning would be viciously circular.

Nor can one argue that the conclusions of proposition 12 can properly be relied on before being demonstratively established, since they were already thoroughly entrenched in the harmonic tradition and in that sense could be treated as known. That would involve using them like the data of empirical harmonics, which are introduced into the arguments as known facts which do not call for proof. Such a strategy would make no sense. Their role is quite unlike that of the empirical data. As proposition 12 shows, they are conclusions – arguably the most important conclusions – which this mathematical harmonics sets out to demonstrate, and they cannot simultaneously be factual assumptions on which those same demonstrations depend, even though both the writer and his readers will no doubt have been confident of their truth before any proofs were devised or presented.

The consequence is that the project of the *Sectio* is fatally flawed. From proposition 12 onwards, every conclusion depends directly or indirectly on proposition 11, and the latter's argumentation cannot be repaired. An analogous and acceptable theorem would have to include a premise of the form 'Every ratio of type T is the ratio of a concord.' The premise must avoid presupposing the conclusions of proposition 12; it must supply the necessary step in an argument to the conclusion that the ratios of the fourth and the fifth are epimoric; and there must be good reasons for thinking it true. The fact that no such premise was ever discovered by Greek theorists is no reflection on their ingenuity, since there is none to be found.[40]

Another difficulty arises from the summary statement with which proposition 12 ends: 'Thus it has been demonstrated, for each of the concords, in what ratio the bounding notes stand to one another.' This claim to comprehensiveness seems unjustified, since one interval regularly accepted as a concord, the octave plus fourth, has not been quantified or even mentioned. The reason for the omission is plain enough, since as I noted earlier, its ratio is 8:3 and is neither epimoric nor multiple. Hence the writer could not include it in his treatment without abandoning his principle about the ratios of concords. As I remarked in the previous chapter (p. 348 above), he deals with the problem by passing over it in silence.

But the problem is genuine and serious. When the *Sectio* treats the intervals listed in proposition 12 as concordant, it is relying on the evidence of musical perception, and the argument can only proceed on that basis. If

[40] On this matter see also pp. 346–8 above.

perception finds the octave plus fourth concordant too, this evidence should arguably be given parity of treatment. It seems very clear that Greek ears did indeed perceive this relation as a concord, and that musicians and non-mathematical theorists were unanimous in accepting it.[41] A writer who uses the evidence of perception and practice to identify the concordant intervals, and who nevertheless commits himself to the thesis that only intervals with epimoric or multiple ratios can be concords, is effectively claiming that in the single case of the octave plus fourth, perception is unreliable. In that case good reasons must be given for regarding it as an exception. The *Sectio*'s author cannot have been unaware of the difficulty, and the fact that he gives no reasons of this sort makes it highly probable that he had none to offer. In later antiquity the problem became notorious, and various strategies were devised to cope with it.[42] None of them is very persuasive, and we have no grounds for reading any of them back into the *Sectio*.

PROPOSITIONS 14–18: INEQUALITIES AND CONTROVERSIAL CONCLUSIONS

We now reach a series of propositions whose role in the enterprise does not spring immediately to the eye. They all follow legitimately (given some further empirical input) from conclusions established or allegedly established previously. But they play no obvious part in the final stage of the work, the division of the *kanōn* in propositions 19–20, and we are entitled to wonder why the author has decided that they, rather than any others, merit demonstration in this treatise.

As a preliminary I shall review them briefly, partly in order to show that they have enough in common to justify my grouping them together. Proposition 14 shows that the octave is less than six tones, proposition 15 that the fourth is less than two and a half tones and the fifth less than three and a half, and proposition 16 that the tone cannot be divided into two or more equal parts. Proposition 17 is not a proof; it sets out a method of construction which is an adjunct to proposition 18. Proposition 18 itself shows that the lower moveable note in an enharmonic tetrachord, its *parhypatē* or *tritē*,

[41] No harmonic scientist who takes the observational data seriously hesitates to accept it. Aristoxenus enunciates the general rule that when an octave is added to any concord the resulting interval is itself concordant (*El. harm.* 45.20–3). Ptolemy, whose approach is primarily mathematical, but who insists that 'rational' principles and conclusions must be consistent with the evidence of perception, elaborates Aristoxenus' rule and defends it vigorously, arguing in particular that it is absurd to deny the octave plus fourth the status of a concord (*Harm.* 13.1–23).

[42] See e.g. Adrastus quoted at Theo Smyrn. 56.12–15, Ptolemy, *Harm.* 15.18–16.12, and cf. Ptolemaïs of Cyrene quoted at Porphyry, *In Ptol. Harm.* 23.25–31.

cannot divide the *pyknon* into equal parts. These conclusions fall neatly into two pairs; propositions 14 and 15 show that each of the principal concords amounts to less than some specified number of tones or half-tones, and propositions 16 and 18 show that certain *diastēmata* cannot be divided into equal sub-intervals or sub-ratios. More broadly, the group is united in its concern with inequalities; the octave is not equal to six tones, the ratios of the two intervals inside the enharmonic *pyknon* are not equal to one another, and so on.

The author's argumentative resources would have allowed him to demonstrate that none of the ratios that are most significant in harmonic analysis can be divided into equal sub-ratios of whole numbers. In the cases of the tone and the lesser concords – the fifth and the fourth – this follows directly from the fact that their ratios are epimoric, together with the Archytan theorem, proposition 3. Proposition 18 shows that the same consideration applies, through slightly more complex reasoning, to the *diastēma* comprising an enharmonic *pyknon*. Though the ratio of the octave is not epimoric but multiple, the argument supporting proposition 3 will cover this case too; all that is needed is a slight rephrasing of the proposition itself.[43] Taken together, these conclusions have an important methodological implication. It is that such intervals cannot be measured as multiples of any smaller interval, and that systems bounded by notes in these fundamental relations cannot be built up additively from units of any one size. There is no elementary unit of harmonic construction.[44]

It is therefore interesting that the *Sectio* does not adopt precisely this approach. It argues that the tone and the enharmonic *pyknon* cannot be divided equally, but in the case of the concords it concentrates on more specific inequalities. It does not claim or even imply that the octave, fifth and fourth cannot be divided into equal parts, only that the octave is less than six tones, that the fifth is less than three and a half tones and that the fourth is less than two and a half. (It leaves open, at this stage, the question whether there are such things as half-tones.)

No one who has studied later essays in this discipline will find this pattern strange; just the same group of contentions is routinely repeated in Platonist and neo-Pythagorean texts of the Roman era. But in the context of the period I am positing for the *Sectio*, we cannot appeal to scholarly routines of this sort, since there is no evidence that such semi-automatic

43 The reasoning of proposition 3 is sufficient to show that there is no mean proportional between terms in any ratio whatever of the form n + 1:n, and applies to the special case of the ratio 2:1 just as firmly as it does to those which can properly be described as epimoric.

44 On this point see also my comments on Aristotle's treatment of the diesis, pp. 350–3 above.

patterns of exposition had yet been set up. The likeliest diagnosis is that the propositions are set in this form because they will then respond most directly to current controversies. We saw in a previous chapter that Aristoxenus dedicated a surprising amount of space to a discussion of the 'size' of the perfect fourth (p. 190 above), and the thesis that it is in fact exactly two and a half tones seems to be one which he shared with his empirically minded predecessors. If they are right, it will follow at once (given the definition of a tone as the difference between a fifth and a fourth, and the agreement that a fifth plus a fourth make up an octave) that the fifth is three and a half tones and the octave six; conversely, if the fourth is not two and a half tones, those quantifications of the fifth and the octave will also fail. Aristoxenus and the empirical *harmonikoi* assume, in addition, that the tone and the enharmonic *pyknon* can be divided into equal parts. They take the same view about other intervals too, but these intervals have special importance. The 'parts' are regularly expressed as unit-fractions of a tone, and the *harmonikoi*, according to Aristoxenus, studied enharmonic systems and no others. These are also his own main focus of attention in Book III of the *El. harm.*

We may reasonably suppose, then, that propositions 14–18 are designed to consolidate a clear line of demarcation between the two harmonic traditions. Their methodological differences had been obvious since Plato's time if not before, and though Aristotle's grasp on the issues may have been uncertain, he plainly knew that they were at odds with one another over at least one of the substantive points set out in the *Sectio*.[45] It is a plausible guess that these points, or some of them, were set out in the work of Archytas. Since the *Sectio* expounds only a small selection of propositions in mathematical harmonics and leaves many topics untouched, and since those that it addresses are established from first principles, they are presumably to be reckoned as in some sense fundamental. The propositions establishing the ratios of the concords and the tone certainly have that status, and the inclusion of the controversial propositions 14–18 implicitly assigns them a similar rank.

The author seems to have thought, then, that mathematical harmonics lacks firm foundations until the theses over which it diverged from empirical harmonics had been brought together as a group and systematically demonstrated. It does not immediately follow that he construed this phase of his work as a full-dress refutation of the empiricist heresy. Aristoxenus, on the other side of the doctrinal fence, refrains from branding the physicists'

[45] Plato, *Rep.* 530–1, Aristotle, *Metaph.* 1053a12–17; cf. p. 350 above.

and mathematical theorists' conclusions as false, a charge he scatters lavishly among his comments on the *harmonikoi*; physics and mathematical harmonics tackle questions quite different from his own, and simply have no bearing on his concerns (see pp. 166–8 above). Since the *Sectio* does not allude directly to the empiricists' procedures, we cannot be certain how its author or his fourth-century sources construed the relation between his conclusions and theirs, but such evidence as we have points to the more extreme interpretation. The *Sectio*'s unadorned statements suggest no scope for compromise or reconciliation; the tone, for example 'will not be divided into two or more equal parts' (proposition 16), and that is that. Since *tonos*, 'tone', is a term shared by theorists on both sides of the chasm, the statement as it stands flatly contradicts Aristoxenus and the *harmonikoi*, and later writers in the mathematical tradition seldom hesitate to conclude that the Aristoxenians and their allies must therefore be wrong. It is hard to imagine that the *Sectio*'s arguments were understood in a more generous sense in their own time. The treatise may indeed have played no small part in entrenching in Greek musicological thought the image of the two approaches as obdurate and irreconcilable rivals.

I have said that propositions 17–18 pose special problems of their own. The problems do not affect the interpretation of the passage's content or reasoning, but they are relevant to the other main issue that has been dogging us through the course of this chapter, since they have been thought to undermine the assumption that these propositions were included in the treatise in its original form. Apart from the fact, which I have already discussed, that they do not appear among Porphyry's quotations from the *Sectio*, two grounds for suspicion have been identified.

The first, however regarded, will not carry very much weight. It is that proposition 17 is presented in a different form from its predecessors. As Barbera notes,[46] all the earlier theorems have been cast as proofs, whereas proposition 17 offers, instead, a method of construction, explaining how to construct a ditone by moves through concordant intervals only. There is no doubt that this shift of approach takes place, but it does not strike me as significant enough to support the hypothesis of a change of author or date. For one thing, the division in propositions 19–20 also amounts to a constructional procedure, not a theorematic proof; but this point will cut no ice with those who reject propositions 17–18, since they typically treat the division, too, as an alien accretion. Barbera (p. 16) comments that proposition 17 marks a transition from mathematical demonstration

[46] Barbera 1991: 16, 24, 60.

to 'the style of a manual', meaning, I think, the style of a work which offers an introductory exposition of musical facts and procedures, designed to help students to find their way, in practice, around musical scales, intervals and so on. But this, I suggest, lays insufficient emphasis on the fact that propositions 17 and 18 belong together, and that the natural way of reading proposition 17 is simply as a preliminary to its sequel. Proposition 18 is, once again, a proof rather than an exposition of method, but we need the method described in proposition 17 if we are confidently to construct the intervals with which it is concerned. Furthermore, as Barbera points out (173, n. 61), even proposition 17 unquestionably displays 'aspects of the Euclidean model' on which its predecessors are patterned. The shift in presentation is not in fact very radical, nor, for the reasons I have given, is it unintelligible or inappropriate in its existing context.

The other objection to propositions 17–18 is more substantial, and has attracted the most discussion. They are concerned exclusively with intervals belonging to a scale in the enharmonic genus. Earlier propositions have examined intervals common to all the genera, and the system constructed in propositions 19–20 is diatonic. What reasons could there possibly be for a systematically minded author to move abruptly into an argument dealing only with the enharmonic, and without even announcing that the transition has occurred? (The name of the genus occurs nowhere in the passage or indeed anywhere else in the treatise.) I have argued above and more fully elsewhere that these propositions may be designed as part of a polemic against a rival school of theorists, the *harmonikoi* who according to Aristoxenus concerned themselves with enharmonic systems and no others, and very possibly also against Aristoxenus himself.[47] The proof with which they conclude purports to show that the small interval which is left in a perfect fourth after the subtraction of a ditone, and which constitutes the enharmonic *pyknon*, cannot be divided into equal intervals; and this would rule out of court any analysis in the manner of these rival theorists, since for them the *pyknon* was divided into equal quarter-tones. In its enunciation of a position incompatible with that of more empirically minded theorists, Aristoxenus included, this proof is of a piece with the *Sectio*'s demonstrations that the octave is not made up of six equal tones, and that the interval of a tone cannot be divided into equal portions (propositions 13 and 16). Further, the proof about the tone cannot be straightforwardly adapted to fit the case of the enharmonic *pyknon*, at least if that miniature structure is conceived as it is in propositions 17–18. The proof dealing with the tone

[47] See Barker 1978a, 1981b.

turns on the thesis that its ratio is epimoric; but the ratio of the enharmonic *pyknon*, as envisaged here, is not epimoric. It is the remainder of a perfect fourth, whose ratio is 4:3, after the subtraction of two whole tones, each in the ratio 9:8. Though the author does not say so, its ratio is 256:243; it is the Platonic or Pythagorean *leimma*. This ratio is epimeric or 'number to number', not epimoric, and nothing has been said to outlaw the equal division of epimeric ratios in general.[48] Hence a rather more elaborate approach is needed, and this is provided by the propositions in question.

I am now inclined to agree with Barbera (1991, 22), however, that the hypothesis of a polemical agenda is not enough by itself to allow us confidently to link propositions 17–18 with the preceding theorems. But there is another reason for believing that these propositions, whatever their relation to the remainder, were composed at an early date. Aristoxenus asserts that his predecessors had analysed systems in the enharmonic genus and no other (*El. harm.* 2.7–25). If that is true, it presumably reflects the currency, in those predecessors' time, of a mode of musical culture in which this genus had a central, privileged status. Support for this conclusion is provided both by the claim made by Aristoxenus – well known as a conservative in matters of musical taste – that it is the 'finest' of all genera, though falling into disuse in his day (*El. harm.* 23.3–23), and by the fact that it is not most commonly referred to as 'the enharmonic genus', but simply as *harmonia*, 'attunement'. There is also a tradition recurrent in later writings, and asserted in one document which probably dates from about 380 BC, that the music of the genre which had the highest status in classical Athens, that is, tragedy, was exclusively enharmonic, at least up to the closing years of the fifth century.[49] The 'ancient scales' described by Aristides Quintilianus (pp. 45–52 above) are based on an enharmonic structure; and if he is right to identify them with the *harmoniai* of Plato's *Republic*, they must also be those discussed by Aristotle in the *Politics*, since Aristotle makes it clear that the systems he is considering and those of the *Republic* are the same. Hence though neither Plato nor Aristotle specifies the *harmoniai* they review as being 'enharmonic', it is rather probable that they were. Neither in fact draws any distinctions recognisable as marking differences of genus. In tacitly assuming an enharmonic scheme, the *Sectio* would seem to be in good company.

[48] Some of them can indeed be 'halved', in the sense intended here, those, that is, both of whose terms have rational square roots.

[49] See Plut. *Quaest. conv.* 645d–e, [Plut.] *De mus.* 1137e–f, Psellus, *De trag.* 5. The probably fourth-century discussion appears in a papyrus fragment, *P.Hib.* 1.13; see West 1992a: 247–8, with the references in his n. 84; cf. also pp. 69–73 above.

With that in mind, we may judge that a treatise bringing together a corpus of significant fourth-century propositions about musical intervals could hardly fail to include some account of those peculiar to enharmonic systems. This rather impressionistic point can be reinforced. I have commented that the text does not actually state that the propositions refer to the enharmonic genus; it is the method of construction and the mathematics involved which assure us that they do.[50] The text names the note which is central to its argument simply as '*lichanos*', not as 'enharmonic *lichanos*', and the relevant structure simply as 'the *pyknon*', without distinguishing it from the various *pykna* of the chromatic genus. But this mode of reference would be unimaginable in a text originating in a period later than the fourth century. It makes sense only in a context where systems are presumed to be enharmonic unless explicitly described otherwise. If that context is assumed, the two propositions can stand, in a way precisely parallel to earlier ones, as orderly systematisations of pre-Euclidean material.

PROPOSITIONS 19–20: THE DIVISION OF THE *KANŌN*

This final stretch of the text describes a procedure for marking out the *kanōn* or 'ruler' under a monochord's string, in such a way that when a bridge is moved successively to each of the positions indicated, it will determine lengths of string that sound each successive note of a two-octave diatonic scale. As Barbera has noted, though the process of division is described in mathematical terms, and offers students no help with the practicalities of such operations as dividing a length into eight equal parts, the writer implies (realistically or otherwise) that the work will be carried out on an instrument, not merely in a diagram. The second sentence of proposition 19 reads: 'Let there be a length of the *kanōn* which is also the length AB of the string'.[51] The construction produces the results claimed for it, and the exposition is not hard to follow. But questions arise about certain details, and there is a problem of a more general sort. One would expect the sequence of steps followed in the division to be governed either by mathematical or by musical considerations, or perhaps by some intelligible combination of the two; and we shall find that the *Sectio*'s procedure has puzzling features on any of these interpretations.

The construction is in two parts. They correspond to the segments numbered as propositions 19 and 20 in modern editions, and though this

[50] The implication is obscured by Jan's emendations to the text of proposition 18, which are unnecessary and misleading.

[51] Barbera 1991: 60–1. The sentence quoted, with different and perhaps preferable punctuation, would read: 'Let there be a length of the *kanōn*, AB, which is also the length of the string.'

numbering has no manuscript authority, a line of demarcation is securely indicated by the text itself. Proposition 19 begins 'To mark out the *kanōn* according to what is called the immutable *systēma*', and ends 'Thus all the notes of the immutable *systēma* will have been found on the *kanōn*'; and proposition 20 begins 'It remains to find the moveable notes'. We can therefore start by considering proposition 19 by itself; discussion will be easier if we have a translation in front of us, along with an illustrative diagram.[52]

To mark out the *kanōn* according to what is called the immutable *systēma*. Let there be a length of the *kanōn* which is also the length AB of the string, and let it be divided into equal parts by C, D, E. Then AB, being the lowest in pitch, will be the bass note. This note AB is in the ratio 4:3 to CB, so that CB will be concordant with AB at the fourth above it. AB is *proslambanomenos*, and CB will therefore be *diatonos hypatōn*. Again, since AB is double BD, it will be in concord with it at the octave, and BD will be *mesē*. Again, since AB is four times EB, EB will be *nētē hyperbolaiōn*. I divided CB in half at F, and CB will be double FB, so that CB is concordant with FB at the octave, and FB will be *nētē synēmmenōn*. I cut off from DB one third, DG. DB will be in the ratio 3:2 to GB, so that DB will be concordant with GB at the fifth. Hence GB will be *nētē diezeugmenōn*. I set out GH, equal to GB, so that HB will be in concord with GB at the octave; thus HB is *hypatē mesōn*. From HB I took away one third, HK. HB will be in the ratio 3:2 to KB, so that KB is *paramesos*.[53] I marked off LK, equal to KB, and LB will become the low *hypatē*.[54] Thus all the notes of the immutable *systēma* will have been found on the *kanōn*.

Considered from a formal perspective, the route taken by this construction is straightforward but not strikingly systematic. After an initial quartering of the whole length, AB, producing four sounding-lengths (whole, three quarters, half and quarter), we halve the three-quarter length to construct a fifth length. We then take two thirds of half of AB to make the sixth length, double the sixth to produce the seventh, take two thirds of the seventh to produce the eighth, and finally double the eighth to construct the ninth.

The last four steps have a certain rhythm to them, with operations involving two thirds alternating with doublings. But by comparison with procedures adopted by some later theorists, the sequence is relatively haphazard. We may compare it, for example, with the one followed by Thrasyllus, some three centuries later. His division too falls into two distinct phases,

[52] A diagram broadly similar to the one given here appears in some MSS. Whether it or any other diagrams appeared in the original text is uncertain.

[53] *Paramesos* is found occasionally as a variant for *paramesē*.

[54] That is, the note usually named as *hypatē hypatōn*. Despite the unfamiliar expression used here, the author seems to have known the 'normal' way of naming it, since he designates the note above it as *parhypatē hypatōn* in proposition 20.

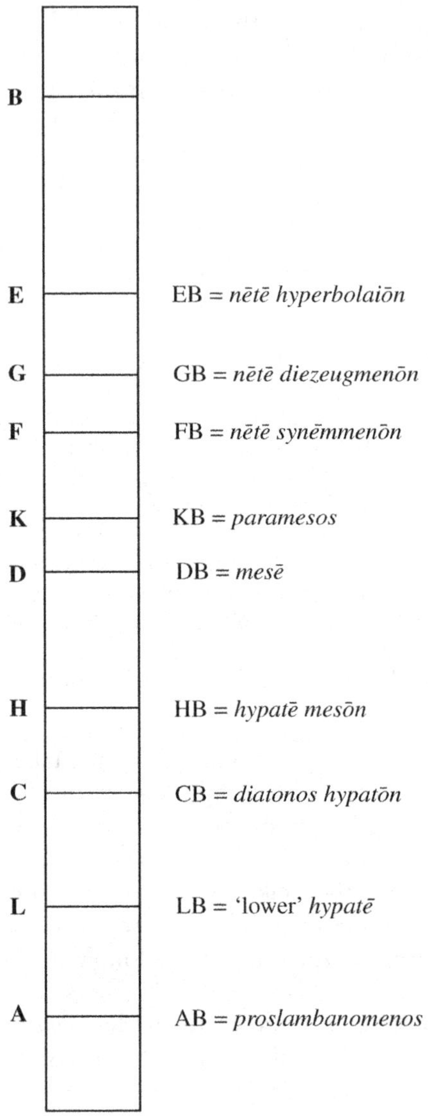

Figure 10 *Sectio canonis* proposition 19

corresponding roughly to the *Sectio*'s propositions 19 and 20; here we need consider only the first. It is remarkably orderly, beginning from the whole string, then halving it, then dividing it into thirds, and finally into quarters. By these means Thrasyllus constructs lengths for only six of the nine notes supplied by the *Sectio*'s proposition 19, leaving the remainder to be found in the second, more complex stage of the division. This seems anomalous, since the missing notes are among the fundamental, fixed notes of the system, and one might reasonably expect that fact to be reflected in the process by which they are mathematically constructed. Thrasyllus has evidently sacrificed musical considerations on the altar of formal orderliness, and has also enshrined in his procedure the symbolic significance of the numbers 1, 2, 3, 4, which make up the Pythagorean *tetraktys* of the decad.[55]

By Thrasyllus' standards the *Sectio*'s procedure is chaotic, and it shows no signs of allegiance to Pythagorean arithmological symbolism. It is less obvious, however, that we can form a sound judgement on it simply by reversing the verdict on Thrasyllus, arguing that at points where the two sets of requirements are in competition it has allowed musical principles free rein, and has made no concessions to systematic neatness for its own sake. That would suggest that it contains no significant musicological oddities, but in fact there are at least three. One of them, on the face of it, is merely terminological. The writer claims that he has constructed all the notes of the 'immutable' (*ametabolon*) *systēma*. If by this he means (in Aristoxenian language) 'all the fixed notes', the usage is intelligible; the system comprising all and only fixed notes is in an acceptable sense 'unchangeable'. But when the expression *ametabolon systēma* occurs in other writings it has a quite different sense, referring to the regular, 'unmodulated' ordering of all the notes within the standard two-octave compass.[56] Secondly, one of the notes constructed through the initial quartering of the string is not normally reckoned a fixed note at all. What the *Sectio* calls *diatonos hypatōn* is in other terminology the diatonic *lichanos hypatōn*, and is elsewhere usually assigned a status no different from that of any other moveable note, whose counterparts in the other genera are at different pitches. The *Sectio*'s treatment seems to award it a more fundamental role. Finally, another note constructed here, *nētē synēmmenōn*, is indeed treated by most theorists as

55 That is, they sum to the perfect number 10, for whose numinous role in Pythagorean lore see e.g. Sextus Empiricus, *Adv. math.* 7.94–5. For Thrasyllus' construction see Theo Smyrn. 87.4–89.9. It is uncertain whether the introductory comment linking the procedure explicitly with the *tetraktys* comes from Thrasyllus or from Theon himself.

56 For examples see Cleonides 201.14–18, Theo Smyrn. 92.26–7, Ptol. *Harm.* 52.11–12, 53.18, Arist. Quint. 14.24–5, and cf. Fig. 5 on p. 17 above.

a fixed note; but when the moveable notes of the system are located in proposition 20, those in the tetrachord whose upper boundary it is, the tetrachord *synēmmenōn*, are nowhere mentioned. *Nētē synēmmenōn* is left strangely isolated from the structure to which it belongs.

Let us take the second problem first. It is unlikely that *diatonos hypatōn* is included here merely because it is bound to emerge from a division of the length into quarters, since that move is itself unnecessary. The other notes constructed through this quartering are *mesē* and *nētē hyperbolaiōn*, and *mesē* could have been found by halving the string, *nētē hyperbolaiōn* by halving the half. The writer seems to have had a genuine reason for wishing to locate *diatonos hypatōn* as well; after *proslambanomenos* it is indeed the first note to be identified. In this respect the case is different from that of Thrasyllus' division, where this note is again constructed in the opening phase, since its appearance there is an inevitable consequence of his unwavering pursuit of a sequence through the terms of the Pythagorean *tetraktys*.

Thrasyllus, however, supplies a useful additional piece of evidence.[57] He refers to this note as '*hyperhypatē*, also known as *diatonos hypatōn*'. The term *hyperhypatē* reappears in the same role in a passage of Boethius derived from Nicomachus;[58] elsewhere it is almost unknown.[59] But the existence of a distinctive name for a note in this position, and one that does not specify it as diatonic, encourages the belief that in some Greek systems it had a role independent of genus, as part of the 'immutable' framework. This hypothesis is supported by Aristides Quintilianus' description of the Dorian *harmonia* he associates with Plato's *Republic*, in which a regularly formed enharmonic octave is supplemented, at the bottom, by the interval of a tone; the lowest note of the system would be this *hyperhypatē*.[60] It is encouraged also by the occurrence of a note in that position, in a non-diatonic context and immediately below a *pyknon*, in two of the surviving musical scores, those of a fragment from Euripides' *Orestes* and the Delphic paean of Limenius. In the *Orestes* the *pyknon* is enharmonic and in the

57 In much of my discussion of this issue I am following the suggestions of Winnington-Ingram 1936: 25; cf. 28, 32, 35–6.

58 Boeth. *Inst. mus.* 1.20. Here the mysteriously named Prophrastus of Pieria is credited with the addition of a ninth string, called *hyperhypatē* and tuned a tone below *hypatē*, to the eight allegedly first used by the equally mysterious Lycaon of Samos (possibly to be identified with Pythagoras). Cf. *Exc. ex Nicom.* 4, 274.3–4, and see Bower 1989: 33 n. 107, 34 nn. 109–10. These attributions are part of an almost wholly fictional reconstruction of progressive additions to the number of strings on the lyre or kithara; but they suggest that Nicomachus had some record of an ancient nine-note system, which included the note *hyperhypatē* lying at the interval of a tone below the regular octave.

59 The usage at Arist. Quint. 8.12, which seems to imply that there are distinct diatonic, chromatic and enharmonic *hyperhypatai*, is probably based on a misunderstanding.

60 Arist. Quint. 18.13–15; see p. 49 above.

paean chromatic, but in neither case does the note in question fit into the pattern standardly assigned to the relevant genus.[61]

Aristides attributes this form of the Dorian *harmonia* to the usage of 'very ancient times' (18.5–6). If the music of the *Orestes* fragment goes back to Euripides himself,[62] it originated in the fifth century; and the Delphic paeans of Athenaeus and Limenius, dating from 127 BC, adopt a deliberately 'antique' style. The various pieces of evidence are consistent, then, with the hypothesis that in representing *diatonos hypatōn* (or *hyperhypatē*) as one of an 'immutable' array of fundamental notes, the *Sectio* is in tune with musical realities, and in particular with ones prevalent in relatively early times.

The hypothesis has a bearing also on the fact that proposition 19 constructs *nētē synēmmenōn*, even though the tetrachord that includes it is ignored in proposition 20. The pertinent evidence I have found is thin, but points in a helpful direction. A passage of the Plutarchan *De musica*, certainly derived from Aristoxenus, reports that music in the very ancient 'libation-style' (*spondeiazōn tropos*) used certain notes in the accompaniment which it avoided in the melody. It also mentions the notes of the melody with which these accompanying notes formed concords or discords. The melodic notes are *parhypatē, lichanos, mesē, paramesē* and *paranētē diezeugmenōn*, and the accompanying notes are *tritē diezeugmenōn, nētē diezeugmenōn* and *nētē synēmmenōn*. (The system is to be construed throughout as enharmonic.) Taken together, these include all except the lowest of the notes of the tetrachord *mesōn*, and all the notes of the tetrachord *diezeugmenōn*; but the tetrachord *synēmmenōn* is represented only by its *nētē*.[63]

The absence of any reference to *paranētē* and *tritē synēmmenōn* might be an accident; but I rather think it is not. *Nētē synēmmenōn* lies a tone below *nētē diezeugmenōn*, and an octave above *diatonos hypatōn* or *hyperhypatē*, and is thus the analogue of the latter in the higher range of the system. (It is noteworthy that the only constructional use which the *Sectio* makes of *diatonos hypatōn* in proposition 19 is to locate *nētē synēmmenōn*, and that the latter is not employed in the construction of any other length.) In the Plutarchan account, *nētē synēmmenōn* appears in an enharmonic context, just as does its counterpart an octave below in Aristides' Dorian and in

[61] For the *Orestes* fragment see Pöhlmann and West 2001: 12–13, where the note appears in bars 2, 8, 9 and 12 of their modern transcription; cf. West 1992a: 284–5. For the relevant part of Limenius' paean see Pöhlmann and West 2001: 80–1, where the note appears several times between figures 25 and 26 of their transcription; cf. West 1992a: 297.

[62] I agree with West 1992a: 270, that there is no reason to doubt it. He provides a short bibliography on the piece on p. 278; see also the extensive bibliography on the scores in Pöhlmann and West 2001.

[63] See [Plut.] *De mus.* 1137b–d.

the *Orestes* fragment; and the *Orestes* score itself includes the symbol for an instrumental (but not a melodic) note at the pitch of *nētē synēmmenōn*, repeated five times, an octave above the note which apparently plays the role of *hyperhypatē*. No other notes of the tetrachord *synēmmenōn* appear.[64]

Taken by themselves, these indications would not add up to solid evidence for the thesis that a note in this position was commonly used, irrespective of the genus of the prevailing *systēma*, as an item unattached to the tetrachord *synēmmenōn*. In the Plutarchan treatise (as in the *Sectio*), the name assigned to it seems to suggest the contrary; it identifies it precisely as the *nētē* of that tetrachord. But the source of that account is Aristoxenus, and he is using the terminology current in his own day; it is the only designation he has for a note at that pitch in an enharmonic context. When linked with the evidence about its counterpart an octave below, the argument for the thesis becomes a good deal stronger; and in both cases the practices in question apparently go back at least to the fifth century and probably further.

There may then be a simple explanation for the first difficulty raised by the *Sectio*'s nineteenth proposition, its anomalous use of the expression *ametabolon* ('immutable') *systēma*. If the hypotheses I have sketched are correct, it does indeed include only 'unchangeable' notes, but within a musical system which by the late fourth century was obsolescent if not extinct. The writer, then, is basing his account on an analysis that was already out-dated, and it would hardly be surprising if it brought with it elements of an earlier terminology. The expression *ametabolon systēma* is perfectly appropriate to the construction of proposition 19, just so long as theorists had not yet appropriated it for other purposes. It is likely, though not certain, that the person responsible for this appropriation was Aristoxenus.[65] We can therefore make sense of the *Sectio*'s usage so long as we attribute it either to a pre-Aristoxenian source, or to a writer for whom Aristoxenian terminology, even if he knew it, did not have canonical authority. The language and strategy of proposition 19 can thus be made musicologically intelligible; and there is something to be said, after all, for the suggestion that it has approached the matter from the opposite direction to Thrasyllus, allowing musical considerations rather than formal

[64] Cf. West 1992a: 206–7.

[65] The expression does not occur in the *El. harm.*, but the fact that it is used to mean 'unmodulated *systēma*' both by Cleonides (201.14–18) and by Aristides Quintilianus in the Aristoxenian phase of his treatise (14.24–5, explicating the phrase in a similar way to Cleonides) makes an Aristoxenian pedigree very likely. Cf. also Bacchius 74 (308.2–7). But by the Roman period the usage was widespread among theorists, regardless of their Aristoxenian or other allegiances. We may note that Thrasyllus signs off his entire division, including the moveable notes, with the comment that he has now 'filled out the entire *ametabolon systēma*', Theo Smyrn. 92.26–7.

or mathematical ones to guide the process of construction. We shall return to these issues shortly, but it will be helpful, first, to glance briefly at the final stage of the division, in proposition 20.

Sounding-lengths are constructed here for all the standardly recognised moveable notes in a two-octave diatonic system (omitting the tetrachord *synēmmenōn*), except two for which lengths were in effect established in proposition 19. (Diatonic *paranētē diezeugmenōn* stands at the same pitch as *nētē synēmmenōn* and need not be constructed independently; and the *lichanos* of the tetrachord *hypatōn* is identical with *diatonos hypatōn* or *hyperhypatē*.) Since there are eight moveable notes in all, two in each tetrachord, proposition 20 deals with six. A diagram of the completed *systēma* (Fig. 11 below) will give some help in following the reasoning.

(i) The first to be constructed is *diatonos hyperbolaiōn*,[66] a tone below the highest note of the system; the latter's sounding-length, EB, is one quarter of the whole length of string. This *diatonos* is found by dividing EB into eighths, and then adding to EB a distance equal to one of those eighths. The resulting length, MB, stands to EB in the ratio of the tone, 9:8. (ii) The lower moveable note in this tetrachord, *tritē hyperbolaiōn*, is a tone below *diatonos*, and is found by treating MB precisely as EB was treated in the previous step, to produce NB.

Once the highest tetrachord is completed, the procedure changes. (iii) Instead of constructing *tritē diezeugmenōn* by repeating the same manoeuvre with FB as its starting point (this is the length for *nētē synēmmenōn*, and therefore for diatonic *paranētē diezeugmenōn*), the author locates it by adding one third of itself to NB, *tritē hyperbolaiōn*. The new length, XB, is in the ratio 4:3 to NB, which corresponds to the fact that the two *tritai* are a perfect fourth apart. (iv) Next, XB is increased by half, giving OB, in the ratio 3:2 to XB. OB therefore sounds a fifth below XB, and is the length for *parhypatē mesōn*. (v) The fifth step amounts to doubling XB (*tritē diezeugmenōn*) to give PB (*parhypatē hypatōn*, an octave lower), though it seems to be conceived and expressed in a needlessly roundabout way.[67] (vi) Finally, CB (*diatonos hypatōn*) is decreased by a quarter to give RB, so that CB:RB = 4:3, and RB is *diatonos* (diatonic *lichanos*) *mesōn*, a fourth above *diatonos hypatōn*.

The procedure works, but there is no clear order in it, and it seems a little makeshift. Once again we can contrast the strategy of Thrasyllus,

[66] The note is usually called diatonic *paranētē hyperbolaiōn*. The name *diatonos hyperbolaiōn* is uncommon, but reappears, for instance, in Thrasyllus' division.

[67] It is done by adding to OX a length equal to itself, OP. Since OX is equal to half XB, this is equivalent to doubling XB to reach PB.

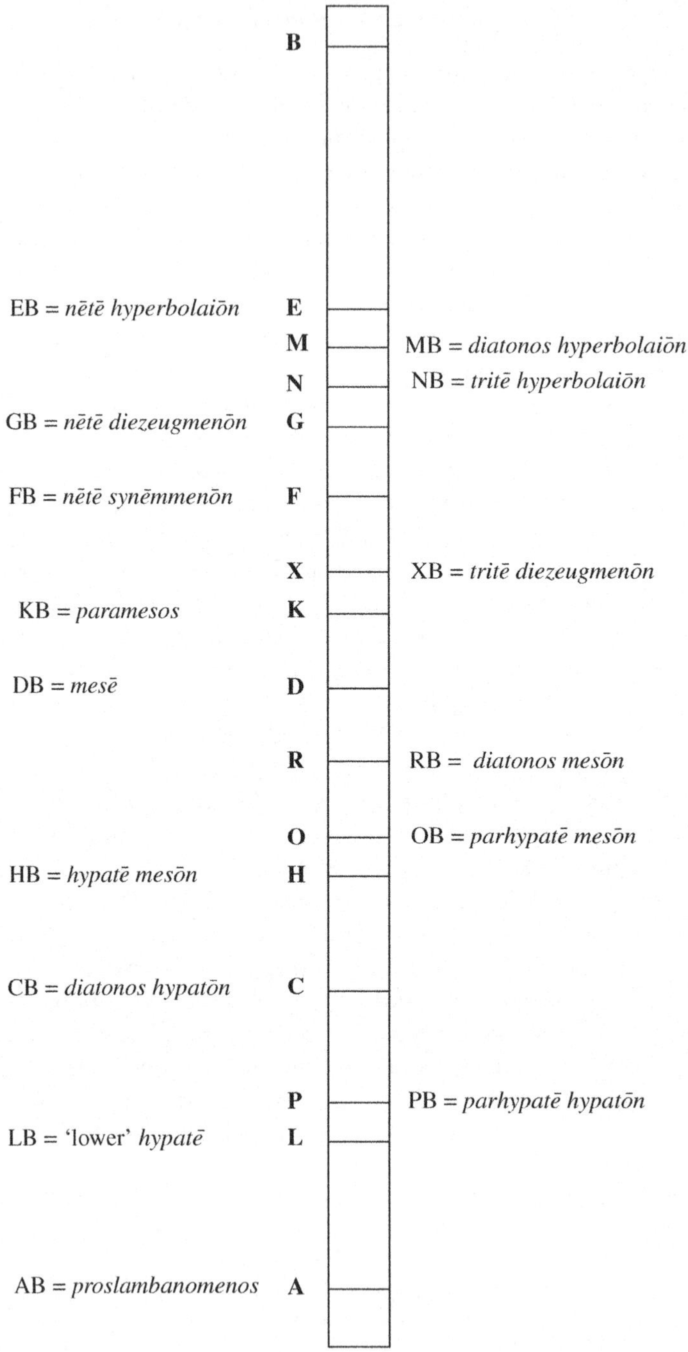

Figure 11 *Sectio canonis* proposition 20
[Notes constructed in prop. 19 are on the left, and those constructed in prop. 20 on the right]

which operates systematically, in every tetrachord, with repetitions of moves analogous to the first two in the *Sectio*.[68] Nor does the route taken in the *Sectio* seem to be governed by principles of a musical sort; the best we can say of it is that it takes the lower moveable notes of the tetrachords one after another in descending order (steps (ii)–(v)), and frames them with the construction of the two upper moveable notes that had not been put on the map in proposition 19. That gives the exposition a certain symmetry. But its basis is not very impressive, and it offers no explanation for the peculiarity of step (v). Since there are several other ways of presenting an equivalent construction in a systematic manner, one must conclude that the writer was unconcerned with such niceties, and perhaps also that his version belongs to a relatively undeveloped stage in the history of canonic division.

It is quite clear that the structure generated by the *Sectio*'s procedure is diatonic, or at least that it would have been so described by Aristoxenus and his successors. This creates a further difficulty for those who suppose, as I do, that the treatise as we have it is a single, fully integrated work, since propositions 17–18 presuppose an enharmonic structure, as we have seen, and nothing has been said to prepare us for the transition to a different system. The final part of my discussion of propositions 17–18 may indeed have made matters worse. If enharmonic is taken as the norm, the automatic point of reference unless we are told to the contrary, why is the division diatonic and not enharmonic, and why, given that it is, does the author not alert us to the change by naming the genus in which his construction is set?

The short answer to the first part of this question is that a complete division in the enharmonic genus is simply impossible on the basis of theorems propounded earlier in the work. There is nothing in them that would ratify any particular way of constructing the position of the middle note of the enharmonic *pyknon*, since the small intervals inside the *pyknon* have not been quantified. It is not merely, as Barbera puts it (1991: 24), that the author fails to provide 'the necessary mathematics to represent these intervals with ratios'. The fact is that there is no way of establishing them on the basis of the principles that the author had at his disposal, and any specific quantification would have been arbitrary. Among fourth-century theorists, only Archytas, so far as we know, offered a mathematical division of the enharmonic; and not only is it based on proportional principles of

[68] The presentation is complicated by Thrasyllus' construction, within the same sequence, of the chromatic *paranētai* and *lichanoi* (the *tritai* and *parhypatai* are identical with their diatonic counterparts). These too are found through repetitions of a single procedure.

which the *Sectio* is innocent, but it posits an enharmonic structure quite different from that of the present treatise. As late as the imperial period, writers who quantify the enharmonic *pyknon* in the same way as the *Sectio*, as the remainder of a perfect fourth after the subtraction of two tones in the ratio 9:8, do not attempt the task of dividing it mathematically.[69]

A diatonic system, by contrast, is easily constructed on the basis of the *Sectio*'s mathematical resources, since they provide a ratio for the tone, and allow us to treat the 'semitone' at the bottom of the tetrachord merely as the residue of the fourth after two tones; no separate quantification is needed. But none of this explains why the shift of genus from enharmonic to diatonic is unannounced, and why the author never explicitly refers to distinctions between genera at all.

The answer, I think, is that he recognised no such distinctions. In the light of the discussion so far this may seem a perverse opinion, but I believe that it both fits the facts and explains them. The first point to be made is terminological. Aristoxenus and later theorists regularly agree that the note with the most important function in determining genus is the higher of the two moveable notes in the tetrachord, which is usually named as *lichanos* in the tetrachords below *mesē*, and as *paranētē* in those above it. These primary names would, where necessary, be qualified by adjectives meaning 'enharmonic', 'chromatic' or 'diatonic'. Now it is these primary names, with no qualifying adjectives, that appear in propositions 17–18 of the *Sectio*. The adjectives' omission, in a context which in Aristoxenian terms is 'enharmonic', indicates that the names will automatically be understood as referring to notes lying roughly a ditone below the highest notes of their tetrachords.[70] When we come to propositions 19–20, where an Aristoxenian would call the upper moveable note of each tetrachord 'diatonic *lichanos*' or 'diatonic *paranētē*', the terms *lichanos* and *paranētē* disappear, and are replaced by the name *diatonos*, 'the note through a tone'.

This terminology is not unique. It is found occasionally in later sources (in Thrasyllus, for example); but I suspect that in those writers the usage is an archaism. Where it properly belongs is in a context where the notion of genus, as Aristoxenus presents it, had not yet taken firm root, and the special characteristic of what we call 'diatonic' systems was conceived less

[69] This is most clearly seen in Thrasyllus' sketchy treatment of the enharmonic at Theo Smyrn. 92.27–93.2.

[70] Traces of this convention survive in Aristoxenus, *El. harm.* Book III, notably at 65.25–68.1, in which (apart from a brief excursus on the diatonic at 66.17–25) the reference is exclusively but inexplicitly to enharmonic systems. In the immediate sequel (68.1–12), by contrast, where attention shifts to diatonic and chromatic, the genera are explicitly named.

as its transference of the notes *lichanos* and *paranētē* to a higher position in the tetrachord than as its omission of those notes, and its inclusion, instead, of a special note of its own, the 'note through a tone'. Structures that we would call 'enharmonic', conversely, are ones in which the 'note through a tone' is omitted, and the note marking off the interval which principally determines the system's musical character is given the special designation *lichanos* or *paranētē*, to distinguish it from the note which stands at the same pitch in a 'diatonic' tetrachord, its *parhypatē* or *tritē*. It was only with Aristoxenus' wholesale re-conceptualisation of musical systems and their relations that the notion of genus became fully articulated, and established the picture of a single note, *lichanos* or *paranētē*, moving up and down in pitch to determine the genus, while still retaining its own identity and hence the same name.[71]

In that case we can no longer insist on the point that propositions 17–18 do not announce themselves as concerned with the enharmonic genus, and propositions 19–20 do not indicate explicitly a shift to the diatonic. The language in which they are couched tells us quite clearly, in each case, which notes are being constructed or discussed. All that they presuppose is that the positions of the relevant notes are known, and that systems of two different sorts existed side by side. They are not distinguished from one another in the Aristoxenian manner, but by the fact that each of them uses a different selection of the notes available in the familiar repertoire.

Evidence of a different sort does a little to commend this hypothesis. The Plutarchan *De musica* records a report by Aristoxenus about the alleged 'invention' or 'discovery' of enharmonic music by the aulete Olympus (1134f–1135b). It is important to note that it is not represented as Aristoxenus' own analysis, but as his account of what other and presumably earlier 'musical experts' (*mousikoi*) have said about the matter (cf. pp. 98–100 above). Olympus is assumed to have been working, initially, within a diatonic structure; and he arrived at the basic outline of the enharmonic not by shifting the 'diatonic *lichanos*' to a lower position, but by omitting it altogether. The remaining note in the diatonic tetrachord, diatonic *parhypatē*,

[71] It is worth recalling that in his analyses of these structures and their relationships, the noun that Aristoxenus uses is *genos*, 'genus'. In musical contexts the usage is new, and is certainly drawn from the terminology of recent, Aristotelian science. If some other noun with the same reference had already been common currency among musicians and musical theorists, Aristoxenus would have been bound to make use of it, if only to connect his analyses with familiar descriptions of the 'facts', and perhaps to announce his intention of replacing this term, for scientific purposes, with another. No ancient text shows any trace of the existence of a pre-Aristoxenian noun of this sort. I conclude that there was none, and correspondingly no clearly articulated conception of the classes of system which Aristoxenus calls *genē*, 'genera'.

can hardly have retained the same name in this new context; in Aristoxenian terms it would have become enharmonic *lichanos*, but the writer does not tell us what it was originally called. This is disappointing but only to be expected, since Aristoxenus has evidently paraphrased the account given by the *mousikoi* into his own terminology; the passage is full of references to the enharmonic and diatonic genera, and it names the note that was 'omitted' as 'diatonic *lichanos*', not, like the *Sectio*, as *diatonos*. The crucial point is that the two systems are characterised by the presence or absence of specific notes, and not by the appearance of the same notes in different positions.

If these suggestions are on the right lines, they go some way towards an explanation of another puzzle in musicological history. There is evidence, as we have seen, that the core of the structures of the *harmoniai* of Plato's *Republic*, including those of which Socrates approves, was – in Aristoxenus' terms – enharmonic. That of the theoretically ideal system of the World-Soul in the *Timaeus*, by contrast, is diatonic, and yet it is to its perfection, not that of some enharmonic construction, that human music ought, we are told, to aspire. In this respect the transition from the *Republic* to the *Timaeus* parallels, on a much larger scale, the one between propositions 18 and 19 in the *Sectio*; and Plato, again like the author of the *Sectio*, nowhere makes any direct allusion to distinctions between the genera. On the hypothesis I am suggesting, Plato's strategy becomes at least a little less puzzling. There are not two radically different systems, nor is there a situation in which notes can move into different locations, of which – on Plato's view of things – only one can be correct. Instead there is a single reservoir of notes, each of which has its correct position, and musicians (or divine soul-makers) can form their harmonic patterns from them by selecting some and omitting others. I do not pretend that my reconstruction of this approach will resolve all the difficulties, either in Plato or in the *Sectio*; but I think it does a good deal to ease them.

CLOSING REFLECTIONS ON THE *SECTIO* AND ITS TARGET READERSHIP

Propositions 19–20 can be understood only by readers armed with a degree of musical knowledge, though it need not be great. They must be familiar with the names of all the notes in the system, and be aware of some (not necessarily all) of the relations that hold between them in a 'diatonic' structure. They must know that the interval between *nētē* and *diatonos hyperbolaiōn* is a tone, for instance, that the interval between *mesē* and *nētē diezeugmenōn* is a fifth, and so on. But the construction calls for knowledge only of intervals whose ratios have previously been established, that is, the concords and the tone, and in particular it neither asserts nor presupposes anything about the interval between a *tritē* or a *parhypatē* and the lowest note of its tetrachord, which will in fact, given the rest of the construction, be a *leimma* in the ratio 256:243. Nothing in the *Sectio* suggests that the construction it describes specifies just one system within the general category 'diatonic'; it points only to the conclusion that in one familiar pattern of attunement the two upper intervals of each tetrachord were commonly referred to as 'tones'. It assumes that this designation is accurate, and that readers of the *Sectio* will be familiar with it. Aside from that, they need know only the position of each named note in the tetrachord to which it belongs, and the (concordant) relations between the tetrachords themselves.

The exercise seems designed, in part, to show how the findings of propositions 12–13 (no others are needed) can be used to give a mathematical interpretation of a complete and well-known musical structure. But that could have been done, on the basis of the physical theory outlined in the introduction, without any reference to a string or to lengths on the *kanōn*. It would simply have specified the system's pattern of intervals as ratios, representing these, as in earlier propositions, as ratios between pitches rather than as ratios between lengths. To put the point another way, the writer could have continued with his original, arithmetical approach (see pp. 380–1 above), instead of shifting into a geometrical mode.

The way he actually goes to work has one obvious advantage. When the relations are constructed geometrically, on a monochord, the system of notes they define can be presented in its musical guise, to the ear as well as the mind.[72] Hence the claim that these mathematical manoeuvres produce the required musical results can be confirmed empirically, and their confirmation will give strong support to the thesis that these mathematical relations are responsible for the system's musical credentials. It would be misleading and anachronistic to suppose that the application of the mathematical formulae to the monochord was undertaken in an 'experimental' spirit, allowing mathematical hypotheses to be subjected to empirical tests. No harmonic theorist before Ptolemy seems to have envisaged the role of 'laboratory instruments' in this way. They were thought of rather as audiovisual aids, whose use would clarify the truths of mathematical harmonics

[72] There are of course practical difficulties to be overcome. Ptolemy argues that while the monochord is suitable for showing how concordant intervals correspond to ratios, it is unsatisfactory when used to display anything more complex, such as the mathematical anatomy of an entire *systēma*. See *Harm.* 66.5–69.12, especially 69.1–5.

and help learners to grasp them. The truths are presented, as it were, to the eyes on the *kanōn* and to the ears through their expression in sound, as well as abstractly to the intellect. There is no suggestion in any pre-Ptolemaic source that the audible results might put the mathematical conclusions at risk.[73]

The division is based, mathematically, on earlier theorems, but there is a logical gap between them which the *Sectio* does nothing to fill. The theorems' transition from pure arithmetic to harmonics is mediated by the physical acoustics of the introduction, which legitimises the treatment of pitch-relations as ratios of numbers. If a note at one pitch is made up of n impulses in a given time, and a note at another of m impulses; the relation between them is the ratio n:m. By itself, however, this does not justify the assumption, essential to propositions 19–20, that the same ratio will be reflected in the relation between the lengths of string from which the notes are produced. As later theorists often point out, the terms of the ratio have to be reversed when one shifts from ratios of pitches to ratios of lengths; the higher pitch, with its supposedly more rapid sequence of impacts or its greater speed of transmission, will attract the larger number, but it is produced by the smaller length. Thus n:m becomes m:n.[74] But this is not the main problem, which is that the *Sectio* has made no effort to show that ratios between pitches are correlated with ratios between lengths in any way at all.

The case for the correlation could evidently be argued on the basis of familiar observations of a kind apparently relied on in early Pythagorean musicology (see pp. 26–7 above). The fact that a string, stopped half-way along its length, produces a note an octave above that sounded by the whole string could also be linked intelligibly with the *Sectio*'s physics. The greater length oscillates more slowly and its impacts are more widely spaced in time; given the physical theory and the observations of a string's behaviour, one could infer that the half-length oscillates twice as rapidly as the whole, and similarly for the other relevant relations. But the *Sectio* is silent about such matters. It seems to take it for granted that its readers are as familiar with these observations as they are with the elementary musical data, and that they are capable of drawing the inference for themselves.

Readers of the treatise must therefore be musically literate, to the extent I have specified; they must have a background in mathematics, since it is assumed that they have mastered propositions not proved in the *Sectio* itself;

[73] See for instance Theo Smyrn. 57.11–58.12, Nicomachus, *Harm.* ch. 10, and cf. Barker 2000a, chapters 10–11.

[74] See for instance Theo Smyrn. 87.9–18, quoting Thrasyllus.

and they must be aware of the links that have been identified in the past between relative pitches and relative lengths of a string. The writer can also assume that they will recognise an implicit allusion to the monochord, and will understand what it is without being told; he mentions a string and the *kanōn*, but gives no description of the instrument to which they belong.[75] Rather few Greeks in 300 BC would have satisfied all these conditions. The *Sectio*, it seems, is appealing to a restricted and specialised audience.

In matching concordant intervals and the tone with the appropriate ratios (propositions 9–13), and in using the correlations to construct a musical *systēma* mathematically (propositions 19–20), the *Sectio* is doing nothing new. The fundamental ratios were known to the fifth-century Pythagoreans; the principles needed for a comparable construction procedure are at least embryonically present in Philolaus and are fully exploited in Plato's *Timaeus*. Some or all of the propositions about inequalities (propositions 14–16, 18) can persuasively be associated with Archytas and were known to Aristotle and Aristoxenus; the method of constructing a ditone which is used in proposition 17 was used also by Aristoxenus, and was probably common among practising musicians. What the *Sectio* contributes is its orderly, formal demonstration of these propositions. Its theorems about musical intervals are systematically arranged, and are dependent in their turn on a coordinated sequence of theorems in pure arithmetic; and they are also neatly integrated with a well-argued exposition of a theory about the physics of sound and pitch. Thus the harmonic propositions become assimilated into a nexus of interlinked explorations in several scientific domains. The *Sectio* is an attempt not only to establish propositions in mathematical harmonics on firm, rational foundations, but also to locate this science among others, fitting it smoothly into their multifaceted enterprise of investigating and understanding the world.

The fact that there are gaps and minor uncertainties in the *Sectio*'s exposition, and one crucial and irremediable logical error, scarcely detracts from its overall coherence or from the excellence of its aspirations. A more cogent general criticism would focus on its limitations. The harmonic theorems seek to demonstrate the ratios of only the most elementary musical relations; they mark off this science's findings from those of empirical harmonics at

[75] The earliest secure reference to the monochord dates from the late fourth century (Duris frag. 23). This passage, like the *Sectio*, mentions it without explaining what it is, and it implies that it had been in use for some time. See the discussion in Creese 2002: 22–5. But even if that is so, it had no public profile. There is no record of its use in musical performance until Roman times (Ptol. *Harm.* II.12); it served only as an accessory to scientific discussion, and only those exposed to such studies are likely to have known about it.

only the most obtrusive points of disagreement; and the concluding division describes only one pattern of attunement among many, from a mathematical perspective the simplest of all. In setting out the division it makes no attempt to ground its form in mathematical principles which might explain why it falls into this pattern rather than any other; it offers no counterpart of the theory of proportions which governs the analyses of Archytas and Plato. In these respects it seems primitive even by the standards of the earlier fourth century, and a comparison with the work of Aristoxenus shows immediately how large and complex a collection of musical relations and structures remains unexplored.

'Primitive', however, is probably the wrong word. There is no good reason to think that the writer was stuck in a time-warp, unaware of ways in which the science had moved on since the time of Philolaus; fairly clearly he was not. A better diagnosis is that he set himself only a determinate and restricted goal. His task was to secure firm foundations for mathematical harmonics, rather than to construct upon them an elaborate edifice to rival Aristoxenus'. That might indeed be a legitimate enterprise, but only when the masonry at the base of the building had been solidly cemented in place. It is notable, too, that the bed-rock upon which he sets it shows no trace of contamination with the deposits of Pythagorean or Platonist metaphysics. Certainly he has borrowed arguments and constructional procedures from those sources, directly or indirectly; but he has relocated them in the setting of a more hard-nosed, more nearly positivistic mathematics and science. In this he differs sharply from his most eminent predecessors,[76] as also from most of those successors who revived mathematical harmonics in the Roman period.

[76] It seems likely, however, though it cannot be proved, that in this respect his approach had precursors among the exponents of mathematical harmonics mentioned by Aristotle in the *Posterior Analytics*.

CHAPTER 15

Quantification under attack: Theophrastus' critique

Theophrastus was born in about 370 BC, succeeded Aristotle as head of the Lyceum in 322 BC and died, full of years, early in the second decade of the third century. Like several other Peripatetics of his generation he wrote copiously; Diogenes Laertius (V.42–50) lists the titles of 224 works, many of them in several books, amounting in all to 232,850 lines and addressing an astonishing range of topics. Later writers refer occasionally to his interest in music, and though he can hardly be regarded as a specialist in the field, the works in Diogenes' catalogue include an *On Music* in three books, and a *Harmonics* and an *On Musicians* in one book each. We know little of their contents. The great majority of his writings are lost, or survive only in fragments quoted by others, and from his publications on music we have only a handful of scraps.[1] Only one of them concerns us here. According to Porphyry, the source who quotes it, it comes from the second book of Theophrastus' *On Music*. It is by far the longest of the fragments on musical topics, running to 126 lines in Fortenbaugh's edition.[2]

The passage begins with a reference to a *kinēma melōidētikon*, a 'movement productive of melody' or a 'melody-making movement' which occurs in the soul. Right at the end of the fragment (130–1) Theophrastus asserts, more bluntly, that the 'nature' of music *is* a movement of the soul, a movement designed to release it from 'the evils caused by the emotions'. The precise relation between these statements is debatable, but the central point is that music and melody originate inside us. They arise from a movement intrinsic to the soul and consist, essentially, in such a movement; when they appear in the public domain as patterns of sounds, these must therefore be secondary manifestations of music, dependent on the first.

[1] The fragments of Theophrastus' writings are collected in the two volumes of Fortenbaugh 1992; the passages on music are in the second volume, fragments 714–726C.

[2] The work from which it is taken is named at Porph. *In Ptol. Harm.* 61.18–20, with the quotation itself at 61.22–65.15; frag. 716 Fortenbaugh. I shall cite it by the line numbers of Fortenbaugh's edition, which includes Porphyry's brief introductory remarks; the excerpt from Theophrastus begins at line 7.

Theophrastus' opening sentence draws attention to the remarkable accuracy with which the soul brings about the transition from psychic movement to audible sound, when it seeks to express its *kinēma melōidētikon* by means of the voice.[3] The soul 'turns' or perhaps 'steers' (*trepei*) the voice just as it wishes, 'to the extent that it is capable of steering something non-rational' (7–9).[4] Talk of the soul and its movement may be alien to modern minds, and a crude paraphrase of his assertion which strips it of its metaphysical associations will no doubt distort his meaning. But he has a point whose force we can easily recognise. A few unfortunates are 'tone dumb', as a friend of mine used to put it (or 'crows' in the patois of a teacher whose tyranny I endured over fifty years ago); but most of us, when we have a tune in our heads, can steer our voices around it with at least tolerable accuracy, singing just the notes we are imagining and no others. Soberly considered it is an extraordinary trick, and one that we perform without having the least idea how we do it.[5] It cries out for an explanation.

Some people, Theophrastus continues, tried to account for the soul's accuracy by reference to numbers, 'saying that the accuracy of the intervals arises in accordance with the ratios of the numbers' (10–12). He goes on to record their view that the ratio of the octave is 2:1, that of the fifth 3:2 and that of the fourth 4:3, and that 'for all the other intervals, in the same way, just as for the other numbers, there is a ratio peculiar to each. Hence they said that music consists in quantity, since the differences exist on the basis of quantity' (12–16).

Here we are on familiar ground, but the use to which, so Theophrastus implies, these theorists put their quantitative conceptions is one we have not previously met. There are, in fact, no other fourth-century allusions to the idea that the soul's ability to create an exact audible counterpart of its inner movements (or, more prosaically, our ability to sing the notes we intend) depends on the existence of precise, quantitative relations between the notes of a melody; and so far as I know there is only one in later literature.[6] The

[3] *Hermēneuein* means 'to express' or 'to interpret'. The related noun *hermēneia* is commonly used of a musician's 'performance' of a composition, and that sense is implicit here in the verb; the soul seeks to give voice to its silent melody in audible performance.

[4] 'Non-rational' translates *alogon*, 'without *logos*'. Sicking 1998: 107–8, is probably right to find a double meaning in it (I am less convinced by his third interpretation); vocal sound is 'something "irrational" and therefore "unmanageable"', and in Theophrastus' view, as we shall see, it is also *alogon* in the sense that it lacks any connection with the ratios (*logoi*) by reference to which some theorists account for the accuracy to which Theophrastus has alluded.

[5] The same point is made by Ptolemy, at the end of his brief account of the physical operations involved in singing. He is not given to exclamatory rhetoric; but he describes the capacity of the soul's ruling part to pick out and execute the appropriate physiological manoeuvres as 'astonishing' (*Harm.* 9.12).

[6] It appears in the same passage of Ptolemy as the remark mentioned in n. 5 above, *Harm.* 9.6–15. Ptolemy's theory is that the pitch of a vocal sound is determined by the distance between a point

thought seems to be that the soul identifies the relations between these notes through a kind of subconscious mental arithmetic, and reproduces them in perceptible form rather in the manner of a craftsman working from a set of measurements, transferring the results of its computations first to some quantifiable feature of the body's vocal apparatus, and from there to the medium of sound.

I do not know whether any fourth-century writer worked out any such hypothesis in detail, and from one point of view it hardly matters. Theophrastus will have none of it, and devotes the bulk of the passage to elaborate arguments designed to refute it. But he says very little more about its psychological aspects, or about the process through which inner movements or imagined melodies are made audible. All his criticisms in this part of the fragment are aimed at the familiar underlying theory that differences in pitch are essentially quantitative, and that each musical interval corresponds to a ratio of numbers. In the course of the discussion we learn a little about his own, strictly qualitative conception of the distinction between high pitch and low; and towards the end he turns his fire briefly on another quantitatively based approach, which seems – at least at first sight – to be that of theorists in the 'empirical' tradition which culminated in Aristoxenus. If that reading of the relevant lines is correct (which will turn out to be questionable), and if his arguments hold water, he will have undermined the basis of every variety of harmonic analysis known in his time.[7] His critique seems to have made little impact on later harmonic scientists. Porphyry, who was clearly impressed by its reasoning, is the only writer who mentions it; and when he says that though many others agree, 'I cannot list them all by name since I am not in possession of their writings,' and that 'Theophrastus will be adequate to stand for me in the place of them all' (1–3), one cannot help suspecting that the 'many others' are

in the windpipe from which an impulse of breath is thrown towards the mouth, and the point at which the impulse 'strikes' the outer air. The latter point is fixed; the former is shifted closer to it or further away through the agency of our 'ruling principles', which 'find and grasp astonishingly and easily, as though with a bridge [i.e. as on a monochord], the places on the windpipe from which the distances to the outer air will produce differences of sounds in proportion to the amounts by which the distances exceed one another'. It is possible that Ptolemy found this theory, like much else in *Harm.* 1.3, in a work of the fourth or third century BC; but there is nothing to prove it.

[7] It is a great pity that his *Harmonics* is lost. Unless it consisted wholly of criticisms of existing approaches to the science, we might expect it to have sketched out a mode of analysis in which quantitative descriptions and comparisons of intervals play no part. The state of the evidence, unfortunately, offers little scope for informed speculation about its procedures or conclusions, or about the terms in which its analyses were expressed. I make a few tentative suggestions at the end of this chapter. For an interesting attempt to set Theophrastus' views on music within a wider philosophical framework see Lippman 1964: 157–66.

a mirage.[8] The passage is nevertheless important. It shows that the whole project of harmonic science, as it stood at the end of the fourth century, could be subjected to critical attack; its status as a reputable discipline was by no means assured. Even if Theophrastus' assault was unprecedented and waited six hundred years to find its first enthusiastic convert, it clearly deserves our attention.

Theophrastus' arguments are complex and often obscurely expressed; there are uncertainties both about their meaning and about the exact nature of the views which some of them are designed to refute. Along with several other commentators, I have tackled these and other minutiae elsewhere, and though we shall have to grapple with some of them in the course of this chapter, a good many will be elided.[9] Instead of following the passage through in precisely Theophrastus' order, I shall consider it under four headings, looking first at one group of arguments against mathematical theorists, then at another, thirdly at arguments apparently directed against an empirical, perhaps Aristoxenian approach, and finally at such hints as we have about Theophrastus' own views, most of which appear in the text just before the end of the second group of arguments against the *mathēmatikoi.*

ARGUMENTS AGAINST MATHEMATICAL THEORISTS: THE FIRST PHASE

The whole of Theophrastus' reasoning seems to be premised on the assumption that attributes fall into two types, quantitative and qualitative, and that these categories are mutually exclusive.[10] The position he initially attacks is encapsulated in the thesis attributed to 'some people' in lines 15–16, that 'music consists in quantity, since the differences exist on the basis of

[8] There are some affinities with Theophrastus' position, perhaps even some distant echoes of it, in a passage from an otherwise unknown mathematical writer, Panaetius the Younger, quoted by Porphyry immediately after the Theophrastan fragment (*In Ptol. Harm.* 65.21–67.10). But Panaetius' views seem a good deal less radical than Theophrastus'.

[9] See Barker 1978b and 1985, and the notes to Barker 1989a: 111–18. There is a valuable, more recent discussion (which includes trenchant comments on some of my earlier opinions) in Sicking 1998. For an older, more summary but still useful account see Lippman 1964: 157–61; cf. also Gottschalk 1998: 293–5.

[10] The terms regularly used in later writings to mark this contrast, *posotēs*, 'quantity', and *poiotēs*, 'quality', are apparently fourth-century philosophical coinages. In Aristotle the former occurs frequently, the latter less often; at *Theaetetus* 182a8–9 Plato indicates that *poiotēs* is an unfamiliar, 'uncouth' expression, perhaps one that he himself has just invented. In the present passage Theophrastus uses *posotēs* freely, but *poiotēs* does not appear; instead we find *idiotēs*, meaning something like 'characteristic property', and designating e.g. the redness of red things or the fluidity of liquids. This word too is rare before the third century (it occurs only once in Plato and not at all in Aristotle), and no earlier writer uses it as Theophrastus does, to contrast with *posotēs* and to refer specifically to non-quantitative attributes.

quantity'. These 'differences', as the context makes clear, are differences in pitch; and the issue on which Theophrastus focuses is whether, when one note is higher or lower in pitch than another, they differ quantitatively.[11]

The mathematical theorists' view leads, so he argues, to absurd consequences. Suppose first that the difference between a high note and a low one is a difference of quantity, just that and nothing else. In that case a difference of quantity, simply as such, must amount to 'a melody or part of a melody' (16–20). It will follow that any two things that differ in quantity, two different-sized patches of colour, for example, will differ in precisely the way in which two notes do, and they too will form 'a melody or part of a melody, if indeed melody and interval are number, and if melody and the [kind of] difference involved in it exist because of number' (20–3). In general, on this hypothesis, 'everything numerable would participate in melody to the extent that it does in number' (25–6). If, for example, I have twelve apples, nine peaches, eight pears and six bananas in my fruit-bowl, it would not merely be the case that the apples:peaches:pears:bananas ratio-pattern mirrors that of an octave divided into two fourths and a tone, but that I have a bowlful of music, humming along on the four fixed notes of a familiar attunement. Which is, alas, ridiculous.

Pitch-differences, then, cannot be merely differences of quantity. Theophrastus now shifts the angle of his attack, homing in on the more plausible idea that they might be quantitative differences attached to subjects of a particular sort, those that are sounds or notes. Suppose, he says, that 'plurality belongs to notes in the same way as it does to colour', in the way, that is, in which the redness of a set of red patches is one thing and anything quantitative about them (such as their number or size) is something quite different. On this approach 'a note is one thing and the plurality related to it is another' (26–7), and we are presented with two possibilities. One is that a high note and a low note differ 'as notes', that is, in some aspect of the feature that makes them notes rather than anything else, just as the colours of a geranium (red) and a delphinium (blue) differ in some

[11] He expresses the view he is criticising in various ways, representing it sometimes as saying that the difference 'depends upon' quantity, sometimes that it 'is' a quantity, sometimes that the items in question differ 'in quantity' (all these locutions can be found in lines 16–21). I do not think that the various forms of words mark significant differences in meaning. Nor do I agree with Sicking 1998: 112, that the shift of terminology in lines 22 ff., where he begins to speak of the thesis that an interval (*diastēma*) is a number (*arithmos*) or a plurality (*plēthos*) indicates that his attention is now directed to a different theory. Lines 17–26 form a single, continuous argument, and despite the variations in Theophrastus' formulations its target is the same throughout. Lines 26–49 pick up the point made in 17–26 and develop out of it a new line of criticism; but here again there is still (in my view) no change in the position under attack. Some reasons for adopting this interpretation will be offered below.

aspect of the feature that makes them colours. Alternatively, they differ 'in respect of their plurality'; their pitch-difference is a relation between quantitative attributes that they possess over and above the qualitative *idiotēs* which constitutes them as notes or sounds (27–9).[12]

These alternatives provide Theophrastus with the makings of a dilemma, on one horn or the other of which the theorists under attack must, so he argues, impale themselves. Suppose first that differently pitched notes differ 'in plurality, and the higher is as it is by being moved in respect of more numbers, the lower by being moved in respect of fewer'. But 'every sound is such as to be grasped either as high or as low, for every sound is higher than some other and lower than some other'. That is, the possession of pitch is essential to anything's being a sound; and if the characteristic of pitch is 'removed', by being treated as a quantitative attribute over and above the specific *idiotēs* of sounds and notes, 'what would be left over,' Theophrastus rhetorically demands, 'to make it a sound?' (29–36). We cannot abstract a sound's pitch from its character as a sound, in the way that we can abstract the size of a colour-patch from its character as a colour, for example; there will not be enough left of the subject to constitute it as a sound at all.

We move, then, to the second horn of the dilemma, and posit that high notes and low ones differ from one another 'as notes'. Here the argument is straightforward, since in that case 'we shall no longer have any need of plurality, for their own intrinsic difference will be sufficient by itself to bring melodies into being'. We are no longer supposing that pitch-differences depend on plurality or quantity; they depend, *ex hypothesi*, on the *idiotēs* of the sounds, and one note will differ from another in pitch just as one colour differs from another in hue, regardless of the quantities involved. If equal amounts of white and black are brought together, their 'numbers' do not differ, but their distinct characters as black and white remain. Just so, on this hypothesis, any quantitative attributes that notes may acquire on particular occasions are irrelevant to their intrinsic characters as high-pitched or low (36–46). Thus if we choose the first horn of the dilemma we reach an impossibility, while if we choose the second the quantitative account of pitch evaporates.

These intriguing and ingenious arguments are scarcely watertight, but it is not my business to criticise them here. It is more important to try

[12] Theophrastus shifts back and forth in this context between speaking of notes (*phthongoi*) and of sound (or 'vocal sound', *phōnē*). It is clear that he intends no distinction that is relevant to his argument, since he treats a note, for present purposes at least, simply as a pitched sound, and asserts at 32–3 that every sound (*phōnē*) is higher than some other and lower than some other; every sound, in short, or at least every sound that he is considering, is a pitched sound and hence a note.

to identify the thesis they are intended to demolish. They characterise it as one that makes pitch-differences depend on quantity or plurality or number; and these various, apparently interchangeable forms of expression, together with the very abstract character of the reasoning, suggest that the position under attack is conceived in rather broad and general terms. It is, perhaps, any kind of acoustic theory whatever that might serve to underpin the representation of pitch-relations as ratios of numbers. As he comes to the end of this stretch of argument, however, Theophrastus seems to indicate that he has either one quite specific version of such a theory in mind, or two, depending on how we read his remark. Looking back with evident satisfaction at the results of his reasoning, he concludes: 'thus the high-pitched sound does not consist of more numerous [parts] or move in accordance with more numbers, and neither does the low one' (46–7).

I have supplied the word 'parts' in the first half of this sentence, but it is hard to see how anything else could be meant. If it is correct, the thesis that is being denied must be that differently pitched notes are composed of different numbers of elements; and the only theory we know that carries such an implication is the one enunciated in the introduction to the *Sectio canonis*, where each note is formed from a number of impacts on the air, and a higher pitch arises from a more rapid and closely packed sequence of such impacts. We need therefore to ask whether Theophrastus is alluding to a different theory in the second half of the sentence, where he denies that a higher-pitched sound 'moves in accordance with more numbers', or is merely referring to the same one again in different terms. If Porphyry had broken off his quotation here, we could have been forgiven for supposing that the phrase refers, though rather obliquely, to the best-known theory of pitch current in the period, the theory adopted by Plato and (with some modifications) by Aristotle, that higher pitch is correlated with the greater speed of a sound's transmission through its medium. 'A higher-pitched sound moves in accordance with more numbers' might plausibly be taken as a roundabout way of saying 'a higher-pitched sound moves faster'. But in fact this cannot be so, since later in the fragment (lines 101–3) the thesis that a lower note 'moves in accordance with fewer numbers' (and by implication, that a higher note moves in accordance with more) is quite explicitly distinguished from the view that 'the high note is distinguished by its speed'; and the two are refuted separately and by different arguments.[13]

[13] Hence I cannot agree with Sicking 1998: 124 (cf. 115), when he asserts that the expression 'moves in accordance with more numbers' 'covers both the vibration theory and the speed theory', or at any rate not in the present context.

Since no third interpretation immediately suggests itself for the phrase 'moves in accordance with more numbers', we might conclude that like the phrase 'consists of more numerous [parts]', it must allude to a theory of the sort spelled out in the *Sectio*; and we might go on to revise our reading of the whole sequence of arguments we have just reviewed. Perhaps it is aimed exclusively at a thesis of that one particular type. But this interpretation also fails. If the next stretch of argument is directed against any one specific theory, it is certainly not that of the *Sectio*, as we shall see. Yet the formula 'moves in accordance with more numbers' reappears again during the course of it (line 69). Further, the way in which Theophrastus negotiates the transition from the arguments we have reviewed into the next, at line 50, clearly indicates that he does not mean us to think that the object of his criticism has changed. Lines 46–9 read as follows. 'Thus the high-pitched sound does not consist of more numerous [parts] or move in accordance with more numbers, and neither does the low one; for one can say this of the latter as well as the former, since there is a characteristic magnitude (*megethos*) that belongs to a low-pitched sound.' Line 50 introduces the new phase of the critique by saying: 'This is clear from the force exerted when people sing.' Hence the contention that a sound of any pitch has a 'characteristic magnitude' is used, in successive lines, first to help undermine the thesis that either high-pitched or low-pitched sounds possess more numerous parts or move in accordance with more numbers, and then to show the untenability of a theory which cannot be identical with the *Sectio*'s.

I shall argue that this new theory is not the 'speed' hypothesis either. But whether it is or not, the points I have raised suggest that the best interpretation of Theophrastus' strategy, both in the first phase of his polemic and (in view of the form of transition) in the second, is, after all, the one I initially sketched. Though at certain points in the discussion he is apparently thinking more of one type of theory than of another, he intends each of his arguments to cut against all of them equally.

ARGUMENTS AGAINST MATHEMATICAL THEORISTS: THE SECOND PHASE

There are arguments of two main sorts in this part of the text (50–80, together with a few lines combining criticism with positive theorising at 100–7). Both are essentially quite simple. In the first (50–64), Theophrastus draws attention to observable facts (or alleged facts) about the power or force required to produce high and low pitches by various different means. He argues that in singing and in playing a stringed instrument, the quantity of power needed to produce a note is the same regardless of its pitch, while in playing a wind instrument it is the lower pitch whose production calls for the greater power. (This last observation provides him with appropriate ammunition because all fourth-century theories of the type under examination associate greater numbers or quantities with higher pitches.) The second argument focuses on the phenomenon of concord. Summarily, if two notes form a concord they blend seamlessly together, and what we hear is the product of their fusion (see pp. 344–5 above); but if one were more forceful than the other it would overpower the weaker and would 'appropriate the perception for itself', so that there could be no experience of a fully blended, unified concordance (64–80).

In both of these arguments the larger numbers or pluralities or quantities of the earlier passage have been replaced by greater power (*dynamis*) or force (*bia*). If any specific theory is now under assault, it is therefore one that correlates higher pitch with the greater vigour of the action needed to produce it, and hence, we may infer, with greater vigour in the movement that constitutes the sound itself. This inference is supported by remarks in the second part of the argument. The theorists in question say, we are told, that 'the higher note is heard at a greater distance, through its travelling further because of the sharpness (*oxytēs*) of its movement, or because of its arising as the result of plurality' (69–72); and from the fact that it is capable of travelling further it is argued that it must be more vigorous (*sphodroteros*, 75–6).

We might expect Theophrastus to say that according to this theory, the higher note travels more swiftly, but in this phase of the discussion he does not. It is of course very probable – indeed virtually certain – that any theorist who associated higher pitch with greater force and power would link it also with greater speed, but we should not ignore the fact that Theophrastus has chosen, so far, to suppress that aspect of the matter, and so to obscure any connection between this theory and the views of Plato and Aristotle. (Plato says nothing about force or power in the relevant passage at *Timaeus* 79e–80b. Aristotle denies the existence of a direct correlation between higher pitch and greater force, in a complex argument at *De gen. an.* 786b–787a, and unless Theophrastus has misunderstood him badly, it cannot be his position that is under attack here.[14]) The only known fourth-century writer who correlates higher pitch directly and explicitly with the greater force exerted by an agent and with the strength of a sound's movement through the air is Archytas, in frag. 1 (see pp. 27–9 above); he also connects it with

[14] For the contrary view see Sicking 1998: 120. It may be my fault, but I fail to follow his reasoning.

greater speed, but speed figures much less prominently in his argument. In other respects too, Archytas' thesis fits smoothly into the target-area of Theophrastus' criticisms. He insists, for example, and Theophrastus denies, that a higher-pitched sound is heard at a greater distance because it travels more strongly and vigorously; and he offers an account of the forces involved in playing high and low notes on the aulos which Theophrastus confronts point by point.

If Theophrastus is deliberately criticising any one particular account of the phenomena, then, it must be that of Archytas (unless, of course, he found it in a source unknown to us), and I have no real doubt that the passage we know as Archytas fragment 1 is at the centre of his attention. But once again there are signs that he intends his comments to have wider application. We saw that the language of the first phase of his argument (lines 17–49) points the finger at a theory akin to that of the *Sectio*, if anywhere; and it certainly does not suggest the Archytan hypothesis. Yet two of its key expressions recur in the present passage, as if the criticisms developed here apply to the thesis or theses they designate just as much as they do to Archytas' position. Theophrastus' contention that two notes which jointly form a concord must be equal in power prompts his rhetorical query, 'for if the high note moved in accordance with more numbers, how could consonance come about?' (66–9, cf. 100–2); and the thesis that a higher note arises 'as a result of plurality' is represented as an alternative to the hypothesis that its movement is 'sharper', one that has relevantly identical implications and is refutable in the same way (69–78). It apparently makes no difference whether the position under bombardment is designated in ways that bring to mind Archytas' approach or some other; the same arguments will demolish it regardless of the language in which it is dressed.

If we suppose that Theophrastus was concerned to distinguish carefully between different types of mathematical theory and to refute them one by one, we shall be hard pressed to explain his treatment of the 'speed' hypothesis adopted, in one variant or another, by Plato, Aristotle and many later writers. It must have been very familiar to members of the Lyceum and associated intellectuals who took an interest in such matters, and it bore the *imprimatur* of the period's most respected philosophical heavyweights. As the dominant hypothesis in the field, it merited careful and extensive critical attention from any reputable opponent, certainly no less than the Archytan theory of force, and only secondarily speed, which it had largely displaced, or the theory of multiple impacts. Yet it surfaces explicitly in Theophrastus' polemic only once, and is dismissed in two short sentences. 'But neither can the high note be distinguished by its speed, for then it would occupy the hearing first, so that a concord would not arise. If it does arise, both [the high note and the low] are of equal speed' (103–5).

It is also relevant that at this point Theophrastus moves to an apparently much more general conclusion. In the preceding lines he has argued first that a lower note does not 'move in accordance with fewer numbers' (100–3), and then, in the sentences quoted above, that a higher note is not 'distinguished by its speed'. 'Hence,' he concludes, in a statement evidently designed to embrace both formulations, 'it is not some unequal numbers that give the explanation of the differences' (105–6). Similarly, as he reaches a peroration near the end of the fragment, he sums up his findings about mathematical theories by saying 'nor are the numbers causes, by the notes' differing from one another in quantity' (125–6). These generalised and capacious formulations seem designed to capture any theories whatever of the mathematical sort. The distinctions between them are unimportant, since each of the arguments that Theophrastus levels against them is intended to be sufficiently broad and flexible to count against them all. In that case he had no need to direct a special barrage of criticism against the 'speed' hypothesis, for all its prominence in contemporary thought. It had been eliminated along with the others, by the cumulative onslaught of his whole succession of arguments. The fact that with one brief exception he avoids formulations that would draw particular attention to Aristotle's theory might possibly be explained in personal or social terms, as prompted by *pietas* or prudence. Since his strategy did not require him to tackle head-on the views of his senior colleague and friend, even though he clearly disagreed with him, he chose to present his critique in a less obtrusively provocative guise. It would be pleasant to believe such a story, but we know almost nothing of the roles, if any, that were played by deference and tact in Aristotle's circle. It is only an agreeable speculation, though it might, after all, be true.

AN ARGUMENT AGAINST ARISTOXENUS?

The short passage we shall consider here (108–25) comes near the end of the fragment, when Theophrastus' critique of mathematical theories is complete, and immediately after an exposition of some positive views of his own which we shall examine in the next section of this chapter. It returns to the critical mode, but it is clearly directed at a new set of opponents. They too construe the relations between pitches in quantitative terms, but not in the manner of those who represent them as ratios. These people assert, Theophrastus says, that the causes of pitch-differences and of *emmeleia*, 'melodiousness', are intervals, *diastēmata*; and when the argument is over he sums up the conclusions of the whole passage preserved in the fragment by saying that the *diastēmata* are not the relevant causes, and 'neither are the numbers causes, by the notes' differing from one another in quantity'. Since this latter statement plainly encapsulates his conclusions about all mathematical theories, the thesis about *diastēmata* must come from a different stable.

Previous commentators, among whom I include myself, have generally identified the view criticised here as that of Aristoxenus, or of the empirically minded predecessors he calls *harmonikoi*, or perhaps of both. I shall argue, however, that unless Theophrastus' comments are very far from the mark indeed, this interpretation needs to be radically qualified. It is of course very tempting to assume that the theory comes from Aristoxenus or the earlier empiricist tradition, since we know of no other school of thought that could appropriately be contrasted with that of the number-theorists; and I accept that in a certain sense the assumption holds. But I shall try to show that Theophrastus' target cannot be Aristoxenus, and that although he may well have believed that he was engaging with the views of earlier empiricists, he relied on only one source for his information about them, one that presents a very peculiar and probably a misleading picture of their approach.

The passage begins as follows. 'Nor is it the intervals (*diastēmata*), as some people say, that are the causes of the differences and hence their principles (*archai*), since even when these are left out the differences remain' (108–9). The 'differences', here as in the earlier arguments, are those between differently pitched sounds or notes; and one might offer Theophrastus' statement a straightforward and moderately plausible interpretation. The distinct pitch of a note cannot be dependent on the intervals or 'distances' by which it is separated from other notes, since it retains that pitch even when it is sounded in complete isolation and all such *diastēmata* are 'left out'. But as the passage unfolds it becomes obvious that this cannot be what Theophrastus means.

For when something comes into being if certain things are left out, these are not the causes of its existence, not [that is] as productive causes, but [only] as not preventing it. For neither is the unmelodic (*ekmeleia*) a cause of the melodic (*emmeleia*), merely because the melodic would not come into being unless the unmelodic were rejected . . .[15] so that neither are the intervals the causes of melody as producing it, but as not preventing it. For if someone were to utter in addition

[15] The sense of the lines omitted here (112–15) is uncertain and the integrity of the text is debatable; see Sicking 1998: 130. On any reading, however, they will add little or nothing to the argument.

the continuous series of intervening positions, wouldn't he emit an unmelodic sound? So while the unmelodic would arise if these were not rejected, the melodic does not arise from their being left out, just because they would prevent it if they were not left out. (111–19)

The abstract theme running through this argument is that if something, X, cannot come into being unless certain other things, Y's, are left out or rejected, then the Y's are not the causes of X. That is, they are not its 'productive' causes, the items that actively bring X into existence; they are causes only in the very etiolated sense that so long as they are left out, they do not prevent its occurrence. The last sentence seems to make a rather stronger point. The fact that the omission of the Y's makes it possible for X to arise does not entail that their omission causes X's existence. Theophrastus apparently means that their omission is merely a necessary condition of X's coming into being, and that something else is needed as a 'productive' cause, to ensure that it actually arises.

In the case under scrutiny, the things that must be left out are the *diastēmata*. Since it is clear that Aristoxenus conceived intervals merely as empty spaces between differently pitched notes, and it seems likely that his predecessors did so too, the sense in which they must be omitted does not leap to the eye. Fortunately, Theophrastus explains what he means quite lucidly in the penultimate sentence. Melody and 'the melodic', *emmeleia*, can exist only when notes are distinctly pitched and the differences between them are clear. If a singer produced audibly not only the relevant notes but also 'all the intervening positions', the notes' pitches would be hopelessly obscured and the result would be unmelodic. What has to be left out between any two musical notes is not, then, the intervening *diastēma* conceived as an empty space, but what we might call its 'content' or 'potential content', that is, the continuous gradient of pitched sound by which the space could be filled.

But it is hard to find a theorist in the empirical tradition who identifies the *diastēmata* with their potentially audible content. Certainly Aristoxenus does not. On his account, 'a *diastēma* is that which is bounded by two notes which do not have the same pitch. To put it straightforwardly, an interval presents itself to perception as a difference between pitches, and a space capable of receiving notes higher than the lower of the pitches that bound the interval and lower than the higher' (*El. harm.* 15.24–31). It is a 'space capable of receiving notes', a *topos dektikos phthongōn*, and it is absolutely not a sound or a continuum of audible pitches. Further, in an elaborate passage at *El. harm.* 8.13–10.20 (see pp. 143–5 above), Aristoxenus argues at length for precisely the thesis that Theophrastus is asserting here, that in singing melodically the voice must sound only the notes bounding the intervals, and must pass silently across the intervals themselves. It therefore seems much more likely that Theophrastus is borrowing from Aristoxenus to fuel his polemic than that he is attempting to criticise him; and in due course we shall find other indications too that he is in Aristoxenus' debt. Nor does Aristoxenus suggest, as those attacked by Theophrastus apparently do, that the *diastēmata* are to be thought of as the 'causes' of differences in pitch. They are merely the spaces between pitches, and play no active role in bringing them into existence. Aristoxenus seems wholly uninterested, in fact, in causal issues of the sort that Theophrastus is addressing.

Nothing in Aristoxenus encourages the view that the predecessors he discusses adopted the strange position conjured up by Theophrastus, and neither does anything said by Aristotle about exponents of harmonics in its empirical form. Neither author tells us straightforwardly that these *harmonikoi* construed intervals precisely as Aristoxenus did, as gaps that could be filled with sound but considered simply as intervals are not; but equally they say nothing to suggest that their view was relevantly different. Even when Aristotle writes of those who treated certain tiny intervals as minimal, and regarded them as constituting units of which all other intervals are measurable as multiples, he does not say that they did so because they thought that there was no space left between their boundaries and that they were 'filled up' with pitched sound; so far as we can tell they were envisaged merely as the smallest 'spaces' or 'distances' that the ear could reliably identify.[16]

We need not suppose, however, that Theophrastus has arbitrarily fathered a ridiculous and imaginary theory on the unnamed objects of his criticism. There is one passage in earlier literature which does indeed seem to allude to a theory of this sort, and though it stands quite alone in this respect, it gains weight and apparent reliability from the high profile of the text and its author. It is in a part of Plato's *Republic* to which we have already returned more than once, in which Glaucon describes the activities of certain theorists to whom he mistakenly supposes that Socrates has just referred (this is the account subsequently embroidered by Socrates with his gruesome images of strings tortured on the rack, beaten with the plectrum, and so on). Glaucon describes these people as behaving absurdly,

[16] See e.g. Aristotle, *Metaph.* 1087b33–1088a7, where Aristotle says that such a *metron* may be indivisible either in form or 'in relation to perception'. There is no sense in which the small interval he calls the *diesis* is indivisible in form, and he must be treating it as an instance of the second sort; see pp. 349–50 above.

naming certain things *pyknōmata* and intently inclining their ears as if trying to detect a voice from next door, some of them saying that they can still just hear a sound in the middle and that this is the smallest interval, by which measurement should be made, while others disagree, claiming that the notes are already sounding at the same pitch; and both are putting their ears ahead of their mind. (*Rep.* 531a4–b1)

These people's quest is for 'the smallest interval, by which measurement is to be made', a project which we discussed in Chapters 1 and 2. Their procedure is to adjust the tensions of two strings until their pitches are so close together that any further adjustment would be heard as bringing them into unison; when that situation is reached, they suppose, the smallest *diastēma* has been found. It is thus to be identified by ear, not on the basis of any kind of argument; and it will therefore be the smallest that human hearing can detect, rather than being the smallest absolutely, in some theoretical sense. This is no doubt as it should be, since the crucial task is to locate an interval which harmonic scientists can pick out directly and use in practice as the unit of measurement.

All this makes sense of a sort, and it locates these theorists squarely in the empirical tradition to which Aristotle and Aristoxenus also refer. But there is one detail in Glaucon's depiction which marks it off from the others. Instead of representing the first group of these disputing investigators as saying that they can still just detect a gap or a space between the notes that the strings emit, he attributes to them the claim that they can still 'just hear a *sound* in the middle', where by 'in the middle' they must mean 'in between the notes in question'. Having said this, they add 'that this is the smallest interval, by which measurement should be made', where 'this' apparently refers to the sound just mentioned.[17]

It seems very strange that they should be claiming to detect a sound (*ēchē*) and not a gap, and if they are indeed identifying this sound itself as the 'smallest interval', that seems stranger still. But if Theophrastus had this passage and only this passage in his mind when he wrote his critique,[18]

[17] One can understand the statement in a slightly different sense, 'that this interval is the smallest . . .'; but on the syntactically most natural reading, 'this interval' will again refer back to the sound. The only way of escaping from this conclusion is to suppose that 'this' or 'this interval' refers to nothing explicitly mentioned before, and simply indicates the interval to which their manipulation of the strings has led them. That is possible, though more awkward. But whichever sense Plato intended, the puzzling reference to the sound remains; and if I am right in thinking that Theophrastus extracted the theory he attacks from this passage, he at least must have interpreted it as identifying the interval (*diastēma*) with the sound.

[18] There is one other passage which refers to a sound that lies between the boundaries of the minimal interval; according to Aristotle at *De sensu* 445a2, 'the note inside the diesis is undetectable'. But this does not carry the same implications as Plato's formulation; in fact it has quite the opposite sense, since Plato's theorists claim to be able to detect the 'sound in between'. The note (*phthongos*) in

he might well have taken it to mean that such theorists supposed the notes bounding an interval to be held apart, as it were, by the sound that occupies the space between them, and that it is this sound, and not an empty space, that constitutes the interval between them. If there is a sound between them, they are distinct notes, and if not, they must be identical in pitch. In that sense this 'sounding interval' can be thought of as the 'cause' of the notes' distinct pitches, as Theophrastus says. It is worth noticing that the word by which Glaucon refers to this sound, *ēchē*, is not the one normally used elsewhere for a musical sound (commonly *phōnē*) or a note (*phthongos*). Its range of application is rather more general than theirs, and can include any kind of audible noise. Theophrastus could plausibly have construed it here as indicating the 'unmelodic' continuum of pitch that makes up the 'content' of an interval.

In that case Theophrastus' criticisms will hit their target squarely enough. The target itself is extremely odd; but it would not really be surprising if, as I am suggesting, he extracted his picture of these theorists' conceptions from this passage and no other. Aristoxenus' discussions of them (assuming that they had already been written and had come to Theophrastus' attention[19]) offer no detailed analysis of their notion of a *diastēma*. Aside from Plato in the *Republic*, the only earlier writer to mention exponents of the empirical branch of harmonics is Aristotle,[20] and his allusions, again, are too slight and allusive to present a critic with a clear view of his quarry. One might expect such a critic to look for living examples of the approach he is criticising, not merely for other writers' reports about them. But it is entirely possible that no theorists of the empirical persuasion were at work in Theophrastus' own time, until their approach was revived and revolutionised by Aristoxenus.

question is not identified with the interval; the point, as Aristotle goes on to explain, is merely that we cannot identify the interval between any such note and either of the boundaries of the diesis; cf. pp. 349–50 above. There is nothing here to encourage an interpretation of the sort that Theophrastus indicates.

19 No firm conclusions can be drawn about the chronological relations between Theophrastus' *On Music*, from which the fragment is taken, and the various books of Aristoxenus' *Elementa harmonica*. It seems certain that Theophrastus knew something of Aristoxenus' work, both for the reason given on pp. 423–4 above, and because elements of his terminology and conceptual apparatus seem to echo those of Aristoxenus; I shall note a few of them below. It does not necessarily follow (though it seems likely) that he had read any part of the *El. harm.*, or even that it had been written; he may, for instance, have engaged in informal discussions with his colleague in the Lyceum or attended some of his lectures. Sicking 1998: 135–8, argues that Theophrastus' critique is directed against the Aristoxenus of *El. harm.* Book I, and that Book II, which was written later, incorporates an attempt to evade Theophrastus' objections. As I have explained, I do not think that the first part of this hypothesis holds water; and it seems to me, more generally, that Sicking's arguments cannot survive detailed exposure to Aristoxenus' text.

20 One should perhaps add the writer of the fragment preserved in *P. Hib.* 1.13 (see pp. 69–73 above); but it says nothing to the point.

The fictional drama of the *Republic* projects its conversations back into the fifth century, and though the dates of Aristoxenus' *harmonikoi* cannot be conclusively fixed, most of them, too, seem to belong to that now distant period. In attempting to reconstruct their position, then, Theophrastus had precious little evidence to go on, and so far as the issues that interested him are concerned, much the fullest and most detailed account available was the *Republic*'s. His reliance on it is perfectly understandable.

His reconstruction hangs, however, on a strenuously literal interpretation of a single, seven-word phrase in Glaucon's speech. The thesis he extracts from it is justifiable to the extent that it does not distort the plain meaning of the text; but it is so bizarre that it is hard to believe that anyone really subscribed to it. We have absolutely no independent evidence that they did. There is plainly some cause for suspicion that when he was writing this colourful and rather over-excited passage, Plato's attention to detail wandered; or perhaps he inserted the slip deliberately, either to make the position he was satirising seem even more absurd, or to underline the muddle-headedness of his unfortunate brother Glaucon, Socrates' willing but confused respondent. However that may be, it seems to me overwhelmingly probable that the doctrine which Theophrastus attacks is a construct developed wholly out of Glaucon's speech, and has no more substantial claim to historical reality than that.

If I am even partly right in my reading of this passage, and even if I am wrong in connecting it so closely to the account in the *Republic*, there is no point in Theophrastus' polemic at which he takes issue with Aristoxenus. This seems remarkable, since Aristoxenus' picture of the relations between pitches is unambiguously quantitative in a sense to which Theophrastus might be expected to object. He discusses, for instance, the 'size' of the perfect fourth, and quantifies the intervals inside tetrachords in each genus in terms of tones, their multiples and their fractions. It makes no difference that in the second and third books of the *El. harm.* Aristoxenus emphasises the need to identify musical intervals, in the context of harmonic science, by reference to the *dynameis* of their bounding notes and not by their sizes. Every actual instance of such an interval still has a determinate size, and any two different pitches (as distinct from musical notes) are still separated by a measurable 'distance'.[21] If Theophrastus wishes to demonstrate that

[21] The Aristoxenian passages cited by Sicking 1998: 127, in support of his suggestion that *El. harm.* Book II is (in part) a response to Theophrastus' criticisms, are unquestionably important, but they do nothing to eliminate this difficulty. We may note, *inter alia*, that the two examples of Aristoxenian quantification which I mentioned above are both included in Book II.

any attempt whatever to quantify relations between pitches is inapposite and misguided, why does he not protest?

The answer, I think, is that he understood Aristoxenus well enough to grasp that a work like the *El. harm.* in fact offers no purchase to his criticisms. Theophrastus' arguments against theories underpinning mathematical harmonics focus on alleged defects in their accounts of the physical basis of pitch, and seek to show that pitch-differences cannot be quantitative when considered as attributes of movements in motion, as they really are 'out there'. His attack on the theory of *diastēmata* seems to be similarly motivated; its essential thesis is that these 'intervals' cannot be the *causes* of differences in pitch. But Aristoxenus, as we know, insists that he is not in the least concerned with such issues. His business is with melodic phenomena as they present themselves to perception, *kata tēn tēs aisthēseōs phantasian*, and it simply does not matter to him how questions about the underlying causes of perceived sounds and their attributes are to be answered.[22] Theophrastus recognised, I suggest, that this way of demarcating the scope of his discussion places Aristoxenus' theories firmly outside the battlefield upon which he himself is embroiled. Aristoxenus contends only that pitched sounds present themselves to the hearing as items separated by quantifiable distances. Underlying or causing the pitches of the sounds we perceive there is, no doubt, some variable attribute of objectively real events; its variations may be quantitative and the differences between its values measurable either as ratios or as quasi-linear distances, or again they may not. But about such matters Aristoxenus' thesis implies nothing at all, and his work is in this respect as irrelevant to those who debate them as theirs is to him.

Theophrastus keeps the spotlight on the doctrine about *diastēmata* for another six lines after the point we have reached (120–5). They introduce no fresh varieties of criticism, however, and the only new ideas they put forward will be more appropriately considered in the course of the next section, where certain ingredients of the preceding argument will also be briefly revisited.

THEOPHRASTUS' EXPOSITION OF HIS POSITIVE VIEWS

We pick up the text at the point where Theophrastus has just rebutted the thesis that it is because higher-pitched sounds move with more vigour and power that they can be heard at a greater distance from their origin than

[22] See particularly *El. harm.* 9.3–11, 12.4–19; cf. 32.20–8 and pp. 141–2, 166–8 above.

lower ones (69–80). He accepts that such sounds are indeed audible from further away, but he explains the phenomenon differently.

But since concordance exists, displaying the equality of the two notes, there is equality in their powers, differing in the specific quality (*idiotēs*) of each of them. For what is higher [lit. 'sharper'] is by its nature more conspicuous, not stronger, and that is why it is apprehended at a greater distance than the lower [lit. 'heavier'], just as white is by comparison with any other colour, or as is anything else that is more strongly apprehended not because the other is less what it naturally is, nor because it does not move in accordance with equal numbers, but because perception focuses more on the one than on the other on account of its unlikeness to its surroundings. Thus the low note penetrates too; but the hearing grasps the high note more readily because of its specific quality (*idiotēs*), not because of the plurality it contains. (78–87)

High notes, then, differ qualitatively from low ones; and the qualitative *idiotēs* of a high note makes it stand out in our perceptual field more vividly than a low one, just as a brightly coloured parrot shows up more conspicuously than a dull brown thrush. That, Theophrastus contends, is enough to account for the perceptibility of higher notes at greater distances without resort to quantitative considerations. But he now seems to make a concession to his opponents, allowing that it may after all be the case that a higher note travels further. If it really does so, however, 'it is not because it is moved in accordance with more numbers, but because of its shape, since a high-pitched sound travels more forwards and upwards, while a low one travels more equally all about' (87–90).

The concept of a movement's shape reappears in a few other writings on acoustics, where it plays various roles, not always in connection with differences of pitch;[23] but I have found no examples earlier than Theophrastus. The link that he makes between shapes and pitches may have been prompted, in part, by the root meanings and associations of the terms designating high pitch and low, *oxytēs* (sharpness) and *barytēs* (heaviness). These attributes, as they exist in sounds, seem to be analogous, says Aristotle, 'to the sharp and the blunt in the field of touch, for the sharp pierces, as it were, and the blunt pushes . . .' (*De an.* 420b1–3). Rather similarly, the writer of *Problem* XIX.8 asserts that a low-pitched, 'heavy' sound is larger and is like an obtuse angle, while a high-pitched, 'sharp' sound is like an

[23] According to Priscian, Theophrastus made a broader connection between sound and shape, asserting that hearing (and so, presumably, sound of any sort) occurs when the air is 'shaped' (*Metaphrasis* 1.30, Theophr. frag. 277A Fortenbaugh); see Gottschalk 1998: 294. (The miniature controversy between Gottschalk and myself mentioned in his n. 29 need not concern us here.) For later instances of connections between sound and shape see [Ar.] *Problemata* XI.20 and 23, XIX.8, [Ar.] *De audib.* 800a, Ptol. *Harm.* 7.8–15.

acute one. A good many other writers too seem to treat the meanings of these words in their tactile contexts as reliable guides to the nature of the acoustic properties to which they are transferred.[24]

But Theophrastus does not argue for a connection between the pitches of sounds and the shapes of their movements on the basis of these semantic associations, which he nowhere makes explicit. Instead, having stated his thesis in lines 87–90, he goes on to present evidence of its truth drawn from empirical observation. If you sing a high note and then a low one while touching your ribs with your hand, he says, you will feel the movement of the low note more in the periphery of your rib-cage than that of the high note. Similar results will be found if you touch the surface of the hollow parts of a lyre. 'For the low note travels everywhere all around, while the high note travels forwards, or in the direction in which the utterer forces it to go' (93–100).

Theophrastus uses his thesis as a weapon against the mathematical theorists, arguing that the greater forward motion of the higher note is balanced out by the wider spread of the lower, so that the latter, therefore, does not 'move in accordance with fewer numbers'. But it is presented as a fact of observation, not as an analysis of what high and low pitch are, and it is far from clear that he means to imply that the movement's shape determines or constitutes a sound's pitch. He need mean no more than that high-pitched sounds, in addition to their essential and constitutive *idiotēs*, have the characteristic of expending their impetus mainly in a narrow band of movement in one direction,[25] whereas low-pitched ones disperse it more evenly around their point of origin. Indeed, if he had intended to represent shape as constitutive of pitch he would have been in danger of falling into the snares of quantitativism himself. By his own account, a high note's forward motion and a low note's motion round about are (at least in principle) measurable, since they can be compared and reckoned to be quantitatively equal (100–1, cf. 126–9); and it would therefore have been tempting and presumably possible (again, in principle) to define any individual pitch by reference to the ratio between its forward and its peripheral movement.

Theophrastus could not afford to go down that road. He would have been better advised to stick firmly to his contention that high pitch and low pitch are qualitative *idiotētes* analogous to colours, and to treat the shapes of their movements merely as collateral attributes. I think it probable that

[24] The most elaborate development of such an approach is perhaps that at Ptol. *Harm.* 7.17–8.2. I discuss the behaviour of *oxys* and *barys* in writings on acoustics more fully in Barker 2002b.

[25] Or conceivably in two; line 89 says 'forwards and upwards', but lines 99 and 100 say only 'forwards'.

he intended to do so, and this interpretation of his position is entirely consistent with the text. It is positively encouraged, if not entailed absolutely, by the statement with which this part of the passage closes. 'Hence it is not some unequal numbers that account for the differences, but the sounds being naturally such as they are, naturally attuned together' (105–7). The regular sense of the word *toiaide*, which I have translated 'such as they are', invites the expanded translation 'such as they are *in quality*' (by contrast with 'in number' or 'in quantity'), and can hardly point to any aspect of the sounds but the attributes previously called *idiotētes* and made analogous to colours.

Theophrastus tells us no more about his notion of pitch. But the expression 'naturally attuned together' takes us into new territory. It has a recognisably Aristoxenian ring about it. Though the verb whose passive participle, *synhērmosmenai*, I have translated 'attuned together' does not occur in the *El. harm.*, we hear a good deal there about 'the nature of *to hērmosmenon* ("that which is attuned")' and similar conceptions. In his attack on the theory that the *diastēmata* are the causes of pitch-difference, which follows immediately, much of Theophrastus' discussion revolves around the relations between two other pivots of Aristoxenus' harmonics, *emmeleia* and *ekmeleia*, the melodic and the unmelodic; and in the remarks that conclude this passage he refers to the 'principles', *archai* of melodic sound. Taken together, these expressions suggest an interest in issues close to those addressed in the *El. harm.* What is it that constitutes the 'natural attunement together' of two pitches? What is the difference between the melodic and the unmelodic, and what are the principles that determine what is melodic and what is not? Another sentence (to which we shall return immediately) speaks of the conditions under which we are able to 'find the notes that are attuned to one another' (120–1). On what basis can we confidently identify them? All these questions are staples of Aristoxenus' enquiry, and we may reasonably conjecture that Theophrastus' lost *Harmonics* had an agenda and a conceptual apparatus not very far removed from his.

In the final phase of the fragment (120–32) Theophrastus sums up the conclusions of his various polemical arguments, of which I shall say no more, and adds a brief and enigmatic remark about the nature of music in general. Its connection with the main body of the passage seems tenuous; but the sentence at 120–1, part of which I quoted above, may do something to give it a context. We need therefore to look at this sentence more closely. Its exact wording, and with it its sense, are unfortunately a little uncertain.

As it stands in the manuscripts, the sentence must be translated roughly as follows. 'It is therefore a great help that *melōidia* revolves around [or "depends upon"] these, enabling us to find the notes that are attuned to one another.' For grammatical reasons, 'these' can only be *emmeleia* and *ekmeleia*, the melodic and the unmelodic, and the suggestion would be that we are helped to find the correctly attuned notes by our sense of what is melodic and what is not. That is intelligible, and could be elaborated from the resources of the *El. harm.* But it is open to a linguistic objection;[26] and even if that problem were ignored or resolved the thought conveyed would be completely isolated. Nothing in the text prepares us for it or explains it, and nothing in the sequel refers back to it. A one-letter emendation to the text[27] gives a different meaning, 'It is therefore a great help to *melōidia* that these are avoided . . .', where 'these' are now the *diastēmata*, conceived in this passage's peculiar way as the sonorous content of the spaces between notes.

The emended version is not problem-free,[28] but it has advantages; and I have come round to the view that it is probably preferable. Unlike the statement made by the MSS text, its thesis that in melody the *diastēmata* must be 'left out' or 'avoided' ties in firmly with the substance of the argument to which it is attached. Further, it gives a much clearer indication of the project in which 'help' is required, and allows us to place it in the context of the issues from which the entire passage began. Avoidance of the *diastēmata* is a great help, we are told, to *melōidia*. More exactly, it is a great help 'towards' *melōidia*; or, to paraphrase, it helps us greatly when *melōidia* is our goal. The word *melōidia*, I think, is carefully chosen. It is not the simple *melos*, 'melody', of lines 115 and 122. Though in some settings the two terms are practically interchangeable, and can refer either to a particular tune or to melody in general, *melōidia* may also be used in a way that connects much more closely with the related verb *melōidein*. In these cases it does not designate a 'thing' such as a tune, but an activity, singing or 'making melody'.[29] This interpretation makes good sense here. Avoidance

[26] The verb *periistasthai* with a dative complement is used nowhere else in the sense required here, to 'revolve around' or 'depend upon'.

[27] Reading *tauta eis* for *tautais*, as proposed by Alexanderson 1969 in his note *ad loc.*

[28] There are clear examples elsewhere of the use of *periistasthai* to mean 'avoid', but they all occur in the Greek of a later period (the earliest is in Philodemus).

[29] See for instance Aristox. *El. harm.* 27.19–20, where *melōidia* and *lexis* designate the activities of singing and speaking respectively; and especially 28.20–8, where it is said to be in the nature of *melōidia*, melodic singing, that the voice places its intervals in certain orders 'in accordance with *melos*', and the allusion to *melōidia* is developed by reference to the notes that the voice is capable of singing, *melōidēsai*. This rather intricate passage makes it clear that *melōidia* is an activity, the voice's actual movement through notes and intervals, whereas *melos* (in this instance) is an abstract 'essence' to whose principles *melōidia* naturally conforms. Cf. also 53.21–5.

of the *diastēmata* is a great help to us in our efforts to make melody with our voices, since it enables us to 'find the notes that are attuned to one another'. Theophrastus does not mean that we can identify the well-attuned notes theoretically, but that we can locate them in vocal practice, hitting the right pitches when we sing.

Clearly this takes us right back to the beginning of the fragment, with its comment on the accuracy with which the soul can bring into being an audible manifestation of its own internal 'melody-making movement'. It was this remarkable accomplishment that the ratio-based theory, with its underpinnings in physical acoustics, was allegedly designed to explain. Theophrastus has argued that its foundations are unsound, and it can therefore explain nothing. In the guise in which he presents it, the *diastēma* theory is equally flawed; and it is not, this sentence tells us, these content-filled *diastēmata* that help us to find the notes we need when translating our psychic movements into audible sound, but their omission. Though the passage is of course only an excerpt from a much longer work, this return to its opening suggests that Porphyry's choices of the places to begin and end his quotation are well judged. It seems to be a complete and well-rounded whole.

But it leaves large uncertainties about Theophrastus' own view of the matter. He may have explained elsewhere just how the omission of the *diastēmata* helps us to produce the right notes, how he construes the concept of 'the melodic', and what he means by saying that some notes are 'naturally attuned to one another', but he does not do so here. The closing lines of the passage offer instead the following compendious statements. 'The nature of music is one. It is the movement of the soul that arises for the sake of release from the evils due to the emotions;[30] and if it were not this, neither would it be the nature of music'[31] (130–2).

This tells us not only that music has its primary mode of existence within the soul, but that its role there is essentially therapeutic. Traces of the latter idea, perhaps initially prompted by some remarks of Aristotle in the *Politics*, can be found in several other reports about Theophrastus' opinions. No other Theophrastan view about music, in fact, is so frequently mentioned

[30] 'For the sake of' translates the preposition *kata* with an accusative. It might mean no more than 'in accordance with' or 'in correspondence with', but the purposive sense seems most appropriate here. Cf. the translation in Sicking 1998: 106, 'with a view to', and for the purposive usage in general see LSJ s.v. *kata* III.

[31] An alternative translation, equally possible from a linguistic perspective, is 'and if it did not exist, neither would the nature of music exist'.

in later antiquity, and it probably figured prominently in the work from which our fragment is taken.[32] Like Plato's views on the ethical influence and significance of music, the details of Theophrastus' ideas about its power to cure both psychic and bodily ills are beyond the scope of this study. But one aspect of them seems to have implications for his approach to the science of harmonics.

His thesis that music is 'a movement of the soul that arises for the sake of release from the evils due to the emotions' is presented as a statement of its essence, almost as a definition. 'If it were not this, neither would it be the nature of music'; that is, nothing amounts to music unless it is constituted by a psychic movement of this sort. The oracular remark prefacing these assertions, 'the nature of music is one', reinforces their definitional effect by assuring us that only one kind of activity or occurrence – the one specified in these lines – can genuinely qualify for the title 'music'. Exponents of mathematical harmonics had located the principles governing musical relations in the domain of numbers; they are mathematical principles first, to be discovered through abstract mathematical reflection, and apply only derivatively to the relations and structures with which harmonics is concerned. Aristoxenus represented *melos* or 'the attuned', *to hērmosmenon*, as having an autonomous nature of its own, and consequently held that the process through which the principles governing its modes of organisation and patterns of movement can be discovered must begin from a meticulous empirical examination of its behaviour. If, however, music's nature is that of a movement of the soul, and specifically that of one capable of releasing it from emotional stress and turmoil, the patterns of movement which constitute music will be marked off from those that do not by criteria of an essentially psychological sort. The principles governing it will be those that determine which inner movements are capable of generating emotional 'release'.

There are affinities between this view and the one set out by Plato in the *Timaeus*, where the human soul, when it is in its best condition, has a musical structure akin to that of the soul of the universe, and music can help us to regain that psychic perfection, and to restore 'the unattuned cycle of our soul to order and concord with itself' (*Tim.* 47d5–6). But the connection with Theophrastus is tenuous. None of our reports about him suggest that a 'healthy' soul is structured on a musical basis; and in fact he

[32] See Aristotle, *Pol.* 1342a4–15. The most relevant reports on Theophrastus appear in Fortenbaugh's edition as frags. 719A–B, 723, 726A–C. Cf. the comments of Sicking 1998: 140–2 (who questions, probably rightly, my earlier attempt to connect Theophrastus' ideas with Damon's), and Lippman 1964: 163–6. I discuss his 'musical therapeutics' more fully in Barker 2005a: 131–41.

could hardly have supposed that it was, given his thesis that music is psychic movement of an essentially therapeutic sort. It is difficult to see how its 'nature' could be identical with that of the movement of an emotionally afflicted soul through which its health is brought about, and at the same time constitute the structure of a soul that is already in a healthy condition. Again, the ideal musical structure of the soul in the *Timaeus* is grounded in principles of mathematical order and perfection which, according to Theophrastus, cannot be applied to musical relations at all, since these are not quantitative.

If Theophrastus' *Harmonics* attempted to follow up the implications of his statements in frag. 716, it must have differed from other known works in the field in at least two very striking respects. Musical relations between notes or pitches must have somehow been represented without recourse to quantification; and the principles governing them must have been derived from theories about the nature and operations of the soul. One cannot easily imagine the contents of such a work, and if it existed it seems to have left no trace in the writings of later theorists.[33] Perhaps, however, Theophrastus laid these inscrutable thoughts aside when composing the *Harmonics*. He cannot have followed the path laid down by mathematical theorists.[34] But I have argued that his non-quantitative view of musical relations in their 'real' or physical aspect (and presumably in the context of psychic movement too) is consistent with a broadly Aristoxenian approach to the phenomena as they present themselves in the field of human perception. In that case a Theophrastan *Harmonics* could engage with the musical 'appearances' in terms similar to those that Aristoxenus used, and could even adopt, at this level, a similar way of quantifying intervals. None of this would compromise his position on the real nature of musical relations; he would merely have to record his conviction that an account of this sort depicts them only as they appear and not as they really are. This, of course, is a distinction that Aristoxenus rejects; music, on his account, exists *only* as an audible phenomenon in the realm of 'appearances'. But Theophrastus can follow an Aristoxenian mode of analysis without subscribing to any

[33] The part of Theophrastus' theory that is not an attempt to define music, but simply credits it with psychotherapeutic effects, did of course leave such traces, and not only in the reports cited in n. 32 above; see for instance Arist. Quint. 57.31–58.32. But that is another matter.

[34] He could in principle have agreed that musical intervals are produced, as those theorists contended, from lengths of string or pipe in the ratios they specify. But a mathematical harmonics that restricts itself to enumerating ratios between the dimensions of sounding bodies is sadly limited. If it is to have a basis on which the results of these measurements can be explained, or to identify principles governing an attunement's mathematical form, it must posit that pitch-relations themselves are quantitative at the most fundamental ontological level; and here Theophrastus will certainly part company with them.

such view. He need only posit that the principles uncovered through such investigations, which govern relations and forms of organisation at the phenomenal level, are not autonomous but derivative, and that audible melodies are no more than external manifestations of the soul's silent dances on an internal stage, whose choreography would demand description in language of quite another sort.

Postscript: the later centuries

Harmonic theory is mentioned by various writers of the period between about 300 BC and the beginning of the Christian era, most of them (Philodemus, Vitruvius and Dionysius of Halicarnassus, for example) in its latter years, but nothing they say suggests that original works in the discipline were still being produced. The fact probably reflects more than an accidental gap in our evidence; there is little to encourage the thought that harmonics continued to flourish during those centuries in the hands of theorists whose writings have been lost. The authors of treatises from the Roman imperial period refer quite often to their predecessors; but some of those they mention are their own near-contemporaries, and nearly all the others belong to the fifth or fourth centuries BC, almost none to the third, second or first.[1] The later theorists seem to have thought of themselves, in fact, as the immediate successors of Philolaus, Archytas, Plato, Aristoxenus and 'Euclid', as if the intervening generations had said nothing at all on the subject, or at any rate nothing interesting and new.

Writings devoted wholly or in large part to harmonics reappear early in the first century AD, and continue, if not in a torrent then at least in a steady trickle, right down to Boethius in the early sixth.[2] Through Boethius' Latin reformulation, a version of Greek mathematical harmonic theory passed into the medieval tradition. Boethius' writings were of course greatly revered in the Middle Ages, especially his *Consolations of Philosophy*,

[1] Minor exceptions include Eratosthenes, whose tetrachordal divisions are recorded in Ptol. *Harm.* II.14, and who is mentioned, briefly and dismissively, at Nicom. *Harm.* 260.15; Archestratus (apparently a slightly unorthodox Aristoxenian, Porph. *In Ptol. Harm.* 26.27–27.16); and perhaps Ptolemaïs of Cyrene, excerpts from whose introductory work *Elements of Pythagorean Music* are quoted at Porph. 22.22–23.12, 23.24–24.6, 25.7–26.5. Her dates are uncertain, but cannot be later than the early first century AD. None of these seems to have contributed much to the discipline, and other examples are hard to find.

[2] We have about a dozen complete or almost complete treatises on harmonics from this period, a similar number of texts (mainly by philosophers) which contain extensive passages on the subject, and allusions to a good many other theorists, sometimes accompanied with quotations from their works. For a survey of the major texts see Mathiesen 1999.

and were read not only by specialists in philosophy or music; elements of the terminology of Greek harmonics sometimes escaped from the theorists' clutches and resurfaced, strangely transformed, in unexpected places. (A curious specimen of this sort is printed at the end of this postscript. Readers may like to amuse themselves by trying to decode it.) The Greek texts themselves were preserved largely by the efforts of Byzantine editors and scholars (some of whom developed musical thought in directions of their own); they enjoyed a vigorous revival when they eventually reached Western Europe, becoming the focus of fervent discussion and controversy among theorists and musicians of the Italian Renaissance.[3]

Two writers are quoted by Porphyry for the distinctions they draw between schools of thought in harmonics. One is Ptolemaïs of Cyrene (n. 1 above). The other is a certain Didymus, who probably lived in the time of Nero. He was evidently a theorist of some importance, and I shall mention him again in another context; it is a great pity that his writings are lost. According to Porphyry, much of the material in Ptolemy's *Harmonics* was borrowed from him without acknowledgement, and though this is certainly a fairly wild exaggeration it must have some basis in fact (it is clear that both Ptolemy and Porphyry had access to Didymus' treatise). Ptolemy himself mentions by name and discusses only three individuals (as distinct from the gaggle of anonymous Pythagoreans) who wrote on harmonics before him, and it is significant that Didymus is one of them (the other two are Archytas and Aristoxenus).[4] He comments at some length on Didymus' novel way of using the monochord and examines his divisions of enharmonic, chromatic and diatonic tetrachords.[5]

Didymus and Ptolemaïs draw very similar distinctions between the various schools of thought, though Didymus' are more elaborately expressed.[6] In both cases they hang on the relative weight assigned by the theorists to reason on the one hand and to perception on the other as 'criteria of judgement',[7] and on differences between the roles in which the various kinds of theorist deploy them. The classifications at which they arrive are far from

[3] See especially Palisca 1985.

[4] Eratosthenes' divisions are set out in the tables of *Harm.* II.14, but are not discussed.

[5] Porphyry's quotations from Didymus on the schools of harmonic theory are at 26.6–29, 27.17–28.26; his comment on Ptolemy's alleged plagiarism is at 5.11–15. Didymus is mentioned again as the source (via Archytas) of information about Pythagorean procedures at 107.15. Ptolemy's expositions and discussions of his ideas are in *Harm.* II.13–14.

[6] It is likely that Porphyry found the excerpts from Ptolemaïs as quotations in Didymus' treatise, and that Didymus used them as the basis of his own remarks. Porphyry says that Ptolemaïs wrote about these matters 'succinctly' (*syntomōs*, 25.1–2), and after quoting from her work he tells us that Didymus 'developed these themes fully' (*exergazomenos tous topous*, 26.6).

[7] On the notion of the 'criterion' see e.g. Long 1989, Blumenthal 1989, Striker 1996.

crude, though some are a little confusing; in effect they place harmonic theorists on a spectrum, one end of which is occupied by people who rely almost wholly on reason, the other by those whose judgements are based as exclusively as possible on the evidence of their ears and make minimal use of reasoning. Both authors describe the former as Pythagoreans (but in one question she asks and answers, it seems that for Ptolemaïs there were other and perhaps more authentic followers of Pythagoras who did not fall under this description; we shall come to this problem shortly). They are said to use perception only as a 'spark' to kindle the fires of reason, and from that initial point onwards they work by reasoning alone. If their conclusions conflict with the evidence of the senses, they argue that it is the latter that must be mistaken.[8] The latter are *organikoi*, 'instrumentalists', to whom Didymus adds *phōnaskikoi*, 'voice-trainers' and others of that ilk.[9] It seems clear that the people he has in mind are scarcely 'harmonic theorists' in any recognisable sense, but are those who in practice give instrumentalists and singers their training and judge the results of their students' efforts entirely by ear.

The two writers agree in placing Aristoxenus and his followers somewhere between these two extremes. Some of them, according to Ptolemaïs, set out as skilful instrumentalists and went on to 'apply themselves to a theoretical science' based largely on perception, but made use of reasoning when they found it necessary. Aristoxenus' own approach is described rather differently. Didymus' account of him is much more extensive and a good deal clearer than that of Ptolemaïs, but the gist of both is the same. For Aristoxenus, reason and perception are equally important, but they have quite different roles. Reasoning is used to extract conclusions from the data provided by perception, and cannot go to work on its own. Reason cannot independently discover or demonstrate the facts whose consequences it draws out, but these are evident to perception; and reason must in that respect rely wholly on the judgements of the ear. It follows that no properly reasoned conclusions can conflict with the evidence of the senses; and it follows equally that the theorist must accept the conclusions, however abstract and remote, to which he is led by reasoning from a starting point in his perceptual impressions. Didymus in fact draws a very competent if partial picture of Aristoxenus' position; and his explicit reference to what he calls 'the first book of the *Harmonic Elements*', together with a careful paraphrase of a passage from what we know as Book II of the *El. harm.*,

[8] Ptolemaïs at Porph. 23.24–31, 25.10–14; Didymus at Porph. 26.15–25.

[9] Ptolemaïs at Porph. 25.14–16, Didymus at Porph. 26.6–15.

strongly suggests that he was familiar at first hand with at least part of Aristoxenus' work.[10]

We return, finally, to the puzzling answer Ptolemaïs gives to one of her questions. 'Who are those who give both criteria equal status? Pythagoras and his followers' (Porph. 25.25–6). At one level this fits well into its context. At 25.15 Ptolemaïs had asked what the difference is between those who privilege both criteria, and one of those who do so has turned out to be Aristoxenus. It is entirely appropriate that she should now consider the other, who must presumably conceive the relation between the criteria in a different way. But there are several difficulties, of which only one need concern us; the account she goes on to give of the approach of 'Pythagoras and his followers' (25.26–26.1) turns out to be equivalent to her previous description of the route taken by 'some of the Pythagoreans' (25.10–14), in which the two criteria are emphatically not given equal weight. One would like to think that the confusion is due to a scribal error, and that the account at 25.26 ff. has mistakenly been transferred into this position from another context; and it has indeed already appeared word for word in Porphyry's text at 23.25–31. In that case we might guess that the original passage which it displaced described an approach which took the evidence of perception more seriously and made genuine efforts to examine the data of real musical practice, perhaps in the manner of Archytas. But this hypothesis is at best debatable, since it is 'Pythagoras and his followers' who are saddled with the extreme 'rationalist' position in the earlier passage as well; and Didymus' exposition, which follows Ptolemaïs very closely, contains no trace of the 'moderate' Pythagoreanism I have postulated. I do not know how the problem should be resolved.

The last sentence which Porphyry quotes from Didymus (28.23–6) shows that the principal line of division he recognises, regardless of any finer distinctions there may be, is that between the Pythagoreans and Aristoxenus. Ptolemy states the same view bluntly in *Harm.* 1.2 (5.24–6.13), again basing the classification mainly (but not entirely), and rather simplistically in this passage, on their different attitudes to the criteria of reason and perception.[11] Most writers of the imperial period, in fact, take this distinction as fundamental; and most of their treatises can be labelled, in a rough and preliminary way, as representatives of one or other of the two major schools

[10] Ptolemaïs at Porph. 24.1–6, 26.1–4, Didymus at Porph. 27.17–28.26.

[11] Ptolemy's account here reads very much like a summary paraphrase of Didymus', to which he has added brief criticisms which he will expand later. It is hard to resist the conclusion that here at least he is behaving in the way that Porphyry asserts, and is surreptitiously borrowing from Didymus' work.

of thought that had established themselves by the end of the fourth century. Some are 'Aristoxenian' and others 'mathematical', the latter generally but not quite invariably portraying themselves as either 'Pythagorean' or 'Platonist' or both at once. But though these writers commonly allude to significant divergences between the Aristoxenian and mathematical approaches and to contradictions between their conclusions, in practice they do not always hold firmly to one tradition or the other.

Nicomachus, for example (writing around AD 100), smuggles into his *Introduction to Harmonics* an account of 'two species of vocal sound' which he attributes to 'those of the Pythagorean school', in line with his own overt commitments, but which (though he does not say so) is very plainly an ingenious adaptation of Aristoxenus' study of the 'intervallic' and 'continuous' movements of the voice.[12] In his final chapter (*Harm.* 12) he sets out a whole series of propositions whose Aristoxenian pedigree is unmistakable, even though he ends by aligning himself with an anti-Aristoxenian position, warning the reader that the 'so-called semitone' is not precisely half a tone. Conversely, the little *Introduction to Harmonics* of Gaudentius (third or fourth century AD) is largely a compendium of Aristoxenian doctrines, but incorporates a sequence of seven chapters (10–16) on the harmonic ratios, without any suggestion that the two pillars of his work rest on foundations which cannot survive in the same space. On a grander scale, the first book of Aristides Quintilianus *De musica* (third century AD) is thoroughly Aristoxenian, whereas the third is devoted to explorations of harmonic ratios and numbers, and of their analogical and symbolic associations with features of the cosmos at large. He seems to think of the Aristoxenian and Pythagorean approaches as simply offering two different 'theoretical' perspectives on music (describing the former as 'technical' and the latter as 'physical', *De mus.* 6.8–18), and that they are complementary rather than being rivals.

The case of Ptolemy is particularly intriguing. He insists that the aim of harmonics is to show that all the patterns of attunement which the ear accepts as musically well formed conform to the postulates of mathematical reason. Though he criticises the theorists he calls 'Pythagoreans' for various errors, his approach and theirs have much in common, whereas he

[12] Nicom. *Harm.* ch. 2; cf. Aristox. *El. harm.* 8.13–10.24, with pp. 143–5 above. It is just possible that Nicomachus did not know the Aristoxenian passage, and that he found the adaptation ready-made in the work of a 'Pythagorean' writer of the first century AD. In that case his attribution was made in good faith; but I think it unlikely. Aristoxenus' discussion was very well known in imperial times, and versions of it appear repeatedly in surviving texts. Nowhere else has it been transmuted into a Pythagorean form.

denounces Aristoxenian harmonics in its entirety, arguing that it is misconceived from the foundations up.[13] Nevertheless, when he sets off to derive the forms of ideally perfect attunements from first principles by mathematical procedures, he adopts alongside his mathematical 'principles of reason' another set of three principles which he explicitly describes as 'based on agreed perception', that is, on criteria of the sort on which Aristoxenus had relied, and which seem to have no legitimate place in this phase of Ptolemy's project. His derivations cannot proceed without them, and they cannot be justified mathematically. Two of the three are in fact taken quite plainly from the Aristoxenian repertoire.[14]

Despite these surprising incursions of ingredients from one or the other tradition into an alien environment (and there are many more), they generally stand out as rather obtrusive and unassimilated accretions. The broad division into 'Aristoxenian' and 'Pythagorean' approaches remains for the most part undisturbed; and if there were theorists who seriously tackled the task of fusing them coherently together, their efforts have left few traces on the record.[15]

Though both styles of harmonic theory remained in play throughout these centuries, their fortunes were very different. The essays in Aristoxenian harmonics (principally those of Cleonides, Bacchius, Gaudentius and the first book of Aristides Quintilianus) are useful to modern scholars, since they preserve, more or less faithfully, elements of Aristoxenus' work which have not survived in his own words. But they are little more than summaries of 'doctrine', shorn of the methodological reflections, the reasoning and the colourful polemics which enliven the *Elementa harmonica* and sustain its procedures and conclusions. They make no attempt to develop Aristoxenus' investigations for themselves beyond the point to which he had taken them, and the only such developments which they attribute to 'more recent Aristoxenians' are slight and unimpressive.[16] The abbreviated question-and-answer format of Bacchius' work marks it unambiguously as a schoolroom text, a 'catechism' designed to drum 'correct' answers to

[13] See especially *Harm.* Book I chapters 9–10 and 12.

[14] The passage is *Harm.* I.15; cf. Barker 2000a: 135–7.

[15] Aristides Quintilianus expounds doctrines from both sides with equal enthusiasm, but in separate parts of his work, as I noted above, and apart from labelling them as representing two distinct and complementary 'parts' of musical theory he makes no attempt to integrate them systematically. There are some grounds for thinking that the harmonic divisions of Eratosthenes (third century BC) and Didymus (first century AD), recorded in Ptol. *Harm.* II.13–14, were designed to 'translate' features of Aristoxenian systems into mathematical form; see Barker 1994a, Barker and Creese 2001. But this remains at best a plausible hypothesis; the evidence is insufficient to justify a firm conclusion.

[16] The only example worth mentioning is the addition by more recent theorists of two extra *tonoi* to the thirteen identified by Aristoxenus; see Arist. Quint. 20.5–9. This departure from Aristoxenus' position seems both musically and theoretically trivial.

set questions into the heads of unfortunate school-children.[17] Cleonides' treatise is more sophisticated and thoughtful and was presumably aimed at more mature readers, and with some qualifications the same is true of Gaudentius. But in the end their aims and those of Bacchius are not very different. Their works are 'text-books' which set out to inculcate the elements of an ancient discipline, not to explore new avenues on their own account. The overall agenda of Aristides Quintilianus' *De musica* is much more ambitious, but in its exposition of Aristoxenian harmonics it too follows the text-book pattern, and has in fact a great deal in common with Cleonides.

Mathematical harmonics had a much more colourful career. We saw earlier in this book that though the fourth-century theorists had much in common, they developed it in two different directions. In the hands of Archytas it generated mathematical analyses of systems of attunement used in practice by contemporary musicians, while on Plato's approach (very partially anticipated by Philolaus) it gave birth to an abstract representation of the divine cosmic order, and of the ideally harmonious mode of integration to which the human soul could aspire. If resemblances remained between this construction and the attunements underlying melodies heard in fourth-century Athens, from Plato's perspective they were incidental; and if no such melodies could be found, the fact would in his view have no bearing on the credentials of his analysis. It is possible though not certain, as I noted above, that Ptolemaïs' classification of theoretical attitudes included a 'moderate' as well as a hyperbolically 'rationalistic' version of Pythagorean harmonics; if it did, by her standards we would expect Archytas to fall into the former category and Plato into the latter.

Among the later theorists whose work survives intact or in substantial quantity, only Ptolemy consistently and deliberately set himself to a project akin to that of Archytas. His evidence about Didymus' treatment of the monochord and his harmonic divisions in *Harm.* II.13, together with Porphyry's insinuations about his unacknowledged 'borrowings' from the earlier theorist, give some reasons for thinking that the latter's enterprise had a similar purpose, but we cannot state it as a fact. Ptolemy insists that when conclusions about the structures of perfectly formed attunements and scales have been reached on the basis of mathematical reasoning, they cannot be accepted as correct until they have been submitted to the judgement of the musical ear. His avowed aim is to show that all the various systems which the ear recognises as musically well formed do indeed conform to

[17] The 'introductory' treatise of Ptolemaïs of Cyrene is presented in a similar manner, and would fit equally smoothly into a pedagogic environment.

mathematical principles; he gives precise and elaborate instructions about the procedures which enquirers should follow and the apparatus they should use in order to exercise their musical judgement on his constructions for themselves; and in two remarkable chapters (*Harm.* I.16 and II.16, cf. II.1 and II.14) he explains in detail how the attunements used in practice by musicians in his own time are related to those generated by his 'rational' mathematics. Although in the third book of the *Harmonics* he moves on to identify corresponding patterns of relations in the heavens and in the human soul, the meticulous arguments and constructions which occupy the bulk of his work are directed exclusively to the analysis of structures underlying the melodies which he and his readers could hear every day.[18]

None of the other surviving writings in mathematical harmonics can be viewed in this light. Many of them are embedded in studies of Plato's philosophy, especially of the passage on the World-Soul in the *Timaeus*; this is patently true, for example, of the work of Theon of Smyrna and his principal sources, Thrasyllus and Adrastus, of Plutarch's intricate essay on the 'generation of the soul' in the *Timaeus*, and of the passages devoted to harmonics in the *Timaeus* commentaries by Calcidius and Proclus.[19] Since the main purpose of all these authors is to bring out meanings, presuppositions and implications locked up in a text of the fourth century BC, it is not surprising that they are more concerned to expound existing harmonic lore in the service of their project than to devise new theories or approaches of their own. But at the same time, Platonism in its various guises was not an archaic curiosity but a living philosophy, all of whose aspects were subject to constant reinterpretation; and these writers sometimes extend and elaborate Plato's harmonic construction in novel ways, or adopt new approaches to its analysis, or even offer views of their own about issues that bear on it.[20] Some of them also set their studies in the context of a wider survey of harmonic theory, defining key terms, discussing the 'physics' underlying the representation of musical intervals as ratios, and offering descriptions of systems in the chromatic and enharmonic genera as well as in the diatonic to which the *Timaeus* system corresponds. But they never go far down that road; the only forms of the chromatic and enharmonic which

[18] I examine his sophisticated methodology in Barker 2000a, where I also argue that he gives us grounds for real confidence in the accuracy of his analyses of contemporary schemes of attunement.

[19] Theon wrote around AD 100, as did Adrastus; Thrasyllus, the editor of Plato's works and astrologer to the emperor Tiberius, died in AD 36. Plutarch was born a little before AD 50 and lived into the 120s. Calcidius belongs to the fourth century AD and Proclus to the fifth.

[20] Thus Thrasyllus, for instance, seems to have devised his own procedure for 'dividing the *kanōn*'; Adrastus expounds a thoroughly idiosyncratic thesis about the relation between pitch and number; and Plutarch explores mathematical subtleties and controversies barely hinted at in the *Timaeus* itself.

they consider are those most closely related to the *Timaeus*' diatonic, those, that is, which can be constructed by simple variations on the procedure used to generate a diatonic tetrachord whose ratios are 9:8, 9:8, 256:243.[21] They show no interest in systems which (if Archytas, Didymus and Ptolemy are to be trusted) corresponded more closely to the attunements used in ordinary human music-making, but could not be built through manipulations of the ratios of the concords and the tone (or, in other terms, by what I have called the 'method of concordance').

Nicomachus was roughly contemporary with Theon and Adrastus, and his writings on harmonics have obvious affinities with those of the Platonists of his day. But they are free-standing treatises rather than commentaries on an existing text, and they present themselves as explorations of ancient Pythagorean rather than Platonic wisdom. (He discusses the *Timaeus* in chapter 8 of his *Introduction to Harmonics*, but evidently regards Plato's dialogue as a record of Pythagorean thought. At the end of chapter 11 he says that in a longer work he will set out the Pythagorean division of the *kanōn* 'completed accurately and in accordance with the intentions of the master [i.e. Pythagoras], not in the manner of Eratosthenes and Thrasyllus, who misunderstood it, but in that of Timaeus of Locri, whom Plato also followed'.) Only his engaging little *Introduction* survives intact, but we need not rely on that alone. It is generally agreed that the longer work which it promises was indeed written, and that it is closely paraphrased in the first three books of Boethius' *Institutio musica*, perhaps also in the fourth. We therefore have a good guide to the contents of this much more elaborate treatise. Other material bearing directly on harmonics can be found in Nicomachus' *Introduction to Arithmetic*; it too was paraphrased into Latin by Boethius, and in this case we are in a position to see that though the paraphrase does not slavishly follow the original, it is related to it very closely, often amounting to quite accurate translation.[22] It was Nicomachus' work in its Boethian version that did most to give medieval Europe its picture of Greek harmonic theory.

[21] There is no great difficulty in constructing by these methods a chromatic tetrachord approximating to Aristoxenus' pattern of tone-and-a-half, semitone, semitone. In the enharmonic, as presented by these theorists, the higher of the tetrachord's moveable notes falls in the same place as the lower moveable note in diatonic. None of them attempts (though another writer did, as we shall see below) to identify the ratios into which the remaining interval of 256:243 is divided to form approximate quarter-tones; see pp. 403–4 above.

[22] Some MSS of the *Introduction to Harmonics* also preserve a short and rather motley sequence of paragraphs under the heading 'From the same Nicomachus'; they are nowadays known as the *Excerpta ex Nicomacho*. But there are serious doubts about their authenticity, and they are probably snippets from several different authors. Nicomachus may possibly be among them, but he is certainly not responsible for them all. For a brief discussion see Zanoncelli 1990: 207–9.

I cannot examine his approach to the discipline in any depth or detail here. The main point I want to make is simple; here again, set in the context of a reconstructed 'Pythagoreanism', the central focus is on the analysis of a diatonic scale of the type familiar from Plato, the *Sectio canonis* and the Platonist writers we have discussed. In the *Introduction* Nicomachus leads up to it with an account of a diatonic seven-note system representing the harmonious relations between the heavenly bodies, and with comments on the properties of musical instruments, designed to convince us that pitch-differences are related 'in accordance with numbers' (chs. 3–4). The climax comes with a description of Pythagoras' alleged achievements in completing the diatonic eight-note octave and discovering the ratios of the concords and tones that define its structure (chs. 5–7).[23] The contents of the rest of the work, including its comments on Plato (ch. 8) and Philolaus (ch. 9), are plainly intended as supplementary material, further explicating Pythagoras' fundamental discoveries. The Boethian essay is vastly more complex, but almost all its intricacies are designed to contribute to the project of exploring the arithmetical properties of the relations between terms in this diatonic structure. (Book IV also calculates the divisions of the *kanōn* through which the associated forms of the chromatic and the enharmonic can be constructed, even including – as the Platonist writers do not – the tiny dieses of the enharmonic *pyknon*.)

There is little in either work to show that Nicomachus was seriously concerned with the attunements of musical practice. It is true that when the Boethian paraphrase represents the study of music through reason as superior to the arts of performers and composers, it does so partly on the grounds that an intellectual understanding of the subject enables a person to judge what composers and performers produce, 'carefully weighing up their rhythms and melodies and the composition in its entirety' (*Inst. mus.* I.34). But it offers no explanation of how the bewildering details of its mathematical computations could possibly be put to use as the basis for such aesthetic judgements. It is hard to avoid the impression that they are studied for their own sake alone; and it seems to me that the brief depiction of Pythagoras' philosophy (or what passed as such in this period) which appears early in Book II is a much better guide to Nicomachus' purposes. Pythagoras held, we are told,

> that philosophy was the knowledge and study of whatever may properly and truly be said 'to be'. Moreover, he considered these things to be those that neither increase under tension nor decrease under pressure, things not changed by any

[23] Ch. 6 contains our earliest record of the legend of Pythagoras and the 'harmonious blacksmith', the details of which are (alas!) impossible.

chance occurrences. These things are forms, magnitudes, qualities, relations, and other things which, considered in themselves, are immutable, but which, joined to material substances, suffer radical change and are altered in many ways because of their relationship with a changeable thing.[24]

The treatise is an extended exposition of what pure reasoning can reveal about the properties of a rationally impeccable and unalterable system of quantitative relations. All it could tell us about the practices of human musicians is that they fail to match up to the perfection of the eternal realities; and though Nicomachus no doubt took that view it is incidental to his main concerns, which centre firmly on the 'philosophical' truths disclosed by harmonic arithmetic.

This chapter does not pretend to offer a complete survey of relevant writings of the period.[25] I shall end with a few remarks on just one other, which does not fit exactly into any of the patterns we have so far identified. Porphyry, the third-century disciple of Plotinus, wrote copiously on philosophical, religious and other topics. Prominent among his works were commentaries on at least seven of Plato's dialogues (including the *Timaeus*) and on five or more treatises of Aristotle; and we have seen that he chose also, for reasons which his preface does not convincingly explain, to compose a commentary on Ptolemy's *Harmonics*. In the form in which it has come down to us it is incomplete, and it seems unlikely that it was ever finished. Whereas it sets off by examining the early chapters of Ptolemy's text in minute detail, embroidering its comments with free-wheeling reflections on philosophical issues and introducing a plethora of quotations from other sources on the matters under discussion, in its later parts it tails off into little more than a rather feeble paraphrase of the original. Porphyry seems to have run out of steam.

The fact is not altogether surprising. It seems not to have been the mathematical, musical or other technical minutiae of the *Harmonics* that captured Porphyry's attention, nor Ptolemy's remarkable studies of the designs of the instruments to be used for 'experimental' purposes. Where he is most at home is with the large philosophical questions raised by Ptolemy's discussions of his science's methodology, of the nature of sound and its attributes, and of various other general issues examined in the first half-dozen chapters of Book I. If we exclude the brief preface, Porphyry's

[24] *Inst. mus.* II.2; translation from Bower 1989. There is a very similar account of 'philosophy' at Nicom. *Arith.* I.1, repeated in Boethius' paraphrase at the beginning of the *Inst. arith.*

[25] Anyone who feels the urge to examine these writings more closely will find a useful starting-point in the later chapters of Mathiesen 1999, along with the more specialised studies cited in his substantial bibliography. The harmonics of later antiquity is not altogether uncharted territory, but there is ample room for further research.

commentary runs to 169 pages in Düring's edition; it omits any reference to *Harm.* I.16, and ends part-way through its treatment of *Harm.* II.7. In all it has covered nearly twenty-two of Ptolemy's chapters, but the first six of them have occupied it for almost two thirds of its length.[26]

In parts of his more expansive sections, Porphyry devotes himself to the commentator's familiar task of explaining, line by line, what he takes his author to mean, and bringing out some of its more recondite implications. Elsewhere he adds quite substantial discussions of his own which he apparently conceives as developments of Ptolemy's line of thought; a particularly interesting example is an epistemological excursus presented in the course of his discussion of the first few lines of the *Harmonics* (Porph. 12.29–15.28).[27] A great deal of space is also given over to quotations from earlier writers, some of them very long, which are always intelligibly (though sometimes rather remotely) connected with the current topic of Ptolemy's text, but by no means always support his views. Porphyry thereby preserves a considerable collection of writings of which we would otherwise have known nothing, and modern scholars have regularly plundered his work for these valuable fragments, as indeed I have done in this book.

But they have tended to neglect his own agenda. His overt reason for including so many quotations is spelled out in his preface, where he claims that 'most, if not virtually all' of what Ptolemy says has been taken from earlier theorists, and that he, Porphyry, will reveal the sources in question, as Ptolemy for the most part does not (Porph. 5.7–16). But he uses some of them also to stoke his own fires, and it turns out that on several crucial points he is thoroughly at odds with Ptolemy's position. The most striking example is in his massively extended discussion of *Harm.* I.3, where he mounts a head-on attack on a doctrine which is fundamental to the *Harmonics*, that a sound's pitch is a quantitative attribute of it, and that when two sounds differ in pitch they differ quantitatively. Porphyry argues at length for the view that pitch is not a quantity (*posotēs*) but a quality (*poiotēs*), though he concedes that changes in pitch may nevertheless depend on quantitative variations in the physical movements that underlie them.[28]

[26] The discussion of those six chapters begins at line 19 of p. 5 of Düring's edition, and ends with the third line of p. 112. It amounts, then, to just over 106 pages, by contrast with a mere 39 pages for the next nine chapters of Book I, and 24 for the first seven chapters of Book II.

[27] It has been suggested that the ideas expounded in this passage are derived from the work of Thrasyllus, from whom Porphyry quotes at 12.21–8. See Tarrant 1993, ch. 5; but cf. also the sceptical comments in Gerson 1993.

[28] See especially 43.7–45.20, 58.5–61.15, and cf. 65.16–20. As additional support for his position he quotes the long passage from Theophrastus (61.22–65.15) which we considered in Ch. 15, and a shorter excerpt from a mathematical work by an otherwise unknown writer whom he names as

What we find in Porphyry, then, aside from an interpretation of Ptolemy's text and a treasury of quotations from elsewhere, is an excellent example of the way in which an essay on harmonics could be used as a spring-board to quite esoteric excursions in philosophy. Porphyry's epistemological reflections are only distantly related to Ptolemy's, and his determined pursuit of intricate arguments about the distinctions between qualities and quantities is probably prompted more by his study of Aristotle's *Categories* (on which he wrote two commentaries) than by anything in the text of the *Harmonics* itself.[29] His work is unique in presenting itself as a commentary on a treatise in harmonic theory, but in another important respect it is characteristic of the period. Mathematical harmonics is no doubt a fascinating discipline in its own right; but in the view of the great majority of these later writers it deserves serious study, as Plato had said long ago, only in so far as it can be of service to a philosopher.

In his passage amang the planetis all,
He herd ane hevinly melody and sound,
Passing all instrumentis musicall,
Causit be rolling of the speris round;
Whilk ermony through all this mapamond
Whill moving cess unite perpetuall,
Whilk of this warld Pluto the saull can call.

Thar leirit he tonis proporcionate,
As dupler, tripler and emetricus,
Enoleus, and eike the quadruplat,
Epodyus riht hard and curiouss;
And of their sex, swet and delicious,
Riht consonant five hevinly symphonyis
Componit ar, as clerkis can devis.

First diatasseroun, full sweit, I wiss,
And diapasoun, simple and duplate,
And diapente, componit with a diss;
This makis five of three multiplicat.
This mery musik and mellifluat,
Complete and full with noumeris od and evin,
Is causit be the moving of the hevin.

Robert Henryson, *The Tale of Orpheus and Eurudices his Quene*, 219–39

'Panaetius the Younger' (65.21–66.15 by Düring's reckoning, but in fact almost certainly extending to 67.10). Porphyry seems to believe that his concession will rescue Ptolemy's project; I think he is wrong, but the matter is too complex to be unravelled here.

29 I should emphasise that these cases are only selected examples; Porphyry provides a great deal more meat for philosophers to chew on.

Bibliography

For a much fuller bibliography, including works on all aspects of ancient music and musical theory and references to earlier bibliographical surveys, see Mathiesen 1999: 669–783. The list I give here identifies all the works mentioned in my notes, and offers a modest selection of others that have a direct bearing on issues with which this book is concerned. It supplements Mathiesen's list only to the extent that it includes some items published since his book appeared (and a very few earlier pieces that he omitted), together with some that are more relevant to the cultural, philosophical and scientific topics I have touched on than to specifically musicological matters.

Alexanderson, B. (1969) *Textual Remarks on Ptolemy's Harmonica and Porphyry's Commentary*. Gothenburg.

Amano, C. (1982) 'La "Division du Canon" et la théorie mathématique des intervalles musicaux', *Historia Scientiarum* 22: 97–115.

Anderson, W. D. (1955) 'The importance of Damonian theory in Plato's thought', *Transactions of the American Philological Society* 86: 88–102.

(1966) *Ethos and Education in Greek Music*. Cambridge, Mass.

(1980) 'Musical development in the school of Aristotle', *Research Chronicle* 16: 78–98.

(1994) *Music and Musicians in Ancient Greece*. Ithaca and London.

Annas, J. (1981) *An Introduction to Plato's Republic*, Oxford.

Anton, J. P. (ed.) (1980) *Science and the Sciences in Plato*. Buffalo, N.Y.

Arnoux, G. (1960) *Musique platonicienne, âme du monde*. Paris.

Avezzù, G. (1994) 'Papyrus Hibeh 1.13: Anonymi fragmentum *De Musica*', *Musica e Storia* 2: 109–38.

Baltes, M. (1972) *Timaios Lokros Über die Natur des Kosmos und der Seele*. Leiden.

Baltussen, H. (2000) *Theophrastus against the Presocratics and Plato*. Leiden.

Barbera, A. (1977) 'Arithmetic and geometric division of the tetrachord', *Journal of Music Theory* 21: 294–323.

(1981) '*Republic* 530c–531c: another look at Plato and the Pythagoreans', *American Journal of Philology* 102: 395–410.

(1984a) 'Placing *Sectio canonis* in historical and philosophical contexts', *Journal of Hellenic Studies* 104: 157–61.

(1984b) 'The consonant eleventh and the expansion of the musical tetraktys: a study of ancient Pythagoreanism', *Journal of Music Theory* 28: 191–223.
(1990) *Music Theory and its Sources: Antiquity and the Middle Ages*. Notre Dame, Ind.
(1991) *The Euclidean Division of the Canon: Greek and Latin Sources*. Lincoln and London.
Barker, A. (1978a) '*Symphōnoi arithmoi*: a note on *Republic* 531c1–4', *Classical Philology* 18: 337–42.
(1978b) '*Hoi kaloumenoi harmonikoi*: the predecessors of Aristoxenus', *Proceedings of the Cambridge Philological Society* 24: 1–21.
(1978c) 'Music and mathematics: Theophrastus against the number-theorists', *Proceedings of the Cambridge Philological Society* 23: 1–15.
(1981a) 'Aristotle on perception and ratios', *Phronesis* 26: 248–66.
(1981b) 'Methods and aims in the Euclidean *Sectio Canonis*', *Journal of Hellenic Studies* 101: 1–16.
(1982a) 'Aristides Quintilianus and constructions in early music theory', *Classical Quarterly* 32: 184–97.
(1982b) 'The innovations of Lysander the kitharist', *Classical Quarterly* 32: 266–9.
(1984a) *Greek Musical Writings I, The Musician and his Art*. Cambridge.
(1984b) 'Aristoxenus' theorems and the foundations of harmonic science', *Ancient Philosophy* 4: 23–64.
(1985) 'Theophrastus on pitch and melody', in *Theophrastus of Eresus: on his Life and Work*, eds. W. Fortenbaugh et al. New Brunswick and Oxford: 289–324.
(1989a) *Greek Musical Writings II, Harmonic and Acoustic Theory*. Cambridge.
(1989b) 'Archita di Taranto e l'armonica pitagorica', in *Tra Sicilia e Magna Grecia*, eds. A. C. Cassio and D. Musti. Pisa: 159–78.
(1991) 'Three approaches to canonic division', in *Peri tōn Mathēmatōn*, ed. I. Mueller, *APEIRON* 24 no. 4: 49–84.
(1994a) 'Greek musicologists in the Roman Empire', in *The Sciences in Greco-Roman Society*, ed. T. D. Barnes, *APEIRON* 27 no. 4: 53–74.
(1994b) 'Ptolemy's Pythagoreans, Archytas, and Plato's conception of mathematics', *Phronesis* 39: 113–35.
(1994c) 'An Oxyrhynchus fragment on harmonic theory', *Classical Quarterly* 44: 75–84.
(1995) '*Heterophonia* and *Poikilia*: accompaniments to Greek melody', in Gentili and Perusino (1995): 41–60.
(1996) 'Plato's *Philebus*: the numbering of a unity', in *Dialogues with Plato*, ed. E. Benitez, *APEIRON* 29 no. 4: 143–64.
(2000a) *Scientific Method in Ptolemy's Harmonics*. Cambridge.
(2000b), 'Timaeus on music and the liver', in Wright (2000): 85–100.
(2001a) 'Armonia', in *Storia della Scienza* 1, ed. S. Petruccioli. Rome: 909–26.
(2001b) 'La musica di Stesicoro', *Quaderni Urbinati di Cultura Classica* 67: 7–20.
(2002a) *Euterpe: Ricerche sulla musica greca e romana*, tr. E. Rocconi. Pisa.
(2002b) 'Words for sounds', in Tuplin and Rihll (2002): 22–35.

(2003) 'Early *Timaeus* commentaries and Hellenistic musicology', in *Ancient Approaches to Plato's Timaeus*, eds. R. W. Sharples and A. Sheppard. London: 73–87.

(2004) 'Theophrastus and Aristoxenus: confusions in musical metaphysics', *Bulletin of the Institute of Classical Studies* 47: 101–17.

(2005a) *Psicomusicologia nella Grecia antica*, tr. A. Meriani. Naples.

(2005b) 'The journeying voice: melody and metaphysics in Aristoxenian harmonic science', *APEIRON* 38: 203–26.

(forthcoming), 'Pseudo-Plutarch *De musica* 1–13', in *Pseudo-Plutarch's De Musica: Proceedings of the Ionian University Seminar on Ancient Greek Music, July 1–10, 2005*, ed. P. Vlagopoulos. Leiden.

Barker, A., and D. E. Creese (2001) 'Eratosthenes', in *Die Musik in Geschichte und Gegenwart*, Kassel 1999–, Personenteil 6: 399–400.

Barnes, J. (1969) 'Aristotle's theory of demonstration', *Phronesis* 14: 123–52.

(1981) 'Proof and the syllogism', in *Aristotle on Science: the Posterior Analytics*, ed. E. Berti. Padua: 17–59.

(1993) *Aristotle, Posterior Analytics* (translation and commentary), 2nd edn. Oxford.

Bélis, A. (1982) 'Les nuances dans le *Traité d'harmonique* d'Aristoxène de Tarente', *Revue des études grecques* 95: 54–73.

(1986) *Aristoxène de Tarente et Aristote: Le Traité d'Harmonique*. Paris.

(1992) 'L'harmonique comme science dans l'antiquité grecque', in Lustgarten (1992): 201–8.

von Blumenthal, A. (1939) *Ion von Chios:die Reste seiner Werke*. Stuttgart and Berlin.

Blumenthal, H. (1989) 'Plotinus and Proclus on the criterion of truth', in Huby and Neal (1989): 257–80.

Boccadoro, B. (2002) *Ethos e Varietas*. Florence.

Bolton, R. (1991) 'Aristotle's method in natural science', in Judson (1991): 1–29.

Bostock, D. (1991) 'Aristotle on continuity in *Physics* VI', in Judson (1991): 179–212.

Bowen, A. C. (1978) 'The minor sixth (8:5) in early Greek harmonic science', *American Journal of Philology* 99: 501–6.

(1982), 'The foundations of early Pythagorean harmonic science: Archytas, fragment 1', *Ancient Philosophy* 2: 79–104.

(ed.) (1991a) *Science and Philosophy in Classical Greece*. New York.

(1991b) 'Euclid's *Sectio Canonis* and the history of Pythagoreanism', in Bowen (1991a): 164–87.

Bowen, A. C., and W. R. Bowen (1997) 'The translator as interpreter: Euclid's *Sectio Canonis* and Ptolemy's *Harmonica* in the Latin tradition', in *Music Discourse from Classical to Early Modern Times: Editing and Translating Texts*, ed. M. R. Maniates. Toronto and Buffalo: 97–148.

Bower, C. M. (1978) 'Boethius and Nicomachus: an essay concerning the sources of *De Institutione Musica*', *Vivarium* 16: 1–45.

(1989) *Boethius, Fundamentals of Music*. New Haven.

Brancacci, A. (1984) 'Aristosseno e lo statuto epistemologico della scienza armonica', in *La scienza ellenistica*, eds. G. Giannantoni and M. Vegetti. Naples: 154–85.

(1988) 'Alcidamante e PHibeh 13 "De musica": musica della retorica e retorica della musica', in *Aristoxenica, Menandrea, fragmenta philosophica*, ed. F. Adorna. Florence: 61–84.

Burkert, W. (1962) *Weisheit und Wissenschaft: Studien zu Pythagoras, Philolaos und Platon*. Nuremberg.

(1972) *Lore and Science in Ancient Pythagoreanism* (translation of Burkert (1962) by E. L. Minar, Jr.). Cambridge, Mass.

Burnett, C. (ed.) (1991) *The Second Sense*. London.

Burnyeat, M. F. (1987) 'Platonism and mathematics: a prelude to discussion', in *Mathematics and Metaphysics in Aristotle*, ed. A. Gaeser. Bern and Stuttgart: 213–40.

(2000) 'Why mathematics is good for the soul', in *Mathematics and Necessity in the History of Philosophy*, ed. T. Smiley, *Proceedings of the British Academy* 103: 1–81.

Caldwell, J. (1981) 'The *De Institutione Arithmetica* and the *De Institutione Musica*', in *Boethius: his Life, Thought and Influence*, ed. M. Gibson. Oxford: 135–54.

Campbell, D. A. (1991) *Greek Lyric* III (Loeb Classical Library). Cambridge Mass. and London.

Cassin, B. (2000) 'Who's afraid of the Sophists? Against ethical correctness', *Hypatia* 15.4: 102–20.

Cassio, A. C. (1988) 'Nicomachus of Gerasa and the dialect of Archytas, fr. 1', *Classical Quarterly* 38: 135–9.

Cassio, A. C., D. Musti and L. E. Rossi (eds.) (2000) *Synaulia: Cultura musicale in Grecia e contatti mediterranei*. Naples.

Chailley, J. (1967) 'Nouvelles remarques sur les deux notations musicales grecques', *Recherches de papyrologie* 4: 210–16.

(1968) 'Nicomaque, Aristote et Terpandre devant la transformation de l'heptachorde grec en octacorde', *Yuval* 1: 132–54.

(1973) 'La notation archaïque grecque d'après Aristide Quintilien', *Revue des études grecques* 36: 17–34.

(1979) *La musique grecque antique*. Paris.

Ciancaglini, C. A. (1991) 'Le teorie acustiche dei Greci', *Atti della Academia nazionale dei Lincei: Classe di Scienze morali, storiche e filologiche* 9 vol. I: 47–77, vol. II: 147–77.

(1998) 'L'acustica in Archita', *Maia* 50: 213–51.

Comotti, G. (1972) 'L'endecacordo di Ione di Chio', *Quaderni urbinati di cultura classica* 13: 54–61.

(1991) 'Pitagora, Ippaso, Laso e il metodo sperimentale', in Wallace and MacLachlan (1991): 20–9.

Creese, D. E. (2002) 'The geometry of sound: the monochord in Greek harmonic science', PhD diss. Birmingham.

Crocker, R. L. (1963–4) 'Pythagorean mathematics and music', *Journal of Aesthetics and Art Criticism* 22: 189–98, 325–35.

(1966) 'Aristoxenus and Greek mathematics', in *Aspects of Medieval and Renaissance Music*, ed. J. La Rue. New York: 96–110.

Crombie, A. C. (1969) 'Mathematics, music and medical science', *Organon* 6: 21–36.

Cuomo, S. (2001) *Ancient Mathematics*. London and New York.

Diels, H., and W. Kranz (1951) *Die Fragmente der Vorsokratiker*, 6th edn. [cited throughout as DK]. Berlin.

D'Ooge, M. L. (1926) *Nicomachus of Gerasa: Introduction to Arithmetic*. New York.

Düring, I. (1930) *Die Harmonielehre des Klaudios Ptolemaios*. Gothenburg.

(1932) *Porphyrios Kommentar zur Harmonielehre des Ptolemaios*. Gothenburg.

(1934) *Ptolemaios und Porphyrios über die Musik*. Gothenburg.

(1945) 'Studies in musical terminology in fifth century literature', *Eranos* 43: 176–97.

Einarson, B., and P. H. De Lacy (1967) *Plutarch, Moralia* vol. XIV (Loeb Classical Library). Cambridge, Mass.

Fletcher, N. H., and T. D. Rossing (1998) *The Physics of Musical Instruments*, 2nd edn. New York.

Fortenbaugh, W. W., et al. (eds.) (1992) *Theophrastus of Eresus: Sources for his Life, Writings, Thought and Influence*, 2 vols. Leiden.

Gamberini, L. (1979) *Plutarcho 'della musica'*. Florence.

(1996) 'Musica e medicina: dal mito d'Orfeo alla biomusica', *Atti dell' Academia Ligure di Scienze e Lettere* 53: 513–78.

Gentili, B., and F. Perusino (eds.) (1995) *Mousike: Metrica, Ritmica e Musica Greca in Memoria di Giovanni Comotti*. Pisa and Rome.

Gentili, B., and R. Pretagostini (eds.) (1988) *La musica in Grecia*. Rome and Bari.

Germain, G. (1954) *Homère et la mystique des nombres*. Paris.

Gerson, L. (1993) Review of Tarrant 1993, *Bryn Mawr Classical Review*, 4.6.12.

Gibson, S. (2005) *Aristoxenus of Tarentum and the Birth of Musicology*. Abingdon and New York.

Gilula, D. (2000) 'Stratonicus, the witty harpist', in *Athenaeus and his World*, eds. D. Braund and J. Wilkins. Exeter: 423–33.

Glucker, J. (1998) 'Theophrastus, the Academy, and the Athenian philosophical atmosphere', in van Ophuijsen and van Raalte (1998): 299–316.

Gombosi, O. J. (1939) *Die Tonarten und Stimmungen der antiken Musik*. Copenhagen.

(1951) 'Key, mode, species', *Journal of the American Musicological Society* 4: 20–6.

Gottschalk, H. B. (1968) 'The *De audibilibus* and Peripatetic acoustics', *Hermes* 96: 435–60.

(1980), *Heraclides of Pontus*. Oxford.

(1998) 'Theophrastus and the Peripatos', in van Ophuijsen and van Raalte (1998): 281–98.

Guthrie, W. K. C. (1962–81) *A History of Greek Philosophy*, 6 vols. Cambridge.

Hagel, S. (2005) 'Twenty-four in auloi', in *Ancient Greek music in Performance*, eds. S. Hagel and C. Harrauer. Vienna: 51–91.

Handschin, J. (1950) 'The Timaeus scale', *Musica Disciplina* 4: 3–42.

Hankinson, R. J. (2005) 'Aristotle on kind-crossing', in *Philosophy and the Sciences in Antiquity*, ed. R. W. Sharples. Aldershot: 23–54.

Heath, T. L. (1921) *A History of Greek Mathematics*, 2 vols. Oxford.

Henderson. M. I. (1942) 'The growth of the Greek *harmoniai*', *Classical Quarterly* 36: 94–103.

(1957) 'Ancient Greek music', in *The New Oxford History of Music* 1, ed. E. Wellesz. London: 336–403.

Hirtler, E. (1995) *Die Musik als Scientia Mathematica von der Spätantike bis zur Barock*. Bern and Frankfurt am Main.

Huby, P., and G. Neal (eds.) (1989) *The Criterion of Truth*. Liverpool.

Huffman, C. A. (1985) 'The authenticity of Archytas fr. 1', *Classical Quarterly* 35: 344–8.

(1988) 'The role of number in Philolaus' philosophy', *Phronesis* 33: 1–30.

(1993) *Philolaus of Croton: Pythagorean and Presocratic*. Cambridge.

(2005) *Archytas of Tarentum: Pythagorean, Philosopher and Mathematician King*. Cambridge.

Izzo, A. (1987) 'Musica e numero da Ippaso ad Archita', in *Forme del sapere nei Presocratici*, eds. A. Capizzi and G. Casertano. Rome: 137–67.

von Jan, K. (1895) *Musici Scriptores Graeci*. Leipzig.

Johnson, C. W. L. (1896) 'Musical pitch and the measurement of intervals among the ancient Greeks', PhD. diss. Johns Hopkins.

(1899) 'The motion of the voice in the theory of ancient music', *Transactions of the Americal Philological Association* 30: 42–55.

Joly. R. (1960) *Recherches sur le traité pseudo-Hippocratique Du Régime*. Paris.

Judson, L. (ed.) (1991) *Aristotle's Physics: a Collection of Essays*. Oxford.

Kahn, C. H. (1974) 'Pythagorean philosophy before Plato', in *The Presocratics: A Collection of Critical Essays*, ed. A. P. D. Mourelatos. Princeton: 161–85.

(2001) *Pythagoras and the Pythagoreans: a Brief History*. Indianapolis and Cambridge.

Kárpáti, A. (1993) 'The musical fragments of Philolaus and the Pythagorean tradition', *Acta Antiqua Academiae Scientiarum Hungaricae* 34: 55–67.

(1994) 'Greek music theory in the fourth century BC', *International Journal of Musicology* 3: 57–88.

Kerferd, G. B. (1981) *The Sophistic Movement*. Cambridge.

Kirk, G. S., J. E. Raven and M. Schofield (1983) *The Presocratic Philosophers*, 2nd edn. Cambridge.

Knobloch, E. (1992) 'Rapports historiques entre musique, mathématique et cosmologie', in Lustgarten (1992): 123–67.

Knorr, W. R. (1975) *The Evolution of the Euclidean Elements*. Dordrecht.

Kucharski, P. (1959) 'Le *Philèbe* et les *Eléments harmoniques* d'Aristoxène', *Revue philosophique* 149: 41–72.

Kytzler, B. (1959) 'Die Weltseele und der musikalische Raum (Platons *Timaios* 35a ff.)', *Hermes* 87: 393–414.

Laloy, L. (1900) 'Un passage d'Euclide mal interprété', *Revue de philologie, de littérature et d'histoire ancienne* 24: 236–41.

(1904) *Aristoxène de Tarente, Disciple d'Aristote, et la Musique de l'Antiquité*. Paris.
Lambropoulou, V. (1995) 'On harmony: etymology, pre-Platonic meanings and elements', *Platon* 47–8: 179–93.
Landels, J. G. (1963) 'The Brauron aulos', *Annual of the British School at Athens* 58: 116–19.
(1968) 'A newly discovered aulos', *Annual of the British School at Athens* 63: 231–8.
(1999) *Music in Ancient Greece and Rome*. London.
Lang, H. S. (1998) *The Order of Nature in Aristotle's Physics*. Cambridge.
Lapini, W. (1994) 'Ancora su Papyrus Hibeh 1.13', *Musica e Storia* 2: 139–70.
Lasserre, F. (1954) *Plutarque de la musique*. Olten and Lausanne.
Lennox, J. G. (1986) 'Aristotle, Galileo and "mixed sciences"', in *Reinterpreting Galileo*, ed. W. A. Wallace. Washington, D.C.: 29–51.
Levin, F. R. (1961) 'The hendecachord of Ion of Chios', *Transactions of the American Philological Association* 92: 295–307.
(1972) '*Synesis* in Aristoxenian theory', *Transactions of the Americal Philological Association* 103: 211–34.
(1990) 'Unity in Euclid's *Sectio Canonis*', *Hermes* 118: 430–43.
Lippman, E. A. (1964) *Musical Thought in Ancient Greece*. New York.
Litchfield, M. (1988) 'Aristoxenus and empiricism: a reevaluation based on his theories', *Journal of Music Theory* 32: 51–73.
Lloyd, G. E. R. (1987) *The Revolutions of Wisdom*. Cambridge.
(1990) 'Plato and Archytas in the Seventh Letter', *Phronesis* 35: 159–74.
(1996) *Aristotelian Explorations*. Cambridge.
Long, A. A. (1989) 'Ptolemy on the criterion: an epistemology for the working scientist', in Huby and Neal (1989): 151–78.
Long, A. A., and D. N. Sedley (1987) *The Hellenistic Philosophers*, 2 vols. Cambridge.
Lustgarten, D. (ed.) (1992) *Quadrivium: musiques et sciences*. Paris.
Maas, M., and J. M. Snyder (1989) *Stringed Instruments of Ancient Greece*. New Haven, Conn.
Macran, H. S. (1902) *The Harmonics of Aristoxenus*. Oxford.
Mathiesen, T. J. (1975) 'An annotated translation of Euclid's division of the monochord', *Journal of Music Theory* 19: 236–58.
(1976) 'Problems of terminology in ancient Greek theory: *harmonia*', in *Harmonia: Festival Essays for Pauline Alderman*, ed. B. Kanon. Boston: 1–17.
(1999) *Apollo's Lyre: Greek Music and Music Theory in Antiquity and the Middle Ages*. Lincoln and London.
Mendell, H. (1998) 'Making sense of Aristotelian demonstration', *Oxford Studies in Ancient Philosophy* 16: 161–225.
Meriani, A. (1995) 'Un "esperimento" di Pitagora (Nicom. *Harm. ench.* 6, pp. 245–248 Jan)', in Gentili and Perusino (1995): 77–92.
(2003a) 'Teoria musicale e antiempirismo', in *Platone, la Repubblica: traduzione e commento*, vol. v, libri VI–VII, ed. M. Vegetti. Naples, 565–602.
(2003b) *Sulla musica greca antica: studi e ricerche*. Naples.

Mountford, J. F. (1923) 'The musical scales of Plato's *Republic*', *Classical Quarterly* 17: 125–36.

Mourelatos, A. P. D. (1980) 'Plato's "Real Astronomy": *Republic* 527d–531d', in Anton (1980): 33–73.

(1981) 'Astronomy and kinematics in Plato's project of rationalist explanation', *Studies in the History and Philosophy of Science* 12: 1–32.

Moutsopoulos, E. (1959) *La musique dans l'oeuvre de Platon*. Paris.

(1963) 'Dialectique musicale et dialectique philosophique chez Platon', *Annales de la Faculté des Lettres et Sciences humaines d'Aix* 37: 159–63.

(1971) 'Science harmonique et empirisme musical chez Platon', *Actes du XXIIème Congrès international d'histoire des sciences* 3A, Paris: 109–12.

Mueller, I. (1980) 'Ascending to problems: astronomy and harmonics in *Republic* VII', in Anton (1980): 103–21.

Musti, D. (2000) 'Musica tra aristocrazia e democrazia', in Cassio, Musti and Rossi (2000): 7–55.

Najock, D. (1996) 'Aristoxenos und die Auloi', in *Worte, Bilde, Töne*, eds. R. Faber and B. Seidensticker. Würzburg: 59–76.

Netz, R. (1997) 'Classical mathematics in the classical Mediterranean', *Mediterranean Historical Review* 12: 1–24.

(1999) *The Shaping of Deduction in Greek Mathematics*. Cambridge.

(2002) 'Greek mathematicians: a group picture', in Tuplin and Rihll (2002): 196–216.

Neubecker, A. J. (1977) *Altgriechischer Musik: eine Einführung*. Darmstadt.

Neugebauer, O. (1957) *The Exact Sciences in Antiquity*, 2nd edn. Providence, N.J.

van Ophuijsen, J. M., and M. van Raalte (eds.) (1998) *Theophrastus: Reappraising the Sources*. New Brunswick and London.

Pagliara, A. (2000) 'Musica e politica nella speculazione platonica: considerazioni intorno all' *ethos* del modo frigio (*Resp.* III.10, 399a–c)', in Cassio, Musti and Rossi (2000): 157–216.

Palisca, C. (1985) *Humanism in Italian Renaissance Musical Thought*. New Haven and London.

Paquette, D. (1984) *L'instrument de musique dans la céramique de la Grèce antique*. Paris.

Pearson, L. (1990) *Aristoxenus, Elementa Rhythmica*. Oxford.

Perusino, F. (1993) *Anonimo (Michele Psello?): La Tragedia greca*. Urbino.

Philip, J. A. (1966) *Pythagoras and Early Pythagoreanism*. Toronto.

Phillips, D. J., and D. Pritchard (eds.) (2003) *Sport and Festivals in the Ancient Greek World*. Swansea.

Pöhlmann, E. (1988) *Beiträge zur antiken und neueren Musikgeschichte*. Frankfurt on Main.

Pöhlmann, E., and M. L. West (2001) *Documents of Ancient Greek Music*. Oxford.

Porter, J. I. (2007) 'Lasus of Hermione, Pindar, and the riddle of *S*', *Classical Quarterly* 57:1–21.

Potiron, H. (1961) 'Les notations d'Aristide Quintilien et les harmonies dites Platoniciennes', *Revue de musicologie* 47: 159–76.

(1964) 'La notation grecque au temps d'Aristoxène', *Revue de musicologie* 50: 222–5.
Privitera, G. A. (1965) *Laso di Hermione*. Rome.
Rankin, H. D. (1983) *Sophists, Socratics and Cynics*. London.
Restani, D. (2001) 'Un manuale per Teoderico', in D. Restani, *Musica per governare*. Ravenna: 57–84.
Richter, L. (1958) 'Die Aufgaben der Musiklehre nach Aristoxenos und Klaudios Ptolemaios', *Archiv für Musikwissenschaft* 15: 209–29 [revised in Richter 2000].
(1961) *Zur Wissenschaftslehre von der Musik bei Platon und Aristoteles*. Berlin.
(2000) 'Die Aufgaben der Musiklehre nach Aristoxenos und Klaudios Ptolemaios', in L. Richter, *Momente der Musikgeschichte Antike und Byzanz*, Anif and Salzburg: 93–115 [a revised version of Richter 1958].
Riethmüller, A. (1985) 'Logos und Diastema in der griechischen Musiktheorie', *Archiv für Musikwissenschaft* 42: 18–36.
Rihll, T. E. (1999) *Greek Science*. Oxford.
Da Rios, R. (1954) *Aristoxeni Elementa Harmonica*. Rome.
Rocconi, E. (1998) '*Harmonia* e teoria dei *gene* musicali nella Grecia antica', *Seminari Romani di cultura greca* 1: 345–63.
(1999a) 'La formazione del lessico musicale nella Grecia antica', Tesi di dottorato. Urbino.
(1999b) 'Terminologia dello "spazio sonoro" negli *Elementa Harmonica* di Aristosseno di Taranto', *Quaderni urbinati di cultura classica* 61: 93–103.
(2002) 'The development of vertical direction in the spatial representation of sound', in *Studien zur Musikarchäologie* 3, ed. E. Hickmann, *Orient-Archäologie* 10: 389–92.
(2003) *Le parole delle Muse*. Rome.
(2004) 'Colours in music: metaphoric musical language in Greek antiquity', in *Studien zur Musikarchäologie* 4, eds. E. Hickmann and R. Eichmann, *Orient-Archäologie* 15: 29–34.
De Romilly, J. (1992) *The Great Sophists in Periclean Athens*. Oxford.
Rossi, L. E. (1988) 'La dottrina dell' "ethos" musicale e il simposio', in Gentili and Pretagostini (1988): 238–45.
(2000) 'Musica e psicologia nel mondo antico e nel mondo moderno', in Cassio, Musti and Rossi (2000): 57–96.
Ruelle, C.-E. (1906) 'Sur l'authenticité probable de la division du canon musical attribuée à Euclide', *Revue des études greques* 19: 318–20.
Sachs, C. (1940) *The History of Musical Instruments*. New York.
Schlesinger, K. (1933) 'Further notes on Aristoxenus and musical intervals', *Classical Quarterly* 27: 88–96.
(1939) *The Greek Aulos*. London.
(1944) 'The harmoniai', *Music Review* 5: 7–39.
Scholz, H. (1975) 'The ancient axiomatic theory', in *Articles on Aristotle* 1, eds. J. Barnes, M. Schofield and R. Sorabji. London: 50–64.

Sicking, C. M. J. (1998) 'Theophrastus on the nature of music', in van Ophuijsen and van Raalte (1998): 97–142.

Sifakis, G. M. (1967) *Studies in the History of Hellenistic Drama*. London.

Smith, R. (1982) 'The relationship between Aristotle's two *Analytics*', *Classical Quarterly* 32: 327–35.

Solomon, J. (1980a) 'Cleonides, Eisagōgē Harmonikē: critical edition, translation and commentary', PhD diss. North Carolina–Chapel Hill.

(1980b) 'Eklysis and ecbole', *Symbolae Osloenses* 55: 111–26.

(1981) 'The diastaltic ethos', *Classical Philology* 76: 93–100.

(1984) 'Towards a history of *tonoi*', *Journal of Musicology* 3: 242–51.

Sorabji, R. (1972) 'Aristotle, mathematics and colour', *Classical Quarterly* 22: 293–308.

(2004) 'Aristotle on colour, light and imperceptibles', *Bulletin of the Institute of Classical Studies* 47: 129–40.

Stephanis, I. E. (1988) *Dionysiakoi Technitai*. Heraklion.

Strasburger, H. (1990) *Studien zur alten Geschichte* III. Hildesheim.

Striker, G. (1996) *Essays on Hellenistic Philosophy and Science*. Cambridge.

Sundberg, J. (1991) *The Science of Musical Sounds*. London.

Tannery, P. (1902a) 'Du rôle de la musique grecque dans le développement de la mathématique pure', *Bibliotheca mathematica* 3: 161–75.

(1902b) 'Sur les intervalles de la musique grecque', *Revue des études grecques* 15: 336–52.

(1904a) 'Inauthenticité de la "Division du Canon" attribuée à Euclide', *Comptes rendus des séances de l'Académie des Inscriptions et Belles-Lettres*: 439–45.

(1904b) 'À propos des fragments philolaïques sur la musique', *Revue de philologie* 28: 233–49.

Tarrant, H. (1993) *Thrasyllan Platonism*. Ithaca, N.Y.

Thesleff, H. (1961) *An Introduction to the Pythagorean Writings of the Hellenistic Period*. Åbo.

(1965) *The Pythagorean Texts of the Hellenistic Period*. Åbo.

Thurn, N. (1998) 'Die siebensaitige Lyra', *Mnemosyne* 51: 411–34.

Tuplin, C. J., and T. E. Rihll (eds.) (2002) *Science and Mathematics in Ancient Greek Culture*. Oxford.

Vogel, M. (1963) *Die Enharmonik der Griechen*, 2 vols. Düsseldorf.

Van der Waerden, B. L. (1943) 'Die Harmonielehre der Pythagoreer', *Hermes* 78: 163–99.

Wallace, R. W. (1991) 'Damone di Oa ed i suoi successori: un' analisi delle fonti', in Wallace and MacLachlan (1991): 30–53.

(1995) 'Music theorists in fourth-century Athens', in Gentili and Perusino (1995): 17–40.

(1997) 'Poet, public and "theatocracy": audience performance in classical Athens', in *Poet, Public and Performance in Ancient Greece*, eds. L. Edmunds and R. W. Wallace. Baltimore, Maryland: 97–111.

Wallace, R. W., and B. MacLachlan (eds.) (1991) *Harmonia Mundi*. Rome.

Walter, H. (1994) 'Logos und Aisthesis: zum Methodenstreit in der antiken Musiktheorie', *International Journal of Musicology* 3: 43–55.

Wehrli, F. (1969) *Die Schule des Aristoteles*, 2nd edn. Basle.

Weil, H., and T. Reinach (1900) *Plutarque: de la musique*. Paris.

West, M. L. (1971) 'Stesichorus', *Classical Quarterly* 21: 302–14.

(1974) *Studies in Early Greek Elegy and Iambus*. Berlin and New York.

(1981) 'The singing of Homer and the modes of early Greek music', *Journal of Hellenic Studies* 101: 113–29.

(1983) *The Orphic Poems*. Oxford.

(1985) 'Ion of Chios', *Bulletin of the Institute of Classical Studies* 32: 71–8.

(1992a) *Ancient Greek Music*. Oxford.

(1992b) 'Analecta Musica', *Zeitschrift für Papyrologie und Epigraphik* 92: 1–54.

Winnington-Ingram, R. P. (1928) 'The Spondeion scale', *Classical Quarterly* 22: 83–91.

(1932) 'Aristoxenus and the intervals of Greek music', *Classical Quarterly* 26: 195–208.

(1936) *Mode in Ancient Greek Music*. Cambridge.

(1965) 'Die Enharmonik der Griechen', *Die Musikforschung* 18: 60–4.

(1973) 'The first notational diagram of Aristides Quintilianus', *Philologus* 117: 243–9.

(1978) 'Two studies in Greek musical notation', *Philologus* 122: 237–48.

Wright, M. R. (ed.) (2000) *Reason and Necessity: Essays on Plato's Timaeus*. London.

Wyller, E. A. (1995) 'Kosmos als Klangfigur: Platons Naturvision im *Timaios*', *Perspektiven der Philosophie* 21: 105–21.

Yamamoto, T. (2001) *Study in Aristoxenus' Harmonica Stoicheia* [in Japanese with English summary]. Ahasikawa.

Zaminer, F. (1980) 'Pythagoras und die Anfänge des musiktheoretischen Denkens bei den Griechen', *Jahrbuch des Staatlichen Instituts für Musikforschung 1979–80*: 203–11.

(1984) 'Hypate, Mese und Nete in frühgriechischen Denken', *Archiv für Musikwissenschaft* 41: 1–26.

Zanoncelli, L. (1990) *La manualistica musicale greca*. Milan.

Zedda, S. (2000) 'How to build a world soul: a practical guide', in Wright (2000): 23–42.

Zhmud, L. J. (1997) *Wissenschaft, Philosophie und Religion im frühen Pythagoreismus*. Berlin.

Index of proper names

General index

Made in the USA
Monee, IL
14 September 2026

63198318R00288